Student Solutions Manual

Intermediate Algebra
Connecting Concepts through Applications

Mark Clark
Palomar College

Cynthia Anfinson
Palomar College

Prepared by

Gina Hayes
Palomar College

Karen Mifflin
Palomar College

Australia • Brazil • Japan • Korea • Mexico • Singapore • Spain • United Kingdom • United States

© 2012 Brooks/Cole, Cengage Learning

ALL RIGHTS RESERVED. No part of this work covered by the copyright herein may be reproduced, transmitted, stored, or used in any form or by any means graphic, electronic, or mechanical, including but not limited to photocopying, recording, scanning, digitizing, taping, Web distribution, information networks, or information storage and retrieval systems, except as permitted under Section 107 or 108 of the 1976 United States Copyright Act, without the prior written permission of the publisher.

For product information and technology assistance, contact us at **Cengage Learning Customer & Sales Support, 1-800-354-9706**

For permission to use material from this text or product, submit all requests online at **www.cengage.com/permissions**
Further permissions questions can be emailed to **permissionrequest@cengage.com**

ISBN-13: 978-0-534-49641-8
ISBN-10: 0-534-49641-5

Brooks/Cole
20 Davis Drive
Belmont, CA 94002-3098
USA

Cengage Learning is a leading provider of customized learning solutions with office locations around the globe, including Singapore, the United Kingdom, Australia, Mexico, Brazil, and Japan. Locate your local office at: **www.cengage.com/global**

Cengage Learning products are represented in Canada by Nelson Education, Ltd.

To learn more about Brooks/Cole, visit **www.cengage.com/brookscole**

Purchase any of our products at your local college store or at our preferred online store **www.cengagebrain.com**

Printed in the United States of America
1 2 3 4 5 6 7 14 13 12 11 10

Table of Contents

Chapter 1: Linear Functions
1.1 Solving Linear Equations — 1
1.2 Using Data to Create Scatterplots — 9
1.3 Fundamentals of Graphing and Slope — 11
1.4 Intercepts and Graphing — 17
1.5 Finding Equations of Lines — 23
1.6 Finding Linear Models — 31
1.7 Functions and Function Notation — 37
CHAPTER 1 REVIEW — **42**
CHAPTER 1 TEST — **50**

Chapter 2: Introduction to Systems of Linear Equations and Inequalities
2.1 Systems of Linear Equations — 54
2.2 Solving Systems of Equations Using the Substitution Method — 63
2.3 Solving Systems of Equations Using the Elimination Method — 72
2.4 Solving Linear Inequalities — 79
2.5 Absolute Value Equations and Inequalities — 84
2.6 Solving Systems of Linear Inequalities — 88
CHAPTER 2 REVIEW — **92**
CHAPTER 2 TEST — **101**
CHAPTERS 1-2 CUMULATIVE REVIEW — **105**

Chapter 3: Exponents, Polynomials, and Functions
3.1 Rules for Exponents — 113
3.2 Combining Functions — 117
3.3 Composing Functions — 125
3.4 Factoring Polynomials — 130
3.5 Special Factoring Techniques — 135
CHAPTER 3 REVIEW — **138**
CHAPTER 3 TEST — **146**

Chapter 4: Quadratic Functions
4.1 Quadratic Functions and Parabolas — 149
4.2 Graphing Quadratics in Vertex Form — 152
4.3 Finding Quadratic Models — 156
4.4 Solving Quadratic Equations by Square Root Property and Completing the Square — 161
4.5 Solving Quadratic Equations by Factoring — 170
4.6 Solving Quadratic Equations Using the Quadratic Formula — 178
4.7 Graphing from Standard Form — 186
CHAPTER 4 REVIEW — **196**
CHAPTER 4 TEST — **207**
CHAPTERS 1-4 CUMULATIVE REVIEW — **211**

Chapter 5: Exponential Functions
5.1 Exponential Functions: Patterns of Growth and Decay	223
5.2 Solving Equations Using Exponent Rules	231
5.3 Graphing Exponential Functions	236
5.4 Finding Exponential Models	242
5.5 Exponential Growth and Decay Rates and Compounding Interest	250
CHAPTER 5 REVIEW	**254**
CHAPTER 5 TEST	**260**

Chapter 6: Logarithmic Functions
6.1 Functions and Their Inverses	263
6.2 Logarithmic Functions	271
6.3 Graphing Logarithm Functions	274
6.4 Properties of Logarithms	277
6.5 Solving Exponential Equations	279
6.6 Solving Logarithmic Equations	287
CHAPTER 6 REVIEW	**292**
CHAPTER 6 TEST	**299**
CHAPTERS 1-6 CUMULATIVE REVIEW	**302**

Chapter 7: Rational Functions
7.1 Rational Functions and Variation	315
7.2 Simplifying Rational Expressions	321
7.3 Multiplying and Dividing Rational Expressions	326
7.4 Adding and Subtracting Rational Expressions	329
7.5 Solving Rational Equations	336
CHAPTER 7 REVIEW	**344**
CHAPTER 7 TEST	**352**

Chapter 8: Radical Functions
8.1 Radical Functions	356
8.2 Simplifying, Adding, and Subtracting Radicals	361
8.3 Multiplying and Dividing Radicals	363
8.4 Solving Radical Equations	367
8.5 Complex Numbers	374
CHAPTER 8 REVIEW	**378**
CHAPTER 8 TEST	**387**
CHAPTERS 1-8 CUMULATIVE REVIEW	**391**

Chapter 9: Conics, Sequences, and Series
9.1 Parabolas and Circles	407
9.2 Ellipses and Hyperbolas	415
9.3 Arithmetic Sequences	421
9.4 Geometric Sequences	427
9.5 Series	433
CHAPTER 9 REVIEW EXERCISES	**437**
CHAPTER 9 TEST	**449**
CHAPTERS 1-9 CUMULATIVE REVIEW	**452**

Appendix A: Basic Algebra Review	**470**
Appendix B: Matrices	**474**

Section 1.1

1.
$2x + 10 = 40$
$2x + 10 - 10 = 40 - 10$
$2x = 30$
$\dfrac{2x}{2} = \dfrac{30}{2}$
$x = 15$

3.
$-4t + 8 = -32$
$-4t + 8 - 8 = -32 - 8$
$-4t = -40$
$\dfrac{-4t}{-4} = \dfrac{-40}{-4}$
$t = 10$

5.
$2.5x + 7.5 = 32.5$
$2.5x + 7.5 - 7.5 = 32.5 - 7.5$
$2.5x = 25$
$\dfrac{2.5x}{2.5} = \dfrac{25}{2.5}$
$x = 10$

7.
$20 = 5.2x - 0.8$
$20 + 0.8 = 5.2x - 0.8 + 0.8$
$20.8 = 5.2x$
$\dfrac{20.8}{5.2} = \dfrac{5.2x}{5.2}$
$4 = x$
$x = 4$

9.
$0.05(x - 200) = 240$
$0.05x - 10 = 240$
$0.05x - 10 + 10 = 240 + 10$
$0.05x = 250$
$\dfrac{0.05x}{0.05} = \dfrac{250}{0.05}$
$x = 5000$

11.

a.
$C = 10h + 20$
$C = 10(1) + 20$
$C = 30$

After 1 hour of training, a new employee can make 30 candies an hour.

b.
$C = 10h = 20$
$C = 10(4) + 20$
$C = 40 + 20 = 60$

After 4 hour of training, a new employee can make 60 candies an hour.

c.
Let $C = 150$
$150 = 10h + 20$
$150 - 20 = 10h + 20 - 20$
$130 = 10h$
$\dfrac{130}{10} = \dfrac{10h}{10}$
$13 = h$

A new employee can make 150 candies an hour after 13 hours of training.

13.

a.
$N = -315.9t + 4809.8$
$N = -315.9(2) + 4809.8$
$N = -631.8 + 4809.8$
$N = 4178$

In 1992, there were 4178 homicides of 15-19 year olds in the United States.

b.
$N = -315.9t + 4809.8$
$N = -315.9(12) + 4809.8$
$N = -3790.8 + 4809.8$
$N = 1019$

In 2002, there were 1019 homicides of 15-19 year olds in the United States.

c.

Let $N = 7337$

$7337 - 4809.8 = -315.9t + 4809.8 - 4809.8$

$2527.2 = -315.9t$

$\dfrac{2527.2}{-315.9} = \dfrac{-315.9t}{-315.9}$

$-8 = t$

In 1982, there were 7337 homicides of 15-19 year olds in the United States.

15.

a.

$P = 1.5t - 300$

$P = 1.5(100) - 300$

$P = 150 - 300$

$P = -150$

There is a loss of $150 for selling only 100 T-shirts.

b.

$P = 1.5t - 300$

$P = 1.5(400) - 300$

$P = 600 - 300$

$P = 300$

There is a profit of $300 for selling 400 T-shirts.

c.

Let $P = 1000$

$1000 = 1.5t - 300$

$1000 + 300 = 1.5t - 300 + 300$

$1300 = 1.5t$

$\dfrac{1300}{1.5} = \dfrac{1.5t}{1.5}$

$866.67 \approx t$

To make $1000 profit, you must sell 867 T-shirts.

17.

a.

$C = 2.50 + 2.0m$

$C = 2.50 + 2.0(25)$

$C = 2.50 + 50.0$

$C = 52.50$

It costs $52.50 to take a 25-mile taxi ride in NYC.

b.

$100 = 2.50 + 2.0m$

$100 - 2.50 = 2.50 - 2.50 + 2.0m$

$97.50 = 2.0m$

$\dfrac{97.50}{2.0} = \dfrac{2.0m}{2.0}$

$48.75 = m$

For $100, you can take about a 48-mile taxi ride in NYC.

19.

a. $P = 3.5$. This too few people. This would mean that only 3500 people live in Kentucky.

b. $P = 4200$. This answer is most reasonable. This would mean that 4,200,000 people live in Kentucky.

c. $P = -210$. This not possible. This would mean that $-210,000$ people live in Kentucky.

21.

a. $T = -50$. This answer is most reasonable.

b. $T = 75$. This temperature is too warm for South Pole temperatures.

c. $T = 82$. This temperature is too warm for South Pole temperatures.

23.

a.

$P = 0.08(s - 1000)$

$P = 0.08(2000 - 1000)$

$P = 0.08(1000) = 80$

On $2000 in sales, you will make $80 in commissions.

b.

$P = 0.08(s - 1000)$

$P = 0.08(50,000 - 1000)$

$P = 0.08(49,000)$

$P = 3920$

On sales of $50,000, you will make $3920 in commissions.

c.

$P = 0.08(s - 1000)$
$500 = 0.08s - 80$
$500 + 80 = 0.08s - 80 + 80$
$580 = 0.08s$
$\dfrac{580}{0.08} = \dfrac{0.08s}{0.08}$
$7250 = s$

To make $500 per week, you will need $7250 in sales each week.

25.

a. $B = 29.95 + 0.55m$

b.

$B = 29.95 + 0.55(75)$
$B = 29.95 + 41.25$
$B = 71.20$

If you drive the 10-foot truck 75 miles, it will cost you $71.20.

c.

$B = 29.95 + 0.55m$
$100 = 29.95 + 0.55m$
$100 - 29.95 = 29.95 - 29.95 + 0.55m$
$70.05 = 0.55m$
$\dfrac{70.05}{0.55} = \dfrac{0.55m}{0.55}$
$127.36 \approx m$

$37 will buy 135 minutes.

For a total of $100, you can rent the 10-foot truck from Budget and drive it 127 miles.

27.

a. $P = 250 + 0.07s$

b.

Let $s = 2000$
$P = 250 + 0.07(2000)$
$P = 250 + 140 = 390$

If you have sales of $2000 in a week, your pay will be $390.

c.

Let $P = 650$
$650 = 250 + 0.07s$
$650 - 250 = 250 - 250 + 0.07s$
$400 = 0.07s$
$\dfrac{400}{0.07} = \dfrac{0.07s}{0.07}$
$5714.29 = s$

To earn $650 per week, you must have $5714.29 in sales each week.

29. Let C be the total cost of a trip to Las Vegas for d days.

a. $C = 125 + 100d$

b.

$C = 125 + 100(3)$
$C = 125 + 300 = 425$

A three day trip to Las Vegas will cost $425.

c.

$\dfrac{\$700}{2} = \$350$
$C = 125 + 100d$
$350 = 125 + 100d$
$350 - 125 = 125 - 125 + 100d$
$225 = 100d$
$\dfrac{225}{100} = \dfrac{100d}{100}$
$2.25 = d$

If you have $700 and gamble half of it, you can stay in Las Vegas for only two days.

31. Let C be the total cost (in dollars) of shooting a wedding, and p be the number of proofs the photographer edits and prints.

a. $C = 5.29p + 400$

b.

$C = 5.29(100) + 400$
$C = 529 + 400$
$C = 929$

If the photographer edits and prints 100 proofs the cost will be $929.

Chapter 1 Linear Functions

c.

Let $C = 1250$

$1250 = 5.29p + 400$

$1250 - 400 = 5.29p + 400 - 400$

$850 = 5.29p$

$\dfrac{850}{5.29} = \dfrac{5.29p}{5.29}$

$160.68 \approx p$

The photographer can edit and print 160 proofs with a budget of $1250.

33.

a. Let C be the total cost (in dollars) for selling s snow cones for a month.

Fixed costs are: $5500 + 1500 = 7000$.

$C = 7000 + 0.45s$

b.

$C = 7000 + 0.45(3000)$

$C = 8350$

The monthly cost for selling 3000 snow cones is $8350.

c.

$10,600 = 7000 + 0.45s$

$10,600 - 7000 = 7000 - 7000 + 0.45s$

$3600 = 0.45s$

$\dfrac{3600}{0.45} = \dfrac{0.45s}{0.45}$

$8000 = s$

For a $10,600 budget, the vendor can sell up to 8000 snow cones.

35.

a. Let C be the total cost (in dollars) for the Squeaky Clean Window Company to clean windows for a day when w windows are cleaned. $C = 1.50w + 230$.

b.

$C = 1.50(60) + 230$

$C = 90 + 230$

$C = 320$

If the Squeaky Clean Window Company cleans 60 windows in a day, it will cost the company $320.

c.

$450 = 1.50w + 230$

$450 - 230 = 1.50w + 230 - 230$

$220 = 1.50w$

$\dfrac{220}{1.50} = \dfrac{1.50w}{1.50}$

$146.7 \approx w$

To stay within a budget of $450, the Squeaky Clean Window Company can clean up to 146 windows.

37. Maria's work is correct. Javier needs a decimal to correctly represent 55 cents per bottle in terms of dollars per bottle.

39.

a. Let C be the total cost (in dollars) for pest management from Enviro-Safe Pest Management when m monthly treatments are done. $C = 150 + 38m$.

b. There are 18 months in 1.5 years.

$C = 150 + 38(18)$

$C = 150 + 684$

$C = 834$

If the house has an initial treatment, and then is treated for an additional 1.5 years, it will cost $834.

41.

a. Let C be the total monthly cost (in dollars) for a manufacturer to produce g sets of golf clubs.

$C = 23,250 + 145g$.

b.

$C = 23,250 + 145(100)$

$C = 23,250 + 14,500$

$C = 37,750$

It costs the manufacturer $37,750 to produce 100 sets of golf clubs.

c.

$20,000 - 23,250 = 23,250 - 23,250 + 145g$

$-3250 = 145g$

$\dfrac{-3250}{145} = \dfrac{145g}{145}$

$-22.41 \approx g$

This is model breakdown. Their costs can never be lower than their fixed costs of $23,250.

d.

$\dfrac{\$37,750}{100 \text{ sets}} = \$377.50$ per set

To break even selling 100 sets of golf clubs per month, the manufacturer must sell each set for $377.50.

43.

a. $C = 1500 + 1.50n$ for $n \le 500$.

b.

$C = 1500 + 1.50(250)$

$C = 1500 + 375$

$C = 1875$

It costs Rockon $1875 to make 250 CDs.

c.

$2000 = 1500 + 1.50n$

$2000 - 1500 = 1500 - 1500 + 1.50n$

$500 = 1.50n$

$\dfrac{500}{1.50} = \dfrac{1.50n}{1.50}$

$333.3 \approx n$

With a budget of $2000, Rockon can order 333 CDs.

d.

$3000 = 1500 + 1.50n$

$3000 - 1500 = 1500 - 1500 + 1.50n$

$1500 = 1.50n$

$\dfrac{1500}{1.50} = \dfrac{1.50n}{1.50}$

$1000 = n$

With a budget of $3000, Rockon can order 1000 CDs.

This is model breakdown. They can only order up to 500 CDs.

45.

$5x + 60 = 2x + 90$

$5x + 60 - 2x = 2x + 90 - 2x$

$3x + 60 = 90$

$3x + 60 - 60 = 90 - 60$

$3x = 30$

$\dfrac{3x}{3} = \dfrac{30}{3}$

$x = 10$

47.

$\dfrac{2}{5}d + 6 = 14$

$\dfrac{2}{5}d + 6 - 6 = 14 - 6$

$\dfrac{2}{5}d = 8$

$\dfrac{5}{2}\left(\dfrac{2}{5}d\right) = \dfrac{5}{2}(8)$

$d = 20$

49.

$\dfrac{1}{3}m + \dfrac{4}{3} = 4$

$3\left(\dfrac{1}{3}m + \dfrac{4}{3}\right) = 3(4)$

$m + 4 = 12$

$m + 4 - 4 = 12 - 4$

$m = 8$

51.

$-3x - 6 = 14 + 8x$

$-3x - 6 - 8x = 14 + 8x - 8x$

$-11x - 6 = 14$

$-11x - 6 + 6 = 14 + 6$

$-11x = 20$

$\dfrac{-11x}{-11} = \dfrac{20}{-11}$

$x = -\dfrac{20}{11}$

53.

$$\frac{5}{7}d - \frac{3}{10} = \frac{4}{7}d + 4$$

$$70\left(\frac{5}{7}d - \frac{3}{10}\right) = 70\left(\frac{4}{7}d + 4\right)$$

$$50d - 21 = 40d + 280$$

$$50d - 21 - 40d = 40d + 280 - 40d$$

$$10d - 21 = 280$$

$$10d - 21 + 21 = 280 + 21$$

$$10d = 301$$

$$\frac{10d}{10} = \frac{301}{10}$$

$$d = \frac{301}{10}$$

55.

$$1.25d - 3.4 = -2.3(5d + 4)$$

$$1.25d - 3.4 = -11.5d - 9.2$$

$$1.25d - 3.4 + 11.5d = -11.5d - 9.2 + 11.5d$$

$$12.75d - 3.4 = -9.2$$

$$12.75d - 3.4 + 3.4 = -9.2 + 3.4$$

$$12.75d = -5.8$$

$$\frac{12.75d}{12.75} = \frac{-5.8}{12.75}$$

$$d \approx -0.45$$

57.

$$3(c + 5) - 21 = 107$$

$$3c + 15 - 21 = 107$$

$$3c - 6 = 107$$

$$3c - 6 + 6 = 107 + 6$$

$$3c = 113$$

$$\frac{3c}{3} = \frac{113}{3}$$

$$c = \frac{113}{3}$$

59.

$$1.7d + 5.7 = 29.7 + 5d$$

$$1.7d + 5.7 - 5d = 29.7 + 5d - 5d$$

$$-3.3d + 5.7 = 29.7$$

$$-3.3d + 5.7 - 5.7 = 29.7 - 5.7$$

$$-3.3d = 24$$

$$\frac{-3.3d}{-3.3} = \frac{24}{-3.3}$$

$$d \approx -7.27$$

61.

$$\frac{3}{7}(2z - 5) = \frac{4}{7}(-3z + 9)$$

$$\frac{6}{7}z - \frac{15}{7} = -\frac{12}{7}z + \frac{36}{7}$$

$$7\left(\frac{6}{7}z - \frac{15}{7}\right) = 7\left(-\frac{12}{7}z + \frac{36}{7}\right)$$

$$6z - 15 = -12z + 36$$

$$6z - 15 + 12z = -12z + 36 + 12z$$

$$18z - 15 = 36$$

$$18z - 15 + 15 = 36 + 15$$

$$18z = 51$$

$$\frac{18z}{18} = \frac{51}{18}$$

$$z = \frac{51}{18}$$

$$z = \frac{17}{6}$$

63.

$$-3(2v + 9) - 3(3v - 7) = 4v + 6(2v - 8)$$

$$-6v - 27 - 9v + 21 = 4v + 12v - 48$$

$$-15v - 6 = 16v - 48$$

$$-15v - 6 - 16v = 16v - 48 - 16v$$

$$-31v - 6 = -48$$

$$-31v - 6 + 6 = -48 + 6$$

$$-31v = -42$$

$$\frac{-31v}{-31} = \frac{-42}{-31}$$

$$v = \frac{42}{31}$$

65.

$$-\frac{8}{9}(3t+5) = \frac{2}{3}t - 12$$
$$-\frac{24}{9}t - \frac{40}{9} = \frac{2}{3}t - 12$$
$$9\left(-\frac{24}{9}t - \frac{40}{9}\right) = 9\left(\frac{2}{3}t - 12\right)$$
$$-24t - 40 = 6t - 108$$
$$-24t - 40 - 6t = 6t - 108 - 6t$$
$$-30t - 40 = -108$$
$$-30t - 40 + 40 = -108 + 40$$
$$-30t = -68$$
$$\frac{-30t}{-30} = \frac{-68}{-30}$$
$$t = \frac{68}{30}$$
$$t = \frac{34}{15}$$

67.

$$F = ma$$
$$\frac{F}{m} = \frac{ma}{m}$$
$$a = \frac{F}{m}$$

69.

$$J = Ft$$
$$\frac{J}{t} = \frac{Ft}{t}$$
$$F = \frac{J}{t}$$

71.

$$\omega = \omega_0 + \alpha t$$
$$\omega - \omega_0 = \omega_0 + \alpha t - \omega_0$$
$$\omega - \omega_0 = \alpha t$$
$$\frac{\omega - \omega_0}{t} = \frac{\alpha t}{t}$$
$$\alpha = \frac{\omega - \omega_0}{t}$$

73.

$$K = \frac{1}{2}I\omega^2$$
$$2(K) = 2\left(\frac{1}{2}I\omega^2\right)$$
$$2K = I\omega^2$$
$$\frac{2K}{\omega^2} = \frac{I\omega^2}{\omega^2}$$
$$I = \frac{2k}{\omega^2}$$

75.

$$K = \frac{1}{2}mv^2$$
$$2(K) = 2\left(\frac{1}{2}mv^2\right)$$
$$2K = mv^2$$
$$\frac{2K}{v^2} = \frac{mv^2}{v^2}$$
$$m = \frac{2K}{v^2}$$

77.

$$ax + by = c$$
$$ax + by - ax = c - ax$$
$$by = c - ax$$
$$\frac{by}{b} = \frac{c - ax}{b}$$
$$y = \frac{c - ax}{b}$$

79.

$$ax + 5 = y$$
$$ax + 5 - 5 = y - 5$$
$$ax = y - 5$$
$$\frac{ax}{a} = \frac{y - 5}{a}$$
$$x = \frac{y - 5}{a}$$

81.

$$b = 2c + 3d$$
$$b - 3d = 2c + 3d - 3d$$
$$b - 3d = 2c$$
$$\frac{b - 3d}{2} = \frac{2c}{2}$$
$$c = \frac{b - 3d}{2}$$

Chapter 1 Linear Functions

83.

$5x^2 + 3y = z$

$5x^2 + 3y - 5x^2 = z - 5x^2$

$3y = z - 5x^2$

$\dfrac{3y}{3} = \dfrac{z - 5x^2}{3}$

$y = \dfrac{z - 5x^2}{3}$

85. Rounding the outside temperature to $73°F$ is appropriate because a difference of $0.4°F$ would not be felt.

87. A result of $236.5725 would be rounded to $236.57 because our monetary units extend to the nearest hundredth.

89. The company would need to wash 313 cars (312.25 rounded up) to make a profit of $400. Anything less would result in a profit of less than $400.

Section 1.2

1. a. The total prize money at Wimbledon in 2002 was 8.5 million British pounds.

b. In 2003, the total prize money at Wimbledon was 9 million British pounds.

c. $(0,7)$. In the year 2000, the winnings were 7 million British pounds

d. Domain: $[2,11]$, range: $[8.5,14]$.

3. (A) and (F)

5. (C) and (E)

7. 50

9. 0.01

11. 50

13. 10,000

15.

a. Dependent variable: C = Cost in dollars for producing chocolate dipped key lime pie bars.

Independent variable: b = Number of bars produced.

b.

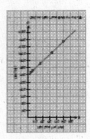

c. When $b = 200$, then $C = \$485$. The cost of producing 200 key lime bars on a stick is about $485.

d. Domain: $0 \leq b \leq 200$, range: $240 \leq C \leq 485$

17.

a. Dependent variable: G = Gross Profit for Quicksilver, Inc. in millions of dollars.

Independent variable: t = years since 2000.

b.

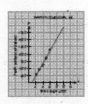

c. When $t = 10$, then $G = 1850$. The gross profit for Quicksilver, Inc. in 2010 will be approximately $1850 million.

d. Domain: $1 \leq t \leq 10$, range: $50 \leq G \leq 1850$.

19. Maria's model fits the data better. There is a smaller variance between the data points and the line of best fit.

21. $0 \leq t \leq 10$ is not a reasonable domain because the model predicts a negative value of cases per 100,000 population for t-values of greater than 8.2. This is model breakdown. Domain: $0 \leq t \leq 7$, range: $0.6 \leq A \leq 4.1$.

23.

a. Dependent variable: P = Population of the United States in millions

Independent variable: t = Years since 1995

b.

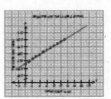

c. When $t = 14$ then $P \approx 308$ The population of the United States in 2009 will be around 308 million people.

d. Domain: $[0,14]$, range: $[266,308]$

e. $(0,266)$

f. In 1995, there were approximately 266 million people in the United States.

25.

a. Let W be the number of deaths of women induced by illegal drugs in the United States t years since 2000.

t	W
0	6583
1	7452
2	9306
3	10,297
4	11,349

Chapter 1 Linear Functions

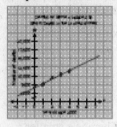

b. When $t = 7$, then $W \approx 15{,}000$. The number of drug-induced deaths of females in the United States in 2007 was around 15,000.

c. The number of drug-induced deaths of females in the United States reached 4000 in 1998.

d. Domain: $-1 \le t \le 7$, range: $5200 \le W \le 15{,}000$

27.

a. Let T be the number of years someone is expected live if they are a years in age.

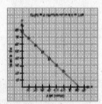

b. When $a = 45$, then $T = 35$. According to the model, a 45-year-old person will live approximately 35 years.

c. Domain: $[0, 80]$, range: $[3, 77]$

d. The graphical model predicts a 90-year-old to live negative years. This is model breakdown.

e. The vertical intercept is $(0, 77)$.

f. At birth, a person will live about 77 years.

29.

a. $(0, 2)$. The point that crosses the y-axis.

b. $(-3, 0)$. The point that crosses the x-axis.

c. $x = 3$ from the point $(3, 4)$.

d. $y = 2.6$ from the point $(1, 2.6)$.

31.

a. $(0, 5)$. The point that crosses the y-axis.

b. $(4, 0)$. The point that crosses the x-axis.

c. The input value is -10 when the output value is 18. That is, $x = -10$ when $y = 18$.

d. The input value is 15 when the output value is -15. That is, $x = 15$ when $y = -15$.

e. The output value is -8 when the input value is 10. That is, $y = -8$ when $x = 10$.

33.

a. $(0, -8)$. The point that crosses the y-axis.

b. $(31, 0)$. The point that crosses the x-axis.

c. $x = 11$ from the point $(11, -5)$.

d. $x = -9$ from the point $(-9, -10)$.

e. $y = -3$ from the point $(20, -3)$.

35.

a. False, vertical intercept $= (0, -2)$

b. True.

c. False, $x = -4.5$

d. False, $x = -1.5$

e. True.

37.

a. $(-15, 0)$. The point that crosses the x-axis.

b. $(0, -10)$ The point that crosses the y-axis.

c. $x = 15$ from the point $(15, -20)$.

d. $x = 7.5$ from the point $(7.5, -15)$.

e. $y = -3$ from the point $(-10, -3)$.

39.

a. $(0, 1)$ The point that crosses the y-axis.

b. $(-2, 0)$ The point that crosses the x-axis.

c. The input value is 1 when the output value is 1.5. That is, $x = 1$ when $y = 1.5$.

d. The input value is -4 when the output value is -1. That is, $x = -4$ when $y = -1$.

e. The output value is 2.5 when the input value is 3. That is, $y = 2.5$ when $x = 3$.

Section 1.3

1.

a. Use two points: $(0,-1)$ & $(2,0)$

Slope $= \dfrac{0-(-1)}{2-0} = \dfrac{1}{2}$

b. Increasing

c. Vertical intercept: $(0,-1)$

d. Horizontal intercept: $(2,0)$

3.

a. Use two points: $(-3,-15)$ & $(3,0)$

Slope $= \dfrac{0-(-15)}{3-(-3)} = \dfrac{15}{6} = \dfrac{5}{2}$

b. Increasing

c. y-intercept: $(0,-7.5)$

d. x-intercept: $(3,0)$

5.

a. Use two points: $(0,5)$ & $(1,0)$

Slope $= \dfrac{0-5}{1-0} = -5$

b. Decreasing

c. Vertical intercept: $(0,5)$

d. Horizontal intercept: $(1,0)$

7. $y = 2x+3$

x	$y = 2x+3$
-3	$2(-3)+3 = -3$
-1	$2(-1)+3 = 1$
0	$2(0)+3 = 3$

9. $x = 5y+4$

x	y
$5(-1)+4 = -1$	-1
$5(0)+4 = 4$	0
$5(1)+4 = 9$	1

11. $y = x^2 + 2$

x	y
-3	$(-3)^2 + 2 = 11$
-2	$(-2)^2 + 2 = 6$
0	$(0)^2 + 2 = 2$
2	$(2)^2 + 2 = 6$
3	$(3)^2 + 2 = 11$

13. $y = \dfrac{2}{3}x + 6$

x	y
-6	$\frac{2}{3}(-6)+6 = 2$
0	$\frac{2}{3}(0)+6 = 6$
3	$\frac{2}{3}(3)+6 = 8$

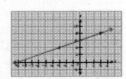

15. $x = \dfrac{2}{3}y - 4$

x	y
$\frac{2}{3}(-3) - 4 = -6$	-3
$\frac{2}{3}(3) - 4 = -2$	3
$\frac{2}{3}(6) - 4 = 0$	6

17. $y = 0.5x - 3$

x	y
-2	$0.5(-2) - 3 = -4$
2	$0.5(2) - 3 = -2$
6	$0.5(6) - 3 = 0$

19. $x = -1.5y + 7$

x	y
$-1.5(2) + 7 = 4$	2
$-1.5(4) + 7 = 1$	4
$-1.5(6) + 7 = -2$	6

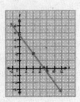

21. $y = -2x^2 + 15$

x	y
-3	$-2(-3)^2 + 15 = -3$
-2	$-2(-2)^2 + 15 = 7$
0	$-2(0)^2 + 15 = 15$
2	$-2(2)^2 + 15 = 7$
3	$-2(3)^2 + 15 = -3$

23. $B = 0.55m + 29.95$

a.

m	B
0	29.95
10	35.45
20	40.95
30	46.45
40	51.95

b.

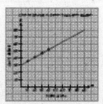

25.

a. Let W be the sales clerk's weekly salary in dollars for selling s dollars of merchandise during the week.

$W = 0.04s + 100$

b.

s	W
0	100
100	104
500	120
1000	140
1500	160

c.

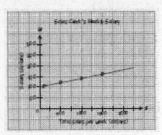

27.

a. Let T be the total tuition and fees Western Washington University charges resident undergrad students (in dollars) for u units taken by the student, up to 10 units.

$T = 137u + 208$

b.

u	T
5	893
6	1030
7	1167
8	1304
9	1441
10	1578

c.

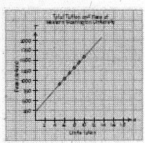

29.

a. $P = -3t + 50$

t	P
0	50
1	47
2	44
3	41
4	38

b.

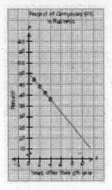

31. The slope is 2.

x	y	slope
0	7	$\dfrac{11-7}{2-0} = \dfrac{4}{2} = 2$
2	11	$\dfrac{15-11}{4-2} = \dfrac{4}{2} = 2$
4	15	$\dfrac{19-15}{6-4} = \dfrac{4}{2} = 2$
6	19	

33. The slope is -7.

x	y	slope
-2	29	$\dfrac{15-29}{0-(-2)} = \dfrac{-14}{2} = -7$
0	15	$\dfrac{1-15}{2-0} = \dfrac{-14}{2} = -7$
2	1	$\dfrac{-13-1}{4-2} = \dfrac{-14}{2} = -7$
4	-13	

Chapter 1 Linear Functions

35. The slope is 4.

x	y	slope
2	-18	$\dfrac{-6-(18)}{5-2} = \dfrac{12}{3} = 4$
5	-6	$\dfrac{6-(-6)}{8-5} = \dfrac{12}{3} = 4$
8	6	$\dfrac{18-6}{11-8} = \dfrac{12}{3} = 4$
11	18	

37. Yes, the points given in the table all lie on a line. The slopes are all equal.

x	y	slope
-3	-14	$\dfrac{-6-(-14)}{0-(-2)} = 4$
0	-6	$\dfrac{2-(-6)}{2-0} = 4$
2	2	$\dfrac{10-2}{4-2} = 4$
4	10	

39. No, the points given in the table do not all lie on a line. The slopes are not all equal.

x	y	slope
-6	12	$\dfrac{10-12}{-3-(-6)} = -\dfrac{2}{3}$
-3	10	$\dfrac{8-10}{0-(-3)} = -\dfrac{2}{3}$
0	8	$\dfrac{5-8}{3-0} = -1$
3	5	

41. The slope is 3. The y-intercept is $(0, 5)$.

The slope is the coefficient of x. The y-intercept is the constant of the equation.

43. The slope is -4. The y-intercept is $(0, 8)$.

The slope is the coefficient of x. The y-intercept is the constant of the equation.

45. The slope is $\dfrac{1}{2}$. The y-intercept is $(0, 5)$.

The slope is the coefficient of x. The y-intercept is the constant of the equation.

47. The slope is 0.4. The y-intercept is $(0, -7.2)$.

The slope is the coefficient of x. The y-intercept is the constant of the equation.

49. The slope is $\dfrac{4}{3}$. The y-intercept is $\left(0, \dfrac{7}{5}\right)$.

The slope is the coefficient of x. The y-intercept is the constant of the equation.

51. No, the y-intercept is $(0, -9)$.

53. Write the equation in the form $y = mx + b$.

The slope is the coefficient of x. The y-intercept is the constant of the equation.

$2x + y = 20$
$2x + y - 2x = 20 - 2x$
$y = -2x + 20$

The slope is -2, and the y-intercept is $(0, 20)$.

55. Write the equation in the form $y = mx + b$.

The slope is the coefficient of x. The y-intercept is the constant of the equation.

$4x - 2y = 20$
$4x - 2y - 4x = 20 - 4x$
$-2y = -4x + 20$
$\dfrac{-2y}{-2} = \dfrac{-4x}{-2} + \dfrac{20}{-2}$
$y = 2x - 10$

The slope is 2, and the y-intercept is $(0, -10)$.

57. Write the equation in the form $y = mx + b$. The slope is the coefficient of x. The y-intercept is the constant of the equation.
$$3x + 5y = 12$$
$$3x + 5y - 3x = 12 - 3x$$
$$5y = -3x + 12$$
$$\frac{5y}{5} = \frac{-3x}{5} + \frac{12}{5}$$
$$y = -\frac{3}{5}x + \frac{12}{5}$$
The slope is $-\frac{3}{5}$, and the y-intercept is $\left(0, \frac{12}{5}\right)$.

59. Write the equation in the form $y = mx + b$. The slope is the coefficient of x. The y-intercept is the constant of the equation.
$$3x + 2y = 6$$
$$3x + 2y - 3x = 6 - 3x$$
$$2y = -3x + 6$$
$$\frac{2y}{2} = \frac{-3x}{2} + \frac{6}{2}$$
$$y = -\frac{3}{2}x + 3$$
No, the slope is $-\frac{3}{2}$, and the y-intercept is $(0, 3)$.

61. The slope is 0.55. The cost of renting a 10 foot Budget truck increases by $0.55 per mile.

63. The slope is 0.40. The sales clerk's salary increases by $0.04 for every one dollar in sales.

65. The slope is 137. The tuition at Western Washington University increases by $137 per unit taken (up to 10 units) by resident undergraduate students.

67. The slope is -3. The percentage of companies still in business decreases by 3 percentage points each year.

69. $y = 2x - 7$. The y-intercept is $(0, -7)$. The slope is $2 = \frac{2}{1}$. Plot a point at the y-intercept, then using the slope, go up 2 and to the right 1.

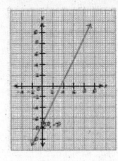

71. $y = \frac{4}{5}x - 6$. The y-intercept is $(0, -6)$. The slope is $\frac{4}{5}$. Plot a point at the y-intercept, then using the slope, go up 4 and to the right 5.

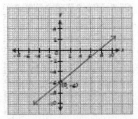

73. $y = -2x + 5$. The y-intercept is $(0, 5)$. The slope is $-2 = \frac{-2}{1}$. Plot a point at the y-intercept, then using the slope, go down 2 and to the right 1.

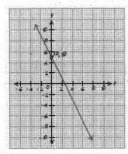

75. $y = -\frac{7}{5}x + 11$. The y-intercept is $(0, 11)$. The slope is $-\frac{7}{5} = \frac{-7}{5}$. Plot a point at the y-intercept, then using the slope, go down 7 and to the right 5.

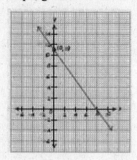

77. The graph has the y-intercept at $(0, 1.5)$, but it should be at $(0, -3)$.

79. The graph uses the slope $\frac{2}{3}$, but it should use the slope $-\frac{2}{3}$.

81. $y = 0.5x + 4$. The y-intercept is $(0, 4)$. The slope is $0.5 = \frac{1}{2}$. Plot a point at the y-intercept, then using the slope, go up 1 and to the right 2.

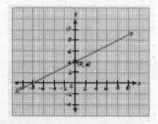

83. $y = -0.25x + 2$. The y-intercept is $(0, 2)$. The slope is $-0.25 = -\frac{1}{4} = \frac{-1}{4}$. Plot a point at the y-intercept, then using the slope, go down 1 and to the right 4.

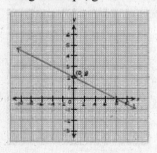

85. $y = -\frac{2}{5}x + \frac{3}{5}$. The y-intercept is $\left(0, \frac{3}{5}\right)$. The slope is $-\frac{2}{5} = \frac{-2}{5}$. Plot a point at the y-intercept, then using the slope, go down 2 and to the right 5.

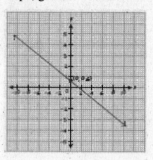

87. $y = \frac{1}{4}x - \frac{3}{4}$. The y-intercept is $\left(0, -\frac{3}{4}\right)$. The slope is $\frac{1}{4}$. Plot a point at the y-intercept, then using the slope, go up 1 and to the right 4.

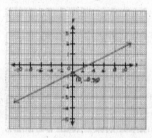

89. $y = -0.75x + 2.5$. The y-intercept is $(0, 2.5)$. The slope is $-0.75 = -\frac{3}{4} = \frac{-3}{4}$. Plot a point at the y-intercept, then using the slope, go down 3 and to the right 4.

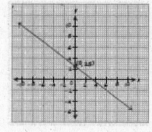

Section 1.4

1.
$y = 5x + 8$
$-5x + y = 5x - 5x + 8$
$-5x + y = 8$
$-1(-5x + y) = -1(8)$
$5x - y = -8$

3.
$y = -4x + 15$
$4x + y = -4x + 4x + 15$
$4x + y = 15$

5.
$y = \frac{2}{3}x - 8$
$3(y) = 3\left(\frac{2}{3}x - 8\right)$
$3y = 2x - 24$
$-2x + 3y = 2x - 2x - 24$
$-1(-2x + 3y) = -1(-24)$
$2x - 3y = 24$

7.
$y = \frac{1}{2}x + \frac{2}{5}$
$10y = 10\left(\frac{1}{2}x + \frac{2}{5}\right)$
$10y = 5x + 4$
$-5x + 10y = 5x - 5x + 4$
$-5x + 10y = 4$
$-1(-5x + 10y) = -1(4)$
$5x - 10y = -4$

9.
$y = -\frac{4}{5}x - \frac{1}{3}$
$15y = 15\left(-\frac{4}{5}x - \frac{1}{3}\right)$
$15y = -12x - 5$
$12x + 15y = -12x + 12x - 5$
$12x + 15y = -5$

11. $B = 0.55m + 29.95$

a. Vertical intercept:
$B = 0.55(0) + 29.95$
$B = 29.95$
$(0, 29.95)$

It will cost you $29.95 to rent a 10-ft truck from Budget and drive it 0 miles.

b. Horizontal intercept:
$0 = 0.55m + 29.95$
$-29.95 = 0.55m + 29.95 - 29.95$
$-29.95 = 0.55m$
$\frac{-29.95}{0.55} = \frac{0.55m}{0.55}$
$-54.5 \approx m$
$(-54.5, 0)$

To rent a 10-ft truck from Budget for $0, you would have to drive it −55 miles. This is model breakdown.

13. $P = 5a + 7$

a. P-intercept
$P = 5(0) + 7$
$P = 7$
$(0, 7)$

7% of 10-year-old girls are sexually active.

b. a-intercept
$0 = 5a + 7$
$-7 = 5a + 7 - 7$
$-7 = 5a$
$\frac{-7}{5} = \frac{5a}{5}$
$-1.4 = a$
$(-1.4, 0)$

0% of 8.6-year-old girls are sexually active.

15.

a. Let M be the total monthly salary a salesperson earns in dollars. Let s be the total dollars in sales per month.
$M = 0.05s + 500$

b. Vertical intercept:

$M = 0.05(0) + 500$
$M = 500$
$(0, 500)$

If a salesperson sells $0 during the month, his/her monthly salary will be $500.

c. Horizontal intercept

$0 = 0.05s + 500$
$0 - 500 = 0.05s + 500 - 500$
$-500 = 0.05s$
$\dfrac{-500}{0.05} = \dfrac{0.05s}{0.05}$
$-10{,}000 = s$
$(-10{,}000, 0)$

A salesperson will need to sell $-\$10{,}000$ in order to make $\$0$ for the month. This is model breakdown.

17. $P = 9.5t + 1277$

a. P-intercept:

$P = 9.5(0) + 1277$
$P = 1277$
$(0, 1277)$

The population of Maine in 2000 was 1277 thousand people.

b. t-intercept:

$0 = 9.5t + 1277$
$-1277 = 9.5t + 1277 - 1277$
$-1277 = 9.5t$
$\dfrac{-1277}{9.5} = \dfrac{9.5}{9.5}t$
$-134.4 \approx t$
$(-134.4, 0)$

The population of Maine was 0 people in 1866. This is model breakdown.

19. $M = 44t + 2798$

a. M-intercept:

$M = 44(0) + 2798$
$M = 2798$
$(0, 2798)$

The number of Florida residents enrolled in Medicare in 2000 was 2798 thousand people.

b. t-intercept:

$0 = 44t + 2798$
$-2798 = 44t + 2798 - 2798$
$-2798 = 44t$
$\dfrac{-2798}{44} = \dfrac{44t}{44}$
$63.6 \approx t$
$(63.6, 0)$

There were 0 residents of Florida that were enrolled in Medicare in 1936. This is model breakdown.

21. $2x + 4y = 8$

vertical intercept:

$2(0) + 4y = 8$
$\dfrac{4y}{4} = \dfrac{8}{4}$
$y = 2$
$(0, 2)$

horizontal intercept:

$2x + 4(0) = 8$
$\dfrac{2x}{2} = \dfrac{8}{2}$
$x = 4$
$(4, 0)$

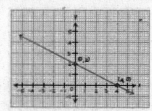

23. $3x - 5y = 15$

vertical intercept:

$3(0) - 5y = 15$
$\dfrac{-5y}{-5} = \dfrac{15}{-5}$
$y = -3$
$(0, -3)$

horizontal intercept:

$3x - 5(0) = 15$
$\dfrac{3x}{3} = \dfrac{15}{3}$
$x = 5$
$(5, 0)$

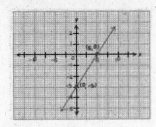

25. $4x + 6y = 30$

vertical intercept:
$4(0) + 6y = 30$
$\dfrac{6y}{6} = \dfrac{30}{6}$
$y = 5$
$(0, 5)$

horizontal intercept:
$4x + 6(0) = 30$
$\dfrac{4x}{4} = \dfrac{30}{4}$
$x = \dfrac{15}{2} = 7.5$
$(7.5, 0)$

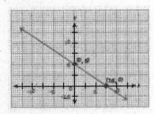

27. $-3x + 2y = 16$

vertical intercept:
$-3(0) + 2y = 16$
$\dfrac{2y}{2} = \dfrac{16}{2}$
$y = 8$
$(0, 8)$

horizontal intercept:
$-3x + 2(0) = 16$
$\dfrac{-3x}{-3} = \dfrac{16}{-3}$
$x = -5\tfrac{1}{3}$
$(-5\tfrac{1}{3}, 0)$

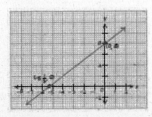

29. $-2x - 4y = -20$

vertical intercept:
$-2(0) - 4y = -20$
$\dfrac{-4y}{-4} = \dfrac{-20}{-4}$
$y = 5$
$(0, 5)$

horizontal intercept:
$-2x - 4(0) = -20$
$\dfrac{-2x}{-2} = \dfrac{-20}{-2}$
$x = 10$
$(10, 0)$

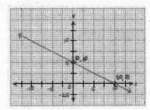

31. $x = 4$

no vertical intercept

horizontal intercept:
$x = 4$
$(4, 0)$

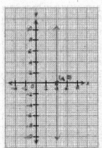

33. $y = 3$

no horizontal intercept

vertical intercept:
$y = 3$
$(0, 3)$

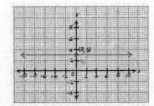

Chapter 1 Linear Functions

35. The graph of the line goes through the point $(0,0)$, but if you plug $x=0$ and $y=0$ into the equation, you get a false statement.

37. This is the graph of the line $y=-1$.

39. $y=\dfrac{3}{7}x-6$. The y-intercept is $(0,-6)$. The slope is $\dfrac{3}{7}$. Plot a point at the y-intercept, then using the slope, go up 3 and to the right 7.

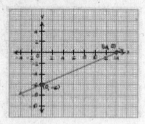

41.
$y-5=3(x-4)$
$y-5=3x-12$
$y-5+5=3x-12+5$
$y=3x-7$

The y-intercept is $(0,-7)$. The slope is $3=\dfrac{3}{1}$. Plot a point at the y-intercept, then using the slope, go up 3 and to the right 1.

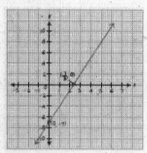

43. $y-4=\dfrac{1}{2}(x+1)$

Let $x=0$ and solve for y:
$y-4=\dfrac{1}{2}(0+1)$
$y-4=\dfrac{1}{2}(1)$
$y-4=\dfrac{1}{2}$
$y=4.5$

Plot a point at the y-intercept $(0,4.5)$.

Let $y=0$ and solve for x:
$0-4=\dfrac{1}{2}(x+1)$
$-4=\dfrac{1}{2}(x+1)$
$\dfrac{2}{1}\cdot(-4)=\dfrac{2}{1}\cdot\dfrac{1}{2}(x+1)$
$-8=x+1$
$x=-9$

Plot a point at the x-intercept $(-9,0)$.
Draw a line through the two points.

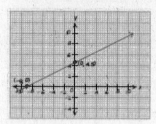

45. The line $y=7$ is a horizontal line through the point $(0,7)$.

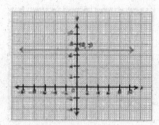

47. The line $x=9$ is a vertical line through the point $(9,0)$.

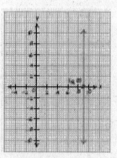

49. The line $y=-4.5$ is a horizontal line through the point $(0,-4.5)$.

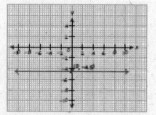

51. The line $x = -8$ is a vertical line through the point $(-8, 0)$.

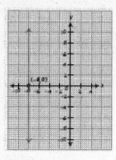

53.
$$y + 3x = 6\left(\frac{1}{2}x - 2\right)$$
$$y + 3x = 3x - 12$$
$$y + 3x - 3x = 3x - 12 - 3x$$
$$y = -12$$

The line $y = -12$ is a horizontal line through the point $(0, -12)$.

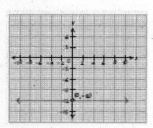

55.
$$5y + x = 2y + 3(y - 10)$$
$$5y + x = 2y + 3y - 30$$
$$5y + x = 5y - 30$$
$$5y + x - 5y = 5y - 30 - 5y$$
$$x = -30$$

The line $x = -30$ is a vertical line through the point $(-30, 0)$.

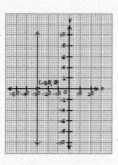

57.
$$y + 2 = 2(x + 4) - 2x$$
$$y + 2 = 2x + 8 - 2x$$
$$y + 2 = 8$$
$$y = 6$$

The line $y = 6$ is a horizontal line through the point $(0, 6)$.

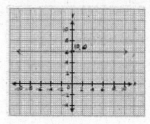

59. $y = \dfrac{2}{15}x - 6$

The y-intercept is $(0, -6)$. The slope is $\dfrac{2}{15}$. Plot a point at the y-intercept, then using the slope, go up 2 and to the right 15.

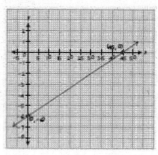

1.4 Exercises

Chapter 1 Linear Functions

61. $10x - 25y = 100$

Let $x = 0$ and solve for y:
$10(0) - 25y = 100$
$0 - 25y = 100$
$-25y = 100$
$y = -4$
Plot a point at the y-intercept $(0, -4)$.

Let $y = 0$ and solve for x:
$10x - 25(0) = 100$
$10x - 0 = 100$
$10x = 100$
$x = 10$
Plot a point at the x-intercept $(10, 0)$.
Draw a line through the two points.

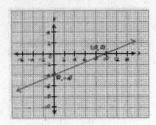

63. $y = 0.001x - 20$

The y-intercept is $(0, -20)$. The slope is $0.001 = \dfrac{1}{1000}$.

Plot a point at the y-intercept, then using the slope, go up 1 and to the right 1000.

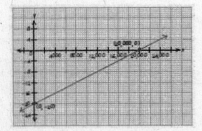

Section 1.5

1.

The y-intercept is $(0,-1)$.

The slope is $\frac{1}{2}$.

Since $m = \frac{1}{2}$ and $b = -1$

$y = \frac{1}{2}x - 1$

3.

Using two points: $(-1,-10)$ and $(3,0)$

Find the slope: $m = \frac{0-(-10)}{3-(-1)} = \frac{10}{4} = \frac{5}{2}$

Using $m = \frac{5}{2}$ and $(3,0)$.

$0 = \frac{5}{2}(3) + b$

$0 = 7.5 + b$

$b = -7.5$

Since $m = \frac{5}{2}$ amd $b = -7.5$

$y = \frac{5}{2}x - 7.5$

5.

The y-intercept is $(0,5)$.

The slope is $\frac{-5}{1} = -5$.

Since $m = -5$ and $b = 5$

$y = -5x + 5$

7.

The y-intercept is $(0,3)$.

The slope is $\frac{-3}{-10} = \frac{3}{10}$.

Since $m = \frac{3}{10}$ and $b = 3$

$y = \frac{3}{10}x + 3$

9.

Using two points: $(2,5)$ and $(6,0)$

Find the slope: $m = \frac{0-5}{6-2} = -\frac{5}{4}$

Using $m = -\frac{5}{4}$ and $(6,0)$.

$0 = -\frac{5}{4}(6) + b$

$0 = -7.5 + b$

$b = 7.5$

Since $m = -\frac{5}{4}$ amd $b = 7.5$

$y = -\frac{5}{4}x + 7.5$

11. The equation for a horizontal line that goes through the point $(0,4)$ is $y = 4$.

13. The equation for a vertical line that goes through the point $(3,0)$ is $x = 3$.

15. Sherry's work is correct. One of the points that Maritza used did not clearly cross the intersection on the graph paper. The point $(0,1.5)$ was incorrect.

17. Let C be the total cost of shirts in dollars, and let n be the total number of shirts purchased.

$(10,110)$ & $(30,280)$

Slope: $m = \frac{280-110}{30-10} = \frac{170}{20} = 8.5$

Use the point slope form of the equation:

$C - 110 = 8.5(n - 10)$

$C - 110 = 8.5n - 85$

$C - 110 + 110 = 8.5n - 85 + 110$

$C = 8.5n + 25$

19. Let P be the total population of Washington State in thousands, and let t be the years since 2000.

$(15,6951)$ & $(25,7996)$

Slope: $m = \frac{7996 - 6951}{25 - 15} = \frac{1045}{10} = 104.5$

Use the point slope form of the equation:

$P - 6951 = 104.5(t - 15)$

$P - 6951 = 104.5t - 1567.5$

$P - 6951 + 6951 = 104.5t - 1567.5 + 6951$

$P = 104.5t + 5383.5$

Chapter 1 Linear Functions

21. Let W be a woman's optimal weight in pounds, and let h be a woman's height in inches above 5 feet.

$(0,100)$ & $(6,130)$

Slope: $m = \dfrac{130-100}{6-0} = \dfrac{30}{6} = 5$

The vertical intercept is $(0,100)$

$W = 5h + 100$

23. Let G be the total number of teenagers who underwent gastric bypass surgery, and let t be years since 2000.

$(3, 771)$ & $(1, 408)$

Slope: $m = \dfrac{771-408}{3-1} = \dfrac{363}{2} = 181.5$

Use the point slope form of the equation:

$G - 408 = 181.5(t-1)$
$G - 408 = 181.5t - 181.5$
$G - 408 + 408 = 181.5t - 181.5 + 408$
$G = 181.5t + 226.5$

25.

Using $m = -2$ and $(3, 7)$.
$7 = -2(3) + b$
$7 = -6 + b$
$b = 13$
Since $m = -2$ amd $b = 13$
$y = -2x + 13$

27.

Using $m = \dfrac{2}{3}$ and $(12, 20)$.

$20 = \dfrac{2}{3}(12) + b$

$20 = 8 + b$

$b = 12$

Since $m = \dfrac{2}{3}$ amd $b = 12$

$y = \dfrac{2}{3}x + 12$

29.

Using two points: $(1, 3)$ and $(4, 12)$

Find the slope: $m = \dfrac{12-3}{4-1} = 3$

Using $m = 3$ and $(1, 3)$.
$y - 3 = 3(x - 1)$
$y - 3 = 3x - 3$
$y - 3 + 3 = 3x - 3 + 3$
$y = 3x$

31.

Using two points: $(7, 6)$ and $(21, -1)$

Find the slope: $m = \dfrac{-1-6}{21-7} = -\dfrac{1}{2}$

Using $m = -\dfrac{1}{2}$ and $(7, 6)$.

$y - 6 = -\dfrac{1}{2}(x - 7)$

$y - 6 = -\dfrac{1}{2}x + 3.5$

$y - 6 + 6 = -\dfrac{1}{2}x + 3.5 + 6$

$y = -\dfrac{1}{2}x + 9.5$

33.

Using two points: $(-4, -5)$ and $(-1, 7)$

Find the slope: $m = \dfrac{7-(-5)}{-1-(-4)} = 4$

Using $m = 4$ and $(-4, -5)$.
$y + 5 = 4(x + 4)$
$y + 5 = 4x + 16$
$y + 5 - 5 = 4x + 16 - 5$
$y = 4x + 11$

35.

Using two points: $(7, -3)$ and $(7, 9)$

Find the slope: $m = \dfrac{9-(-3)}{7-7} = \dfrac{12}{0}\Big\}$ undefined

If the slope is undefined, then the line is a vertical line. The vertical line that goes through both $(7, -3)$ and $(7, 9)$ is $x = 7$.

37.

Using two points: $(2,8)$ and $(4,8)$

Find the slope: $m = \dfrac{8-8}{4-2} = \dfrac{0}{2} = 0$

If the slope is zero, then the line is a horizontal line. The horizontal line that goes through both $(2,8)$ and $(4,8)$ is $y = 8$.

39.

Using two points: $(0,7)$ and $(2,11)$

Find the slope: $m = \dfrac{11-7}{2-0} = 2$

Since $m = 2$ and $b = 7$

$y = 2x + 7$

41.

Using two points: $(2,1)$ and $(4,-13)$

Find the slope: $m = \dfrac{-13-1}{4-2} = -7$

Using $m = -7$ and $(2,1)$.

$y - 1 = -7(x - 2)$
$y - 1 = -7x + 14$
$y - 1 + 1 = -7x + 14 + 1$
$y = -7x + 15$

43.

Using two points: $(8,6)$ and $(11,18)$

Find the slope: $m = \dfrac{18-6}{11-8} = 4$

Using $m = 4$ and $(8,6)$.

$y - 6 = 4(x - 8)$
$y - 6 = 4x - 32$
$y - 6 + 6 = 4x - 32 + 6$
$y = 4x - 26$

45.

$y = 3x + 5 \;\rightarrow\; m_1 = 3$
$y = 3x - 7 \;\rightarrow\; m_2 = 3$

Since $m_1 = m_2$ the lines are parallel.

47.

$2x + 3y = 15$
$2x + 3y - 2x = 15 - 2x$
$3y = -2x + 15$
$\dfrac{3y}{3} = \dfrac{-2x}{3} + \dfrac{15}{3}$
$y = -\dfrac{2}{3}x + 5 \;\rightarrow\; m_1 = -\dfrac{2}{3}$

$y = \dfrac{3}{2}x + 4 \;\rightarrow\; m_2 = \dfrac{3}{2}$

Since $m_1 = -\dfrac{1}{m_2}$ the lines are perpendicular.

49.

$2x - 5y = 40$
$2x - 5y - 2x = 40 - 2x$
$-5y = -2x + 40$
$\dfrac{-5y}{-5} = \dfrac{-2x}{-5} + \dfrac{40}{-5}$
$y = \dfrac{2}{5}x - 8 \;\rightarrow\; m_1 = \dfrac{2}{5}$

$-4y = 10x + 10$
$\dfrac{-4y}{-4} = \dfrac{10x}{-4} + \dfrac{10}{-4}$
$y = -\dfrac{5}{2}x - 2.5 \;\rightarrow\; m_2 = -\dfrac{5}{2}$

Since $m_1 = -\dfrac{1}{m_2}$, the lines are perpendicular.

51.

$4x - 3y = 20$
$4x - 3y - 4x = 20 - 4x$
$-3y = -4x + 20$
$\dfrac{-3y}{-3} = \dfrac{-4x}{-3} + \dfrac{20}{-3}$
$y = \dfrac{4}{3}x - \dfrac{20}{3} \;\rightarrow\; m_1 = \dfrac{4}{3}$

$12x - 9y = 30$
$12x - 9y - 12x = 30 - 12x$
$-9y = -12x + 30$
$\dfrac{-9y}{-9} = \dfrac{-12x}{-9} + \dfrac{30}{-9}$
$y = \dfrac{4}{3}x - \dfrac{10}{3} \;\rightarrow\; m_2 = \dfrac{4}{3}$

Since $m_1 = m_2$ the lines are parallel.

Chapter 1 Linear Functions

53. Perpendicular, any horizontal line is perpendicular to any vertical line.

55.
$2x+5y=20$
$2x+5y-2x=20-2x$
$5y=-2x+20$
$\dfrac{5y}{5}=\dfrac{-2x}{5}+\dfrac{20}{5}$
$y=-\dfrac{2}{5}x+4 \rightarrow m_1=-\dfrac{2}{5}$

$10x-4y=20$
$10x-4y-10x=20-10x$
$-4y=-10x+20$
$\dfrac{-4y}{-4}=\dfrac{-10x}{-4}+\dfrac{20}{-4}$
$y=\dfrac{5}{2}x-5 \rightarrow m_2=\dfrac{5}{2}$

Since $m_1=-\dfrac{1}{m_2}$, the lines are perpendicular.

57.
$5x+y=7$
$5x+y-5x=7-5x$
$5x+y=-5x+7$
$y=-5x+7 \rightarrow m_1=-5$

$2y=10x-9$
$\dfrac{2y}{2}=\dfrac{10x}{2}-\dfrac{9}{2}$
$y=5x-4.5 \rightarrow m_2=5$

Since $m_1 \neq m_2$ the lines are not parallel. Since $m_1 \neq -\dfrac{1}{m_2}$, the lines are not perpendicular. The lines are neither parallel nor perpendicular.

59. Since the lines are parallel, $m_1=m_2$.

$y=4x-13$
Using $m=4$ and the point $(2,8)$
$y-8=4(x-2)$
$y-8=4x-8$
$y-8+8=4x-8+8$
$y=4x$

61. Since the lines are parallel, $m_1=m_2$.
$10x-15y=-12$
$10x-15y-10x=-12-10x$
$-15y=-10x-12$
$\dfrac{-15y}{-15}=\dfrac{-10x}{-15}-\dfrac{12}{-15}$
$y=\dfrac{2}{3}x+\dfrac{4}{5}$

Using $m=\dfrac{2}{3}$ and the point $(-6,8)$
$y-8=\dfrac{2}{3}(x+6)$
$y-8=\dfrac{2}{3}x+4$
$y=\dfrac{2}{3}x+12$

63. Since the lines are perpendicular, $m_1=-\dfrac{1}{m_2}$.

$y=2x-1$
Using $m=-\dfrac{1}{2}$ and the point $(1,7)$
$y-7=-\dfrac{1}{2}(x-1)$
$y-7=-\dfrac{1}{2}x+\dfrac{1}{2}$
$y-7+7=-\dfrac{1}{2}x+\dfrac{1}{2}+7$
$y=-\dfrac{1}{2}x+\dfrac{15}{2}$

65. Since the lines are perpendicular, $m_1=-\dfrac{1}{m_2}$.

$4x+y=5$
$y=-4x+5$

Using $m=\dfrac{1}{4}$ and the point $(5,1)$
$y-1=\dfrac{1}{4}(x-5)$
$y-1=\dfrac{1}{4}x-1.25$
$y-1+1=\dfrac{1}{4}x-1.25+1$
$y=\dfrac{1}{4}x-0.25$

67. Since the lines are perpendicular, $m_1 = -\dfrac{1}{m_2}$.

$y = \dfrac{1}{5}x - 8$

Using $m = -5$ and the point $(2,3)$

$y - 3 = -5(x - 2)$

$y - 3 = -5x + 10$

$y - 3 + 3 = -5x + 10 + 3$

$y = -5x + 13$

69. Since the two lines are perpendicular, and the first line is horizontal, then the other line is vertical. The vertical line that goes through the point $(4,3)$ is $x = 4$.

71. The student needs to take the negative reciprocal of the original slope to find the slope of the line perpendicular.

$m = -\dfrac{3}{2}$.

73. The slopes must be the same if the two lines are parallel. To find the slope of the original line, put it into slope-intercept form.

$2x + 5y = 14$

$2x + 5y - 2x = 14 - 2x$

$5y = -2x + 14$

$\dfrac{5y}{5} = \dfrac{-2x}{5} + \dfrac{14}{5}$

$y = -\dfrac{2}{5}x + \dfrac{14}{5}$

$m = -\dfrac{2}{5}$

75. The point $(2,6)$ is not the y-intercept, so b is not 6.

$y - 6 = 4(x - 2)$

$y - 6 = 4x - 8$

$y - 6 + 6 = 4x - 8 + 6$

$y = 4x - 2$

77. $P = 75.8t + 5906.2$

a. Slope $= 75.8$. The population of Washington state increases 75.8 thousand each year.

b. The vertical intercept is $(0, 5906.2)$. In 2000, the population of Washington state was 5906.2 thousand.

c. Horizontal intercept:

$0 = 75.8t + 5906.2$

$-5906.2 = 75.8t$

$\dfrac{-5906.2}{75.8} = \dfrac{75.8t}{75.8}$

$-77.9 \approx t$

The horizontal intercept is $(-77.9, 0)$. In 1922, the population of Washington State was zero. This is model breakdown.

79. $M = 4.02t + 25.8$

a. Slope $= 4.02$. The annual amount of public expenditure on medical research in the United States increases $4.02 billion each year.

b. The vertical intercept is $(0, 25.8)$. The annual amount of public expenditure on medical research in the United States in 2000 was $25.8 billion.

c. Horizontal intercept:

$0 = 4.02t + 25.8$

$-25.8 = 4.02t$

$\dfrac{-25.8}{4.02} = \dfrac{4.02t}{4.02}$

$-6.42 \approx t$

The horizontal intercept is $(-6.42, 0)$. The annual amount of public expenditure on medical research in the United States was $0 in 1994. This is model breakdown.

81. $P = 500c - 6000$

a. Slope $= 500$. The monthly profit for a small used car lot increases by $500 for every car that is sold.

b. The vertical intercept is $(0, -6000)$. If the small used car lot sells 0 cars, it will lose $6000.

Chapter 1 Linear Functions

c. Horizontal intercept:
$$0 = 500c - 6000$$
$$6000 = 500c$$
$$\frac{6000}{500} = \frac{500c}{500}$$
$$12 = c$$

The horizontal intercept is (12, 0). The small used car lot will need to sell 12 cars in order to break even in their monthly profit.

83.

a. Let P the total profit in dollars Dan earns teaching surfing lessons, and let s be the number of one-hour surf lessons Dan gives. This can be represented by the equation $P = 30s - 700$.

b. Slope = 30. Dan's earnings increase by $30 for every one-hour surf lesson he gives.

c. The vertical intercept is $(0, -700)$. If Dan doesn't give any surf lessons, he will lose $700 in profit.

d. Horizontal intercept:
$$0 = 30s - 700$$
$$700 = 30s$$
$$\frac{700}{30} = \frac{30s}{30}$$
$$23.33 \approx c$$

The horizontal intercept is (23.33, 0). Dan will need to give 24 one-hour surf lessons in order to break even.

85.

a. Let T be the total amount of money raised by the PTA for a new track at Mission Meadows Elementary School in dollars, and let a be the total pounds of aluminum cans the PTA recycles. $T = 1.24a + 2000$.

b. Slope = 1.24. The funds raised by the PTA at Mission Meadows Elementary School increases by $1.24 for every pound of aluminum cans they recycle.

c. The vertical intercept is (0, 2000). The PTA at Mission Meadows Elementary School will have $2000 to put toward a new track if they recycle 0 pounds of aluminum cans.

d. Horizontal intercept:
$$0 = 1.24a + 2000$$
$$-2000 = 1.24a$$
$$\frac{-2000}{1.24} = \frac{1.24a}{1.24}$$
$$-1612.9 \approx a$$

The horizontal intercept is $(-1612.9, 0)$. The PTA at Mission Meadows Elementary School will need to recycle -1612.9 pounds of aluminum cans in order to earn $0 for a new track. This is model breakdown.

87.

a. Let P be the total percentage of Americans who have been diagnosed with diabetes, and let t be years since 2000. Use $(2, 4.8)$ and $(4, 5.1)$:

$$\text{Slope} = \frac{5.1 - 4.8}{4 - 2} = \frac{0.3}{2} = 0.15$$

Use the point slope form of the equation:
$$P - 4.8 = 0.15(t - 2)$$
$$P - 4.8 = 0.15t - 0.3$$
$$P - 4.8 + 4.8 = 0.15t - 0.3 + 4.8$$
$$P = 0.15t + 4.5$$

b.
$$P = 0.15(10) + 4.5$$
$$P = 6$$

This implies 6% of Americans will be diagnosed with diabetes in 2010.

c. Slope = 0.15. The percentage of Americans diagnosed with diabetes increases 0.15 percentage points each year.

89. Let C be the total cost of shirts in dollars, and let n be the total number of shirts purchased.
$$C = 8.5n + 25$$

a.
$$C = 8.5(50) + 25$$
$$C = 425 + 25$$
$$C = 450$$

It will cost $450 for 50 shirts.

b. Slope = 8.5. The cost per shirt increases by $8.50 per shift.

c. The vertical intercept is (0, 25). It costs $25 if 0 shirts are purchased. This is model breakdown.

d.
$$0 = 8.5n + 25$$
$$-25 = 8.5n$$
$$\frac{-25}{8.5} = \frac{8.5n}{8.5}$$
$$-2.94 \approx n$$

The horizontal intercept is $(-2.94, 0)$. To pay $0, you would have to purchase -3 shirts. This is model breakdown.

91. Let P be the total population of Washington state in thousands, and let t be the years since 2000.
$$P = 104.5t + 5383.5$$

a.
$$P = 104.5(30) + 5383.5$$
$$P = 3135 + 5383.5$$
$$P = 8518.5$$

In 2030, the population of Washington state will be about 8518.5 thousand.

b. Slope = 104.5. The population of Washington state increases by 104.5 thousand each year.

c. The vertical intercept is (0, 5383.5). In 2000, the population of Washington state was 5383.5 thousand.

d.
$$0 = 104.5t + 5383.5$$
$$-5383.5 = 104.5t$$
$$\frac{-5383.5}{104.5} = \frac{104.5t}{104.5}$$
$$-51.52 \approx t$$

The horizontal intercept is $(-51.52, 0)$. In 1949, the population of Washington State was zero. This is model breakdown.

93. $W = 5h + 100$, where W is the weight of the woman, and h is the number of inches past 5 feet.

a.
6 feet is 12 inches over 5 feet.
Use $h = 12$:
$$W = 5(12) + 100$$
$$W = 60 + 100$$
$$W = 160$$

The optimal weight of a woman who is 6 feet tall is 160 pounds.

b. Slope = 5. For every inch taller than 5 feet, a woman's optimal weight increases 5 pounds.

c. The vertical intercept is (0, 100). A woman's optimal weight at 5 feet tall is 100 pounds.

d.
$$0 = 5h + 100$$
$$-100 = 5h$$
$$\frac{-100}{5} = \frac{5h}{5}$$
$$-20 = h$$

The horizontal intercept is $(-20, 0)$. At 3 feet 4 inches, a woman's optimal weight is 0 pounds. This is model breakdown.

95. Let G be the total number of teenagers who underwent gastric bypass surgery t years since 2000.
$$G = 181.5t + 226.5$$

a.
$$G = 181.5(5) + 226.5$$
$$G = 907.5 + 226.5$$
$$G = 1134$$

The number of teenagers who had gastric bypass surgery in 2005 was 1134.

b. Slope = 181.5. The number of teenagers who undergo gastric bypass surgery increases by about 182 each year.

c. The vertical intercept is (0, 226.5). In 2000, 227 teenagers underwent gastric bypass surgery.

Chapter 1 Linear Functions

d.

$0 = 181.5t + 226.5$

$-181.5t = 226.5$

$\dfrac{-181.5t}{-181.5} = \dfrac{226.5}{-181.5}$

$t \approx -1.25$

The horizontal intercept is $(-1.25, 0)$ (-1.25, 0). In 1998, 0 teenagers underwent gastric bypass surgery. This is model breakdown.

Section 1.6

1. b

3. c

5. c

7. a

9. x-scl = 1, y-scl = 5

11. x-scl = 1, y-scl = 0.1

13. The student switched the min and max for the x-values.

```
WINDOW
 Xmin=-45
 Xmax=-5
 Xscl=1
 Ymin=-20
 Ymax=35
 Yscl=1
↓Xres=■
```

15.

a. Let R be the revenue in thousands of dollars for Quick Tire Repair, Inc. and t be the years since 2000.

t	R
6	608
7	611
8	616
9	620
10	625

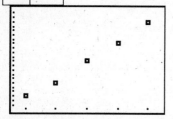

$R = mt + b$

Using two points: $(6, 608)$ and $(9, 620)$

Find the slope: $m = \dfrac{620 - 608}{9 - 6} = \dfrac{12}{3} = 4$

$R = 4t + b$

$b = 608 - 4(6)$

$b = 584$

$R = 4t + 584$

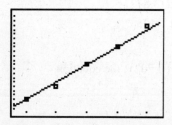

b. Domain [2,14]

$R = 4(2) + 584 = 592$

$R = 4(14) + 584 = 640$

Range [592, 640]

c.

$700 = 4t + 584$

$700 - 584 = 4t + 584 - 584$

$116 = 4t$

$\dfrac{116}{4} = \dfrac{4t}{4}$

$29 = t$

In 2029, the revenue for Quick Tire Repair, Inc. will be $700 thousand.

d. $S = 4$. Quick Tire Repair, Inc. revenue is increasing about $4 thousand per year.

17. $y = 2.5x + 8$; The graph must shift up so b needs to be increased.

19. $y = -0.75x + 20$; The slope needs to be less negative so m needs to be increased.

21.

a. Let M be millions of metric tons of beef and pork, and let t be years since 2000.

t	M
0	132.1
1	133.2
2	137.7
3	139.0
4	141.2
5	144.5

$M = mt + b$

Using two points: $(0, 132.1)$ and $(3, 139.0)$

Find the slope: $m = \dfrac{139.0 - 132.1}{3 - 0} = 2.3$

Chapter 1 Linear Functions

Since the vertical intercept is $(0,132.1)$, the equation becomes

$M = 2.3t + 132.1$

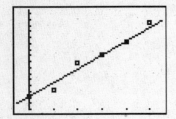

b.

$M = 2.3(6) + 132.1$

$M = 145.9$

In 2006, there were about 145.9 million metric tons of beef and pork produced worldwide.

c.

$175 = 2.3t + 132.1$

$175 - 132.1 = 2.3t + 132.1 - 132.1$

$42.9 = 2.3t$

$\dfrac{42.9}{2.3} = \dfrac{2.3t}{2.3}$

$18.65 \approx t$

In 2018, the worldwide production of beef and pork will be about 175 million metric tons.

d. Domain $[-3, 10]$

$M = 2.3(-3) + 132.1 = 125.2$

$M = 2.3(10) + 132.1 = 155.1$

Range $[125.2, 155.1]$

e. Slope = 2.3. There is an increase of 2.3 million metric tons per year in the production of beef and pork worldwide.

23. The slope is incorrect. The numerator should be the difference in cost and denominator should be the difference in number of bars. The slope should be

$m = \dfrac{365 - 265}{100 - 20} = 1.25$. This made the value of b incorrect

as well. The calculation should be

$265 = 1.25(20) + b$

$265 = 25 + b$

$240 = b$

$C = 1.25b + 240$

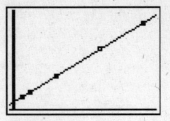

25. The slope is incorrect. The order of subtraction is reversed in the numerator. The slope should be

$m = \dfrac{1105 - 90}{3000 - 100} = 0.35$. This made the value of b incorrect

as well. The calculation should be:

$90 = 0.35(100) + b$

$90 = 35 + b$

$55 = b$

$C = 0.35p + 55$

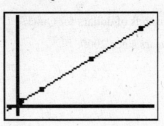

27.

a. Let C be the number of reported Chlamydia cases in thousands, and t be the number of years since 2000.

t	C
0	702
1	783
2	835
3	877
4	929

$C = mt + b$

Using two points: $(0, 702)$ and $(3, 877)$

Find the slope: $m = \dfrac{877 - 702}{3 - 0} = \dfrac{175}{3} \approx 58.3$

Since the vertical intercept is $(0, 702)$, the equation becomes:

$C = 58.3t + 702$

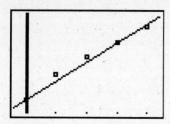
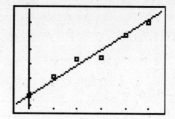

b. Slope = 58.3. The reported number of Chlamydia cases increases by 58.3 thousand cases per year.

c.

$C = 58.3(6) + 702$
$C = 349.8 + 702$
$C = 1051.8$

There is 1,051,800 estimated Chlamydia cases for 2006.

29.

a. Let E be the total egg production in the United States (in billions), and let t be years since 2000.

t	E
0	84.7
1	86.1
2	87.3
3	87.5
4	89.1
5	90.0

$E = mt + b$

Using two points: $(0, 84.7)$ and $(4, 89.1)$

Find the slope: $m = \dfrac{89.1 - 84.7}{4 - 0} = \dfrac{4.4}{4} = 1.1$

Since the vertical intercept is $(0, 84.7)$, the equation becomes

$E = 1.1t + 84.7$

b.

$E = 1.1(8) + 84.7$
$E = 8.8 + 84.7$
$E = 93.5$

In 2008, the total egg production in the United States was 93.5 billion.

c. Domain $[-4, 10]$

$E = 1.1(-4) + 84.7 = 80.3$
$E = 1.1(10) + 84.7 = 95.7$

Range $[80.3, 95.7]$

d. Slope = 1.1. The egg production is increased by 1.1 billion eggs per year in the United States.

31. Let U, S, and F be the number of Internet users in the United Kingdom, Spain, and Finland in millions, respectively, and let t be years since 1990.

t	U	S	F
8	6.7	0.6	1.4
9	12.5	2.8	1.7
10	18.0	5.4	1.9
11	24.0	7.7	2.2
14	38.9	14.3	2.9

United Kingdom : $U = mt + b$

Using two points: $(9, 12.5)$ and $(10, 18.0)$

$m = \dfrac{18.0 - 12.5}{10 - 9} = 5.5$

$b = 18.0 - 5.5(10) = 37$

$\boxed{U = 5.5t + 37}$

Chapter 1 Linear Functions

Spain: $S = mt + b$

Using two points: $(8, 0.6)$ and $(11, 7.4)$

$m = \dfrac{7.4 - 0.6}{11 - 8} = \dfrac{6.8}{3} \approx 2.27$

$b = 7.4 - 2.27(11) = -17.57 \approx -17.6$

$\boxed{S = 2.27t - 17.6}$

Finland: $F = mt + b$

Using two points: $(8, 1.4)$ and $(14, 2.9)$

$m = \dfrac{2.9 - 1.4}{14 - 8} = \dfrac{1.5}{6} = 0.25$

$b = 1.4 - 0.25(8) = -0.6$

$\boxed{F = 0.25t - 0.6}$

33. By comparing the slopes of the models we can see that the United Kingdom has the greatest growth.

United Kingdom $(5.5) >$ Spain $(2.27) >$ Finland (0.25)

35.

a. Let R be total revenue for FedEx in millions of dollars, and let t be the number of years since 2000.

t	R
0	18,257
1	19,629
2	20,607
3	22,487
4	24,710
5	29,363
6	32,294
7	35,214

$R = mt + b$

Using two points: $(1, 19629)$ and $(5, 29363)$

Find the slope: $m = \dfrac{29{,}363 - 19{,}629}{5 - 1} = 2433.5$

$R = 2433.5t + b$

$b = 19{,}629 - 2433.5(1)$

$b = 17{,}195.5$

$R = 2433.5t + 17{,}195.5$

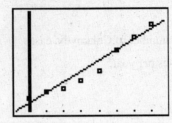

b. Domain $[-3, 10]$

$R = 2433.5(-3) + 17{,}195.5 = 9895$

$R = 2433.5(10) + 17{,}195.5 = 41{,}531$

Range $[9895, 41531]$

c. The slope is 2433.5. The total revenue for FedEx increases by $2433.5 million each year.

d.

$R = 2433.5(10) + 17{,}195.5$

$R = 41{,}530.5$

In 2010, the total revenue for FedEx will be $41530.5 million.

37.

a. Let M be the total consumption of milk in the United States in millions of gallons, and let t be years since 2000.

t	M
0	9297
2	9244
3	9192
4	9141
5	9119

$M = mt + b$

Using two points: $(3, 9192)$ and $(5, 9119)$

Find the slope: $m = \dfrac{9119 - 9192}{5 - 3} = -36.5$

$M = -36.5t + b$
$b = 9192 - (-36.5)(3)$
$b = 9301.5$
$M = -36.5t + 9301.5$

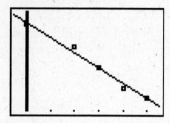

t	H
6	251
7	266
8	281
9	299
10	317
11	335

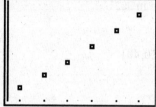

b. Domain $[-3, 10]$

$M = -36.5(-3) + 9301.5 = 9411$
$M = -36.5(10) + 9301.5 = 8936.5$

Range $[8936.5, 9411]$

c. The vertical intercept is (0, 9301.5). In 2000, the total consumption of milk in the United States was 9301.5 million gallons of milk.

d.
$9000 = -36.5t + 9301.5$
$9000 - 9301.5 = -36.5t + 9301.5 - 9301.5$
$-301.5 = -36.5t$
$\dfrac{-301.5}{-36.5} = \dfrac{-36.5t}{-36.5}$
$8.3 \approx t$

In 2008, Americans consumed approximately 9000 million gallons of milk.

e.
$0 = -36.5t + 9301.5$
$36.5t = 9301.5$
$\dfrac{36.5t}{36.5} = \dfrac{9301.5}{36.5}$
$t \approx 254.8$

The horizontal intercept is (254.8, 0). In 2255, Americans will consume 0 gallons of milk. This is model breakdown.

39.

a. Let H be the total amount spent by individuals on health care expenses in the United States in billions of dollars, and let t be the years since 2000.

$H = mt + b$

Using two points: (7, 266) and (10, 317)

Find the slope: $m = \dfrac{317 - 266}{10 - 7} = 17$

$H = 17t + b$
$b = 266 - 17(7)$
$b = 147$
$H = 17t + 147$

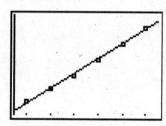

b. Domain [3, 14]

$H = 17(3) + 147 = 198$
$H = 17(14) + 147 = 385$

Range [198, 385]

c. The slope is 17. The amount of money Americans spend on health care expenses increases $17 billion each year.

d.
$H = 17(15) + 147$
$H = 402$

In 2015, Americans will spend $402 billion on health care expenses.

Chapter 1 Linear Functions

e.

$500 = 17t + 147$

$500 - 147 = 17t + 147 - 147$

$353 = 17t$

$\dfrac{353}{17} = \dfrac{17t}{17}$

$20.8 \approx t$

The amount that Americans will spend on health care expenses will reach $500 billion in 2021.

41.

Using two points: $(4, 4)$ and $(10, 40)$

Find the slope: $m = \dfrac{40 - 4}{10 - 4} = 6$

$4 = 6(4) + b$

$4 = 24 + b$

$b = -20$

Since $m = 6$ and $b = -20$

$y = 6x - 20$

43.

Using two points: $(1, 5)$ and $(5, -7)$

Find the slope: $m = \dfrac{5 - (-7)}{1 - 5} = -3$

$5 = -3(1) + b$

$5 = -3 + b$

$b = 8$

Since $m = -3$ and $b = 8$

$y = -3x + 8$

45.

Using two points: $(3, -4)$ and $(12, -1)$

Find the slope: $m = \dfrac{-1 - (-4)}{12 - 3} = \dfrac{1}{3}$

$-4 = \dfrac{1}{3}(3) + b$

$-4 = 1 + b$

$b = -5$

Since $m = \dfrac{1}{3}$ and $b = -5$

$y = \dfrac{1}{3}x - 5$

47.

Using two points: $(-8, 4.167)$ and $(7, 1.333)$

Find the slope: $m = \dfrac{1.333 - 4.167}{7 - (-8)} = -\dfrac{2.834}{11} \approx -0.189$

$4.167 = -0.189(-8) + b$

$4.167 = 1.512 + b$

$b \approx 2.66$

Since $m \approx -0.19$ and $b = 2.66$

$y = -0.19x + 2.66$

49.

Using two points: $(10, 492)$ and $(19, 627)$

Find the slope: $m = \dfrac{627 - 492}{19 - 10} = 15$

$492 = 15(10) + b$

$492 = 150 + b$

$b = 342$

Since $m = 15$ and $b = 342$

$y = 15x + 342$

51.

Using two points: $(7, -3755)$ and $(22, -4370)$

Find the slope: $m = \dfrac{-4370 - (-3755)}{22 - 7} = -41$

$-3755 = -41(7) + b$

$-3755 = -287 + b$

$b = -3468$

Since $m = -41$ and $b = -3468$

$y = -41x - 3468$

Section 1.7

1. This is not a function because two children the same age may not be in the same grade.
Input: a = years old.
Output: G = grade level of student.

3. This is not a function because two children the same age may not be the same height.
Input: a = years old.
Output: H = heights in inches of children attending Mission Meadows Elementary School.

5. This is a function.
Input: t = years after starting the investment.
Output: I = interest earned from an investment in dollars.

7. This is not a function because each year there will be several songs at the top of the pop charts.
Input: y = year.
Output: S = song at the top of the pop charts.

9. This is a function.
Input: t = year.
Output: T = amount of taxes you paid in dollars.

11. This is a function. Every input value is unique.

13. This is not a function, because the input of 1 hour of poker play is related to two different amounts of winnings.

15. This is a function. Every input value is unique.

17. This is not a function, because the input of Monday is related to two different amounts of money spent on lunch.

19. Domain = {Jan, Feb, May, June}
Range = {5689.35, 7856.12, 2689.15, 1005.36}

21. Domain = {1, 2, 3, 4, 5}
Range = $\{-650, -100, -150, 60, 125, 200, 300\}$

23. The death rate from HIV for 20-year-olds is 0.5 per 100,000.

25. In 2002, 14,304 small business loans were made to minority-owned small businesses.

27. The amount spent on lunch on Tuesday was $5.95 and $6.33.

29. Yes, this is a function. The graph passes the Vertical line test.

31. No, this is not a function. The graph fails the Vertical line test.

33. Yes, this is a function. The graph passes the Vertical line test.

35. Yes, this is a function. The graph passes the Vertical line test.

37. Yes, this is a function. The graph passes the Vertical line test.

39. No, this is not a function. The graph fails the Vertical line test.

41.
a. $W(0) = 86.5$. At the start of a diet, a person's weight will be 86.5 kg.
b. $W(10) = 82$. Ten days after starting a diet, a person's weight will be 82 kg.
c. $W = 75$ when $d = 30$. Thirty days after starting a diet, a person's weight will be 75 kg.
d. $W(100) = 88$. One hundred days after starting a diet, a person's weight will be 88 kg.

43.
a. $C(5) = 8$. Eight ounces of chocolate were consumed in the Clark household on the fifth day of the month.
b. $C = 20$ when $d = 15$. Twenty ounces of chocolate were consumed in the Clark household on the 15th day of the month.
c. $C(30) = 28$. Twenty-eight ounces of chocolate were consumed in the Clark household on the 30th day of the month.

45.
a. $P(\text{Ohio}) = 11.48$
b. $P(\text{Texas}) = 23.507783$
c. $P(\text{Wyoming}) = 0.515004$

47. Let $B(t)$ be the average monthly Social Security benefit in dollars t years since 2000.

a.

t	$B(t)$
2	834
3	862
4	894
5	938
6	978

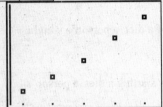

$B(t) = mt + b$

Using two points: $(3, 862)$ and $(5, 938)$

Find the slope: $m = \dfrac{938 - 862}{5 - 3} = 38$

$B(t) = 38t + b$
$b = 862 - 38(3)$
$b = 748$

$B(t) = 38t + 748$

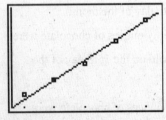

b. $B(9) = 38(9) + 748 = 1090$ In 2009, the average social security benefit for retired workers was $1090.

c.

$2000 = 38t + 748$
$2000 - 748 = 38t + 748 - 748$
$1252 = 38t$
$\dfrac{1252}{38} = \dfrac{38t}{38}$
$33 \approx t$

The average monthly Social Security benefit will be $2000 in 2033.

d. Domain: $[-1, 10]$

$B(-1) = 38(-1) + 748 = 710$
$B(10) = 38(10) + 748 = 1128$

Range: $[710, 1128]$

e. $(0, 748)$. In 2000, the average monthly Social Security benefit was $748.

49. Let $M(t)$ be the number of Medicare enrollees t years since 2000.

a.

t	$M(t)$
2	40.5
3	41.2
4	41.9
5	42.5

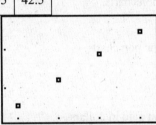

$M(t) = mt + b$

Using two points: $(2, 40.5)$ and $(5, 42.5)$

Find the slope: $m = \dfrac{42.5 - 40.5}{5 - 2} \approx 0.67$

$M(t) = 0.67t + b$
$b = 40.5 - 0.67(2)$
$b = 39.16$

$M(t) = 0.67t + 39.16$

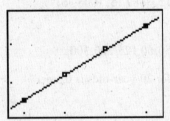

b. Domain: $[-3, 10]$

$M(-3) = 0.67(-3) + 39.16 = 37.15$
$M(10) = 0.67(10) + 39.16 = 45.86$

Range: $[37.15, 45.86]$

c. $(0, 39.16)$ In 2000, there were 39.16 million people enrolled in Medicare.

d.

$M(10) = 0.67(10) + 39.16$

$M(10) = 45.86$

In 2010, there were 45.86 million people enrolled in Medicare.

e.

$50 = 0.67t + 39.16$

$50 - 39.16 = 0.67t + 39.16 - 39.16$

$10.84 = 0.67t$

$\dfrac{10.84}{0.67} = \dfrac{0.67t}{0.67}$

$16.18 \approx t$

In 2016, there will be approximately 50 million people enrolled in Medicare.

51.

a.

$f(x) = 2x - 7$

$f(\ ) = 2(\ ) - 7$

$f(5) = 2(5) - 7$

$f(5) = 10 - 7$

$f(5) = 3$

b.

$f(x) = 2x - 7$

$f(\ ) = 2(\ ) - 7$

$f(-10) = 2(-10) - 7$

$f(-10) = -20 - 7$

$f(-10) = -27$

c.

$f(x) = 2x - 7$

$-1 = 2x - 7$

$-1 + 7 = 2x - 7 + 7$

$6 = 2x$

$x = 3$

d. Domain: All real numbers, range: All real numbers

53.

a.

$h(x) = \dfrac{2}{3}x + \dfrac{1}{3}$

$h(\ ) = \dfrac{2}{3}(\ ) + \dfrac{1}{3}$

$h(15) = \dfrac{2}{3}(15) + \dfrac{1}{3}$

$h(15) = \dfrac{30}{3} + \dfrac{1}{3}$

$h(15) = \dfrac{31}{3}$

b.

$h(x) = \dfrac{2}{3}x + \dfrac{1}{3}$

$h(\ ) = \dfrac{2}{3}(\ ) + \dfrac{1}{3}$

$h(-9) = \dfrac{2}{3}(-9) + \dfrac{1}{3}$

$h(-9) = \dfrac{-18}{3} + \dfrac{1}{3}$

$h(-9) = \dfrac{-17}{3}$

c.

$h(x) = \dfrac{2}{3}x + \dfrac{1}{3}$

$4 = \dfrac{2}{3}x + \dfrac{1}{3}$

$3(4) = 3\left(\dfrac{2}{3}x + \dfrac{1}{3}\right)$

$12 = 2x + 1$

$12 - 1 = 2x + 1 - 1$

$11 = 2x$

$\dfrac{11}{2} = \dfrac{2x}{2}$

$x = \dfrac{11}{2}$

d. Domain: All real numbers, range: All real numbers

55.

a.

$g(2) = -18$

b.

$g(-11) = -18$

c. Domain: All real numbers, range: $\{-18\}$

Chapter 1 Linear Functions

57.

The student set $x = 20$ and solved for $f(20)$.

The correct solution is:

$f(x) = 4x + 8$
$20 = 4x + 8$
$20 - 8 = 4x + 8 - 8$
$12 = 4x$
$x = 3$

59.

The student restricted the domain and range.

The correct solution is: Domain: All real numbers, range: All real numbers

61.

a.

$f(x) = 3.2x - 4.8$
$f(\ ) = 3.2(\ ) - 4.8$
$f(2) = 3.2(2) - 4.8$
$f(2) = 6.4 - 4.8$
$f(2) = 1.6$

b.

$f(x) = 3.2x - 4.8$
$f(\ ) = 3.2(\ ) - 4.8$
$f(-14) = 3.2(-14) - 4.8$
$f(-14) = -44.8 - 4.8$
$f(-14) = -49.6$

c.

$f(x) = 3.2x - 4.8$
$-10 = 3.2x - 4.8$
$-10 + 4.8 = 3.2x - 4.8 + 4.8$
$-5.2 = 3.2x$
$\dfrac{-5.2}{3.2} = \dfrac{3.2x}{3.2}$
$x = -1.625$

d. Domain: All real numbers, range: All real numbers

63.

a.

$h(x) = 14x + 500$
$h(\ ) = 14(\ ) + 500$
$h(105) = 14(105) + 500$
$h(105) = 1470 + 500$
$h(105) = 1970$

b.

$h(x) = 14x + 500$
$-140 = 14x + 500$
$-140 - 500 = 14x + 500 - 500$
$-640 = 14x$
$\dfrac{-640}{14} = \dfrac{14x}{14}$
$x = -\dfrac{320}{7}$

c. Domain: All real numbers, range: All real numbers

65.

a. $f(3) = 2$

b. $f(-2) = -8$

c. $x = -1$

d. Domain: All real numbers, range: All real numbers

e. Vertical intercept: $(0, -4)$. This is the point the graph crosses the y-axis.

Horizontal intercept: $(2, 0)$. This is the point the graph crosses the x-axis.

67.

a. $f(25) = 50$

b. $f(100) = 275$

c. $x = 75$

d. Domain: All real numbers, range: All real numbers

e. Vertical intercept $(0, -25)$. This is the point the graph crosses the y-axis.

Horizontal intercept $(8, 0)$. This is the point the graph crosses the x-axis.

69.

a. $h(2) = 3$

b. $h(5) = -4$

c. $x = -0.2$ and $x = 4.2$

d. Domain: All real numbers, range: $(-\infty, 3]$

e. Vertical intercept $(0,0)$. This is the point the graph crosses the *y*-axis.

Horizontal intercepts $(0,0)$ and $(4,0)$. These are the points the graph crosses the *x*-axis.

Chapter 1 Linear Functions
Chapter 1 Review Exercises

1.
$$2x+5=7(x-8)$$
$$2x+5=7x-56$$
$$2x+5-7x=7x-56-7x$$
$$-5x+5=-56$$
$$-5x+5-5=-56-5$$
$$-5x=-61$$
$$\frac{-5x}{-5}=\frac{-61}{-5}$$
$$x=\frac{61}{5}$$

2.
$$\frac{1}{3}x-\frac{2}{3}=5$$
$$3\left(\frac{1}{3}x-\frac{2}{3}\right)=3(5)$$
$$x-2=15$$
$$x-2+2=15+2$$
$$x=17$$

3.
$$0.4t+2.6=0.8(t-8.2)$$
$$0.4t+2.6=0.8t-6.56$$
$$0.4t+2.6-0.8t=0.8t-6.56-0.8t$$
$$-0.4t+2.6=-6.56$$
$$-0.4t+2.6-2.6=-6.56-2.6$$
$$-0.4t=-9.16$$
$$\frac{-0.4t}{-0.4}=\frac{-9.16}{-0.4}$$
$$t=22.9$$

4.
$$-2(x+3.5)+4(2x-1)=7x-5(2x+3)$$
$$-2x-7+8x-4=7x-10x-15$$
$$6x-11=-3x-15$$
$$6x-11+3x=-3x-15+3x$$
$$9x-11=-15$$
$$9x-11+11=-15+11$$
$$9x=-4$$
$$\frac{9x}{9}=\frac{-4}{9}$$
$$x=-\frac{4}{9}$$

5. $h=0.75d+4.5$

a.
$$h=0.75(7)+4.5$$
$$h=9.75$$

The height after 1 week (7 days) is 9.75 inches.

b.
$$12=0.75d+4.5$$
$$12-4.5=0.75d+4.5-4.5$$
$$7.5=0.75d$$
$$\frac{7.5}{0.75}=\frac{0.75d}{0.75}$$
$$10=d$$

The course grass attendants should cut the grass in the rough 10 days before the tournament.

6.

a. Let C be the cost of the satellite phone service in dollars, and m be the number of minutes used.
$$C=10m+10$$

b.
$$C=10(3)+10$$
$$C=40$$

A 3-minute call cost $40.

c.
$$300=10m+10$$
$$300-10=10m+10-10$$
$$290=10m$$
$$\frac{290}{10}=\frac{10m}{10}$$
$$29=m$$

The satellite call can last 29 minutes for $300.

7. Let C be the cost in dollars of renting a Bobcat for h hours.

a. $C=40h+15$

b. $C=40(2)+15$
$C=95$

Renting a Bobcat for 2 hours will cost $95.

c.
$$(3 \text{ days})\left(\frac{8 \text{ hours}}{\text{day}}\right)=24 \text{ hours}$$
$$C=40(24)+15$$
$$C=975$$

A three-day rental of a Bobcat will cost $975.

8.

a. Let C be the cost in dollars to produce n putters.

n	C
5	600.00
10	725.00
20	950.00
25	1100.00
50	1700.00

t	P
1	4199
2	4779
3	5310
4	6018
5	6605
6	7406
7	7950

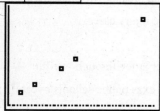

$C = mt + b$

Using two points: $(5, 600)$ and $(50, 1700)$

Find the slope: $m = \dfrac{1700 - 600}{50 - 5} \approx 24.44$

$C = 24.44t + b$
$b = 600 - 24.44(5)$
$b = 477.80$
$C = 24.44n + 477.80$

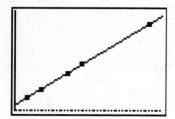

b. The vertical intercept is $(0, 477.80)$. It cost $477.80 to produce 0 putters.

c.

$C = 24.44(100) + 477.80$
$C = 2921.80$

The cost to produce 100 putters is $2921.80.

d. Domain: $[0, 100]$,

$C = 24.44(100) + 477.80 = 2921.80$

Range: $[477.80, 2921.80]$

e. The slope is 24.44. For every putter produced the cost increases by $24.44.

9.

a. Let P be the profit for Costco Wholesale Corporation in millions of dollars, t years since 2000.

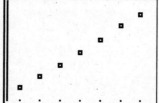

$P = mt + b$

Using two points: $(2, 4779)$ and $(7, 7950)$

Find the slope: $m = \dfrac{7950 - 4779}{7 - 2} = 634.2$

$P = 634.2t + b$
$b = 4779 - 634.2(2)$
$b = 3510.6$
$P = 634.2t + 3510.6$

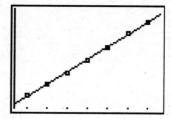

b.

$9000 = 634.2t + 3510.6$
$9000 - 3510.6 = 634.2t + 3510.6 - 3510.6$
$5489.4 = 634.2t$
$\dfrac{5489.4}{634.2} = \dfrac{634.2t}{634.2}$
$8.66 \approx t$

The gross profit for Costco Wholesale will be $9 billion in 2008.

c.

$P = 634.2(9) + 3510.6$
$P = 9218.4$

The gross profit for Costco Wholesale in 2009 will be $9.2184 billion.

Chapter 1 Linear Functions

d. Domain [0,9]

$P = 634.2(9) + 3510.6 = 9218.4$

Range [3510.6, 9218.4]

e. The slope is 634.2. The gross profit for Costco Wholesale increases by $634.2 million per year.

10.

a. Let P be the percent of full-time workers in private industry who filed an injury case t years since 2000.

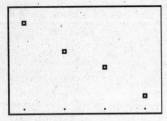

$P(t) = mt + b$

Using two points: $(3, 4.7)$ and $(6, 4.2)$

Find the slope: $m = \dfrac{4.2 - 4.7}{6 - 3} \approx -0.167$

$P(t) = -0.167t + b$
$b = 4.7 - (-0.167)(3)$
$b \approx 5.2$

$P(t) = -0.167t + 5.2$

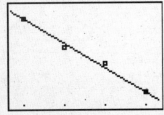

b. The slope is -0.167. The percent of full-time workers in private industry who filed an injury case is decreasing by about 0.167 percentage points per year.

c.

$P(3) = -0.167(3) + 5.2$
$P(3) = 4.699$

In 2003, about 4.7 percent of full-time workers in private industry filed an injury case.

d. Vertical intercept is $(0, 5.2)$. In 2000, about 5.2% of full-time workers in private industry filed an injury case.

e.

$0 = -0.167t + 5.2$
$0.167t = 5.2$
$\dfrac{0.167t}{0.167} = \dfrac{5.2}{0.167}$
$t \approx 31.14$

The horizontal intercept is $(31.14, 0)$. In 2031, there will be no injury cases filed by full-time workers in private industry. This is probably model breakdown since there will most likely always be some injury cases.

11.

a. Let S be the number of kindergarten through twelfth grade students, in millions, in Texas public schools t years since 2000.

t	S
0	4.00
1	4.07
2	4.16
3	4.26
4	4.33
5	4.40
6	4.52

$S(t) = mt + b$

Using two points: $(0, 4.00)$ and $(4, 4.33)$

Find the slope: $m = \dfrac{4.33 - 4.00}{4 - 0} \approx 0.08$

Since the vertical intercept is $(0, 4.00)$, the equation becomes

$S(t) = 0.08t + 4$

b. The slope is 0.08. The number of kindergarten through twelfth grade students in millions in Texas public schools is increasing by about 0.08 million (80 thousand) per year.

c.

$S(11) = 0.08(11) + 4$
$S(11) = 4.88$

In 2011, Texas can expect to have about 4.88 million kindergarten through twelfth grade students in public schools.

12. $C(t) = 0.56t + 4.3$

a.

$2015 \to t = 5$

$C(5) = 0.56(5) + 4.3$

$C(5) = 7.1$

7.1 million candles will be produced in 2015.

b.

$C(3) = 0.56(3) + 4.3$

$C(3) = 5.98$

5.98 million candles will be produced in 2013

c.

$10 = 0.56t + 4.3$

$10 - 4.3 = 0.56t + 4.3 - 4.3$

$5.7 = 0.56t$

$\dfrac{5.7}{0.56} = \dfrac{0.56t}{0.56}$

$10.18 \approx t$

10 million candles will be produced in 2020.

13.

$V = bT$

$\dfrac{V}{b} = \dfrac{bT}{b}$

$T = \dfrac{V}{b}$

14.

$F = ma$

$\dfrac{F}{a} = \dfrac{ma}{a}$

$m = \dfrac{F}{a}$

15.

$ax + by = c$

$ax + by - by = c - by$

$ax = c - by$

$\dfrac{ax}{a} = \dfrac{c - by}{a}$

$x = \dfrac{c - by}{a}$

16.

$2x - ay = b$

$2x - ay - 2x = b - 2x$

$-ay = b - 2x$

$\dfrac{-ay}{-a} = \dfrac{b - 2x}{-a}$

$y = \dfrac{b - 2x}{-a}$

$y = \dfrac{-1}{-1} \cdot \dfrac{(b - 2x)}{(-a)}$

$y = \dfrac{-b + 2x}{a}$

$y = \dfrac{2x - b}{a}$

17.

$y = 3x - 4$

Since $b = -4$ the y-intercept is $(0, -4)$.

To find the x-intercept let $y = 0$ and solve for x:

$0 = 3x - 4$

$4 = 3x$

$x = \dfrac{4}{3}$

Plot a point at the x-intercept $\left(\dfrac{4}{3}, 0\right)$.

Draw a line through the two points.

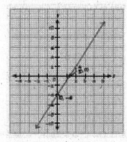

18.

$y = 2x + 3$

Since $b = 3$ the y-intercept is $(0, 3)$.

To find the x-intercept let $y = 0$ and solve for x:

$0 = 2x + 3$

$-3 = 2x$

$x = -1.5$

Plot a point at the x-intercept $(-1.5, 0)$.

Draw a line through the two points.

Chapter 1 Linear Functions

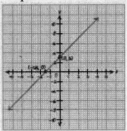

19.

$2x + 3y = 24$

To find the y-intercept let $x = 0$ and solve for y:

$2(0) + 3y = 24$

$0 + 3y = 24$

$3y = 24$

$y = 8$

Plot a point at the y-intercept $(0, 8)$.

To find the x-intercept let $y = 0$ and solve for x:

$2x + 3(0) = 24$

$2x + 0 = 24$

$2x = 24$

$x = 12$

Plot a point at the x-intercept $(12, 0)$.

Draw a line through the two points.

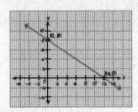

20.

$5x - 6y = 42$

To find the y-intercept let $x = 0$ and solve for y:

$5(0) - 6y = 42$

$0 - 6y = 42$

$-6y = 42$

$y = -7$

Plot a point at the y-intercept $(0, -7)$.

To find the x-intercept let $y = 0$ and solve for x:

$5x - 6(0) = 42$

$5x - 0 = 42$

$5x = 42$

$x = 8.4$

Plot a point at the x-intercept $(8.4, 0)$.

Draw a line through the two points.

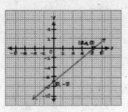

21.

$y = x^2 + 5$

To find the y-intercept let $x = 0$ and solve for y:

$y = (0)^2 + 5$

$y = 0 + 5$

$y = 5$

Plot a point at the y-intercept $(0, 5)$.

Plot additional points to graph the curve.

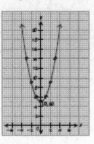

22.

$y = x^2 - 6$

To find the y-intercept let $x = 0$ and solve for y:

$y = (0)^2 - 6$

$y = 0 - 6$

$y = -6$

Plot a point at the y-intercept $(0, -6)$.

To find the x-intercept let $y = 0$ and solve for x:

$0 = x^2 - 6$

$6 = x^2$

$x = \pm\sqrt{6}$

$x \approx \pm 2.4$

Plot the points at the x-intercepts $(-2.4, 0)$ and $(2.4, 0)$.

Plot additional points to graph the curve.

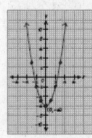

23.

$y = \dfrac{1}{2}x + 7$

Since $b = 7$ the y-intercept is $(0, 7)$.

To find the x-intercept let $y = 0$ and solve for x:

$0 = \dfrac{1}{2}x + 7$

$-7 = \dfrac{1}{2}x$

$2(-7) = 2\left(\dfrac{1}{2}x\right)$

$x = -14$

Plot a point at the x-intercept $(-14, 0)$.
Draw a line through the two points.

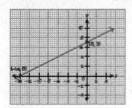

24.

$y = -\dfrac{2}{7}x + 4$

Since $b = 4$ the y-intercept is $(0, 4)$.

To find the x-intercept let $y = 0$ and solve for x:

$0 = -\dfrac{2}{7}x + 4$

$-4 = -\dfrac{2}{7}x$

$-4 = -\dfrac{2}{7}x$

$x = 14$

Plot a point at the x-intercept $(14, 0)$.
Draw a line through the two points.

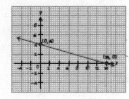

25.

Using two points: $(2, 7)$ and $(7, 27)$

Find the slope: $m = \dfrac{27 - 7}{7 - 2} = \dfrac{20}{5} = 4$

Using $m = 4$ and $(2, 7)$.

$7 = 4(2) + b$

$7 = 8 + b$

$b = -1$

Since $m = 4$ amd $b = -1$

$y = 4x - 1$

26.

Using two points: $(4, 9)$ and $(-3, 23)$

Find the slope: $m = \dfrac{23 - 9}{-3 - 4} = \dfrac{14}{-7} = -2$

Using $m = -2$ and $(4, 9)$.

$y - 9 = -2(x - 4)$

$y - 9 = -2x + 8$

$y - 9 + 9 = -2x + 8 + 9$

$y = -2x + 17$

27.

Since the lines are parallel $m_1 = m_2$.

$-3x + y = 12$

$y = 3x + 12$

Using $m = 3$ and the point $(4, 10)$

$y - 10 = 3(x - 4)$

$y - 10 = 3x - 12$

$y = 3x - 2$

28.

Since the lines are parallel $m_1 = m_2$.

$y = -0.5x + 4$

Using $m = -0.5$ and the point $(2, 16)$

$y - 16 = -0.5(x - 2)$

$y - 16 = -0.5x + 1$

$y = -0.5x + 17$

Chapter 1 Linear Functions

29.

Since the lines are perpendicular $m_1 = -\dfrac{1}{m_2}$.

$y = -4x + 7$

Using $m = \dfrac{1}{4}$ and the point $(3, 7)$

$y - 7 = \dfrac{1}{4}(x - 3)$

$y - 7 = \dfrac{1}{4}x - \dfrac{3}{4}$

$y - 7 + 7 = \dfrac{1}{4}x - \dfrac{3}{4} + 7$

$y = \dfrac{1}{4}x + \dfrac{25}{4}$

30.

Since the lines are perpendicular $m_1 = -\dfrac{1}{m_2}$.

$-2x + 5y = 35$

$5y = 2x + 35$

$y = \dfrac{2}{5}x + 7$

Using $m = -\dfrac{5}{2}$ and the point $(-2, 9)$

$y - 9 = -\dfrac{5}{2}(x + 2)$

$y - 9 = -\dfrac{5}{2}x - 5$

$y = -\dfrac{5}{2}x + 4$

31.

a. $(0, 4)$ is the point where the graph crosses the y-axis.

b. $(1.3, 0)$ is the point where the graph crosses the x-axis.

c.

Use two points from the graph $(0, 4)$ and $(4, -8)$.

slope $= \dfrac{-8 - 4}{4 - 0} = \dfrac{-12}{4} = -3$

d. $f(4) = -8$ Since when $x = 4$ $y = -8$.

e. $x = -4$ when $f(x) = 16$

f.

Since the y-intercept is $(0, 4)$, then $b = 4$.

Since the slope is -3, then $m = -3$.

Use $y = mx + b$ where $m = -3$ and $b = 4$.

The equation of the line is $f(x) = -3x + 4$.

32.

a. $(0, -5)$ is the point where the graph crosses the y-axis.

b. $(3.3, 0)$ is the point where the graph crosses the x-axis.

c.

Use two points from the graph $(-2, -8)$ and $(2, -2)$.

slope $= \dfrac{-2 - (-8)}{2 - (-2)} = \dfrac{6}{4} = \dfrac{3}{2}$

d. $h(-2) = -8$ since when $x = -2$, $y = -8$.

e. $x = 2$ when $h(x) = -2$

f.

Since the y-intercept is $(0, -5)$, then $b = -5$.

Since the slope is $\dfrac{3}{2}$, then $m = \dfrac{3}{2}$.

Use $y = mx + b$ where $m = \dfrac{3}{2}$ and $b = -5$.

The equation of the line is $h(x) = \dfrac{3}{2}x - 5$.

33.

a.

Using two points: $(0, 6)$ and $(3, 8)$

Find the slope: $m = \dfrac{8 - 6}{3 - 0} = \dfrac{2}{3}$

b. $(0, 6)$

c. $y = \dfrac{2}{3}x + 6$

34.

a.

Using two points: $(0, 3)$ and $(3, 1.65)$

Find the slope: $m = \dfrac{1.65 - 3}{3 - 0} = 0.55$

b. $(0, 3)$

c. $y = 0.55x + 3$

35. The electricity cost for using holiday lights 6 hours a day is $13.

36. $P(t) = 30t + 9979$

a.

$P(10) = 30(10) + 9979$

$P(10) = 10,279$

In 2010, the population of Michigan will be about 10,279 thousand.

b.
$$11{,}000 = 30t + 9979$$
$$11{,}000 - 9979 = 30t + 9979 - 9979$$
$$1021 = 30t$$
$$\frac{1021}{30} = \frac{30t}{30}$$
$$34.03 \approx t$$

The population of Michigan will reach 11,000 thousand in about 2034. This may be model breakdown if the current population growth does not continue.

37. $f(x) = 2x - 8$

a.
$$f(10) = 2(10) - 8$$
$$f(10) = 12$$

b.
$$0 = 2x - 8$$
$$8 = 2x$$
$$\frac{8}{2} = \frac{2x}{2}$$
$$4 = x$$

c. Domain: All real numbers, range All real numbers

38. $h(x) = -\frac{2}{3}x + 14$

a.
$$h(12) = -\frac{2}{3}(12) + 14$$
$$h(12) = -8 + 14$$
$$h(12) = 6$$

b
$$-8 = -\frac{2}{3}x + 14$$
$$-8 - 14 = -\frac{2}{3}x + 14 - 14$$
$$-22 = -\frac{2}{3}x$$
$$\left(-\frac{3}{2}\right)(-22) = \left(-\frac{3}{2}\right)\left(-\frac{2}{3}x\right)$$
$$33 = x$$

c. Domain: All real numbers, range All real numbers

39. $g(x) = 12$

a. $g(3) = 12$

b. $g(-20) = 12$

c. Domain: All real numbers, range $\{12\}$

40. $f(x) = 1.25x + 4.5$

a.
$$f(5) = 1.25(5) + 4.5$$
$$f(5) = 10.75$$

b.
$$-8.5 = 1.25x + 4.5$$
$$-8.5 - 4.5 = 1.25x + 4.5 - 4.5$$
$$-13 = 1.25x$$
$$\frac{-13}{1.25} = \frac{1.25x}{1.25}$$
$$-10.4 = x$$

c. Domain: All real numbers, range All real numbers

Chapter 1 Linear Functions
Chapter 1 Test

1.

a. Let C be the number of work-related injury cases in thousands in U.S. private industry t years since 2000.

t	C
3	4095
4	4008
5	3972
6	3857

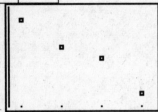

$C(t) = mt + b$

Using two points: $(3, 4095)$ and $(6, 3857)$

Find the slope: $m = \dfrac{3857 - 4095}{6 - 3} \approx -79.3$

$C(t) = -79.3t + b$

$b = 4095 - (-79.3)(3)$

$b \approx 4333$

$C(t) = -79.3t + 4333$

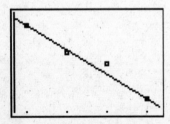

b.

$C(9) = -79.3(9) + 4333$

$C(9) = 3619.3$

In 2009, there were about 3619.3 thousand work-related injury cases in the United States private industry.

c. The slope is -79.3. The number of work-related injury cases in U.S. private industry is decreasing about 79.3 thousand cases per year.

d. Domain: $[0, 10]$

$C(0) = -79.3(0) + 4333 = 4333$

$C(10) = -79.3(10) + 4333 = 3540$

Range: $[3540, 4333]$

e.

$5000 = -79.3t + 4333$

$5000 - 4333 = -79.3t + 4333 - 4333$

$667 = -79.3t$

$\dfrac{667}{-79.3} = \dfrac{-79.3t}{-79.3}$

$-8.4 \approx t$

In 1991, there were about 5 million work-related injury cases in U.S. private industry.

2. $y = 4x - 2$

Since $b = -2$ the y-intercept is $(0, -2)$.

To find the x-intercept let $y = 0$ and solve for x:

$0 = 4x - 2$

$2 = 4x$

$x = \dfrac{1}{2}$

Plot a point at the x-intercept $\left(\dfrac{1}{2}, 0\right)$.

Draw a line through the two points.

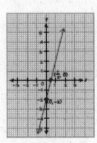

3. $y = -\dfrac{2}{3}x + 5$

Since $b = 5$ the y-intercept is $(0, 5)$.

To find the x-intercept let $y = 0$ and solve for x:

$0 = -\dfrac{2}{3}x + 5$

$-5 = -\dfrac{2}{3}x$

$\left(-\dfrac{3}{2}\right) \cdot (-5) = \left(-\dfrac{3}{2}\right) \cdot \left(-\dfrac{2}{3}x\right)$

$x = 7.5$

Plot a point at the x-intercept $(7.5, 0)$.

Draw a line through the two points.

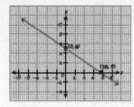

4. $2x - 4y = 10$

To find the y-intercept let $x = 0$ and solve for y:
$2(0) - 4y = 10$
$0 - 4y = 10$
$-4y = 10$
$y = -2.5$

Plot a point at the y-intercept $(0, -2.5)$.

To find the x-intercept let $y = 0$ and solve for x:
$2x - 4(0) = 10$
$2x - 0 = 10$
$2x = 10$
$x = 5$

Plot a point at the x-intercept $(5, 0)$.
Draw a line through the two points.

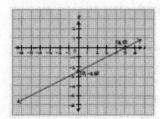

5.
$ax - by = c$
$ax - by + by = c + by$
$ax = c + by$
$\dfrac{ax}{a} = \dfrac{c + by}{a}$
$x = \dfrac{c + by}{a}$

6.

Using two points: $(-4, 8)$ and $(6, 10)$

Find the slope: $m = \dfrac{10 - 8}{6 - (-4)} = \dfrac{2}{10} = 0.2$

Using $m = 0.2$ and $(6, 10)$.
$10 = 0.2(6) + b$
$10 = 1.2 + b$
$b = 8.8$

Since $m = 0.2$ and $b = 8.8$
$y = 0.2x + 8.8$

7.
Since the lines are parallel $m_1 = m_2$.
$y = 2x - 7$
Using $m = 2$ and the point $(5, 8)$
$y - 8 = 2(x - 5)$
$y - 8 = 2x - 10$
$y = 2x - 2$

8. $P = 2m + 30$

a.
$P = 2(6) + 30$
$P = 42$

Six months after starting to sell paintings, John Clark will sell approximately 42 paintings.

b.
$50 = 2m + 30$
$50 - 30 = 2m + 30 - 30$
$20 = 2m$
$\dfrac{20}{2} = \dfrac{2m}{2}$
$10 = m$

John Clark will sell 50 painting during the 10th month after starting to sell painting.

9. Let R be the total revenue in millions of dollars for Apple Inc. t years since 2000.

a.

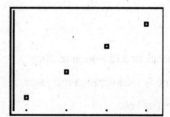

$R(t) = mt + b$

Using two points: $(4, 8279)$ and $(6, 19315)$

Find the slope: $m = \dfrac{19,315 - 8279}{6 - 4} = 5518$

$R(t) = 5518t + b$
$b = 8279 - 5518(4)$
$b = -13,793$

$R(t) = 5518t - 13,793$

Chapter 1 Linear Functions

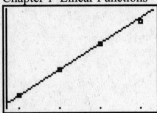

b. Vertical intercept is $(0, -13,793)$. In 2000, Apple, Inc. had total revenue of $-\$13,793$ million. This is model breakdown, since Apple did not have a negative revenue in 2000.

c. Domain: $[3,9]$

$R(3) = 5518(3) - 13,793 = 2761$

$R(9) = 5518(9) - 13,793 = 35,869$

Range: $[2761, 35869]$

d.

$R(10) = 5518(10) - 13,793$

$R(10) = 41,387$

In 2010, Apple, Inc. had a total revenue of about $41,387 million.

e.

$30,000 = 5518t - 13,793$

$30,000 + 13,793 = 5518t - 13,793 + 13,793$

$43,793 = 5518t$

$\dfrac{43,793}{5518} = \dfrac{5518t}{5518}$

$8 \approx t$

Apple, Inc.'s total revenue was about $30 billion in 2008.

f. The slope is 5518. Apple, Inc.'s total revenue increases by about $5518 million dollars per year.

10.

a.

Using two points from the chart $(0,12)$ and $(5,8)$

$slope = \dfrac{8-12}{5-0} = \dfrac{-4}{5} = -\dfrac{4}{5}$

b. The y-intercept is $(0,12)$ because when $x=0$, $y=12$.

c. The x-intercept is $(15,0)$ because when $y=0$, $x=15$.

d.

Since the y-intercept is $(0,12)$, then $b=12$.

Since the slope is $-\dfrac{4}{5}$, then $m = -\dfrac{4}{5}$.

Use $y = mx + b$ where $m = -\dfrac{4}{5}$ and $b = 12$.

The equation of the line is $y = -\dfrac{4}{5}x + 12$.

11.

a.

Using two points from the chart $(-4,0)$ and $(6,20)$

$slope = \dfrac{20-0}{6-(-4)} = \dfrac{20}{10} = 2$

b. The vertical intercept is $(0,8)$. That is the point where the graph crosses the y-axis.

c. The horizontal intercept is $(-4,0)$. That is the point where the graph crosses the x-axis.

d.

Since the y-intercept is $(0,8)$ then $b = 8$.

Since the slope is 2 then $m = 2$.

Use $y = mx + b$ where $m = 2$ and $b = 8$.

The equation of the line is $y = 2x + 8$.

12. Domain: All real numbers, range: All real numbers

13.

$W = ht^2$

$\dfrac{W}{t^2} = \dfrac{ht^2}{t^2}$

$h = \dfrac{W}{t^2}$

14.

$1.5(x+3) = 4x + 2.5(4x-7)$

$1.5x + 4.5 = 4x + 10x - 17.5$

$1.5x + 4.5 = 14x - 17.5$

$1.5x + 4.5 - 4.5 = 14x - 17.5 - 4.5$

$1.5x = 14x - 22$

$1.5x - 14x = 14x - 22 - 14x$

$-12.5x = -22$

$\dfrac{-12.5x}{-12.5} = \dfrac{-22}{-12.5}$

$x = 1.76$

15. In 2010, the population of New York will be about 19.4 million.

16. $H(p) = 0.20p + 5$

a.
$H(70) = 0.20(70) + 5$
$H(70) = 19$

To earn a 70% on the exam, you need to study about 19 hours.

b.
$H(100) = 0.20(100) + 5$
$H(100) = 25$

To earn a 100% on the exam, you need to study about 25 hours.

17.

a.
$f(x) = 7x - 3$
$f(\ ) = 7(\ ) - 3$
$f(4) = 7(4) - 3$
$f(4) = 28 - 3$
$f(4) = 25$

b.
$f(x) = 7x - 3$
$-31 = 7x - 3$
$-31 + 3 = 7x - 3 + 3$
$-28 = 7x$
$x = -4$

c. Domain: All real numbers, range: All real numbers

18.

a.
$h(x) = \dfrac{4}{7}x + 6$
$h(\ ) = \dfrac{4}{7}(\ ) + 6$
$h(35) = \dfrac{4}{7}(35) + 6$
$h(35) = 20 + 6$
$h(35) = 26$

b.
$h(x) = \dfrac{4}{7}x + 6$
$4 = \dfrac{4}{7}x + 6$
$4 - 6 = \dfrac{4}{7}x + 6 - 6$
$-2 = \dfrac{4}{7}x$
$\dfrac{7}{4} \cdot (-2) = \dfrac{7}{4} \cdot \left(\dfrac{4}{7}x\right)$
$x = -\dfrac{7}{2}$

c. Domain: All real numbers, range: All real numbers

19.

a.
$g(x) = -9$
$g(\ ) = -9$
$g(6) = -9$

b. Domain: All real numbers, range: $\{-9\}$

20. A relation may not be a function if the input value was paired with more than one output value.

Chapter 2 Introduction to Systems of Linear Equations and Inequalities
Section 2.1

1. $(3,2)$. This is a consistent system with independent lines. The point where the two lines intersect is the solution to the system.

3. $(-7,3)$. This is a consistent system with independent lines. The point where the two lines intersect is the solution to the system.

5. No solution. This is an inconsistent system. The two lines are parallel, therefore they will never intersect.

7.

a. At 15%, the percentage of people under 18 years was the same as the percentage of people 65 years and over with family income below the poverty level in 1974.

b. At 13%, the percentage of people 65 years-and-over with family income below the poverty level was first equal to the percentage of people 18-64 years old in 1993.

9. $(3,10)$. This is a consistent system with independent lines.

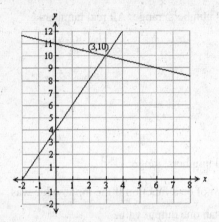

11. $(-6,-8)$, consistent, independent

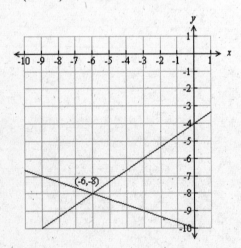

13. No solution. This is an inconsistent system. The two lines are parallel, therefore they will never intersect.

The 1st line is: $y = 0.5x + 2$
The 2nd line is: $x - 2y = 8$

$$\frac{-x \quad\quad -x}{\frac{-2y}{-2} = \frac{-x+8}{-2}}$$
$$y = 0.5x - 4$$

Since both lines have the same slope, and different y intercepts, they are parallel lines.

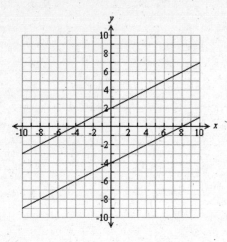

15. The solution is every point (m,n) on the line $n = \frac{2}{3}m + 7$. This is a consistent system with dependent lines.

The 1st line is: $n = \frac{2}{3}m + 7$

The 2nd line is: $6m - 9n = -63$

$$\underline{-6m \qquad\quad -6m}$$

$$-9n = -6m - 63$$

$$\frac{-9n}{-9} = \frac{-6m}{-9} - \frac{63}{-9}$$

$$n = \frac{2}{3}m + 7$$

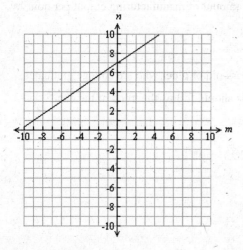

17. Let U and N be the total amount of manufacturing output per hour in dollars by the United States and Norway, respectively, and let t be years since 2000. Note: Answers may vary, but your models should follow the trend of the data.

a.

t	U	N
0	147.7	105.9
3	175.5	121.6
4	187.8	128.8
5	194.0	132.4

Using two points: (0, 147.7) and (5, 194.0)

Find the slope: $m = \dfrac{194.0 - 147.7}{5 - 0} = 9.26$

$U(t) = 9.26t + 147.7$

Using two points: (0, 105.9) and (5, 132.4)

Find the slope: $m = \dfrac{132.4 - 105.9}{5 - 0} = 5.3$

$N(t) = 5.3t + 105.9$

b. Graph both models on the same calculator screen.

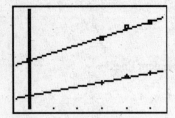

c.

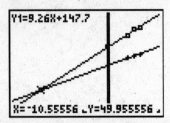

The amount of manufacturing output per hour by the United States was equal to the amount of manufacturing output per hour by Norway in 1989.

19. Let M and F be the total percentages of white male and female Americans 25-years-old or older who are college graduates, respectively, and let t be years since 1980. Note: Answers may vary, but your models should follow the trend of the data.

a.

t	M	F
0	21.3	12.3
5	24.0	16.3
10	25.3	19.0
15	27.2	21.0
20	28.5	23.9
25	29.4	26.8

Using two points: (0, 21.3) and (20, 28.5)

Find the slope: $m = \dfrac{28.5 - 21.3}{20 - 0} = 0.36$

$M(t) = 0.36t + 21.3$

Using two points: (0, 12.3) and (25, 26.8)

Find the slope: $m = \dfrac{26.8 - 12.3}{25 - 0} = 0.58$

$F(t) = 0.58t + 12.3$

b. Graph both models on the same calculator screen.

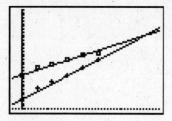

c.

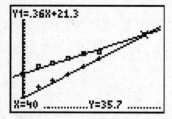

The percentage of white American males with college degrees will be the same as the percentage of white American women with college degrees in 2020.

21.

a.

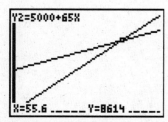

b. Hope's Pottery will break even after producing and selling 56 vases. At this point she will have expenses of $8640 and revenue of $8680. Before this point she will lose money.

23.

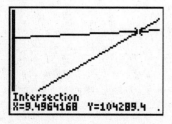

a. *La Opinion* had approximately the same circulation as the *Long Beach Press Telegram* in 1999 with a circulation of about 104,289.

b. The *La Opinion* model has a slope of 7982, which means its circulation is growing at a rate of 7982 new subscribers per year. The *Long Beach Press Telegram* model has a slope of 726, which means its circulation is growing at a rate of 726 new subscribers per year. This means that the circulation of *La Opinion* is increasing much faster than that of the *Long Beach Press Telegram*.

25. Since both options share the same slope and different vertical intercepts, these salary options will never be the same. Salary option number 1 will always pay more than salary option number 2.

27.

$$\begin{cases} x+y=-6 \\ -2x+y=3 \end{cases}$$

Solve for y to graph:
$x+y=-6$
$x+y-x=-6-x$
$y=-x-6$

Solve for y to graph:
$-2x+y=3$
$-2x+y+2x=3+2x$
$y=2x+3$

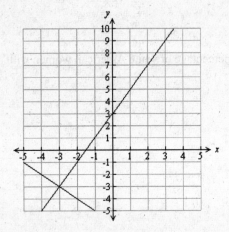

$(-3,-3) \to x=-3, y=-3$. This is a consistent system with independent lines.

29.

$$\begin{cases} p=2.5t+6 \\ p=\dfrac{5}{2}t-6 \end{cases}$$

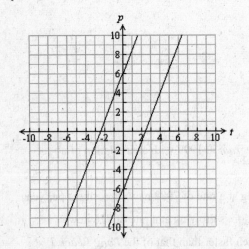

No solution. This is an inconsistent system. The two lines are parallel.

31.

$$\begin{cases} y = \frac{1}{3}x + 5 \\ 2x - 6y = -30 \end{cases}$$

Solve for y in the bottom equation to graph:

$2x - 6y = -30$

$2x - 6y - 2x = -30 - 2x$

$-6y = -2x - 30$

$\dfrac{-6y}{-6} = \dfrac{-2x - 30}{-6}$

$y = \dfrac{1}{3}x + 5$

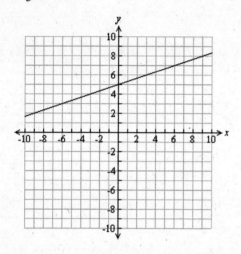

There are an infinite number of solutions. This is a consistent system with dependent lines.

33.

$$\begin{cases} R = 2.75t + 6.35 \\ R = -1.5t + 12.45 \end{cases}$$

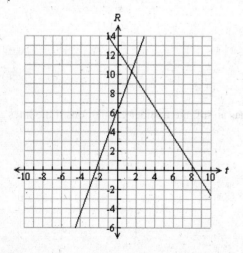

$(1.4, 10.3) \rightarrow t = 1.4, R = 10.3$. This is a consistent system with independent lines.

Chapter 2 Introduction to Systems of Linear Equations and Inequalities

35.
$$\begin{cases} 4x + 5y = -7 \\ 3x - 7y = 27 \end{cases}$$

Solve for y to graph:

$4x + 5y = -7$

$4x + 5y - 4x = -7 - 4x$

$5y = -4x - 7$

$\dfrac{5y}{5} = \dfrac{-4x - 7}{5}$

$y = -\dfrac{4}{5}x - \dfrac{7}{5}$

Solve for y to graph:

$3x - 7y = 27$

$3x - 7y - 3x = 27 - 3x$

$-7y = -3x + 27$

$\dfrac{-7y}{-7} = \dfrac{-3x + 27}{-7}$

$y = \dfrac{3}{7}x - \dfrac{27}{7}$

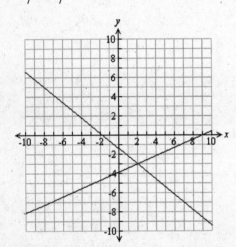

$(2, -3) \rightarrow x = 2, y = -3$. This is a consistent system with independent lines.

37. The two lines are parallel. The system is inconsistent. No solution to the system.

39. A consistent system has exactly one solution if the lines are independent, but it has an infinite number of solutions if the lines are dependent.

41. Many possible answers. Should be a graph with two lines intersecting at the point $(2,4)$.

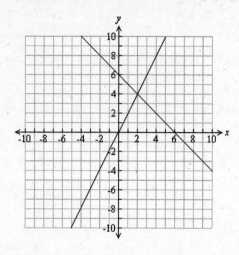

43. Many possible answers. Graph two parallel lines.

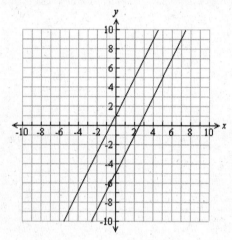

45. $(2,11) \rightarrow x = 2, y = 11$. This is a consistent system with independent lines.

47. No solution. This is an inconsistent system. The two lines are parallel.

49. Infinite number of solutions. This is a consistent system with dependent lines.

51. No solution. This is an inconsistent system. The two lines are parallel.

53. $(9,35) \rightarrow t = 9, p = 35$. This is a consistent system with independent lines.

X	Y1	Y2
5	23	27
7	29	31
8	32	33
9	35	35

X=

55. No Solution. This is an inconsistent system. The two lines are parallel.

X	Y1	Y2
1	52	20.667
2	57	25.667
3	62	30.667
4	67	35.667
5	72	40.667
6	77	45.667
7	82	50.667

X=7

Chapter 2 Introduction to Systems of Linear Equations and Inequalities

57. Infinite number of solutions. This is a consistent system with dependent lines.

X	Y1	Y2
5	81	81
10	88	88
15	95	95

X=

59. $(15,10) \to x = 15, y = 10$. This is a consistent system with independent lines.

X	Y1	Y2
0	-20	70
5	-10	50
10	0	30
15	10	10

X=

Section 2.2

1. Using the substitution method, set the equations equal to each other.

$\begin{cases} y = x+7 \\ y = 2x+3 \end{cases}$

$x+7 = 2x+3$
$x+7-7 = 2x+3-7$
$x = 2x-4$
$x-2x = 2x-4-2x$
$-x = -4$
$x = 4$

Substituting x back into either equation solve for y:
$y = (4)+7$
$y = 11$

The solution to the system is $(4, 11)$.
This is a consistent system with independent lines.

3. Using the substitution method, set the equations equal to each other.

$\begin{cases} P = -5t+20 \\ P = 2t-8 \end{cases}$

$2t-8 = -5t+20$
$2t-8+8 = -5t+20+8$
$2t = -5t+28$
$2t+5t = -5t+28+5t$
$7t = 28$
$t = 4$

Substituting t back into either equation solve for P:
$P = 2(4)-8$
$P = 8-8$
$P = 0$

The solution to the system is $(4, 0)$.
This is a consistent system with independent lines.

5. Using the substitution method, set the equations equal to each other.

$\begin{cases} y = 2x-5 \\ y = \dfrac{1}{3}x+5 \end{cases}$

$2x-5 = \dfrac{1}{3}x+5$

$3(2x-5) = 3\left(\dfrac{1}{3}x+5\right)$

$6x-15 = x+15$
$6x-15+15 = x+15+15$
$6x = x+30$
$6x-x = x+30-x$
$5x = 30$
$x = 6$

Substituting x back into either equation solve for y:
$y = 2(6)-5$
$y = 12-5$
$y = 7$

The solution to the system is $(6, 7)$.
This is a consistent system with independent lines.

7. Using the substitution method, set the equations equal to each other.

$\begin{cases} H = k+8 \\ H = 0.5k+14 \end{cases}$

$k+8 = 0.5k+14$
$2(k+8) = 2(0.5k+14)$
$2k+16 = k+28$
$2k+16-16 = k+28-16$
$2k = k+12$
$2k-k = k+12-k$
$k = 12$

Substituting k back into either equation solve for H:
$H = (12)+8$
$H = 20$

The solution to the system is $(12, 20)$.
This is a consistent system with independent lines.

Chapter 2 Introduction to Systems of Linear Equations and Inequalities

9. Using the substitution method, set the equations equal to each other.

$$\begin{cases} y = \dfrac{1}{2}x + \dfrac{3}{2} \\ y = \dfrac{3}{2}x - \dfrac{7}{2} \end{cases}$$

$$\dfrac{3}{2}x - \dfrac{7}{2} = \dfrac{1}{2}x + \dfrac{3}{2}$$

$$2\left(\dfrac{3}{2}x - \dfrac{7}{2}\right) = 2\left(\dfrac{1}{2}x + \dfrac{3}{2}\right)$$

$$3x - 7 = x + 3$$
$$3x - 7 + 7 = x + 3 + 7$$
$$3x = x + 10$$
$$3x - x = x + 10 - x$$
$$2x = 10$$
$$x = 5$$

Substituting x back into either equation solve for y:

$$y = \dfrac{1}{2}(5) + \dfrac{3}{2}$$

$$y = \dfrac{5}{2} + \dfrac{3}{2}$$

$$y = 4$$

The solution to the system is $(5, 4)$.

This is a consistent system with independent lines.

11.

$$\begin{cases} R(v) = 155v \\ C(v) = 5000 + 65v \end{cases}$$

$$R(v) = C(v)$$
$$155v = 5000 + 65v$$
$$155v - 65v = 5000 + 65v - 65v$$
$$90v = 5000$$
$$\dfrac{90v}{90} = \dfrac{5000}{90}$$
$$v \approx 55.6$$

Hope's Pottery will need to produce and sell 56 vases in order to break even.

13.

$$\begin{cases} R(a) = 175a \\ C(a) = 25a + 7800 \end{cases}$$

$$R(a) = C(a)$$
$$175a = 25a + 7800$$
$$175a - 25a = 25a - 25a + 7800$$
$$150a = 7800$$
$$\dfrac{150a}{150} = \dfrac{7800}{150}$$
$$a = 52$$

The break even point for Optimum Traveling Detail is 52 automotive details.

15.

$$\begin{cases} A(t) = 8.61t + 641.25 \\ M(t) = 12.50t + 543.07 \end{cases}$$

$$A(t) = M(t)$$
$$8.61t + 641.25 = 12.50t + 543.07$$
$$\underline{-12.50t - 641.25 \quad -12.50t - 641.25}$$
$$-3.89t = -98.18$$
$$\dfrac{-3.89t}{-3.89} = \dfrac{-98.18}{-3.89}$$
$$t \approx 25.2$$

In 2026, the number of associate's degrees conferred will be the same as the number of master's degrees.

17.

a. Let $P(t)$ be the student teacher ratio for public schools in the United States, and let $R(t)$ be the total student teacher ratio for private schools in the United States, t years since 2000. Note: Your model may differ, but it must follow the trend of the data.

t	$P(t)$	$R(t)$
-1	16.1	14.7
0	16.0	14.5
2	15.9	14.1
4	15.8	13.8

Using two points: $(0, 16.0)$ and $(4, 15.8)$

Find the slope: $m = \dfrac{15.8 - 16.0}{4 - 0} = -0.05$

$$P(t) = -0.05t + 16.0$$

Using two points: $(0, 14.5)$ and $(4, 13.8)$

Find the slope: $m = \dfrac{13.8 - 14.5}{4 - 0} = -0.175$

$$R(t) = -0.175t + 14.5$$

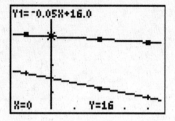

b.

$\begin{cases} P(t) = -0.05t + 16.0 \\ R(t) = -0.175t + 14.5 \end{cases}$

$P(t) = R(t)$

$-0.05t + 16.0 = -0.175t + 14.5$

$\underline{+0.175t - 16.0 \quad +0.175t - 16.0}$

$0.125t = -1.5$

$\dfrac{0.125t}{0.125} = \dfrac{-1.5}{0.125}$

$t = -12$

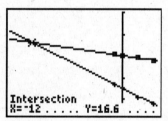

In 1988, the student teacher ratio was the same for public and private schools.

19.

t	H(t)	A(t)
0	10.6	16.5
3	11.4	17.3
4	12.1	17.6
6	12.4	18.5

a. Let $H(t)$ be the percentage of Hispanics 25-years-old or older who have a college degree and let t be years since 2000. Note: Your model may differ, but it must follow the trend of the data.

Using two points: $(0, 10.6)$ and $(6, 12.4)$

Find the slope: $m = \dfrac{12.4 - 10.6}{6 - 0} = 0.3$

$$H(t) = 0.3t + 10.6$$

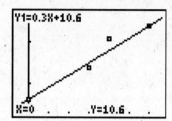

b. The percentage rate of Hispanics who have a college degree is increasing at a rate of 0.3% each year.

c. Let $A(t)$ be the percentage of African Americans 25-years-old or older who have a college degree, and let t be years since 2000. Note: Your model may differ, but it must follow the trend of the data.

Using two points: $(0, 16.5)$ and $(6, 18.5)$

Find the slope: $m = \dfrac{18.5 - 16.5}{6 - 0} \approx 0.33$

$$A(t) = 0.33t + 16.5$$

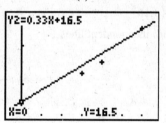

d. The percentage rate of African Americans who have a college degree is increasing at a rate of 0.33% each year.

e.

$A(t) = H(t)$

$0.33t + 16.5 = 0.3t + 10.6$

$\underline{-0.3t - 16.5 \quad -0.3t - 16.5}$

$0.03t = -5.9$

$\dfrac{0.03t}{0.03} = \dfrac{-5.9}{0.03}$

$t \approx -196.7$

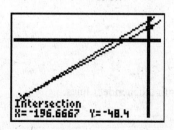

Chapter 2 Introduction to Systems of Linear Equations and Inequalities

In 1803, the percentage of Hispanics who have a college degree was the same as the percentage of African Americans who have a college degree. This is model breakdown.

21.

$\begin{cases} y = 2x - 5 \\ 3x + y = 10 \end{cases}$

Substitute $2x - 5$ for y in the second equation and solve for x.

$3x + (2x - 5) = 10$
$3x + 2x - 5 = 10$
$5x - 5 = 10$
$5x - 5 + 5 = 10 + 5$
$5x = 15$
$x = 3$

Substituting x back into either equation solve for y:

$y = 2(3) - 5$
$y = 6 - 5$
$y = 1$

The solution to the system is $(3, 1)$.
This is a consistent system with independent lines.

23.

$\begin{cases} H = 3k + 6 \\ 2k + 4H = -4 \end{cases}$

Substitute $3k + 6$ for H in the second equation and solve for k.

$2k + 4(3k + 6) = -4$
$2k + 12k + 24 = -4$
$14k + 24 = -4$
$14k + 24 - 24 = -4 - 24$
$14k = -28$
$k = -2$

Substituting k back into either equation solve for H:

$H = 3(-2) + 6$
$H = -6 + 6$
$H = 0$

The solution to the system is $(-2, 0)$.
This is a consistent system with independent lines.

25.

$\begin{cases} G = 4a - 5 \\ 5a - 2G = 1 \end{cases}$

Substitute $4a - 5$ for G in the second equation and solve for a.

$5a - 2(4a - 5) = 1$
$5a - 8a + 10 = 1$
$-3a + 10 = 1$
$-3a + 10 - 10 = 1 - 10$
$-3a = -9$
$a = 3$

Substituting a back into either equation solve for G:

$G = 4(3) - 5$
$G = 12 - 5$
$G = 7$

The solution to the system is $(3, 7)$.
This is a consistent system with independent lines.

27.

$\begin{cases} P = \dfrac{2}{3}t + 5 \\ t + 3P = 21 \end{cases}$

Substitute $\dfrac{2}{3}t + 5$ for P in the second equation and solve for t.

$t + 3\left(\dfrac{2}{3}t + 5\right) = 21$
$t + 2t + 15 = 21$
$3t + 15 = 21$
$3t + 15 - 15 = 21 - 15$
$3t = 6$
$t = 2$

Substituting t back into either equation solve for P:

$P = \dfrac{2}{3}(2) + 5$
$P = \dfrac{4}{3} + 5$
$P = \dfrac{19}{3}$

The solution to the system is $\left(2, \dfrac{19}{3}\right)$.
This is a consistent system with independent lines.

29.

$\begin{cases} 3x+4y=21 \\ x=6y-4 \end{cases}$

Substitute $6y-4$ for x in the first equation and solve for y.

$3(6y-4)+4y=21$
$18y-12+4y=21$
$22y-12=21$
$22y-12+12=21+12$
$22y=33$
$y=1.5$

Substituting y back into either equation solve for x:
$x=6(1.5)-4$
$x=9-4$
$x=5$

The solution to the system is $x=5$ and $y=1.5$.
This is a consistent system with independent lines.

31. For s dollars of sales, the two salary options are:

$O_1 = 700 + 0.05s$
$O_2 = 1300 + 0.035s$

$O_1 = O_2$
$700 + 0.05s = 1300 + 0.035s$
$\underline{-700 - 0.035s \quad -700 - 0.035s}$
$\qquad 0.015s = 600$
$\qquad \dfrac{0.015s}{0.015} = \dfrac{600}{0.015}$
$\qquad s = 40,000$

Sales of $40,000 worth of high end fashion will result in the two options having the same monthly salary.

33.

a. Let A be the amount of money Damian invests into the account paying 5% simple interest, and let B be the amount of money Damian invests into the account paying 7.2% simple interest.

$A + B = 150,000 \to$ Total amount invested.
$0.05A + 0.072B = 9600 \to$ Total interest earned.

b.

Solve for A in the first equation and use it for A in the second equation.

$A = 150,000 - B$
$0.05(150,000 - B) + 0.072B = 9600$

Solve for B.

$7500 - 0.05B + 0.072B = 9600$
$7500 + 0.022B = 9600$
$7500 - 7500 + 0.022B = 9600 - 7500$
$0.022B = 2100$
$\dfrac{0.022B}{0.022} = \dfrac{2100}{0.022}$
$B = 95,454.55$

Now use this value to solve for A.

$A = 150,000 - 95,454.55$
$A = 54,545.45$

Damian needs to invest $54,545.45 in the account paying 5% simple interest, and $95,454.55 in the account paying 7.2% simple interest.

35.

a. Let A be the amount of money Henry invests into the account paying 5% simple interest, and let B be the amount of money Henry invests into the account paying 8% simple interest.

1,500,000   Settlement
−125,000   Debts
1,375,000   Left over to invest

$A + B = 1,375,000 \to$ Total amount invested.
$0.05A + 0.08B = 87,500 \to$ Total interest earned.

Chapter 2 Introduction to Systems of Linear Equations and Inequalities

b.

Solve for A in the first equation and use it for A in the second equation.

$A = 1,375,000 - B$
$0.05(1,375,000 - B) + 0.08B = 87,500$

Solve for B.
$68,750 - 0.05B + 0.08B = 87,500$
$68,750 + 0.03B = 87,500$
$68,750 - 68,750 + 0.03B = 87,500 - 68,750$
$0.03B = 18,750$
$\dfrac{0.03B}{0.03} = \dfrac{18,750}{0.03}$
$B = 625,000$

Now use this value to solve for A.
$A = 1,375,000 - 625,000$
$A = 750,000$

Henry needs to invest \$750,000 in the account paying 5% simple interest, and \$625,000 in the account paying 8% simple interest.

37. Let A be the amount of money Joan invests into the account paying 9% simple interest, and let B be the amount of money Joan invests into the account paying 5% simple interest.

$A + B = 175,000 \to$ Total amount invested.
$0.09A + 0.05B = 12,000 \to$ Total interest earned.

Solve for B in the first equation and use it for B in the second equation.
$B = 175,000 - A$
$0.09A + 0.05(175,000 - A) = 12,000$

Solve for A.
$0.09A + 8750 - 0.05A = 12,000$
$0.04A + 8750 = 12,000$
$0.04A + 8750 - 8750 = 12,000 - 8750$
$0.04A = 3250$
$\dfrac{0.04A}{0.04} = \dfrac{3250}{0.04}$
$A = 81,250$

Now use this value to solve for B.
$B = 175,000 - 81,250$
$B = 93,750$

Joan needs to invest \$93,750 in the account paying 5% simple interest, and \$81,250 in the account paying 9% simple interest.

39. Let S be the amount of money Truong invests in stocks paying 11% simple interest, and let B be the amount of money Truong invests in bonds paying 9% simple interest.

$S + B = 40,000 \to$ Total amount invested.
$0.11S + 0.09B = 4180 \to$ Total interest earned.

Solve for B in the first equation and use it for B in the second equation.
$B = 40,000 - S$
$0.11S + 0.09(40,000 - S) = 4180$

Solve for S.
$0.11S + 3600 - 0.09S = 4180$
$0.02S + 3600 = 4180$
$0.02S + 3600 - 3600 = 4180 - 3600$
$0.02S = 580$
$\dfrac{0.02S}{0.02} = \dfrac{580}{0.02}$
$S = 29,000$

Now use this value to solve for B.
$B = 40,000 - 29,000$
$B = 11,000$

Truong invested \$11,000 in bonds with a return of 9%, and \$29,000 in stocks with a return of 11%.

41. Let p be the number of people on the tour.

$500 + 25p = 700 + 20p$
$\underline{-500 - 20p \quad -500 - 20p}$
$\qquad 5p = 200$
$\qquad \dfrac{5p}{5} = \dfrac{200}{5}$
$\qquad p = 40$

A total of 40 people would result in the same price from either of the two tour companies.

43. Tom did the work correctly. Matt wrote that the system is consistent, which is false. The system is inconsistent because the variables were both eliminated and the remaining statement was false.

45.

$$\begin{cases} 7d + 4r = 8.17 \\ 2r = 8d + 0.98 \end{cases}$$

Solve for r in the second equation:

$\dfrac{2r}{2} = \dfrac{8d}{2} + \dfrac{0.98}{2}$

$r = 4d + 0.49$

Plug this into the first equation:

$7d + 4(4d + 0.49) = 8.17$

$7d + 16d + 1.96 = 8.17$

$23d + 1.96 = 8.17$

$23d = 6.21$

$d = 0.27$

Substituting d back into either equation solve for r:

$r = 4(0.27) + 0.49 = 1.57$

The solution to the system is $(0.27, 1.57)$.

This is a consistent system with independent lines.

47.

$$\begin{cases} p = \dfrac{1}{4}r + \dfrac{7}{4} \\ p = 0.25r + 1.75 \end{cases}$$

Set the equations equal to each other and solve for r.

$\dfrac{1}{4}r + \dfrac{7}{4} = 0.25r + 1.75$

$4\left(\dfrac{1}{4}r + \dfrac{7}{4}\right) = 4(0.25r + 1.75)$

$r + 7 = r + 7$

$r + 7 - 7 = r + 7 - 7$

$r = r$

$r - r = r - r$

$0 = 0$

Infinite number of solutions.

This is a consistent system with dependent lines.

49.

$$\begin{cases} b = \dfrac{3}{10}a + \dfrac{4}{5} \\ b = 0.3a + 2.8 \end{cases}$$

Set the equations equal to each other and solve for a.

$\dfrac{3}{10}a + \dfrac{4}{5} = 0.3a + 2.8$

$10\left(\dfrac{3}{10}a + \dfrac{4}{5}\right) = 10(0.3a + 2.8)$

$3a + 8 = 3a + 28$

$3a + 8 - 8 = 3a + 28 - 8$

$3a = 3a + 20$

$3a - 3a = 3a + 20 - 3a$

$0 \neq 20$

No solution.

This is an inconsistent system.

51.

$$\begin{cases} y = 4(x - 7) + 2 \\ y = -3(x + 2) + 5 \end{cases}$$

Set the equations equal to each other and solve for x.

$4(x - 7) + 2 = -3(x + 2) + 5$

$4x - 28 + 2 = -3x - 6 + 5$

$4x - 26 = -3x - 1$

$4x - 26 + 26 = -3x - 1 + 26$

$4x = -3x + 25$

$4x + 3x = -3x + 25 + 3x$

$7x = 25$

$x = \dfrac{25}{7}$

Substituting x back into either equation solve for y:

$y = 4\left(\dfrac{25}{7} - 7\right) + 2 = -\dfrac{82}{7}$

The solution to the system is $\left(\dfrac{25}{7}, -\dfrac{82}{7}\right)$.

This is a consistent system with independent lines.

Chapter 2 Introduction to Systems of Linear Equations and Inequalities

53.

$$\begin{cases} w = 2(t+5)-12 \\ w = 2(t-3)+4 \end{cases}$$

Set the equations equal to each other and solve for t.

$2(t+5)-12 = 2(t-3)+4$
$2t+10-12 = 2t-6+4$
$2t-2 = 2t-2$
$2t-2+2 = 2t-2+2$
$2t = 2t$
$2t-2t = 2t-2t$
$0 = 0$

Infinite number of solutions.
This is a consistent system with dependent lines.

55.

$$\begin{cases} 6x-15y = 24 \\ y = \dfrac{2}{5}x - \dfrac{8}{5} \end{cases}$$

Substitute $\dfrac{2}{5}x - \dfrac{8}{5}$ for y in the first equation and solve for x.

$6x - 15\left(\dfrac{2}{5}x - \dfrac{8}{5}\right) = 24$
$6x - 6x + 24 = 24$
$24 = 24$

Infinite number of solutions.
This is a consistent system with dependent lines.

57.

$$\begin{cases} 10x-4y = 30 \\ y = \dfrac{5}{2}x - 14 \end{cases}$$

Substitute $\dfrac{5}{2}x - 14$ for y in the first equation and solve for x.

$10x - 4\left(\dfrac{5}{2}x - 14\right) = 30$
$10x - 10x + 56 = 30$
$56 \neq 30$

No solution.
This is an inconsistent system.

59.

$$\begin{cases} g + 5h = 37 \\ 3g + 4h = 34 \end{cases}$$

Solve for g in the first equation:
$g + 5h - 5h = 37 - 5h$
$g = 37 - 5h$

Plug this into the second equation:
$3(37-5h) + 4h = 34$
$111 - 15h + 4h = 34$
$111 - 11h = 34$
$111 - 11h - 111 = 34 - 111$
$-11h = -77$
$h = 7$

Substituting h back into either equation solve for g:
$g = 37 - 5(7) = 2$

The solution to the system is $(2, 7)$.
This is a consistent system with independent lines.

61.
$$\begin{cases} \dfrac{2}{3}x + \dfrac{1}{2}y = -\dfrac{15}{2} \\ \dfrac{3}{4}x - y = \dfrac{5}{2} \end{cases}$$

Solve for y in the second equation:

$\dfrac{3}{4}x - y + y = \dfrac{5}{2} + y$

$\dfrac{3}{4}x = \dfrac{5}{2} + y$

$\dfrac{3}{4}x - \dfrac{5}{2} = \dfrac{5}{2} + y - \dfrac{5}{2}$

$\dfrac{3}{4}x - \dfrac{5}{2} = y$

Plug this into the first equation:

$\dfrac{2}{3}x + \dfrac{1}{2}\left(\dfrac{3}{4}x - \dfrac{5}{2}\right) = -\dfrac{15}{2}$

$\dfrac{2}{3}x + \dfrac{3}{8}x - \dfrac{5}{4} = -\dfrac{15}{2}$

$24\left(\dfrac{2}{3}x + \dfrac{3}{8}x - \dfrac{5}{4}\right) = 24\left(-\dfrac{15}{2}\right)$

$16x + 9x - 30 = -180$

$25x - 30 = -180$

$25x = -150$

$x = -6$

Substituting x back into either equation solve for y:

$y = \dfrac{3}{4}(-6) - \dfrac{5}{2} = -7$

The solution to the system is $(-6, -7)$.

This is a consistent system with independent lines.

Chapter 2 Introduction to Systems of Linear Equations and Inequalities
Section 2.3

1.

$\begin{cases} x+4y=11 \\ 5x-4y=7 \end{cases}$

$\begin{cases} x\cancel{+4y}=11 \\ 5x\cancel{-4y}=7 \end{cases}$

$6x=18$
$x=3$

Substituting x back into either equation solve for y:
$(3)+4y=11$
$3+4y=11$
$4y=8$
$y=2$

The solution to the system is $(3,2)$.

This is a consistent system with independent lines.

3.

$\begin{cases} 2g-3h=9 \\ -2g-4h=26 \end{cases}$

$\begin{cases} \cancel{2g}-3h=9 \\ \cancel{-2g}-4h=26 \end{cases}$

$-7h=35$
$h=-5$

Substituting h back into either equation solve for g:
$2g-3(-5)=9$
$2g+15=9$
$2g=-6$
$g=-3$

The solution to the system is $(-3,-5)$.

This is a consistent system with independent lines.

5.

$\begin{cases} 8x-2y=36 \\ 3x+4y=23 \end{cases}$

$\begin{cases} 2(8x-2y)=2(36) \\ 3x+4y=23 \end{cases}$

$\begin{cases} 16x-4y=72 \\ 3x+4y=23 \end{cases}$

$\begin{cases} 16x\cancel{-4y}=72 \\ 3x\cancel{+4y}=23 \end{cases}$

$19x=95$
$x=5$

Substituting x back into either equation solve for y:
$3(5)+4y=23$
$15+4y=23$
$4y=8$
$y=2$

The solution to the system is $(5,2)$.

This is a consistent system with independent lines.

7.

$\begin{cases} 7k-4H=16 \\ 2k+8H=32 \end{cases}$

$\begin{cases} 2(7k-4H)=2(16) \\ 2k+8H=32 \end{cases}$

$\begin{cases} 14k-8H=32 \\ 2k+8H=32 \end{cases}$

$\begin{cases} 14k\cancel{-8H}=32 \\ 2k\cancel{+8H}=32 \end{cases}$

$16k=64$
$k=4$

Substituting k back into either equation solve for H:
$2(4)+8H=32$
$8+8H=32$
$8H=24$
$H=3$

The solution to the system is $(4,3)$.

This is a consistent system with independent lines.

9.

$\begin{cases} 3t+8k=20 \\ 5t+6k=26 \end{cases}$

$\begin{cases} 5(3t+8k)=5(20) \\ -3(5t+6k)=-3(26) \end{cases}$

$\begin{cases} 15t+40k=100 \\ -15t-18k=-78 \end{cases}$

$\begin{cases} \cancel{15t}+40k=100 \\ \cancel{-15t}-18k=-78 \end{cases}$

$22k=22$
$k=1$

Substituting k back into either equation solve for t:
$3t + 8(1) = 20$
$3t + 8 = 20$
$3t = 12$
$t = 4$
The solution to the system is $(4,1)$.
This is a consistent system with independent lines.

11. Let x be the number of ml of 5% HCl solution, and y be the number of ml of 50% HCl solution used to get the 20 ml of a 15% HCl solution.
$$\begin{cases} x + y = 20 \\ 0.05x + 0.50y = (0.15)(20) = 3 \end{cases}$$
$$\begin{cases} x + y = 20 \\ (-20)(0.05x + 0.50y) = (3)(-20) \end{cases}$$
$$\begin{cases} \cancel{x} + y = 20 \\ \cancel{-x} - 10y = -60 \end{cases}$$
$-9y = -40$
$\dfrac{-9y}{-9} = \dfrac{-40}{-9}$
$y \approx 4.444$

Now solve for x:
$x = 20 - 4.444$
$x = 15.556$

The chemistry student will need 15.556 ml of the 5% HCl solution, and 4.444 ml of the 50% HCl solution, in order to make 20 ml of a 15% HCl solution.

13. Let Kristy mix x ml of 2% NaCl solution with y ml of 10% NaCl solution to get 25 ml of a 5% NaCl solution.
$$\begin{cases} x + y = 25 \\ 0.02x + 0.10y = (0.05)(25) = 1.25 \end{cases}$$
$$\begin{cases} x + y = 25 \\ (-50)(0.02x + 0.10y) = (1.25)(-50) \end{cases}$$
$$\begin{cases} \cancel{x} + y = 25 \\ \cancel{-x} - 5y = -62.5 \end{cases}$$
$-4y = -37.5$
$\dfrac{-4y}{-4} = \dfrac{-37.5}{-4}$
$y = 9.375$

Now solve for x:
$x = 25 - 9.375$
$x = 15.625$

Kristy will need 15.625 ml of the 2% NaCl solution, and 9.375 ml of the 10% NaCl solution, in order to make 25 ml of a 5% NaCl solution.

15. Let x be the number of pounds of the 5% premade mix, and y be the number of pounds of the 20% premade mix, used to get the 100 pounds of mix with 10% peanuts.
$$\begin{cases} x + y = 100 \\ 0.05x + 0.20y = (0.10)(100) = 10 \end{cases}$$
$$\begin{cases} x + y = 100 \\ (-20)(0.05x + 0.20y) = (10)(-20) \end{cases}$$
$$\begin{cases} \cancel{x} + y = 100 \\ \cancel{-x} - 4y = -200 \end{cases}$$
$-3y = -100$
$\dfrac{-3y}{-3} = \dfrac{-100}{-3}$
$y = 33.33$

Now solve for x:
$x = 100 - 33.33$
$x = 66.67$

The store manager must use 33.33 pounds of the 20% premade mix with 66.67 pounds of the 5% premade mix to make a 100 pounds 10% of mix with 10% peanuts.

17.
$$\begin{cases} w + 8s = -42 \\ -9w - s = -48 \end{cases}$$
$$\begin{cases} 9(w + 8s) = 9(-42) \\ -9w - s = -48 \end{cases}$$
$$\begin{cases} 9w + 72s = -378 \\ -9w - s = -48 \end{cases}$$
$$\begin{cases} \cancel{9w} + 72s = -378 \\ \cancel{-9w} - s = -48 \end{cases}$$
$71s = -426$
$s = -6$

Chapter 2 Introduction to Systems of Linear Equations and Inequalities

Substituting s back into either equation solve for w:

$w + 8(-6) = -42$

$w - 48 = -42$

$w = 6$

The solution to the system is $(6, -6)$.

This is a consistent system with independent lines.

19.

$\begin{cases} y = \dfrac{2}{3}x - 9 \\ -2x + 3y = -27 \end{cases}$

$\begin{cases} 3(y) = 3\left(\dfrac{2}{3}x - 9\right) \\ -2x + 3y = -27 \end{cases}$

$\begin{cases} 3y - 2x = 2x - 27 - 2x \\ -2x + 3y = -27 \end{cases}$

$\begin{cases} -1(-2x + 3y) = -1(-27) \\ -2x + 3y = -27 \end{cases}$

$\begin{cases} 2x - 3y = 27 \\ -2x + 3y = -27 \end{cases}$

$\begin{cases} \cancel{2x} - 3y = 27 \\ \cancel{-2x} + 3y = -27 \end{cases}$

$0 = 0$

Infinite number of solutions.

This is a consistent system with dependent lines.

21.

$\begin{cases} 2.5x + y = 4 \\ x + 0.4y = -5 \end{cases}$

$\begin{cases} 2.5x + y = 4 \\ -2.5(x + 0.4y) = -2.5(-5) \end{cases}$

$\begin{cases} 2.5x + y = 4 \\ -2.5x - y = 12.5 \end{cases}$

$\begin{cases} \cancel{2.5x} + y = 4 \\ \cancel{-2.5x} - y = 12.5 \end{cases}$

$0 \neq 16.5$

No solution.

This is an inconsistent system.

23.

$\begin{cases} W = -3.2c + 4.1 \\ W = 2.4c - 8.8 \end{cases}$

$\begin{cases} 3.2c + W = 4.1 \\ -2.4c + W = -8.8 \end{cases}$

$\begin{cases} -1(3.2c + W) = -1(4.1) \\ -2.4c + W = -8.8 \end{cases}$

$\begin{cases} -3.2c - W = -4.1 \\ -2.4c + W = -8.8 \end{cases}$

$\begin{cases} -3.2c \cancel{-W} = -4.1 \\ -2.4c \cancel{+W} = -8.8 \end{cases}$

$-5.6c = -12.9$

$c \approx 2.3$

Substituting c back into either equation solve for W:

$W \approx 2.4(2.3) - 8.8$

$W \approx 5.52 - 8.8$

$W \approx -3.28$

The solution to the system is $(2.3, -3.28)$.

This is a consistent system with independent lines.

25.

$\begin{cases} p = \dfrac{7}{4}t + 9 \\ 4p - 7t = 36 \end{cases}$

$\begin{cases} 4(p) = 4\left(\dfrac{7}{4}t + 9\right) \\ 4p - 7t = 36 \end{cases}$

$\begin{cases} 4p = 7t + 36 \\ 4p - 7t = 36 \end{cases}$

$\begin{cases} 4p - 7t = 36 \\ 4p - 7t = 36 \end{cases}$

$\begin{cases} -1(4p - 7t) = -1(36) \\ 4p - 7t = 36 \end{cases}$

$\begin{cases} -4p + 7t = -36 \\ 4p - 7t = 36 \end{cases}$

$\begin{cases} \cancel{-4p} + 7t = -36 \\ \cancel{4p} - 7t = 36 \end{cases}$

$0 = 0$

Infinite number of solutions.

This is a consistent system with dependent lines.

27.

a. Let A be option 1 for admission and ride tickets at a local fair in dollars, let B be option 2 for admission and ride tickets at a local fair in dollars, and let r be the total number of rides purchased.

$A(r) = 0.50r + 22$ and $B(r) = 0.75r + 15$

$$A(r) = B(r)$$
$$\left.\begin{aligned}0.50r + 22 &= 0.75r + 15 \\ -0.75r - 22 &= -0.75r - 22\end{aligned}\right\} \text{add}$$
$$-0.25r = -7$$
$$\frac{-0.25r}{-0.25} = \frac{-7}{-0.25}$$
$$r = 28$$

Therefore, 28 rides will result in the same cost for both options for admission and ride tickets at a local fair.

b. Option 1 would be a better deal if you are planning on riding a larger number of rides at the fair because the cost per ride is less.

29. Let x be the cost of a plank in dollars, and y be the cost of a 4x4 in dollars.

$$\begin{cases} 150x + 8y = 222.92 \to \text{order \#1} \\ 45x + 2y = 65.18 \to \text{order \#2} \end{cases}$$

$$\begin{cases} 150x + 8y = 222.92 \\ (-4)(45x + 2y) = (65.18)(-4) \end{cases}$$

$$\left.\begin{aligned}150x + 8y &= 222.92 \\ -180x - 8y &= -260.72\end{aligned}\right\} \text{add}$$
$$-30x = -37.8$$
$$\frac{-30x}{-30} = \frac{-37.8}{-30}$$
$$x = 1.26$$

Substitute $x = 1.26$ into one of the original equations to solve for y.

$45(1.26) + 2y = 65.18$
$56.7 + 2y = 65.18$
$2y = 65.18 - 56.7$
$2y = 8.48$
$\frac{2y}{2} = \frac{8.48}{2}$
$y = 4.24$

Each plank has a cost of $1.26 and each 4x4 has a cost of $4.24.

31. Let x be the number of regular hours worked, and y be the number of overtime hours worked.

$$\begin{cases} x + y = 165 \to \text{Total hours worked} \\ 8x + 12y = 1420 \to \text{Total pay} \end{cases}$$

$$\begin{cases} (2)(x + y) = (165)(2) \\ \left(\frac{-1}{4}\right)(8x + 12y) = (1420)\left(\frac{-1}{4}\right) \end{cases}$$

$$\left.\begin{aligned}-2x - 3y &= -355 \\ 2x + 2y &= 330\end{aligned}\right\} \text{add}$$
$$y = -25$$

Now solve for x:
$x = 165 - 25 = 140$

Ana worked a total of 140 regular hours, and 25 over time hours.

33. Substitution, because H is already isolated.

35. Substitution, because T is already isolated.

37. When solving an inconsistent system using the elimination or substitution method, both variables will be eliminated, and the remaining statement will be false.

39. Use a table or graph the two equations and see where the lines intersect.

41.

$$\begin{cases} \frac{5}{3}d - \frac{3}{5}g = 52 \\ \frac{3}{5}d + \frac{5}{3}g = -66 \end{cases}$$

$$\begin{cases} 375\left(\frac{5}{3}d - \frac{3}{5}g\right) = 375(52) \\ 135\left(\frac{3}{5}d + \frac{5}{3}g\right) = 135(-66) \end{cases}$$

$$\begin{cases} 625d - 225g = 19500 \\ 81d + 225g = -8910 \end{cases}$$

$$\begin{cases} 625d \cancel{-225g} = 19500 \\ 81d \cancel{+225g} = -8910 \end{cases}$$

$706d = 10590$

$d = 15$

Substituting d back into either equation solve for g:

$81(15) + 225g = -8910$

$1215 + 225g = -8910$

$225g = -10125$

$g = -45$

The solution to the system is $(15, -45)$.

This is a consistent system with independent lines.

43.

$$\begin{cases} -2x + 11y = 2 \\ 11(x - y) = -2 \end{cases}$$

$$\begin{cases} -2x + 11y = 2 \\ 11x - 11y = -2 \end{cases}$$

$$\begin{cases} -2x \cancel{+11y} = 2 \\ 11x \cancel{-11y} = -2 \end{cases}$$

$9x = 0$

$x = 0$

Substituting x back into either equation solve for y:

$-2(0) + 11y = 2$

$0 + 11y = 2$

$11y = 2$

$y = \frac{2}{11}$

The solution to the system is $\left(0, \frac{2}{11}\right)$.

This is a consistent system with independent lines.

45.

$$\begin{cases} -3.5x + y = -16.95 \\ y = -2.4x + 4.88 \end{cases}$$

$-3.5x + (-2.4x + 4.88) = -16.95$

$-3.5x - 2.4x + 4.88 = -16.95$

$-5.9x + 4.88 = -16.95$

$-5.9x = -21.83$

$x = 3.7$

Substituting x back into either equation solve for y:

$y = -2.4(3.7) + 4.88$

$y = -8.88 + 4.88$

$y = -4$

The solution to the system is $(3.7, -4)$.

This is a consistent system with independent lines.

47.

$$\begin{cases} W = \frac{3}{7}d + 6 \\ W = \frac{2}{7}d + 8.8 \end{cases}$$

$\frac{3}{7}d + 6 = \frac{2}{7}d + 8.8$

$7\left(\frac{3}{7}d + 6\right) = 7\left(\frac{2}{7}d + 8.8\right)$

$3d + 42 = 2d + 61.6$

$3d + 42 - 2d = 2d + 61.6 - 2d$

$d + 42 = 61.6$

$d = 19.6$

Substituting d back into either equation solve for W:

$W = \frac{3}{7}(19.6) + 6$

$W = 8.4 + 6$

$W = 14.4$

The solution to the system is $(19.6, 14.4)$.

This is a consistent system with independent lines.

49.

$$\begin{cases} T = 0.5g + 8.5 \\ g - 2T = -17 \end{cases}$$

$g - 2(0.5g + 8.5) = -17$

$g - g - 17 = -17$

$-17 = -17$

Infinite number of solutions.

This is a consistent system with dependent lines.

51. Frank did not multiply the 30 by 2.

53.

$\begin{cases} 3.7x + 3.5y = 5.3 \\ y = 3.57x + 4.21 \end{cases}$

$3.7x + 3.5(3.57x + 4.21) = 5.3$

$3.7x + 12.495x + 14.735 = 5.3$

$16.195x + 14.735 = 5.3$

$16.195x = -9.435$

$x \approx -0.58$

Substituting x back into either equation solve for y:

$y \approx 3.57(-0.58) + 4.21$

$y \approx -2.0706 + 4.21$

$y \approx 2.1394$

The solution to the system is $(-0.58, 2.1394)$.

This is a consistent system with independent lines.

55.

$\begin{cases} -9c - 7d = 8 \\ c - 7d = 8 \end{cases}$

$\begin{cases} -9c - 7d = 8 \\ -1(c - 7d) = -1(8) \end{cases}$

$\begin{cases} -9c - 7d = 8 \\ -c + 7d = -8 \end{cases}$

$\begin{cases} -9c \cancel{-7d} = 8 \\ -c \cancel{+7d} = -8 \end{cases}$

$-10c = 0$

$c = 0$

Substituting c back into either equation solve for d:

$(0) - 7d = 8$

$0 - 7d = 8$

$-7d = 8$

$d = -\dfrac{8}{7}$

The solution to the system is $\left(0, -\dfrac{8}{7}\right)$.

This is a consistent system with independent lines.

57.

$\begin{cases} 3x + 4y = -15 \\ y = x - 9 \end{cases}$

$3x + 4(x - 9) = -15$

$3x + 4x - 36 = -15$

$7x - 36 = -15$

$7x = 21$

$x = 3$

Substituting x back into either equation solve for y:

$y = (3) - 9$

$y = 3 - 9$

$y = -6$

The solution to the system is $(3, -6)$.

This is a consistent system with independent lines.

59.

$\begin{cases} \dfrac{4}{3}x + y = 13 \\ x - \dfrac{7}{6}y = 4 \end{cases}$

$\begin{cases} 21\left(\dfrac{4}{3}x + y\right) = 21(13) \\ 18\left(x - \dfrac{7}{6}y\right) = 18(4) \end{cases}$

$\begin{cases} 28x + 21y = 273 \\ 18x - 21y = 72 \end{cases}$

$\begin{cases} 28x \cancel{+21y} = 273 \\ 18x \cancel{-21y} = 72 \end{cases}$

$46x = 345$

$x = 7.5$

Substituting x back into either equation solve for y:

$28(7.5) + 21y = 273$

$210 + 21y = 273$

$21y = 63$

$y = 3$

The solution to the system is $(7.5, 3)$.

This is a consistent system with independent lines.

Chapter 2 Introduction to Systems of Linear Equations and Inequalities

61.

$\begin{cases} -8.4x - 2.8y = -58.8 \\ 3x + y = 21 \end{cases}$

$\begin{cases} -8.4x - 2.8y = -58.8 \\ 2.8(3x + y) = 2.8(21) \end{cases}$

$\begin{cases} -8.4x - 2.8y = -58.8 \\ 8.4x + 2.8y = 58.8 \end{cases}$

$\begin{cases} -8.4x - \cancel{2.8y} = -58.8 \\ 8.4x + \cancel{2.8y} = 58.8 \end{cases}$

$0 = 0$

Infinite number of solutions.

This is a consistent system with dependent lines.

63.

$\begin{cases} 3c - b = -17 \\ b = 3c + 18 \end{cases}$

$3c - (3c + 18) = -17$

$3c - 3c - 18 = -17$

$-18 \neq -17$

No solution.

This is an inconsistent system.

65.

$\begin{cases} -2.4x + y = 2.5 \\ 4.1x + y = -2 \end{cases}$

$\begin{cases} -2.4x + y = 2.5 \\ -1(4.1x + y) = -1(-2) \end{cases}$

$\begin{cases} -2.4x + y = 2.5 \\ -4.1x - y = 2 \end{cases}$

$\begin{cases} -2.4x + \cancel{y} = 2.5 \\ -4.1x - \cancel{y} = 2 \end{cases}$

$-6.5x = 4.5$

$x \approx -0.7$

Substituting x back into either equation solve for y:

$4.1(-0.7) + y \approx -2$

$-2.87 + y \approx -2$

$y \approx 0.87$

The solution to the system is $(-0.7, 0.87)$.

This is a consistent system with independent lines.

Section 2.4

1.
$$5x+7 > 37$$
$$5x+7-7 > 37-7$$
$$5x > 30$$
$$\frac{5x}{5} > \frac{30}{5}$$
$$x > 6$$

3.
$$\frac{P}{4} \geq 2$$
$$(4)\left(\frac{P}{4}\right) \geq (2)(4)$$
$$P \geq 8$$

5.
$$8V + 4 < 5V - 20$$
$$-5V - 4 \quad -5V - 4$$
$$3V < -24$$
$$\frac{3V}{3} < \frac{-24}{3}$$
$$V < -8$$

7.
$$2t + 12 < 5t + 39$$
$$-5t - 12 \quad -5t - 12$$
$$-3t < 27$$
$$\frac{-3t}{-3} > \frac{27}{-3}$$
$$t > -9$$

9.
$$\frac{K}{-4} + 7 \leq 21$$
$$\frac{K}{-4} + 7 - 7 \leq 21 - 7$$
$$\frac{K}{-4} \leq 14$$
$$(-4)\left(\frac{K}{-4}\right) \geq (14)(-4)$$
$$K \geq -56$$

11. Michelle's work is correct. Amy reversed the inequality symbol for subtracting 8 from both sides, which is incorrect.

13. Let $F(t)$ be the number of full-time faculty and let $P(t)$ be the number of part-time faculty in U.S. higher education institutions t years since 1990. Note: Your models might differ, but they must follow the trend of the data.

a.

t	$F(t)$	$P(t)$
9	523	437
11	618	495
13	632	543
15	676	615

Full-time faculty in U.S. higher education institutions:
$$F(t) = mt + b$$
$$m = \frac{676-632}{15-13} = 22$$
$$b = 632 - (22)(13) = 346$$
$$F(t) = 22t + 346$$

Part-time faculty in U.S. higher education institutions:
$$P(t) = mt + b$$
$$m = \frac{615-495}{15-11} = 30$$
$$b = 615 - (30)(15) = 165$$
$$P(t) = 30t + 165$$

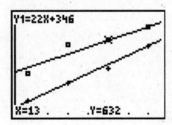

b.

$$P(t) > F(t)$$
$$30t + 165 > 22t + 346$$
$$\underline{-22t - 165 \quad -22t - 165}$$
$$8t > 181$$
$$\frac{8t}{8} > \frac{181}{8}$$
$$t > 22.625$$

The number of part-time faculty will be greater than the number of full-time faculty after 2012.

15. Let $A(t)$ be the total population of Africa in millions (represented by □ on the graph). Let $E(t)$ be the total population of Europe in millions (represented by + on the graph). Let t be time in years since 1970.

a.

t	A(t)	E(t)
0	361	656
10	472	694
20	626	721
30	805	729

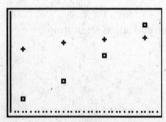

Africa's population in millions:

$$A(t) = mt + b$$
$$m = \frac{626 - 361}{90 - 70} = 13.25$$
$$b = 361$$
$$A(t) = 13.25t + 361$$

Europe's population in millions:

$$E(t) = mt + b$$
$$m = \frac{729 - 656}{100 - 70} \approx 2.43$$
$$b = 656$$
$$E(t) = 2.43t + 656$$

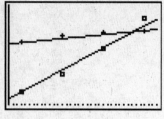

b.

$$A(t) > E(t)$$
$$13.25t + 361 > 2.43t + 656$$
$$\underline{-2.43t - 361 \quad -2.43t - 361}$$
$$10.82t > 295$$
$$\frac{10.82t}{10.82} > \frac{295}{10.82}$$
$$t > 27.3$$

Africa has a greater population than Europe after 1997.

17. Cheaper than: <

19. At least: ≥

21. More expensive than: >

23. Lower than: <

25.

$$R(c) \geq C(c)$$
$$450c \geq 280c + 20,000$$
$$\underline{-280c \quad -280c}$$
$$170c \geq 20,000$$
$$\frac{170c}{170} \geq \frac{20,000}{170}$$
$$c \geq 117.65$$

The cabinet manufacturer must sell at least 118 cabinets to break even or make a profit.

27.

$$R(a) \geq C(a)$$
$$175a \geq 25a + 7800$$
$$\underline{-25a \quad -25a}$$
$$150a \geq 7800$$
$$\frac{150a}{150} \geq \frac{7800}{150}$$
$$a \geq 52$$

Optimum Traveling Detail must detail at least 52 cars to break even or make a profit.

2.4 Exercises

29.
a.
Let $F(m)$ be the cost in dollars for the *Freedom* plan from Uptown Wireless, when used for m minutes.
$$F(m) = 69.99 + .3m$$
Let $N(m)$ be the cost in dollars for the *NO-Strings* plan from Go Wireless, when used for m minutes.
$$N(m) = 69.99 + .35m$$
Let $Y(m)$ be the cost in dollars for the *Your Free* plan from U-R Mobile, when used for m minutes.
$$Y(m) = 79 + .25m$$

b. Since Uptown Wireless is always cheaper than Go Wireless we need only compare Uptown Wireless with U-R Mobile.

$$Y(m) < F(m)$$
$$\cancel{79} + 0.25m < 69.99 \cancel{+0.30m}$$
$$\underline{\cancel{79} - 0.30m \quad -79 \cancel{-0.30m}}$$
$$-0.05m < -9.01$$
$$\frac{-0.05m}{-0.05} > \frac{-9.01}{-0.05}$$
$$m > 180.2$$

U-R Mobile will have the cheapest plan if used for more than 180 minutes.

31.
$$U(t) > K(t)$$
$$0.71t \cancel{+37.1} > 1.3t + 44.7$$
$$\underline{-1.3t \cancel{-37.1} \quad -1.3t - 37.1}$$
$$-0.59t > 7.6$$
$$\frac{-0.59t}{-0.59} < \frac{7.6}{-0.59}$$
$$t < 12.88$$

In 2017, the percentage of births to unmarried women in the United States will be greater than the percentage of births to unmarried women in the United Kingdom.

33. Let $P(s)$ be the monthly salary for s dollars of sales.
$$P(s) = 600 + 0.21s$$
$$600 + 0.21s \geq 1225$$
$$600 - 600 + 0.21s \geq 1225 - 600$$
$$0.21s \geq 625$$
$$\frac{0.21s}{0.21} \geq \frac{625}{0.21}$$
$$s \geq 2976.19$$

Sales need to be at least $2976.19 for the salesperson to earn at least $1225.

35. $Y1 < Y2$ when $x < 2$

37. $Y1 > Y2$ when $x < -7$

39. $x > 5$

X	Y1	Y2
2	17	47
3	25	45
4	33	43
5	41	41
6	49	39
7	57	37
8	65	35

X=8

41. $x \geq 4$

X	Y1	Y2
1	-3.5	-12.5
2	1	-5
3	5.5	2.5
4	10	10
5	14.5	17.5
6	19	25
7	23.5	32.5

X=7

43. $x \leq 3.5$

X	Y1	Y2
1	16.5	6.5
2	15	9
3	13.5	11.5
3.5	12.75	12.75
4	12	14
5	10.5	16.5
6	9	19

X=6

45. $x < -3.33333$

X	Y1	Y2
-6	-3.5	-1.5
-5	-3.25	-2
-4	-3	-2.5
-3.5	-2.875	-2.75
-3.4	-2.85	-2.8
-3.33333	-2.833	-2.833
-3.2	-2.8	-2.9

X=-3.33333

47. $x > -5$

49. $x \leq -3$

51. $x > 2$

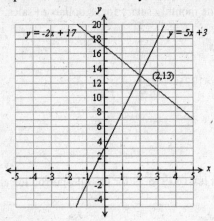

53. $x < 12$

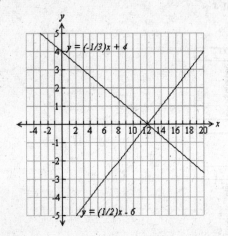

55.

$5 + \dfrac{3x}{2} \leq -3$

$2\left(5 + \dfrac{3x}{2}\right) \leq 2(-3)$

$10 + 3x \leq -6$

$10 + 3x - 10 \leq -6 - 10$

$3x \leq -16$

$x \leq -\dfrac{16}{3}$

57.

$5x + 3 \geq 3(x - 2)$

$5x + 3 \geq 3x - 6$

$5x + 3 - 3x \geq 3x - 6 - 3x$

$2x + 3 \geq -6$

$2x + 3 - 3 \geq -6 - 3$

$2x \geq -9$

$x \geq -\dfrac{9}{2}$

59.

$\dfrac{-7d}{3} + 5 < 3$

$\dfrac{-7d}{3} + 5 - 5 < 3 - 5$

$\dfrac{-7d}{3} < -2$

$\left(-\dfrac{3}{7}\right)\left(\dfrac{-7d}{3}\right) > \left(-\dfrac{3}{7}\right)(-2)$

$d > \dfrac{6}{7}$

61.

$2.7v + 3.69 > 1.5v - 6.5$

$2.7v + 3.69 - 1.5v > 1.5v - 6.5 - 1.5v$

$1.2v + 3.69 > -6.5$

$1.2v + 3.69 - 3.69 > -6.5 - 3.69$

$1.2v > -10.19$

$\dfrac{1.2v}{1.2} > \dfrac{-10.19}{1.2}$

$v > -8.49$

63.

$3.2 + 2.7(1.5k - 3.1) \geq 9.43k - 17.5$

$3.2 + 4.05k - 8.37 \geq 9.43k - 17.5$

$4.05k - 5.17 \geq 9.43k - 17.5$

$4.05k - 5.17 - 9.43k \geq 9.43k - 17.5 - 9.43k$

$-5.38k - 5.17 \geq -17.5$

$-5.38k - 5.17 + 5.17 \geq -17.5 + 5.17$

$-5.38k \geq -12.33$

$\dfrac{-5.38k}{-5.38} \leq \dfrac{-12.33}{-5.38}$

$k \leq 2.3$

65.

$\dfrac{2}{5}(w - 20) \leq -\dfrac{3}{7}(4w - 9)$

$\dfrac{2}{5}w - 8 \leq -\dfrac{12}{7}w + \dfrac{27}{7}$

$35\left(\dfrac{2}{5}w - 8\right) \leq 35\left(-\dfrac{12}{7}w + \dfrac{27}{7}\right)$

$14w - 280 \leq -60w + 135$

$14w - 280 + 60w \leq -60w + 135 + 60w$

$74w - 280 \leq 135$

$74w - 280 + 280 \leq 135 + 280$

$74w \leq 415$

$w \leq \dfrac{415}{74}$

67.

$2.35x + 7.42 < 1.3x - 4.75$

$2.35x + 7.42 - 1.3x < 1.3x - 4.75 - 1.3x$

$1.05x + 7.42 < -4.75$

$1.05x + 7.42 - 7.42 < -4.75 - 7.42$

$1.05x < -12.17$

$\dfrac{1.05x}{1.05} < \dfrac{-12.17}{1.05}$

$x < -11.59$

Chapter 2 Introduction to Systems of Linear Equations and Inequalities
Section 2.5

1.
$|x| = 12$
$x = 12$ or $x = -12$
$x = \pm 12$

3.
$|h+7| = 15$
$h+7 = 15$ or $h+7 = -15$
$h = 8$ or $h = -22$

5.
$|b-12| = 8$
$b-12 = 8$ or $b-12 = -8$
$b = 20$ or $b = 4$

7.
$|k+10| = -3$
There is no solution, because an absolute value cannot equal a negative number.

9.
$2|r+11| = 36$
$\dfrac{2|r+11|}{2} = \dfrac{36}{2}$
$|r+11| = 18$
$r+11 = 18$ or $r+11 = -18$
$r = 7$ or $r = -29$

11.
$|g-9| + 12 = 8$
$|g-9| + 12 - 12 = 8 - 12$
$|g-9| = -4$

There is no solution, because an absolute value cannot equal a negative number.
absolute value cannot equal a negative number.

13.
$-2|d-8| + 1 = 11$
$-2|d-8| + 1 - 1 = 11 - 1$
$-2|d-8| = 10$
$\dfrac{-2|d-8|}{-2} = \dfrac{10}{-2}$
$|d-8| = -5$

There is no solution, because an absolute value cannot equal a negative number.

15.
$-3|x-12| = -5$
$\dfrac{-3|x-12|}{-3} = \dfrac{-5}{-3}$
$|x-12| = \dfrac{5}{3}$
$x-12 = \dfrac{5}{3}$ or $x-12 = -\dfrac{5}{3}$
$x = \dfrac{41}{3}$ or $x = \dfrac{31}{3}$

17.
$-\dfrac{1}{3}|s-60| + 25 = 20$
$-\dfrac{1}{3}|s-60| + 25 - 25 = 20 - 25$
$-\dfrac{1}{3}|s-60| = -5$
$(-3)\left(-\dfrac{1}{3}\right)|s-60| = (-5)(-3)$
$|s-60| = 15$
$s-60 = 15$ or $s-60 = -15$
$s = 75$ or $s = 45$

Ricardo should travel at an average speed of 45 mph or 75 mph to get gas mileage of 20 mpg during his trip.

19.
$|700 - 60t| = 50$
$700 - 60t = 50$ or $700 - 60t = -50$
$-60t = -650$ or $-60t = -750$
$\dfrac{-60t}{-60} = \dfrac{-650}{-60}$ or $\dfrac{-60t}{-60} = \dfrac{-750}{-60}$
$t \approx 10.83$ or $t = 12.5$

Ricardo will be fifty miles away from Fresno after about 10.83 hours and again after about 12.5 hours.

21. $-6 < x < 6$

23. $x \leq -3$ or $x \geq 3$

25. $x < -5$ or $x > -2$

27. $-3 \leq x \leq 8$

29. $x < -6$ or $x > 5$

31.
a. For $f(x) < g(x): -4 < x < 4$
b. For $f(x) > g(x): x < -4$ or $x > 4$

33.

a. For $f(x) \leq g(x): 0 \leq x \leq 4$

b. For $f(x) \geq g(x): x \leq 0$ or $x \geq 4$

35.

a. For $f(x) < g(x): -5 < x < 5$

b. For $f(x) > g(x): x < -5$ or $x > 5$

37.

$|x| < 5$

$-5 < x < 5$

39.

$|h-3| \leq 4$

$-4 \leq h-3 \leq 4$

$-4+3 \leq h-3+3 \leq 4+3$

$-1 \leq h \leq 7$

41.

$|b-1| < 6$

$-6 < b-1 < 6$

$-6+1 < b-1+1 < 6+1$

$-5 < b < 7$

43.

$|L-3.00| \leq 0.001$

$-0.001 \leq L-3.00 \leq 0.001$

$-0.001+3.00 \leq L-3.00+3.00 \leq 0.001+3.00$

$2.999 \leq L \leq 3.001$

The acceptable lengths of this coupler are from a low of 2.999 cm, to a high of 3.001 cm.

45.

$|P-20| \leq 0.5$

$-0.5 \leq P-20 \leq 0.5$

$-0.5+20 \leq P-20+20 \leq 0.5+20$

$19.5 \leq P \leq 20.5$

The acceptable pressures for this experiment are from a low of 19.5 psi, to a high of 20.5 psi.

47.

$|x| > 3$

$x < -3$ or $x > 3$

49.

$|p-4| \geq 2$

$p-4 \leq -2$ or $p-4 \geq 2$

$p-4+4 \leq -2+4$ or $p-4+4 \geq 2+4$

$p \leq 2$ or $p \geq 6$

51.

$|y+6| > 3$

$y+6 < -3$ or $y+6 > 3$

$y+6-6 < -3-6$ or $y+6-6 > 3-6$

$y < -9$ or $y > -3$

53.

$\left|\dfrac{h-69}{2.5}\right| > 2$

$\dfrac{h-69}{2.5} < -2$ or $\dfrac{h-69}{2.5} > 2$

$2.5\left(\dfrac{h-69}{2.5}\right) < (-2)(2.5)$ or $2.5\left(\dfrac{h-69}{2.5}\right) > (2)(2.5)$

$h-69 < -5$ or $h-69 > 5$

$h-69+69 < -5+69$ or $h-69+69 > 5+69$

$h < 64$ or $h > 74$

Men's heights that are less than 64 inches or greater than 74 inches are considered unusual.

55.

a.

$\left|\dfrac{S-100}{16}\right| > 2$

$\dfrac{S-100}{16} < -2$ or $\dfrac{S-100}{16} > 2$

$16\left(\dfrac{S-100}{16}\right) < (-2)(16)$ or $16\left(\dfrac{S-100}{16}\right) > (2)(16)$

$S-100 < -32$ or $S-100 > 32$

$S-100+100 < -32+100$ or $S-100+100 > 32+100$

$S < 68$ or $S > 132$

IQ scores that are below 68 or above 132 would be considered unusual.

b. Yes, Albert Einstein's IQ was unusual. His IQ was greater thatn 132.

57.

$2|r+6| < 8$

$\dfrac{2|r+6|}{2} < \dfrac{8}{2}$

$|r+6| < 4$

$-4 < r+6 < 4$

$-4-6 < r+6-6 < 4-6$

$-10 < r < -2$

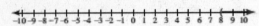

59.

$4x - 20 > 12$

$4x - 20 + 20 > 12 + 20$

$4x > 32$

$x > 8$

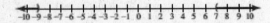

61.

$|d-5| + 10 \le 7$

$|d-5| + 10 - 10 \le 7 - 10$

$|d-5| \le -3$

There is no solution, because an absolute value cannot equal a negative number.

63.

$|m+1| - 2 > 6$

$|m+1| - 2 + 2 > 6 + 2$

$|m+1| > 8$

$m + 1 < -8$ or $m + 1 > 8$

$m + 1 - 1 < -8 - 1$ or $m + 1 - 1 > 8 - 1$

$m < -9$ or $m > 7$

65.

$\dfrac{2}{3}x - 8 > \dfrac{5}{6}x - 4$

$6\left(\dfrac{2}{3}x - 8\right) > 6\left(\dfrac{5}{6}x - 4\right)$

$4x - 48 > 5x - 24$

$4x - 48 - 5x > 5x - 24 - 5x$

$-x - 48 > -24$

$-x - 48 + 48 > -24 + 48$

$-x > 24$

$\dfrac{-x}{-1} < \dfrac{24}{-1}$

$x < -24$

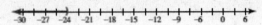

67.

$-6|p-4| + 3 \le 16$

$-6|p-4| + 3 - 3 \le 16 - 3$

$-6|p-4| \le 13$

$\dfrac{-6|p-4|}{-6} \ge \dfrac{13}{-6}$

$|p-4| \ge -\dfrac{13}{6}$

An absolute value will always be greater than a negative number. Therefore, the solution for p is all real numbers.

69.

$-|x-4| \ge -2$

$\dfrac{-|x-4|}{-1} \le \dfrac{-2}{-1}$

$|x-4| \le 2$

$-2 \le x - 4 \le 2$

$-2 + 4 \le x - 4 + 4 \le 2 + 4$

$2 \le x \le 6$

71.

$5 + 2|2x - 4| > 11$

$5 + 2|2x - 4| - 5 > 11 - 5$

$2|2x - 4| > 6$

$\dfrac{2|2x - 4|}{2} > \dfrac{6}{2}$

$|2x - 4| > 3$

$2x - 4 < -3$ or $2x - 4 > 3$

$2x - 4 + 4 < -3 + 4$ or $2x - 4 + 4 > 3 + 4$

$2x < 1$ or $2x > 7$

$x < \dfrac{1}{2}$ or $x > \dfrac{7}{2}$

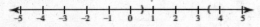

73.

$12 + 8|3d - 8| \leq 10$

$12 + 8|3d - 8| - 12 \leq 10 - 12$

$8|3d - 8| \leq -2$

$\dfrac{8|3d - 8|}{8} \leq \dfrac{-2}{8}$

$|3d - 8| \leq -\dfrac{1}{4}$

There is no solution, because an absolute value cannot equal a negative number.

75.

$1.4x + 3 \geq 4.4x - 1$

$1.4x + 3 - 4.4x \geq 4.4x - 1 - 4.4x$

$-3x + 3 \geq -1$

$-3x + 3 - 3 \geq -1 - 3$

$-3x \geq -4$

$\dfrac{-3x}{-3} \leq \dfrac{-4}{-3}$

$x \leq \dfrac{4}{3}$

77.

$\left|\dfrac{1}{2}x + 3\right| \leq 5$

$-5 \leq \dfrac{1}{2}x + 3 \leq 5$

$-5 - 3 \leq \dfrac{1}{2}x + 3 - 3 \leq 5 - 3$

$-8 \leq \dfrac{1}{2}x \leq 2$

$2(-8) \leq 2\left(\dfrac{1}{2}x\right) \leq 2(2)$

$-16 \leq x \leq 4$

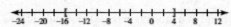

Chapter 2 Introduction to Systems of Linear Equations and Inequalities

Section 2.6

1. A 61" tall person weighing 105 pounds falls into the normal range.

3. A 5'9"(69") tall person is considered overweight if he weighs above 172 pounds.

5. A person who weighs 150 pounds must be between 5'5" and 6'3" to fall in the normal range.

7. A person who is 67" tall has a normal weight range of 120 pounds to 160 pounds.

9. They can build at most 370 mountain bikes per month.

11. Yes, they can build 400 cruisers and 100 mountain bikes per month. This is at the 500 bike per month limit and under the cost limitations for a month.

13.

a. Let C be the total number of cruisers Bicycles Galore produces monthly and let M be the total number of mountain bikes Bicycles Galore produces monthly.

$C + M \leq 500$

$65C + 120M \geq 40000$

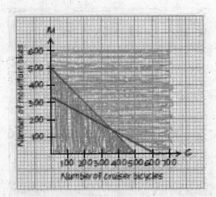

b. Bicycles Galore will be able to meet a demand by the board of directors for a profit of $40,000 per month. Any combination of mountain bikes and cruisers in the overlapping shaded region will meet the board's demand.

c.
$$\begin{cases} C + M = 500 \\ 65C + 120M = 40,000 \end{cases}$$

Multiply the first equation by -65 and add the result to the second equation.

$(-65)(C+M) = (500)(-65)$

$-65C - 65M = -32,500$

$\left.\begin{array}{r}65C + 120M = 40,000 \\ -65C - 65M = -32,500\end{array}\right\}$ add

$55M = 7500$

$\dfrac{55M}{55} = \dfrac{7500}{55}$

$M \approx 136.36$

Now solve for C:

$C = 500 - 136.36$

$C \approx 363.64$

They must make 137 mountain bikes and 363 cruisers to make $40,000 profit. If they make more mountain bikes they will make more profit.

15. $y > 3x + 2$

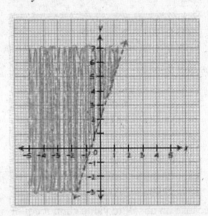

17. $y > \dfrac{5}{4}x - 2$

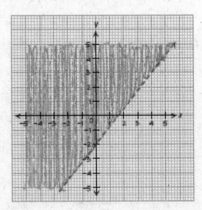

19. $y \geq \dfrac{2}{3}x + 6$

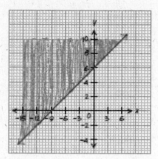

21. $y \leq -2x + 6$

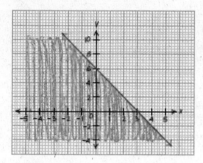

23.
$2x + 5y > 10$
$2x + 5y - 2x > 10 - 2x$
$5y > -2x + 10$
$\dfrac{5y}{5} > \dfrac{-2x + 10}{5}$
$y > -\dfrac{2}{5}x + 2$

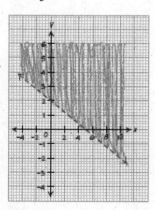

25.
$12x - 4y > 8$
$12x - 4y - 12x > 8 - 12x$
$-4y > -12x + 8$
$\dfrac{-4y}{-4} < \dfrac{-12x + 8}{-4}$
$y < 3x - 2$

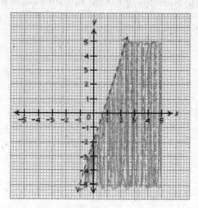

27.

Using the points $(3,3)$ and $(0,-3)$:

$m = \dfrac{3 - (-3)}{3 - 0} = \dfrac{6}{3} = 2$

$b = -3$

Since the solid line is shaded below use the $\leq$ symbol.

$y \leq 2x - 3$

29.

Using the points $(4,0)$ and $(0,6)$:

$m = \dfrac{6 - 0}{0 - 4} = \dfrac{6}{-4} = -\dfrac{3}{2}$

$b = 6$

Since the dasshed line is shaded above use the $>$ symbol.

$y > -\dfrac{3}{2}x + 6$

31.

Using the points $(0,9)$ and $(6,6)$:

$m = \dfrac{9 - 6}{0 - 6} = \dfrac{3}{-6} = -\dfrac{1}{2}$

$b = 9$

Since the dashed line is shaded above use the $>$ symbol.

$y > -\dfrac{1}{2}x + 9$

33.

Using the points $(1,0)$ and $(3,-2)$:

$$m = \frac{-2-0}{3-1} = \frac{-2}{2} = -1$$

Using $m = -1$ and $(1,0)$:

$$0 = -1(1) + b$$
$$0 = -1 + b$$
$$1 = b$$

Since the solid line is shaded below use the $\leq$ symbol.

$$y \leq -x + 1$$

35. Let A be the amount (in dollars) Juanita invests in the account paying 4.5% interest. Let B be the amount (in dollars) invested in the account paying 3.75% interest.

$$A + B \leq 750,000$$
$$0.045A + 0.0375B \geq 30,000$$

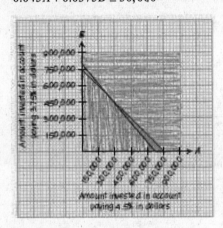

37.

a. Let P be the number of power bars, and let D be the number of sports drinks that need to be consumed by the athletes.

$$240P + 300D \geq 2000$$
$$30P + 70D \geq 350$$

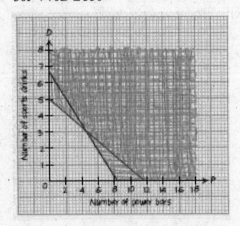

b. The point of intersection gives the least amount of each. Each athlete can carry a minimum of 3 drinks and 5 power bars. This combination will meet the minimum calorie and carbohydrate requirements.

c. If a racer carries 4 drinks, he or she must also carry 4 power bars.

39.

$$\begin{cases} y > 2x + 4 \\ y < -3x + 7 \end{cases}$$

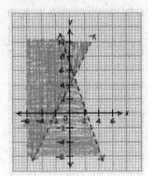

41.

$$\begin{cases} y < \frac{2}{5}x - 3 \\ y < -\frac{3}{4}x + 6 \end{cases}$$

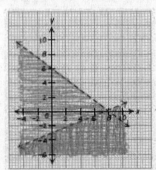

43.

$$\begin{cases} y \leq 2x + 5 \\ y \geq 2x + 8 \end{cases}$$

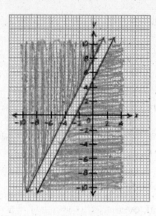

No solution because there is no overlap.

45.

$\begin{cases} 2x + 4y \leq 5 \\ 2x - 4y \leq 5 \end{cases}$

Solve for y in both equations to graph:

$2x + 4y \leq 5$, $2x - 4y \leq 5$

$4y \leq -2x + 5$, $-4y \leq -2x + 5$

$\dfrac{4y}{4} \leq \dfrac{-2x+5}{4}$, $\dfrac{-4y}{-4} \geq \dfrac{-2x+5}{-4}$

$y \leq -\dfrac{1}{2}x + \dfrac{5}{4}$, $y \geq \dfrac{1}{2}x - \dfrac{5}{4}$

$\begin{cases} y \leq -\dfrac{1}{2}x + \dfrac{5}{4} \\ y \geq \dfrac{1}{2}x - \dfrac{5}{4} \end{cases}$

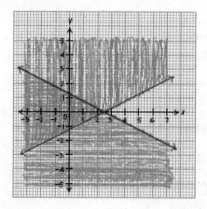

47.

$\begin{cases} 5x + 3y > 7 \\ 4x + 2y < 10 \end{cases}$

Solve for y in both equations to graph:

$5x + 3y > 7$, $4x + 2y < 10$

$3y > -5x + 7$, $2y < -4x + 10$

$\dfrac{3y}{3} > \dfrac{-5x+7}{3}$, $\dfrac{2y}{2} < \dfrac{-4x+10}{2}$

$y > -\dfrac{5}{3}x + \dfrac{7}{3}$, $y < -2x + 5$

$\begin{cases} y > -\dfrac{5}{3}x + \dfrac{7}{3} \\ y < -2x + 5 \end{cases}$

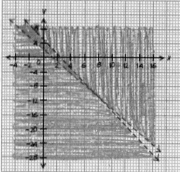

49.

$\begin{cases} y > \dfrac{2}{3}x - 12 \\ 6x - 4y > 12 \end{cases}$

Solve for y in the second equation to graph:

$6x - 4y - 6x > 12 - 6x$

$-4y > -6x + 12$

$\dfrac{-4y}{-4} < \dfrac{-6x+12}{-4}$

$y < \dfrac{3}{2}x - 3$

$\begin{cases} y > \dfrac{2}{3}x - 12 \\ y < \dfrac{3}{2}x - 3 \end{cases}$

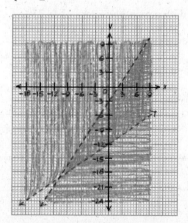

Chapter 2 Introduction to Systems of Linear Equations and Inequalities
Chapter 2 Review Exercises

1.

a.

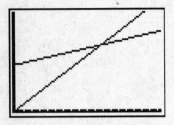

b.

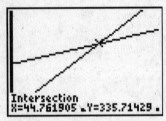

Frank's shoe repair will break even when they repair 45 pairs of shoes a month.

2.

a.

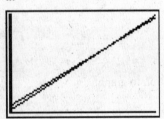

b.

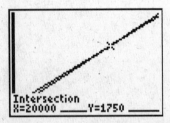

These two salary options will be the same when the sales reach $20,000 per month, with a monthly salary of $1750.

3. Let $R(d)$ be the revenue generated by selling d computer software disks, and let $C(d)$ be the cost of producing d computer software disks.

$$\begin{cases} R(d) = 9d \\ C(d) = 3232 + 3.95d \end{cases}$$

$R(d) = C(d)$

$9d = 3232 + 3.95d$

$9d - 3.95d = 3232 + 3.95d - 3.95d$

$5.05d = 3232$

$\dfrac{5.05d}{5.05} = \dfrac{3232}{5.05}$

$d = 640$

You must sell 640 copies to break even.

4. Let C be the number of pounds of cashews in the mix and let P be the number of pounds of peanuts in the mix.

$$\begin{cases} C + P = 100 \\ 4.5C + 2P = (3.00)(100) \end{cases}$$

$$\begin{cases} (-2)(C + P) = (100)(-2) \\ 4.5C + 2P = 300 \end{cases}$$

$\left. \begin{array}{l} -2C - 2P = -200 \\ 4.5C + 2P = 300 \end{array} \right\}$ add

$2.5C = 100$

$\dfrac{2.5C}{2.5} = \dfrac{100}{2.5}$

$C = 40$

Now solve for P:

$P = 100 - 40$

$P = 60$

John needs to mix 60 pounds of peanuts and 40 pounds of cashews to make 100 pounds of mixed nuts that cost $3.00 per pound.

5.

a. Let $R(a)$ be the residual value in dollars of a Mazda RX-8 that is a years old. Let $C(a)$ be the residual value in dollars of a Mini Cooper S that is a years old.

a	$R(a)$	$C(a)$
0	27,200	20,900
3	16,900	13,000
5	12,800	10,900
8	3300	4000

Using two points: $(0, 27200)$ and $(8, 3300)$

Find the slope: $m = \dfrac{3300 - 27,200}{8 - 0} = -2987.5$

$R(a) = -2987.5a + 27,200$

Using two points: $(0, 20900)$ and $(8, 4000)$

Find the slope: $m = \dfrac{4000 - 20900}{8 - 0} = -2112.5$

$C(a) = -2112.5a + 20{,}900$

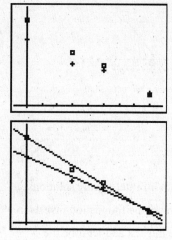

b.

$R(a) = C(a)$

$-2987.5a + 27{,}200 = -2112.5a + 20{,}900$

$+2112.5a - 27{,}200 \quad +2112.5a - 27{,}200$

$\qquad -875a = -6300$

$\qquad \dfrac{-875a}{-875} = \dfrac{-6300}{-875}$

$\qquad a = 7.2$

The two cars will have the same residual value of approximately $6200.00 when they are both about seven years old.

6.

a. Let $G(t)$ be the percent of births to unmarried women in Germany t years since 2000.

t	G(t)
1	25.0
2	26.0
3	27.0
4	27.9
5	29.2

Percent of births to unmarried women in Germany:

$G(t) = mt + b$

$m = \dfrac{27 - 25}{3 - 1} = 1$

$b = 27 - (1)(3) = 24$

$G(t) = t + 24$

b. Slope = 1. The percentage of births in Germany to unmarried women is increasing by one percentage point per year.

c. Let $U(t)$ be the percent of births to unmarried women in the United Kingdom t years since 2000.

t	U(t)
1	40.1
2	40.6
3	41.5
4	42.3
5	42.9

Percent of births to unmarried women in United Kingdom:

$U(t) = mt + b$

$m = \dfrac{42.9 - 40.1}{5 - 1} = 0.7$

$b = 40.1 - (0.7)(1) = 39.4$

$U(t) = 0.7t + 39.4$

d.

$G(t) > U(t)$

$t + 24 > 0.7t + 39.4$

$-0.7t - 24 \quad -0.7t - 24$

$0.3t > 15.4$

$\dfrac{0.3t}{0.3} > \dfrac{15.4}{0.3}$

$t > 51.3$

The percentage of births to unmarried women in the United Kingdom will be greater than the percentage in Germany after 2051. This is probably model breakdown.

7. The solution is $(10, 1)$. This is a consistent system with independent lines.

8.

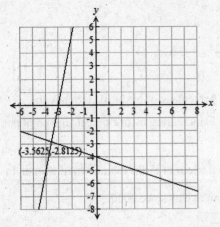

The solution is $\left(-\dfrac{57}{16}, -\dfrac{45}{16}\right)$ or $(-3.5625, -2.8125)$. This is a consistent system with independent lines.

9.

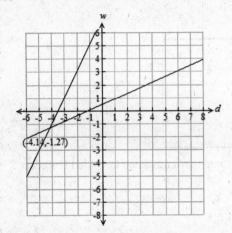

The solution is $\left(-\dfrac{91}{22}, -\dfrac{14}{11}\right)$ or $(-4.14, -1.27)$. This is a consistent system with independent lines.

10.

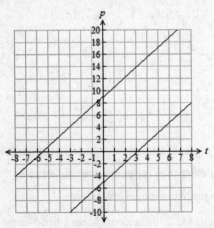

This is an inconsistent system with no solution.

11.

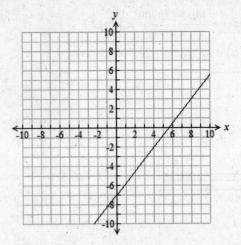

This is a consistent system with infinitely many solutions.

12. Let A be the amount of money the foundation invests in the account paying 7% simple interest and let B be the amount of money the foundation invests in the account paying 11% simple interest.

$A + B = 3,000,000 \to$ Total amount invested.
$0.07A + 0.11B = 260,000 \to$ Total interest earned.

Solve for B in the first equation and substitute for B in the second equation.
$B = 3,000,000 - A$
$0.07A + 0.11(3,000,000 - A) = 260,000$
$0.07A + 330,000 - 0.11A = 260,000$
$-0.04A + 330,000 = 260,000$

Solve for A.
$-0.04A + 330,000 - 330,000 = 260,000 - 330,000$
$-0.04A = -70,000$
$\dfrac{-0.04A}{-0.04} = \dfrac{-70,000}{-0.04}$
$A = 1,750,000$

Now solve for B.
$B = 3,000,000 - 1,750,000$
$B = 1,250,000$

The foundation should invest \$1,750,000 in the account that pays 7% simple interest, and \$1,250,000 in the account that pays 11% simple interest.

13. Let x be the number of gallons of 5% test chemical AX-14 solution, and let y be the number of gallons of 15% test chemical AX-14 solution used to get the 150 gallons of a 12% test chemical AX-14 solution.

$$\begin{cases} x+y=150 \\ 0.05x+0.15y=(0.12)(150)=18 \end{cases}$$

To eliminate x multiply each equation as follows:

$$\begin{cases} (-5)(x+y)=(150)(-5) \\ (100)(0.05x+0.15y)=(18)(100) \end{cases}$$

$\left.\begin{array}{r} -5x-5y=-750 \\ 5x+15y=1800 \end{array}\right\}$ add

$10y=1050$

$\dfrac{10y}{10}=\dfrac{1050}{10}$

$y=105$

Now solve for x:

$x=150-105$

$x=45$

Brian should use 105 gallons of the 15% solution and 45 gallons of the 5% solutions to make the 150 gallons of the 12% solution of test chemical AX-14.

14.

$$\begin{cases} 3x+4y=-26 \\ y=x-3 \end{cases}$$

$3x+4(x-3)=-26$

$3x+4x-12=-26$

$7x-12=-26$

$7x-12+12=-26+12$

$7x=-14$

$x=-2$

Substituting x back into either equation solve for y:

$y=(-2)-3$

$y=-2-3$

$y=-5$

The solution is $(-2,-5)$. This is a consistent system with independent lines.

15.

$$\begin{cases} 2w-5t=-1 \\ 3w-4t=2 \end{cases}$$

To eliminate w multiply each equation as follows:

$$\begin{cases} -3(2w-5t)=-3(-1) \\ 2(3w-4t)=2(2) \end{cases}$$

$\left.\begin{array}{r} -6w+15t=3 \\ 6w-8t=4 \end{array}\right.$

$7t=7$

$t=1$

Substituting t back into either equation solve for w:

$3w-4(1)=2$

$3w-4=2$

$3w-4+4=2+4$

$3w=6$

$w=2$

The solution to the system is $w=2$ and $t=1$.

The solution is $(2,1)$. This is a consistent system with independent lines.

16.

$$\begin{cases} 2.35d+4.7c=4.7 \\ c=-7.05d-21.15 \end{cases}$$

$2.35d+4.7(-7.05d-21.15)=4.7$

$2.35d-33.135d-99.405=4.7$

$-30.785d-99.405=4.7$

$-30.785d-99.405+99.405=4.7+99.405$

$\dfrac{-30.785d}{-30.785}=\dfrac{104.105}{-30.785}\approx -3.38168$

$d\approx -3.38$

Substituting d back into either equation solve for c:

$c\approx -7.05(-3.38168)-21.15$

$c\approx 23.84-21.15$

$c\approx 2.69$

The solution is $(-3.38, 2.69)$. This is a consistent system with independent lines.

17.

$$\begin{cases} \dfrac{5}{6}m+n=25 \\ m-\dfrac{4}{5}n=5 \end{cases}$$

$$\begin{cases} 12\left(\dfrac{5}{6}m+n\right)=12(25) \\ 15\left(m-\dfrac{4}{5}n\right)=15(5) \end{cases}$$

Chapter 2 Introduction to Systems of Linear Equations and Inequalities

$\begin{cases} 10m + 12n = 300 \\ 15m - 12n = 75 \end{cases}$
$25m = 375$
$m = 15$

Substituting m back into either equation solve for n:
$10(15) + 12n = 300$
$150 + 12n = 300$
$12n = 150$
$n = 12.5$

The solution is $(15, 12.5)$. This is a consistent system with independent lines

18.
$\begin{cases} y = 4.1x - 2.2 \\ y = -2.9x - 7.1 \end{cases}$
$4.1x - 2.2 = -2.9x - 7.1$
$+2.9x + 2.2 \quad +2.9x + 2.2$
$7x = -4.9$
$\dfrac{7x}{7} = \dfrac{-4.9}{7}$
$x = -0.7$

Substituting x back into either equation solve for y:
$y = 4.1(-0.7) - 2.2$
$y = -2.87 - 2.2$
$y = -5.07$

The solution is $(-0.7, -5.07)$. This is a consistent system with independent lines.

19.
$\begin{cases} -7x + 7y = -3 \\ 7(x - y) = 3 \end{cases}$
$\begin{cases} -7x + 7y = -3 \\ 7x - 7y = 3 \end{cases}$
$0 = 0$

This is a consistent system with dependent lines. There are infinitely many solutions.

20.
$\begin{cases} 4.1w + 3.7t = 5.1 \\ t = 4.43w + 4.63 \end{cases}$
$4.1w + 3.7(4.43w + 4.63) = 5.1$
$4.1w + 16.391w + 17.131 = 5.1$
$20.491w + 17.131 = 5.1$

$20.491w + 17.131 - 17.131 = 5.1 - 17.131$
$20.491w = -12.031$
$\dfrac{20.491w}{20.491} = \dfrac{-12.031}{20.491}$
$w \approx -0.587$

Substituting w back into either equation solve for t:
$t \approx 4.43(-0.587) + 4.63$
$t \approx -2.600 + 4.63$
$t \approx 2.03$

The solution is $(-0.587, 2.03)$. This is a consistent system with independent lines.

21.
$\begin{cases} 2f + 2g = -22 \\ g = -f - 9 \end{cases}$
$2f + 2(-f - 9) = -22$
$2f - 2f - 18 = -22$
$-18 \neq -22$

No solution.
This is an inconsistent system.

22.
$\begin{cases} -6x + 15y = 5 \\ 15(x - y) = -5 \end{cases}$
$\begin{cases} -6x + 15y = 5 \\ 15x - 15y = -5 \end{cases}$
$9x = 0$
$x = 0$

Substituting x back into either equation solve for y:
$-6(0) + 15y = 5$
$0 + 15y = 5$
$15y = 5$
$\dfrac{15y}{15} = \dfrac{5}{15}$
$y = \dfrac{1}{3}$

The solution is $\left(0, \dfrac{1}{3}\right)$. This is a consistent system with independent lines.

23.
$\begin{cases} \dfrac{w}{2} + \dfrac{2z}{3} = 32 \\ \dfrac{w}{4} - \dfrac{5z}{9} = 40 \end{cases}$

$$\begin{cases} 30\left(\dfrac{w}{2}+\dfrac{2z}{3}\right)=30(32) \\ 36\left(\dfrac{w}{4}-\dfrac{5z}{9}\right)=36(40) \end{cases}$$

$$\begin{cases} 15w+20z=960 \\ 9w-20z=1440 \end{cases}$$

$24w = 2400$

$w = 100$

Substituting w back into either equation solve for z:

$15(100) + 20z = 960$

$1500 + 20z = 960$

$1500 + 20z - 1500 = 960 - 1500$

$20z = -540$

$z = -27$

The solution is $(100, -27)$. This is a consistent system with independent lines.

24.

$$\begin{cases} -2.7x + y = -13.61 \\ y = -4.28x + 5.69 \end{cases}$$

$-2.7x + (-4.28x + 5.69) = -13.61$

$-2.7x - 4.28x + 5.69 = -13.61$

$-6.98x + 5.69 = -13.61$

$-6.98x + 5.69 - 5.69 = -13.61 - 5.69$

$-6.98x = -19.3$

$\dfrac{-6.98x}{-6.98} = \dfrac{-19.3}{-6.98}$

$x \approx 2.765$

Substituting x back into either equation solve for y:

$y \approx -4.28(2.765) + 5.69$

$y \approx -11.834 + 5.69$

$y \approx -6.144$

The solution is $(2.765, -6.144)$. This is a consistent system with independent lines.

25.

$L(t) < D(t)$

$-0.67t + 16.94 < -0.46t + 12.35$

$+0.46t - 16.94 \quad +0.46t - 16.94$

$-0.21t < -4.59$

$\dfrac{-0.21t}{-0.21} > \dfrac{-4.59}{-0.21}$

$t > 21.86$

The percentage of births that are to teenage mothers in Louisiana will be less than the percentage in Delaware after 2021. This is probably model breakdown.

26. $f(x) < g(x)$ when $x > 12$

27. $Y_1 > Y_2$ when $x > 7$

28.

$7 - \dfrac{9x}{11} \leq -8$

$-\dfrac{9x}{11} \leq -15$

$\left(-\dfrac{11}{9}\right)\left(-\dfrac{9x}{11}\right) \geq (-15)\left(-\dfrac{11}{9}\right)$

$x \geq \dfrac{55}{3}$

29.

$-5x + 4 \geq 7(x-1)$

$-5x + 4 \geq 7x - 7$

$-7x + 4 \quad -7x - 4$

$-12x \geq -11$

$\dfrac{-12x}{-12} \leq \dfrac{-11}{-12}$

$x \leq \dfrac{11}{12}$

30.

$3t + 4 > -6(4t + 2)$

$3t + 4 > -24t - 12$

$+24t - 4 \quad +24t - 4$

$27t > -16$

$\dfrac{27t}{27} > \dfrac{-16}{27}$

$t > -\dfrac{16}{27}$

31.

$\dfrac{3d}{5} + 7 < 4$

$\dfrac{3d}{5} < -3$

$\left(\dfrac{5}{3}\right)\left(\dfrac{3d}{5}\right) < (-3)\left(\dfrac{5}{3}\right)$

$d < -5$

32.

$-1.5v + 2.84 > -3.2v - 1.48$
$\underline{+3.2v - 2.84 \quad +3.2v - 2.84}$
$\qquad 1.7v > -4.32$
$\qquad \dfrac{1.7v}{1.7} > \dfrac{-4.32}{1.7}$
$\qquad\qquad v > -2.54$

33.

$1.85 + 1.34(2.4k - 5.7) \geq 3.25k - 14.62$
$1.85 + 3.216k - 7.638 \geq 3.25k - 14.62$
$3.216k - 5.788 \geq 3.25k - 14.62$
$\underline{-3.25k + 5.788 \quad -3.25k + 5.788}$
$\qquad -0.034k \geq -8.832$
$\qquad \dfrac{-0.034k}{-0.034} \leq \dfrac{-8.832}{-0.034}$
$\qquad\qquad k \leq 259.76$

34.

$|x - 7| = 20$
$x - 7 = 20 \text{ or } x - 7 = -20$
$x = 27 \text{ or } x = -13$

35.

$|2x + 3| - 10 = 40$
$|2x + 3| - 10 + 10 = 40 + 10$
$|2x + 3| = 50$
$2x + 3 = 50 \text{ or } 2x + 3 = -50$
$2x = 47 \text{ or } 2x = -53$
$x = 23.5 \text{ or } x = -26.5$

36.

$|x + 19| = 30$
$x + 19 = 30 \text{ or } x + 19 = -30$
$x = 11 \text{ or } x = -49$

37.

$5|x - 7| + 10 = 95$
$5|x - 7| + 10 - 10 = 95 - 10$
$5|x - 7| = 85$
$\dfrac{5|x - 7|}{5} = \dfrac{85}{5}$
$|x - 7| = 17$
$x - 7 = 17 \text{ or } x - 7 = -17$
$x = 24 \text{ or } x = -10$

38.

$|x| < 8.5$
$-8.5 < x < 8.5$

39.

$|x| > 6.5$
$x < -6.5 \text{ or } x > 6.5$

40.

$|x + 3| \geq 12$
$x + 3 \leq -12 \text{ or } x + 3 \geq 12$
$x \leq -15 \text{ or } x \geq 9$

41.

$|x - 7| \leq 3$
$-3 \leq x - 7 \leq 3$
$-3 + 7 \leq x - 7 + 7 \leq 3 + 7$
$4 \leq x \leq 10$

42.

$2|x + 5| - 4 > 16$
$2|x + 5| - 4 + 4 > 16 + 4$
$2|x + 5| > 20$
$\dfrac{2|x + 5|}{2} > \dfrac{20}{2}$
$|x + 5| > 10$
$x + 5 < -10 \text{ or } x + 5 > 10$
$\quad x < -15 \text{ or } x > 5$

43.

$-3|x + 5| + 7 \leq 4$
$-3|x + 5| + 7 - 7 \leq 4 - 7$
$-3|x + 5| \leq -3$
$\dfrac{-3|x + 5|}{-3} \geq \dfrac{-3}{-3}$
$|x + 5| \geq 1$
$x + 5 \leq -1 \text{ or } x + 5 \geq 1$
$\quad x \leq -6 \text{ or } x \geq -4$

44. $y \leq 4x - 10$

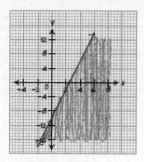

45. $y \geq 1.5x - 5$

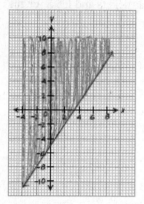

46.
$2x + 5y > 15$
$2x + 5y - 2x > 15 - 2x$
$5y > -2x + 15$
$\dfrac{5y}{5} > \dfrac{-2x + 15}{5}$
$y > -\dfrac{2}{5}x + 3$

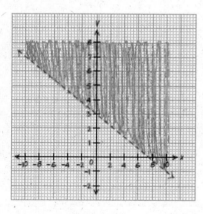

47.
$-3x - 4y > 12$
$-3x - 4y + 3x > 12 + 3x$
$-4y > 3x + 12$
$\dfrac{-4y}{-4} < \dfrac{3x + 12}{-4}$
$y < -\dfrac{3}{4}x - 3$

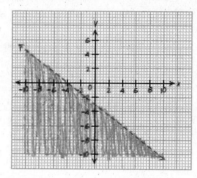

48.
$\begin{cases} y > 2x + 5 \\ y < -x + 7 \end{cases}$

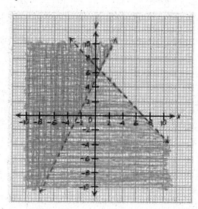

49.
$\begin{cases} y > 1.5x + 2 \\ 4.5x - 3y > 6 \end{cases}$

Solve for y in the second equation to graph:
$4.5x - 3y - 4.5x > 6 - 4.5x$
$-3y > -4.5x + 6$
$\dfrac{-3y}{-3} < \dfrac{-4.5x + 6}{-3}$
$y < 1.5x - 2$

$\begin{cases} y > 1.5x + 2 \\ y < 1.5x - 2 \end{cases}$

These are parallel lines so there is no solution to this system.

Chapter 2 Introduction to Systems of Linear Equations and Inequalities

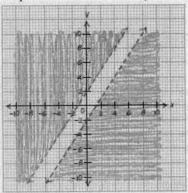

50.

$\begin{cases} 4x+5y \le 12 \\ 2x+8y \ge 5 \end{cases}$

Solve for y in both equations to graph:

$4x+5y \le 12$, $2x+8y \ge 5$

$5y \le -4x+12$, $8y \ge -2x+5$

$\dfrac{5y}{5} \le \dfrac{-4x+12}{5}$, $\dfrac{8y}{8} \ge \dfrac{-2x+5}{8}$

$y \le -\dfrac{4}{5}x + \dfrac{12}{5}$, $y \ge -\dfrac{1}{4}x + \dfrac{5}{8}$

$\begin{cases} y \le -\dfrac{4}{5}x + \dfrac{12}{5} \\ y \ge -\dfrac{1}{4}x + \dfrac{5}{8} \end{cases}$

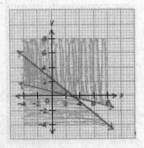

51.

$\begin{cases} y > \dfrac{2}{3}x - 7 \\ y < \dfrac{2}{3}x + 5 \end{cases}$

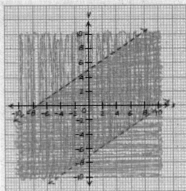

52. Let R be the number of regular tickets and let S be the number of student tickets.

$S + R \le 12000$

$9S + 15R \ge 140000$

53.

$(0,-6)$ and $(6,-4)$

$m = \dfrac{-4-(-6)}{6-0} = \dfrac{2}{6} = \dfrac{1}{3}$

This is a solid line, shaded below:

$y \le \dfrac{1}{3}x - 6$

54.

$(0,6)$ and $(4,4)$

$m = \dfrac{4-6}{4-0} = \dfrac{-2}{4} = -\dfrac{1}{2}$

This is a dashed line, shaded above:

$y > -\dfrac{1}{2}x + 6$

Chapter 2 Test

1.

a. Let $C(t)$ be the average hourly earnings in dollars per hour, of production workers in manufacturing industries in California t years since 2000. Let $M(t)$ be the average hourly earnings in dollars per hour, of production workers in manufacturing industries in Massachusetts t years since 2000. Note: Your models may differ, but they must follow the trend of the data.

a.

t	$C(t)$	$M(t)$
3	15.04	16.53
4	15.36	16.89
5	15.70	17.66
6	15.95	18.26

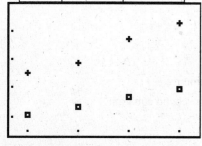

California average hourly earnings:
$C(t) = mt + b$
$m = \dfrac{15.95 - 15.04}{6 - 3} = 0.303$
$b = 15.04 - (0.303)(3) \approx 14.13$
$C(t) = 0.303t + 14.13$

Massachusetts average hourly earnings:
$M(t) = mt + b$
$m = \dfrac{18.26 - 17.66}{6 - 5} = 0.60$
$b = 17.66 - (0.60)(5) = 14.66$
$M(t) = 0.60t + 14.66$

b.

$C(t) = M(t)$
$0.303t + 14.13 = 0.60t + 14.66$
$-0.60t - 14.13 -0.60t - 14.13$
$-0.297t = 0.53$
$\dfrac{-0.297t}{-0.297} = \dfrac{0.53}{-0.297}$
$t \approx -1.8$

In about 1998, the average hourly earnings for production workers in manufacturing industries in California were approximately the same as those in Massachusetts.

2.

a. Let A be the amount in dollars invested in the account paying 12%, and let B be the amount in dollars invested in the account paying 7%.

$A + B = 500,000 \rightarrow$ Total amount invested.
$0.12A + 0.07B = 44,500 \rightarrow$ Total interest earned.

b.

Solve for A in the first equation and substitute for A in the second equation.
$A = 500,000 - B$
$0.12(500,000 - B) + 0.07B = 44,500$
Solve for B.
$60,000 - 0.12B + 0.07B = 44,500$
$60,000 - 0.05B = 44,500$
$60,000 - 60,000 - 0.05B = 44,500 - 60,000$
$-0.05B = -15,500$
$\dfrac{-0.05B}{-0.05} = \dfrac{-15,500}{-0.05}$
$B = 310,000$
Now solve for A.
$A = 500,000 - 310,000$
$A = 190,000$

Christine should invest $190,000 in the account paying 12% interest, and $310,000 in the account paying 7% interest.

3. Let M be the number of hours per week Georgia needs to study for her math class, and let H be the number of hours per week Georgia needs to study for her history class.

$M + H \leq 20$
$M \geq 12$

Chapter 2 Introduction to Systems of Linear Equations and Inequalities

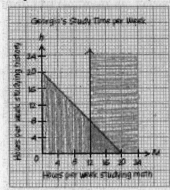

4. Let x be the number of liters of 5% HCl solution, and let y be the number of liters of 20% HCl solution used to get the 2 liters of an 8% HCl solution.

$$\begin{cases} x+y=2 \\ 0.05x+0.20y=(0.08)(2) \end{cases}$$

Multiply the second equation by -20 and add the result to the first equation.

$(-20)(0.05x+0.20y)=(0.16)(-20)$
$-x-4y=-3.2$

$\left.\begin{array}{r} x+y=2 \\ -x-4y=-3.2 \end{array}\right\}$ add

$-3y=-1.2$

$\dfrac{-3y}{-3}=\dfrac{-1.2}{-3}$

$y=0.4$

Now solve for x:
$x=2-0.4$
$x=1.6$

Wendy should use 1.6 liters of the 5% HCl solution and 0.4 liters of the 20% HCl solution to make two liters of an 8% HCl solution.

5.
$5x+7<12x-8$
$5x+7-12x<12x-8-12x$
$-7x+7<-8$
$-7x+7-7<-8-7$
$-7x<-15$
$\dfrac{-7x}{-7}>\dfrac{-15}{-7}$
$x>\dfrac{15}{7}$

6.
$3.2m+4.5\ge 5.7(2m+3.4)$
$3.2m+4.5\ge 11.4m+19.38$
$\underline{-11.4m-4.5-11.4m-4.5}$
$-8.2m\ge 14.88$
$\dfrac{-8.2m}{-8.2}\le\dfrac{14.88}{-8.2}$
$m\le -1.8$

7.
$-4.7+6.5(a+2.5)\le 2.43a-5$
$-4.7+6.5a+16.25\le 2.43a-5$
$6.5a+11.55\le 2.43a-5$
$\underline{-2.43a-11.55-2.43a-11.55}$
$4.07a\le -16.55$
$\dfrac{4.07a}{4.07}\le\dfrac{-16.55}{4.07}$
$a\le -4.1$

8.
$\begin{cases} x+7y=-2 \\ 3x+y=34 \end{cases}$

Multiply the first equation by -3 and add the result to the second equation.

$-3(x+7y)=-3(-2)$
$-3x-21y=6$

$\left.\begin{array}{r} -3x-21y=6 \\ 3x+y=34 \end{array}\right\}$ add

$-20y=40$
$y=-2$

Substitute y back into either equation to solve for x:
$x+7(-2)=-2$
$x-14=-2$
$x-14+14=-2+14$
$x=12$

The solution is $(12,-2)$. This is a consistent system with independent lines.

9.

$\begin{cases} 0.4375w + 4t = 22 \\ -2.4t = 0.2625w - 13.2 \end{cases}$

Solve for t in the second equation:

$\dfrac{-2.4t}{-2.4} = \dfrac{0.2625w}{-2.4} - \dfrac{13.2}{-2.4}$

$t = -0.109375w + 5.5$

Plug this into the first equation:

$0.4375w + 4(-0.109375w + 5.5) = 22$

$0.4375w - 0.4375w + 22 = 22$

$22 = 22$

There are infinitely many solutions. This is a consistent system with dependent lines.

10.

$\begin{cases} 5c + 3d = -15 \\ d = -\dfrac{5}{3}c - 12 \end{cases}$

$5c + 3\left(-\dfrac{5}{3}c - 12\right) = -15$

$5c - 5c - 36 = -15$

$-36 \neq -15$

There is no solution. This is an inconsistent system.

11.

$\begin{cases} 2.68g - 3.45f = 23.87 \\ 4.75g + 6.9f = -12.47 \end{cases}$

$\begin{cases} 6.9(2.68g - 3.45f) = 6.9(23.87) \\ 3.45(4.75g + 6.9f) = 3.45(-12.47) \end{cases}$

$\begin{cases} 18.492g - 23.805f = 164.703 \\ 16.3875g + 23.805f = -43.0215 \end{cases}$

$34.8795g = 121.6815$

$\dfrac{34.8795g}{34.8795} = \dfrac{121.6815}{34.8795}$

$g \approx 3.5$

Substitute g back into either equation to solve for f:

$4.75(3.5) + 6.9f \approx -12.47$

$16.625 + 6.9f \approx -12.47$

$16.625 + 6.9f - 16.625 \approx -12.47 - 16.625$

$6.9f \approx -29.095$

$\dfrac{6.9f}{6.9} \approx \dfrac{-29.095}{6.9}$

$f \approx -4.2$

The solution is $(3.5, -4.2)$. This is a consistent system with independent lines

12.

$f(x) \geq g(x)$

$x \leq 8$

13. Let $C(h)$ be the cost (in dollars) to make h hammock stands. Let $R(h)$ be the revenue (in dollars) from selling h hammock stands.

$\begin{cases} C(h) = 7500 + 395h \\ R(h) = 550h \end{cases}$

$R(h) = C(h)$

$550h = 7500 + 395h$

$550h - 395h = 7500 + 395h - 395h$

$155h = 7500$

$\dfrac{155h}{155} = \dfrac{7500}{155}$

$h \approx 48.39$

Scott needs to sell 49 hammock stands to break even.

14. $y > -\dfrac{4}{5}x + 6$

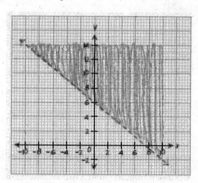

15.

$C(t) > L(t)$

$6428.29t - 11{,}553 > 3804.32t + 84{,}252.07$

$-3804.32t + 11{,}553 \quad -3804.32t + 11{,}553$

$2623.97t > 95{,}805.07$

$\dfrac{2623.97t}{2623.97} > \dfrac{95{,}805.07}{2623.97}$

$t > 36.51$

The revenue for cellular phone providers will be greater than that of local telephone providers after 2026.

16. $Y_1 < Y_1$ when $x < -4$

17.

$|x+5|+3=17$

$|x+5|+3-3=17-3$

$|x+5|=14$

$x+5=14$ or $x+5=-14$

$\quad x=9$ or $x=-19$

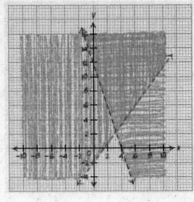

18.

$|x-4|-8<5$

$|x-4|-8+8<5+8$

$|x-4|<13$

$-13<x-4<13$

$-13+4<x-4+4<13+4$

$-9<x<17$

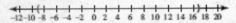

19.

$\begin{cases} y \geq 2x-5 \\ y \leq \dfrac{1}{3}x+4 \end{cases}$

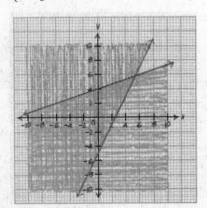

20.

$\begin{cases} 4x-3y<6 \\ 3x+y>12 \end{cases}$

Solve for y in both equations to graph:

$4x-3y<6 \qquad 3x+y>12$

$-3y<-4x+6 \qquad y>-3x+12$

$\dfrac{-3y}{-3}>\dfrac{-4x+6}{-3}$

$y>\dfrac{4}{3}x-2$

$\begin{cases} y>\dfrac{4}{3}x-2 \\ y>-3x+12 \end{cases}$

Chapter 1-2 Cumulative Review

1.
a. Let P the population of the United States in millions, and let t be the years since 2000.
$P = mt + b$
Using two points: $(4, 293)$ and $(7, 301)$
find the slope: $m = \dfrac{301 - 293}{7 - 4} \approx 2.7$
$P = 2.7t + b$
$b = 301 - 2.7(7)$
$b = 282.1$
$P = 2.7t + 282.1$

b.
$P = 2.7(12) + 282.1$
$P = 314.5$
The population of the United States will be 314.5 million in 2012.

c. Domain $[0, 13]$
$P = 2.7(13) + 282.1$
$P = 317.2$
Range $[282.1, 317.2]$

d. Vertical intercept: $(0, 282.1)$. The population of the United States was about 282.1 million in 2000.

e.
$325 = 2.7t + 282.1$
$\underline{-282.1 \quad\quad -282.1}$
$42.9 = 2.7t$
$\dfrac{42.9}{2.7} = \dfrac{2.7t}{2.7}$
$15.9 \approx t$
The U.S. population will reach 325 million in about 2016.

2. $y = 3x - 4$

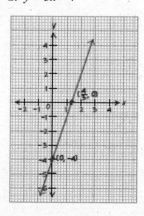

3. $y = -\dfrac{1}{3}x + 2$

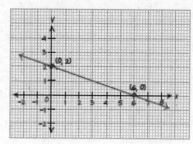

4. $5x - 6y = 9$

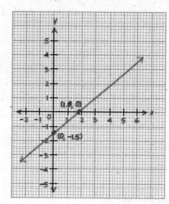

5.
$abc + b = a$
$abc = a - b$
$\dfrac{abc}{ab} = \dfrac{a-b}{ab}$
$c = \dfrac{a-b}{ab}$

6.
$P = \dfrac{1}{2}mn^2 + 3$
$P - 3 = \dfrac{1}{2}mn^2 = \dfrac{mn^2}{2}$
$\left(\dfrac{2}{n^2}\right)(P - 3) = \left(\dfrac{2}{n^2}\right)\left(\dfrac{mn^2}{2}\right)$
$m = \dfrac{2(P-3)}{n^2}$

7.
$y = mx + b$
Using the points: $(-2, 6)$ and $(5, 20)$
$m = \dfrac{20 - 6}{5 + 2} = \dfrac{14}{7} = 2$
$b = 6 - (2)(-2) = 10$
$y = 2x + 10$

Chapter 2 Introduction to Systems of Linear Equations and Inequalities

8.

$3x + 4y = 7$

$4y = -3x + 7$

$\dfrac{4y}{4} = \dfrac{-3x+7}{4}$

$y = -\dfrac{3}{4}x + \dfrac{7}{4}$

Use the slope $m = -\dfrac{3}{4}$, and the point $(6,2)$.

$b = y - mx$

$b = 2 - \left(-\dfrac{3}{4}\right)(6)$

$b = 2 + \dfrac{9}{2} = \dfrac{13}{2}$

$y = -\dfrac{3}{4}x + \dfrac{13}{2}$

9.

a.

$R = 1.5(4) + 5$

$R = 11$

Steve plans to run eleven miles in week four.

b.

$20 = 1.5w + 5$

$15 = 1.5w$

$\dfrac{15}{1.5} = \dfrac{1.5w}{1.5}$

$w = 10$

Steve plans to run twenty miles in week ten.

10. Let $R(t)$ be the total revenue in billions of dollars for Cisco Systems, Inc t years since 2000.

a..

$R(t) = mt + b$

Using two points: $(5, 24.8)$ and $(8, 39.5)$

Find the slope: $m = \dfrac{39.5 - 24.8}{8 - 5} = 4.9$

$b = 39.5 - 4.9(8)$

$b = 0.3$

$R(t) = 4.9t + 0.3$

b. Vertical intercept: $(0, 0.3)$. The total revenue for Cisco Systems, Inc. in 2000, was about $300 million.

c. Domain $[3,8]$ (due to the economic downturn)

$R(3) = 4.9(3) + 0.3 = 15$

$R(8) = 4.9(8) + 0.3 = 39.5$

Range $[15, 39.5]$

d.

$R(10) = 4.9(10) + 0.3 = 49.3$

The model predicts Cisco Systems, Inc.'s revenue at $49.3 billion in 2010.

e.

$4.9t + 0.3 = 64$

$4.9t + 0.3 - 0.3 = 64 - 0.3$

$4.9t = 63.7$

$\dfrac{4.9t}{4.9} = \dfrac{63.7}{4.9}$

$t = 13$

The model predicts Cisco Systems, Inc.'s revenue at $64 billion in 2013.

f. The slope is 4.9, which means Cisco Systems, Inc.'s revenue is increasing at a rate of $4.9 billion per year.

11.

a.

Use two points to find the slope: $(-6, 0)$ and $(0, 4)$.

$m = \dfrac{4 - 0}{0 - (-6)} = \dfrac{4}{6} = \dfrac{2}{3}$

b. The y-intercept is $(0, 4)$.

c. The x-intercept is $(-6, 0)$.

d. The y-intercept is $(0, 4)$, therefore $b = 4$.

Since $m = \dfrac{2}{3}$ and $b = 4$, the line is $y = \dfrac{2}{3}x + 4$.

12.

a.

Use two points to find the slope: $(0, -5)$ and $(4, 10)$.

$m = \dfrac{10 - (-5)}{4 - 0} = \dfrac{15}{4}$

b. The vertical intercept is $(0, -5)$.

c. The horizontal intercept is $\left(\dfrac{4}{3}, 0\right)$.

d. The vertical intercept is $(0, -5)$, therefore $b = -5$.

Since $m = \dfrac{15}{4}$ and $b = -5$, the line is $y = \dfrac{15}{4}x - 5$.

13.

Domain: All real numbers, range: All real numbers

14.

Domain: All real numbers, range: $\{20\}$

15.

$$4a + 6 = 2a - 18$$
$$\underline{-2a - 6 \quad -2a - 6}$$
$$2a = -24$$
$$\frac{2a}{2} = \frac{-24}{2}$$
$$a = -12$$

16.

$$1.5(x+3) = 4x + 2.5(4x - 7)$$
$$1.5x + 4.5 = 4x + 10x - 17.5$$
$$1.5x + 4.5 = 14x - 17.5$$
$$\underline{-14x - 4.5 \quad -14x - 4.5}$$
$$-12.5x = -22$$
$$\frac{-12.5x}{-12.5} = \frac{-22}{-12.5}$$
$$x = 1.76$$

17.

$$\frac{1}{2}x + 10 = \frac{3}{4}x - 6$$
$$4\left(\frac{1}{2}x + 10\right) = 4\left(\frac{3}{4}x - 6\right)$$
$$2x + 40 = 3x - 24$$
$$\underline{-3x - 40 \quad -3x - 40}$$
$$-x = -64$$
$$\frac{-x}{-1} = \frac{-64}{-1}$$
$$x = 64$$

18.

$$p + 3\left(\frac{1}{4}p - 4\right) = \frac{2}{5}$$
$$p + \frac{3}{4}p - 12 = \frac{2}{5}$$
$$20\left(p + \frac{3}{4}p - 12\right) = 20\left(\frac{2}{5}\right)$$
$$20p + 15p - 240 = 8$$
$$35p - 240 = 8$$
$$35p - 240 + 240 = 8 + 240$$
$$35p = 248$$
$$\frac{35p}{35} = \frac{248}{35}$$
$$p = \frac{248}{35}$$

19. $H(6) = 7$

On the 6th day of the month, a person sleeps 7 hours.

20. $P(d) = -3d + 100$

a.
$$P(14) = -3(14) + 100$$
$$P(14) = 58$$

58 people out of 100 remain on the plan 14 days after starting the diet.

b. The slope is -3. The number of people remaining on the diet plan decreases by three people per day.

21. $f(x) = 9x - 45$

a.
$$f(3) = 9(3) - 45$$
$$f(3) = 27 - 45$$
$$f(3) = -18$$

b.
$$9x - 45 = 54$$
$$9x = 99$$
$$\frac{9x}{9} = \frac{99}{9}$$
$$x = 11$$

c.
Domain: All real numbers, range: All real numbers

22. $h(x) = -\frac{5}{8}x + 11$

a.
$$h(32) = -\frac{5}{8}(32) + 11$$
$$h(32) = -20 + 11$$
$$h(32) = -9$$

b.
$$-\frac{5}{8}x + 11 = \frac{13}{8}$$
$$8\left(-\frac{5}{8}x + 11\right) = 8\left(\frac{13}{8}\right)$$
$$-5x + 88 = 13$$
$$-5x + 88 - 88 = 13 - 88$$
$$-5x = -75$$
$$\frac{-5x}{-5} = \frac{-75}{-5}$$
$$x = 15$$

c) Domain: All real numbers, range: All real numbers

23. $g(x) = 7$

a. $g(10) = 7$

b. Domain: All real numbers, range: $\{7\}$

24.

a. $f(6) = 8$

b.
$f(x) = 0$
$f(-6) = 0$
$x = -6$

c. Domain: All real numbers, range: All real numbers

25.

$\begin{cases} 2x + 5y = 26 \\ 10x - 3y = 18 \end{cases}$

$\begin{cases} 3(2x + 5y) = 3(26) \\ 5(10x - 3y) = 5(18) \end{cases}$

$\begin{cases} 6x + \cancel{15y} = 78 \\ 50x - \cancel{15y} = 90 \end{cases}$

$56x = 168$
$x = 3$

Substitute x back into either equation to solve for y.

$2(3) + 5y = 26$
$6 + 5y = 26$
$6 + 5y - 6 = 26 - 6$
$5y = 20$
$y = 4$

The solution is $(3, 4)$. This is a consistent system with independent lines.

26.

$\begin{cases} c = 3d + 10 \\ c = 2d + 14 \end{cases}$

$3d + 10 = 2d + 14$
$\underline{-2d - 10 \quad -2d - 10}$
$d = 4$

Substitute d back into either equation to solve for c.

$c = 2(4) + 14$
$c = 8 + 14$
$c = 22$

The solution is $(4, 22)$. This is a consistent system with independent lines.

27.

$\begin{cases} \frac{2}{5}g + \frac{3}{8}h = 2 \\ 16g + 15h = 80 \end{cases}$

$\begin{cases} -40\left(\frac{2}{5}g + \frac{3}{8}h\right) = -40(2) \\ 16g + 15h = 80 \end{cases}$

$\begin{cases} -16g - \cancel{15h} = -80 \\ 16g + \cancel{15h} = 80 \end{cases}$

$0 = 0$

Infinite number of solutions.

These are dependent lines.

There are an infinite number of solutions. This is a consistent system with dependent lines.

28.

$\begin{cases} 4a + 18b = 8 \\ 6a = 7b + \frac{2}{3} \end{cases}$

$\begin{cases} \left(\frac{1}{2}\right)(4a + 18b) = \left(\frac{1}{2}\right)(8) \\ (3)(6a - 7b) = (3)\left(\frac{2}{3}\right) \end{cases}$

$\begin{cases} 2a + 9b = 4 \\ 18a - 21b = 2 \end{cases}$

$\begin{cases} (-9)(2a + 9b) = (-9)(4) \\ 18a - 21b = 2 \end{cases}$

$\left. \begin{array}{r} -18a - 81b = -36 \\ \underline{18a - 21b = 2} \end{array} \right\}$ add

$-102b = -34$
$\dfrac{-102b}{-102} = \dfrac{-34}{-102}$
$b = \dfrac{1}{3}$

Solve for a in the first equation:

$4a + 18\left(\dfrac{1}{3}\right) = 8$
$4a + 6 = 8$
$4a = 2$
$\dfrac{4a}{4} = \dfrac{2}{4}$
$a = \dfrac{1}{2}$

The solution is $\left(\dfrac{1}{2}, \dfrac{1}{3}\right)$. This is a consistent system with independent lines.

29.
$\begin{cases} 20x + 25y = -1 \\ 35x + 40y = 1 \end{cases}$

$\begin{cases} -40(20x + 25y) = -40(-1) \\ 25(35x + 40y) = 25(1) \end{cases}$

$\begin{cases} -800x - 1000y = 40 \\ 875x + 1000y = 25 \end{cases}$

$75x = 65$

$\dfrac{75x}{75} = \dfrac{65}{75}$

$x = \dfrac{13}{15}$

Substitute x back into either equation to solve for y.

$35\left(\dfrac{13}{15}\right) + 40y = 1$

$\dfrac{91}{3} + 40y = 1$

$3\left(\dfrac{91}{3} + 40y\right) = 3(1)$

$91 + 120y = 3$

$91 + 120y - 91 = 3 - 91$

$120y = -88$

$\dfrac{120y}{120} = \dfrac{-88}{120}$

$y = -\dfrac{11}{15}$

The solution is $\left(\dfrac{13}{15}, -\dfrac{11}{15}\right)$. This is a consistent system with independent lines.

30.
$\begin{cases} m = 3n - 12 \\ 4m - 12n = 10 \end{cases}$

$4(3n - 12) - 12n = 10$

$12n - 48 - 12n = 10$

$-48 \neq 10$

There is no solution. This is an inconsistent system.

31. Let x be the amount of money Don invests in the account paying 5% simple interest, and let y be the amount of money Don invests in the account paying 3.5% simple interest.

a.
$x + y = 1,200,000 \rightarrow$ Total amount invested.
$0.05x + 0.035y = 53,250 \rightarrow$ Total interest earned.

b.
$y = 1,200,000 - x$
$0.05x + 0.035(1,200,000 - x) = 53,250$
$0.05x + 42,000 - 0.035x = 53,250$
$0.015x + 42,000 = 53,250$
$ -42,000 -42,000$
$0.015x = 11,250$
$\dfrac{0.015x}{0.015} = \dfrac{11,250}{0.015}$
$x = 750,000$
$y = 1,200,000 - 750,000$
$y = 450,000$

Don needs to invest \$750,000 in the account paying 5% simple interest, and \$450,000 in the account paying 3.5% simple interest.

32. Let x be the number of liters of 6% HCl solution and y be the number of liters of 24% HCl solution used to get the 4 liters of a 12% HCl solution that Hamid needs.

$\begin{cases} x + y = 4 \\ 0.06x + 0.24y = (0.12)(4) = 0.48 \end{cases}$

Solve for x in the first equation and use the result for x in the second equation.

$x = -y + 4$
$0.06(-y + 4) + 0.24y = 0.48$

Solve for y.
$0.06(-y + 4) + 0.24y = 0.48$
$-0.06y + 0.24 + 0.24y = 0.48$
$0.18y + 0.24 = 0.48$
$ -0.24 -0.24$
$0.18y = 0.24$
$\dfrac{0.18y}{0.18} = \dfrac{0.24}{0.18}$
$y = \dfrac{4}{3}$

Now solve for x:
$x = -\dfrac{4}{3} + 4 = 2\dfrac{2}{3}$

Hamid should mix $2\dfrac{2}{3}$ liters of 6% HCL solution with $1\dfrac{1}{3}$ liters of 24% HCL solution to get 4 liters of 12% HCL solution.

Chapter 2 Introduction to Systems of Linear Equations and Inequalities

33.
$$6x + 20 > 15x - 16$$
$$\underline{-15x - 20 \quad -15x - 20}$$
$$-9x > -36$$
$$\frac{-9x}{-4} < \frac{-36}{-4}$$
$$x < 4$$

34.
$$\frac{1}{4}m + \frac{2}{3}(m-5) \leq \frac{1}{3}(4m+7)$$
$$\frac{1}{4}m + \frac{2}{3}m - \frac{10}{3} \leq \frac{4}{3}m + \frac{7}{3}$$

Clear the fractions by multiplying both sides of the inequality by the least common denominator 12.

$$(12)\left(\frac{1}{4}m + \frac{2}{3}m - \frac{10}{3}\right) \leq (12)\left(\frac{4}{3}m + \frac{7}{3}\right)$$
$$3m + 8m - 40 \leq 16m + 28$$
$$11m - 40 \leq 16m + 28$$
$$\underline{-16m + 40 \quad -16m + 40}$$
$$-5m \leq 68$$
$$\frac{-5m}{-5} \geq \frac{68}{-5}$$
$$m \geq -\frac{68}{5} = -13.6$$
$$m \geq -\frac{68}{5} \text{ or } m \geq -13.6$$

35.
$$-1.4a + 2.34a + 6.1 \geq -0.2a + 8.4$$
$$0.94a + 6.1 \geq -0.2a + 8.4$$
$$\underline{+0.2a - 6.1 \quad +0.2a - 6.1}$$
$$1.14a \geq 2.3$$
$$\frac{1.14a}{1.14} \geq \frac{2.3}{1.14}$$
$$a \geq \frac{2.3}{1.14} = \frac{230}{114}$$
$$a \geq \frac{115}{57}$$

36. $f(x) \geq g(x)$ when $x \leq 3$.

37. Let $P(s)$ be Sandra's weekly salary for s dollars of sales.

$$P(s) = 400 + 0.06s$$
$$400 + 0.06s \geq 565$$
$$400 - 400 + 0.06s \geq 565 - 400$$
$$0.06s \geq 165$$
$$\frac{0.06s}{0.06} \geq \frac{165}{0.06}$$
$$s \geq 2750$$

Sandra must have $2750 or more per week in sales to earn at least $565 per week.

38. $y > -\frac{2}{3}x + 9$

Use a dashed line with shading above.

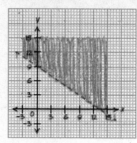

39.
$$2x - 4y \geq 11$$
$$-4y \geq -2x + 11$$
$$\frac{-4y}{-4} \leq \frac{-2x + 11}{-4}$$
$$y \leq \frac{1}{2}x - \frac{11}{4}$$

Use a solid line with shading below.

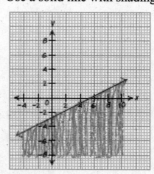

40.
$$6x + y < 12$$
$$y < -6x + 12$$

Use a dashed line with shading below.

110

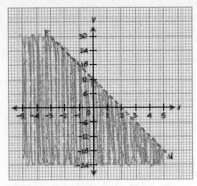

41.
$$S(t) > C(t)$$
$$2t + 17 > -2t + 82$$
$$\underline{+2t - 17 \quad +2t - 17}$$
$$4t > 65$$
$$\frac{4t}{4} > \frac{65}{4}$$
$$t > 16.25$$

The percentage of Canadian TV subscribers that use satellite services will be greater than those who use cable services after 2016.

42. $Y_1 < Y_2$ when $x < 4$

43.
$$|x + 2| + 6 = 18$$
$$|x + 2| + 6 - 6 = 18 - 6$$
$$|x + 2| = 12$$
$$x + 2 = 12 \text{ or } x + 2 = -12$$
$$x = 10 \text{ or } x = -14$$

44.
$$3|a - 8| - 4 = 44$$
$$3|a - 8| - 4 + 4 = 44 + 4$$
$$3|a - 8| = 48$$
$$\frac{3|a - 8|}{3} = \frac{48}{3}$$
$$|a - 8| = 16$$
$$a - 8 = 16 \text{ or } a - 8 = -16$$
$$a = 24 \text{ or } a = -8$$

45.
$$|4n + 9| + 20 = 7$$
$$|4n + 9| + 20 - 20 = 7 - 20$$
$$|4n + 9| = -13$$

There is no solution, because an absolute value cannot equal a negative number.

46.
$$-2|3r + 5| + 15 = 7$$
$$-2|3r + 5| + 15 - 15 = 7 - 15$$
$$-2|3r + 5| = -8$$
$$\frac{-2|3r + 5|}{-2} = \frac{-8}{-2}$$
$$|3r + 5| = 4$$
$$3r + 5 = 4 \text{ or } 3r + 5 = -4$$
$$3r = -1 \text{ or } 3r = -9$$
$$r = -\frac{1}{3} \text{ or } r = -3$$

47.
$$|x - 2| < 7$$
$$-7 < x - 2 < 7$$
$$-7 + 2 < x - 2 + 2 < 7 + 2$$
$$-5 < x < 9$$

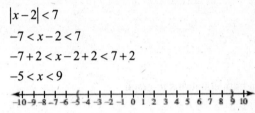

48.
$$|2x + 3| \leq 15$$
$$-15 \leq 2x + 3 \leq 15$$
$$-15 - 3 \leq 2x + 3 - 3 \leq 15 - 3$$
$$-18 \leq 2x \leq 12$$
$$\frac{-18}{2} \leq \frac{2x}{2} \leq \frac{12}{2}$$
$$-9 \leq x \leq 6$$

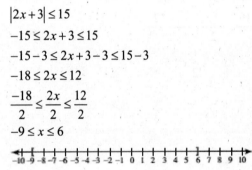

49.
$$|x + 4| \geq 6$$
$$x + 4 \leq -6 \text{ or } x + 4 \geq 6$$
$$x \leq -10 \text{ or } x \geq 2$$

50.
$$|3x - 7| + 10 > 15$$
$$|3x - 7| + 10 - 10 > 15 - 10$$
$$|3x - 7| > 5$$
$$3x - 7 < -5 \text{ or } 3x - 7 > 5$$
$$3x < 2 \text{ or } 3x > 12$$
$$x < \frac{2}{3} \text{ or } x > 4$$

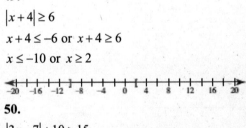

51.

$$\begin{cases} y \geq x-5 \\ y \leq \dfrac{1}{5}x+2 \end{cases}$$

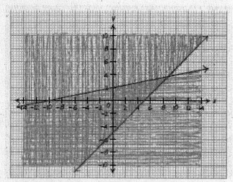

52.

$$\begin{cases} 3x-5y<10 \\ x+2y>12 \end{cases}$$

Solve for y in both equations to graph:

$3x-5y<10$, $x+2y>12$

$-5y<-3x+10$, $2y>-x+12$

$\dfrac{-5y}{-5}>\dfrac{-3x+10}{-5}$, $\dfrac{2y}{2}>\dfrac{-x+12}{2}$

$y>\dfrac{3}{5}x-2$, $y>-\dfrac{1}{2}x+6$

$$\begin{cases} y>\dfrac{3}{5}x-2 \\ y>-\dfrac{1}{2}x+6 \end{cases}$$

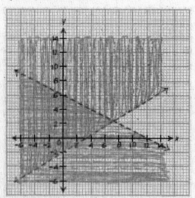

Section 3.1

1.
$2^5 + 3^4 = 32 + 81$
$ = 113$

3.
$2^4(2^5) = 2^{4+5}$
$ = 2^9 = 512$

5. $w^2 w^5 w^3 = w^{2+5+3} = w^{10}$

7.
$\dfrac{7^{23}}{7^{20}} = 7^{23-20}$
$\phantom{\dfrac{7^{23}}{7^{20}}} = 7^3 = 343$

9. $\dfrac{z^{12}}{z^8} = z^{12-8} = z^4$

11.
$3^{245}(3^{-242}) = 3^{245+(-242)}$
$\phantom{3^{245}(3^{-242})} = 3^3 = 27$

13. $s^5 s^{-3} = s^{5+(-3)} = s^2$

15. $4r^5 t^{-4} = \dfrac{4r^5}{t^4}$

17. $-9a^3 b^{-7} = \dfrac{-9a^3}{b^7}$

19. $\dfrac{3}{b^{-2}} = 3b^2$

21.
$\dfrac{x^{-3} y^2}{3x^5 y^{-7}} = \dfrac{x^{-3-5} y^{2-(-7)}}{3}$
$\phantom{\dfrac{x^{-3} y^2}{3x^5 y^{-7}}} = \dfrac{x^{-8} y^9}{3} = \dfrac{y^9}{3x^8}$

23.
$\dfrac{-4x^{-4} y^2}{x^2 y^{-3}} = -4x^{-4-2} y^{2-(-3)}$
$\phantom{\dfrac{-4x^{-4} y^2}{x^2 y^{-3}}} = -4x^{-6} y^5 = \dfrac{-4y^5}{x^6}$

25. The student added the bases, and then added the exponents. The student should have just followed the order of operations.
$5^3 + 2^6 = 125 + 64$
$ = 189$

27. The student treated -7 as a negative exponent and incorrectly moved it to the denominator.
$-7x^2 y^{-3} = \dfrac{-7x^2}{y^3}$

29. The student multiplied the exponents instead of adding the exponents.
$m^3 m^4 = m^{3+4} = m^7$

31.
$\dfrac{200 a^5 bc^3}{25 a^2 bc} = \dfrac{8 a^{5-2} \cancel{b} c^{3-1}}{\cancel{b}}$
$\phantom{\dfrac{200 a^5 bc^3}{25 a^2 bc}} = 8 a^3 c^2$

33.
$3x^2 y(5x^4 y^3) = 15 x^{2+4} y^{1+3}$
$ = 15 x^6 y^4$

35.
$\left(\dfrac{2}{3} a^3 b^7 c\right)^4 = \dfrac{2^4}{3^4} a^{3\times 4} b^{7\times 4} c^{1\times 4}$
$\phantom{\left(\dfrac{2}{3} a^3 b^7 c\right)^4} = \dfrac{16}{81} a^{12} b^{28} c^4$

37.
$(2x^2 y^3)^{-2} = \dfrac{1}{(2x^2 y^3)^2}$
$\phantom{(2x^2 y^3)^{-2}} = \dfrac{1}{2^2 x^{2\times 2} y^{3\times 2}}$
$\phantom{(2x^2 y^3)^{-2}} = \dfrac{1}{4 x^4 y^6}$

39.
$(a + 5b)^2 = (a+5b)(a+5b)$
$ = a^2 + 5ab + 5ab + 25b^2$
$ = a^2 + 10ab + 25b^2$

41.
$\left(\dfrac{5w^3 v^7 x^{-4}}{17 wx^3}\right)^0 = 1$ for $v, w, x \neq 0$

43.
$\left(\dfrac{3}{5}\right)^{-2} = \left(\dfrac{5}{3}\right)^2 = \dfrac{5^2}{3^2} = \dfrac{25}{9}$

Chapter 3 Exponents, Polynomials, and Functions

45.

$$\left(\frac{1}{3}x^3y^{-2}\right)^{-4} = \left(\frac{1}{3}\right)^{-4}(x^3)^{-4}(y^{-2})^{-4}$$

$$= \left(\frac{3}{1}\right)^4 x^{3\times-4} y^{-2\times-4}$$

$$= 3^4 x^{-12} y^8$$

$$= \frac{81y^8}{x^{12}}$$

47.

$$\frac{(7x^2y)^{10}}{(7x^2y)^8}$$

a. The base for the exponent 10 is $7x^2y$.

b. The base for the exponent 8 is $7x^2y$.

c. Yes, as long as the bases are the same, the quotient rule for exponents applies.

d.

$$\frac{(7x^2y)^{10}}{(7x^2y)^8} = (7x^2y)^{10-8}$$

$$= (7x^2y)^2$$

$$= 7^2 x^{2\times 2} y^2$$

$$= 49x^4y^2$$

49.

a. No, because you cannot apply the exponent over addition.

b.

$$(2x+6)^2 = (2x+6)(2x+6)$$

$$= 4x^2 + 12x + 12x + 36$$

$$= 4x^2 + 24x + 36$$

51.

$$\frac{(x+9y)^7}{(x+9y)^5} = (x+9y)^{7-5}$$

$$= (x+9y)^2$$

$$= (x+9y)(x+9y)$$

$$= x^2 + 9xy + 9xy + 81y^2$$

$$= x^2 + 18xy + 81y^2$$

53.

$$\left(\frac{2x^3y^{-4}}{5xy^5}\right)\left(\frac{15xy^2}{7x^5y^{-3}}\right) = \left(\frac{2x^{3-1}y^{-4-5}}{5}\right)\left(\frac{15x^{1-5}y^{2-(-3)}}{7}\right)$$

$$= \left(\frac{2x^2y^{-9}}{5}\right)\left(\frac{15x^{-4}y^5}{7}\right) = \left(\frac{2x^2}{5y^9}\right)\left(\frac{15y^5}{7x^4}\right)$$

$$= \frac{30x^2y^5}{35x^4y^9} = \frac{30x^{2-4}y^{5-9}}{35} = \frac{30x^{-2}y^{-4}}{35}$$

$$= \frac{30}{35x^2y^4} = \frac{6}{7x^2y^4}$$

55.

$$\left(\frac{2g^{-2}h^{-3}}{5gh^{-6}}\right)^2 = \left(\frac{2g^{-2-1}h^{-3-(-6)}}{5}\right)^2$$

$$= \left(\frac{2g^{-3}h^3}{5}\right)^2 = \left(\frac{2h^3}{5g^3}\right)^2$$

$$= \frac{2^2(h^3)^2}{5^2(g^3)^2} = \frac{4h^6}{25g^6}$$

57.

$$(2x^3y^{-4})^{-3}(3x^2y^{-6})^2 = 2^{-3}(x^3)^{-3}(y^{-4})^{-3} \cdot 3^2(x^2)^2(y^{-6})^2$$

$$= 2^{-3}x^{-9}y^{12} \cdot 9x^4y^{-12}$$

$$= 2^{-3}x^{-9+4}y^{12+(-12)} \cdot 9$$

$$= 2^{-3}x^{-5}y^0 \cdot 9$$

$$= 2^{-3}x^{-5} \cdot 1 \cdot 9$$

$$= 2^{-3}x^{-5} \cdot 9$$

$$= \frac{9}{2^3 x^5}$$

$$= \frac{9}{8x^5}$$

59.

$$\left(\frac{1}{5}a^{-2}b^3c\right)^{-2}\left(\frac{2}{3}a^4b^{-6}c\right)^{-1} = \left(\frac{1}{5}\right)^{-2}(a^{-2})^{-2}(b^3)^{-2}c^{-2} \cdot \left(\frac{2}{3}\right)^{-1}(a^4)^{-1}(b^{-6})^{-1}c^{-1}$$

$$= \frac{1^{-2}}{5^{-2}}a^4b^{-6}c^{-2} \cdot \frac{2^{-1}}{3^{-1}}a^{-4}b^6c^{-1}$$

$$= 5^2 a^{4+(-4)}b^{-6+6}c^{-2+(-1)} \cdot \frac{3}{2}$$

$$= 25a^0b^0c^{-3} \cdot \frac{3}{2}$$

$$= 25c^{-3} \cdot \frac{3}{2}$$

$$= \frac{25}{c^3} \cdot \frac{3}{2}$$

$$= \frac{75}{2c^3}$$

61.
a. The power rule.
b. The product rule.
c. Negative exponents.

63.
$\sqrt{x} = x^{\frac{1}{2}}$

65.
$\sqrt[5]{m} = m^{\frac{1}{5}}$

67.
$\sqrt{c^3} = (c^3)^{\frac{1}{2}}$
$= c^{\frac{3}{2}}$

69.
$(\sqrt[3]{t})^2 = (t^{\frac{1}{3}})^2$
$= t^{\frac{2}{3}}$

71.
$\sqrt{5xy} = (5xy)^{\frac{1}{2}}$
$= 5^{\frac{1}{2}} x^{\frac{1}{2}} y^{\frac{1}{2}}$

73.
$\sqrt[7]{4m^3 n^6 p^2} = (4m^3 n^6 p^2)^{\frac{1}{7}}$
$= 4^{\frac{1}{7}} (m^3)^{\frac{1}{7}} (n^6)^{\frac{1}{7}} (p^2)^{\frac{1}{7}}$
$= 4^{\frac{1}{7}} m^{\frac{3}{7}} n^{\frac{6}{7}} p^{\frac{2}{7}}$

75. $r^{\frac{1}{3}} = \sqrt[3]{r}$

77. $n^{\frac{2}{3}} = \sqrt[3]{n^2}$

79. $x^{\frac{1}{3}} y^{\frac{1}{3}} = (xy)^{\frac{1}{3}} = \sqrt[3]{xy}$

81.
$r^{\frac{1}{5}} s^{\frac{2}{5}} = (rs^2)^{\frac{1}{5}}$
$= \sqrt[5]{rs^2}$

83. $(xy^3 z)^{\frac{1}{2}} = \sqrt{xy^3 z}$

85.
$27^{\frac{1}{3}} = 27\wedge(1/3)$
$= 3$

87.
$1,000,000^{\frac{1}{6}} = 1,000,000\wedge(1/6)$
$= 10$

89.
$(-64)^{\frac{1}{3}} = -64\wedge(1/3)$
$= -4$

91.
$25^{-\frac{1}{2}} = 25\wedge(-1/2)$
$= \frac{1}{5}$

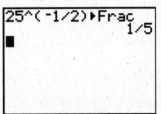

93.
$(8x^3 y^9)^{\frac{1}{3}} = 8^{\frac{1}{3}} x^{\frac{3}{3}} y^{\frac{9}{3}}$
$= 2xy^3$

95.
$(64 m^6 n^{12})^{\frac{1}{2}} = 64^{\frac{1}{2}} m^{\frac{6}{2}} n^{\frac{12}{2}}$
$= 8 m^3 n^6$

97.
$(121 a^{-6} b^8)^{\frac{1}{2}} = (121)^{\frac{1}{2}} a^{\frac{-6}{2}} b^{\frac{8}{2}}$
$= 11 a^{-3} b^4$
$= \frac{11 b^4}{a^3}$

99.
$\left(\frac{49 x^5 y^3}{25 xy^9}\right)^{\frac{1}{2}} = \left(\frac{49 x^{5-1} y^{3-9}}{25}\right)^{\frac{1}{2}} = \left(\frac{49 x^4 y^{-6}}{25}\right)^{\frac{1}{2}}$
$= \left(\frac{49 x^4}{25 y^6}\right)^{\frac{1}{2}} = \frac{49^{\frac{1}{2}} x^{\frac{4}{2}}}{25^{\frac{1}{2}} y^{\frac{6}{2}}} = \frac{7 x^2}{5 y^3}$

Chapter 3 Exponents, Polynomials, and Functions

101.

$$\left(\frac{16m^3n^6p}{mn^{-2}p^3}\right)^{\frac{1}{2}} = \left(16m^{3-1}n^{6-(-2)}p^{1-3}\right)^{\frac{1}{2}}$$

$$= \left(16m^2n^{6+2}p^{-2}\right)^{\frac{1}{2}}$$

$$= \left(16m^2n^8p^{-2}\right)^{\frac{1}{2}}$$

$$= 16^{\frac{1}{2}}m^{\frac{2}{2}}n^{\frac{8}{2}}p^{\frac{-2}{2}}$$

$$= 4mn^4p^{-1}$$

$$= \frac{4mn^4}{p}$$

103.

$$\left(100ab^3c^2\right)^{-\frac{1}{2}}\left(a^6b^{-2}\right)^{\frac{1}{4}} = \left(100^{-\frac{1}{2}}a^{-\frac{1}{2}}b^{-\frac{3}{2}}c^{-\frac{2}{2}}\right)\left(a^{\frac{6}{4}}b^{\frac{-2}{4}}\right)$$

$$= \left(100^{-\frac{1}{2}}a^{\frac{-1}{2}}b^{\frac{-3}{2}}c^{-1}\right)\left(a^{\frac{3}{2}}b^{\frac{-1}{2}}\right)$$

$$= 100^{-\frac{1}{2}}a^{\frac{-1}{2}+\frac{3}{2}}b^{\frac{-3}{2}+\frac{-1}{2}}c^{-1}$$

$$= 100^{-\frac{1}{2}}a^{\frac{2}{2}}b^{\frac{-4}{2}}c^{-1}$$

$$= 100^{-\frac{1}{2}}ab^{-2}c^{-1}$$

$$= \frac{a}{10b^2c}$$

Section 3.2

1. $5x+9 \to$ Two terms.

$5x$: variable term, coefficient 5

9: constant term

3. $-3x^2+2x-8 \to$ Three terms.

$-3x^2$: variable term, coefficient -3

$2x$: variable term, coefficient 2

-8: constant term

5. $12a^3b^2c+3abc^2-8bc \to$ Three terms.

$12a^3b^2c$: variable term, coefficient 12

$3abc^2$: variable term, coefficient 3

$-8bc$: variable term, coefficient -8

7. $4x+9$; Yes, it is a polynomial.

9. $\dfrac{4}{x}+20x-10 = 4x^{-1}+20x-10$; No, it is not a polynomial because the first term has a negative exponent on the variable.

11. $14a^2b+10ab^3-9a$; Yes, it is a polynomial.

13. $5\sqrt{d}+2 = 5d^{\frac{1}{2}}+2$; No, it is not a polynomial because the first term has a rational exponent on the variable.

15. 108; Yes, it is a polynomial.

17. $2x^5$ has degree 5; -7 has degree 0; the polynomial has degree 5.

19. $5p^2$ has degree 2; $4p$ has degree 1; -87 has degree 0; the polynomial has degree 2.

21. $4m$ has degree 1; 8 has degree 0; the polynomial has degree 1.

23. $2a^2b^3$ has degree 5; $3ab^2$ has degree 3; $-8b$ has degree 1; the polynomial has degree 5.

25. $\dfrac{2}{3}gh$ has degree 2; $\dfrac{1}{4}g^3h^5$ has degree 8; $-\dfrac{2}{9}g$ has degree 1; 7 has degree 0; the polynomial has degree 8.

27.

$(5x+6)+(2x+8) = 5x+6+2x+8$
$= 7x+14$

29.

$(7x+6)-(3x+2) = 7x+6-3x-2$
$= 4x+4$

31.

$(5x^2-6x-12)-(-3x^2-10x+8)$
$= 5x^2-6x-12+3x^2+10x-8$
$= 8x^2+4x-20$

33.

$(5x^3z^2-4x^2z+6z)-(3x^3z^2-17xz+3z)$
$= 5x^3z^2-4x^2z+6z-3x^3z^2+17xz-3z$
$= 2x^3z^2-4x^2z+17xz+3z$

35. Chie distributed the subtraction sign and then turned the addition problem into a multiplication problem.

$(2x+7)-(6x-9) = 2x+7-6x+9$
$= -4x+16$

37. Gordon distributed the exponent across the two terms in the binomial.

$(5x-8)^2 = (5x-8)(5x-8)$
$= 25x^2-40x-40x+64$
$= 25x^2-80x+64$

39.

$(5ab-8b)(3a+4b)$
$= 15a^2b+20ab^2-24ab-32b^2$

41.

$(2x+5)(3x^2-5x+7)$
$= 6x^3-10x^2+14x+15x^2-25x+35$
$= 6x^3+5x^2-11x+35$

43.

$(3x+7y)^2 = (3x+7y)(3x+7y)$
$= 9x^2+21xy+21xy+49y^2$
$= 9x^2+42xy+49y^2$

45.

$(2x^2+3x-9)+(7x+2)^2$
$= 2x^2+3x-9+(7x+2)(7x+2)$
$= 2x^2+3x-9+49x^2+14x+14x+4$
$= 51x^2+31x-5$

47.

$7x(3y^2+5y-6)+8xy$
$= 21xy^2+35xy-42x+8xy$
$= 21xy^2+43xy-42x$

Chapter 3 Exponents, Polynomials, and Functions

49.

a. $T(0) = -4.5(0) + 712.3 = 712.3$. In the year 2000, the average American ate 712.3 pounds of fruits and vegetables.

b. $F(0) = -3.5(0) + 289.6 = 289.6$. In the year 2000, the average American ate 289.6 pounds of fruit.

c. $T(0) - F(0) = 712.3 - 289.6 = 422.7$. In the year 2000, the average American ate 422.7 pounds of vegetables.

d. Let $V(y)$ be the average number of pounds of vegetables each American eats per year y years since 2000.

$$V(y) = T(y) - F(y)$$
$$= (-4.5y + 712.3) - (-3.5y + 289.6)$$
$$= -4.5y + 712.3 + 3.5y - 289.6$$
$$V(y) = -y + 422.7$$

e. $V(0) = 422.7$. In the year 2000, the average American ate 422.7 pounds of vegetables.

f.

$V(10) = -10 + 422.7 = 412.7$
$V(15) = -15 + 422.7 = 407.7$
$V(20) = -20 + 422.7 = 402.7$

The average American will eat approximately 412.7 pounds of vegetables in 2010, 407.7 pounds of vegetables in 2015, and 402.7 pounds of vegetables in 2020.

51.

a. Let $O(t)$ be the per capita consumption of milk products other than whole milk in the U.S. (in gallons per person) t years since 2000.

$$O(t) = M(t) - W(t)$$
$$= (-0.29t + 22.42) - (-0.21t + 8.09)$$
$$= -0.29t + 22.42 + 0.21t - 8.09$$
$$O(t) = -0.08t + 14.33$$

b. $W(5) = -0.21(5) + 8.09 = 7.04$. In 2005, 7.04 gallons of whole milk were consumed per person in the United States.

c.

$O(5) = -0.08(5) + 14.33 = 13.93$
$O(10) = -0.08(10) + 14.33 = 13.53$

In the United States, the per capita consumption of milk products other than whole milk was 13.93 gallons in 2005 and 13.53 gallons in 2010.

d. Slope $= -0.21$. The amount of whole milk consumed per person in the United States is decreasing by approximately 0.21 gallons per year.

e. M-intercept: (0, 22.42). In the year 2000, the per capita consumption of milk products in the United States was 22.42 gallons.

53. $U(t) - M(t)$ is the number of women in the United States in millions t years since 1900.

55. The total distance can be found by multiplying $\dfrac{\text{miles}}{\text{hour}} \times \dfrac{\text{hour}}{1} = \text{miles}$. Let $S(d)T(d) = ST(d)$ be the number of miles driven on day d of a cross country trip.

57. The average amount of national debt per person is given by $\dfrac{D(t)}{U(t)} = \dfrac{D}{U}(t)$ dollars per person, t years since 1900.

59. To get the personal dept use $(\#\text{persons})\left(\dfrac{\text{dollars}}{\text{person}}\right)$. Let $U(t) \cdot D(t)$ be the total amount of personal debt in dollars in the United States t years since 1900.

61.

a. $M(t) + F(t) = 100$, because the percent must add up to 100 if it is to represent all of the jail inmates.

b.

$M(t) + F(t) = 100$
$(-0.24t + 88.64) + F(t) = 100$
$F(t) = 100 - (-0.24t + 88.64)$
$F(t) = 100 + 0.24t - 88.64$
$F(t) = 0.24t + 11.36$

c. $F(5) = 0.24(5) + 11.36 = 12.56$. In 2005, approximately 12.56% of jail inmates in the U.S. federal and state prisons were female.

d. Slope $= 0.24$. The percentage of jail inmates who are female in U.S. federal and state prisons is increasing by approximately 0.24 percentage points per year.

e. M intercept $= (0, 88.64)$. The percentage of jail inmates who are male in U.S federal and state prisons was 88.64% in 2000.

63.

a.

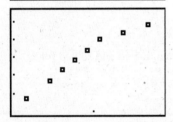

Using two points: $(7, 65.7)$ and $(13, 68.3)$

Find the slope: $m = \dfrac{68.3 - 65.7}{13 - 7} \approx 0.433$

$O(t) = 0.433t + b$

$65.7 = 0.433(7) + b$

$65.7 = 3.031 + b$

$65.7 - 3.031 = 3.031 - 3.031 + b$

$62.669 = b$

$O(t) = 0.433t + 62.67$

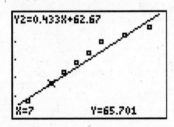

b.

$R(t) = 100 - O(t)$

$R(t) = 100 - (0.433t + 62.67)$

$R(t) = 100 - 0.433t - 62.67$

$R(t) = -0.433t + 37.33$

c. $R(20) = -0.433(20) + 37.33 = 28.67$. In 2010, approximately 28.67% of occupied housing units should be renter occupied.

d. Slope $= -0.433$. The number of housing units that are renter occupied is decreasing by 0.433 percentage point each year.

65.

a. Let $N(w)$ be the total area in square inches of a Norman window whose rectangular part has a height of 75 inches and a width of w inches.

Area of window = Area of half circle + Area of rectangle

$N(w) = \dfrac{1}{2} \cdot \pi \left(\dfrac{1}{2} w\right)^2 + 75w$

$N(w) = \dfrac{\pi}{8} w^2 + 75w$

b.

$N(36) = \dfrac{\pi}{8}(36)^2 + 75(36)$

$N(36) = 3208.94 \; in^2$

A Norman window whose rectangular part has a height of 75 inches and a width of 36 inches will have a total area of 3208.94 square inches.

67. $f(x) = 3x + 8$ and $g(x) = -6x + 9$

a.

$f(x) + g(x) = (3x + 8) + (-6x + 9)$
$= 3x + 8 + (-6x) + 9$
$= -3x + 17$

Chapter 3 Exponents, Polynomials, and Functions

b.

$$f(x)-g(x)=(3x+8)-(-6x+9)$$
$$=3x+8+6x-9$$
$$=9x-1$$

c.

$$f(x)g(x)=(3x+8)(-6x+9)$$
$$=-18x^2+27x-48x+72$$
$$=-18x^2-21x+72$$

d.

$$\frac{f(x)}{g(x)}=\frac{3x+8}{-6x+9}$$

69. $f(x)=\frac{1}{2}x+\frac{1}{5}$ and $g(x)=2x-\frac{3}{5}$

a.

$$f(x)+g(x)=\left(\frac{1}{2}x+\frac{1}{5}\right)+\left(2x-\frac{3}{5}\right)$$
$$=\frac{1}{2}x+\frac{1}{5}+2x-\frac{3}{5}$$
$$=\frac{5}{2}x-\frac{2}{5}$$

b.

$$g(x)-f(x)=\left(2x-\frac{3}{5}\right)-\left(\frac{1}{2}x+\frac{1}{5}\right)$$
$$=2x-\frac{3}{5}-\frac{1}{2}x-\frac{1}{5}$$
$$=\frac{3}{2}x-\frac{4}{5}$$

c.

$$f(x)g(x)=\left(\frac{1}{2}x+\frac{1}{5}\right)\left(2x-\frac{3}{5}\right)$$
$$=x^2-\frac{3}{10}x+\frac{2}{5}x-\frac{3}{25}$$
$$=x^2+\frac{1}{10}x-\frac{3}{25}$$

d.

$$\frac{f(x)}{g(x)}=\frac{\frac{1}{2}x+\frac{1}{5}}{2x-\frac{3}{5}}$$

71. $f(x)=3x+5$ and $g(x)=x^2+4x+10$

a.

$$f(x)+g(x)=(3x+5)+(x^2+4x+10)$$
$$=3x+5+x^2+4x+10$$
$$=x^2+7x+15$$

b.

$$g(x)-f(x)=(x^2+4x+10)-(3x+5)$$
$$=x^2+4x+10-3x-5$$
$$=x^2+x+5$$

c.

$$f(x)g(x)=(3x+5)(x^2+4x+10)$$
$$=3x^3+12x^2+30x+5x^2+20x+50$$
$$=3x^3+17x^2+50x+50$$

d.

$$\frac{f(x)}{g(x)}=\frac{3x+5}{x^2+4x+10}$$

73. Jose set the expression equal to zero and solved for x.

$$f(x)+g(x)=(3x-10)+(7x+8)$$
$$=3x-10+7x+8$$
$$=10x-2$$

75.

a.

$$R(10)=3.1(10)^2-33.7(10)+180.9$$
$$=310-337+180.9$$
$$=153.9$$

In 2010, IBM had revenue of $153.9 billion for goods sold.

$$C(10)=2.5(10)^2-28.9(10)+136.1$$
$$=250-289+136.1$$
$$=97.1$$

In 2010, IBM had a cost of $97.1 billion for selling goods.

b. Let $P(t)$ be the profit of IBM in billions of dollars t years since 2000.

$$P(t) = R(t) - C(t)$$
$$= (3.1t^2 - 33.7t + 180.9) - (2.5t^2 - 28.9t + 136.1)$$
$$= 3.1t^2 - 33.7t + 180.9 - 2.5t^2 + 28.9t - 136.1$$
$$P(t) = 0.6t^2 - 4.8t + 44.8$$

c.

$$P(10) = 0.6(10)^2 - 4.8(10) + 44.8$$
$$= 60 - 48 + 44.8$$
$$= 56.8$$

According to the model, IBM's profit in 2010 was approximately $56.8 billion.

77.

a.

$$R(11) = 164.1(11)^2 - 1353.9(11) + 6380.0$$
$$= 11,343.2$$
$$P(11) = 87.9(11)^2 - 708.9(11) + 3322.8$$
$$= 6160.8$$

$R(11) = 11343.2$. The revenue earned by Pearson Publishing Co. in 2011, is approximately 11,343.2 million British pounds. $P(11) = 6160.8$. The profit earned by Pearson Publishing Co. in 2011, is approximately 6160.8 million British pounds.

b. Let $C(t)$ be the publishing costs at Pearson Publishing in millions of British pounds t years since 2000.

$$C(t) = (164.1t^2 - 1353.9t + 6380.0) - (87.9t^2 - 708.9t + 3322.8)$$
$$= 164.1t^2 - 1353.9t + 6380.0 - 87.9t^2 + 708.9t - 3322.8$$
$$C(t) = 76.2t^2 - 645t + 3057.2$$

c. $C(11) = 76.2(11)^2 - 645(11) + 3057.2 = 5182.4$. The costs at Pearson Publishing Co. for 2011 are approximately 5182.4 million British pounds.

79.

a. $S(100) = 0.009(100)^2 - 0.5(100) + 20 = 60$. Car stereo producers are willing to supply 60 thousand car stereos if the price is set at $100.

b. Let $R(p)$ be the revenue in thousands of dollars from selling the supplied car stereos at a price of p dollars.

Revenue = Number of items × Price
$$R(p) = S(p) \times p$$
$$R(p) = 0.009p^3 - 0.5p^2 + 20p$$

c. $R(90) = 0.009(90)^3 - 0.5(90)^2 + 20(90) = 4311$. If car stereos are sold for $90 each, the projected revenue will be about $4,311 thousand.

81.

a. $S(25) = 0.11(25)^2 - 3.2(25) + 45 = 33.75$. Car manufacturers are willing to supply about 33,750 minivans if they can be sold for $25,000 each.

b. Let $R(p)$ be the revenue in millions of dollars from selling minivans at a price of p thousand dollars.

Revenue = Number of items × Price
$$R(p) = S(p) \times p$$
$$R(p) = 0.11p^3 - 3.2p^2 + 45p$$

c. $R(31) = 0.11(31)^3 - 3.2(31)^2 + 45(31) = 1596.81$. If minivans sell for $31,000 each, the projected revenue will be about $1.6 billion.

83.

a.

$$I(5) = -17,854(5)^2 + 212,449(5) - 501,784$$
$$I(5) = 114,111$$

There were about 114,111 immigrants admitted to the United States as permanent residents under refugee acts in 1995.

b.

$$E(5) = -6432(5)^2 + 74,006(5) - 161,816$$
$$E(5) = 47,414$$

There were about 47,414 European immigrants admitted to the United States as permanent residents under refugee acts in 1995.

Chapter 3 Exponents, Polynomials, and Functions

c. Let $N(t)$ be the number of non-European immigrants admitted to the United States as permanent residents under refugee acts t years since 1990.

$N(t) = I(t) - E(t)$
$= -17{,}854t^2 + 212{,}449t - 501{,}784$
$\underline{-(-6432t^2 + 74{,}006t - 161{,}816)}$

Distribute the negative to each of the terms in $E(t)$.

$N(t) = I(t) - E(t)$
$= -17{,}854t^2 + 212{,}449t - 501{,}784$
$\underline{+6432t^2 \ -74{,}006t \ +161{,}816}$
$N(t) = -11{,}442t^2 + 138{,}443t - 339{,}968$

d.

$N(8) = -11{,}422(8)^2 + 138{,}443(8) - 339{,}968$
$N(8) = 36{,}568$

There were about 36,568 non-European immigrants admitted to the United States as permanent residents under refugee acts in 1998.

85.

a. Let $U(t)$ be the number of United States residents in millions who are Caucasian and under 18 years old, t years since 1990.

$U(t) = C(t) - O(t)$
$= \ \ \ \ \ \ \ \ \ \ 1.693t + 209.107$
$\underline{-(0.014t^2 + 1.169t + 157.528)}$

Distribute the negative to each of the terms in $O(t)$.

$U(t) = C(t) - O(t)$
$= \ \ \ \ \ \ \ \ \ \ 1.693t + 209.107$
$\underline{-0.014t^2 - 1.169t - 157.528}$
$U(t) = -0.014t^2 + 0.524t + 51.579$

b.

$U(12) = -0.014(12)^2 + 0.524(12) + 51.579$
$U(12) = 55.851$

In 2002, there were approximately 56 million Caucasian United States residents who were under 18 years old.

87. $f(x) = 3x + 8$ and $g(x) = -6x + 9$

a.

$f(5) + g(5) = [3(5) + 8] + [-6(5) + 9]$
$= (23) + (-21)$
$= 2$

b.

$f(5) - g(5) = [3(5) + 8] - [-6(5) + 9]$
$= (23) - (-21)$
$= 44$

c.

$f(5)g(5) = [3(5) + 8][-6(5) + 9]$
$= (23)(-21)$
$= -483$

d.

$\dfrac{f(5)}{g(5)} = \dfrac{3(5) + 8}{-6(5) + 9}$
$= -\dfrac{23}{21}$

89. $f(x) = \dfrac{1}{2}x + \dfrac{1}{5}$ and $g(x) = 2x - \dfrac{3}{5}$

a.

$f(2) + g(2) = \left[\dfrac{1}{2}(2) + \dfrac{1}{5}\right] + \left[2(2) - \dfrac{3}{5}\right]$
$= \left(1 + \dfrac{1}{5}\right) + \left(4 - \dfrac{3}{5}\right)$

Use a common denominator of 5.

$= \dfrac{5}{5} + \dfrac{1}{5} + \dfrac{20}{5} - \dfrac{3}{5}$
$= \dfrac{23}{5} = 4.6$

b.

$g(2) - f(2) = \left[2(2) - \dfrac{3}{5}\right] - \left[\dfrac{1}{2}(2) + \dfrac{1}{5}\right]$
$= \left(4 - \dfrac{3}{5}\right) - \left(1 + \dfrac{1}{5}\right)$

Use a common denominator of 5.

$= \dfrac{20}{5} - \dfrac{3}{5} - \dfrac{5}{5} - \dfrac{1}{5}$
$= \dfrac{11}{5} = 2.2$

c.

$$f(2)+g(2)=\left[\frac{1}{2}(2)+\frac{1}{5}\right]\left[2(2)-\frac{3}{5}\right]$$

$$=\left(1+\frac{1}{5}\right)\left(4-\frac{3}{5}\right)$$

$$=4-\frac{3}{5}+\frac{4}{5}-\frac{3}{25}$$

Use a common denominator of 25.

$$=4\left(\frac{25}{25}\right)-\frac{3}{5}\left(\frac{5}{5}\right)+\frac{4}{5}\left(\frac{5}{5}\right)-\frac{3}{25}$$

$$=\frac{100}{25}-\frac{15}{25}+\frac{20}{25}-\frac{3}{25}$$

$$=\frac{102}{25}=4.08$$

d.

$$\frac{f(2)}{g(2)}=\frac{\frac{1}{2}(2)+\frac{1}{5}}{2(2)-\frac{3}{5}}$$

$$=\frac{1+\frac{1}{5}}{4-\frac{3}{5}}$$

$$=\frac{1.2}{3.4}$$

$$\approx 0.35$$

91. $f(x)=3x+5$ and $g(x)=x^2+4x+10$

a.

$$f(5)+g(5)=[3(5)+5]+\left[(5)^2+4(5)+10\right]$$

$$=(20)+(55)$$

$$=75$$

b.

$$g(5)-f(5)=\left[(5)^2+4(5)+10\right]-[3(5)+5]$$

$$=(55)-(20)$$

$$=35$$

c.

$$f(5)g(5)=[3(5)+5]\left[(5)^2+4(5)+10\right]$$

$$=(20)(55)$$

$$=1100$$

d.

$$\frac{f(5)}{g(5)}=\frac{3(5)+5}{(5)^2+4(5)+10}$$

$$=\frac{20}{55}$$

$$=\frac{4}{11}$$

93. c

95. a

97. $f(x)=5x+7$ and $g(x)=4x+1$

a.

$$g(x)+f(x)=(4x+1)+(5x+7)$$

$$=4x+1+5x+7$$

$$=9x+8$$

b.

$$g(2)+f(2)=[4(2)+1]+[5(2)+7]$$

$$=(9)+(17)$$

$$=26$$

c.

$$f(x)g(x)=(5x+7)(4x+1)$$

$$=20x^2+5x+28x+7$$

$$=20x^2+33x+7$$

d.

$$f(8)g(8)=[5(8)+7][4(8)+1]$$

$$=(47)(33)$$

$$=1551$$

99. $f(x)=x+2$ and $g(x)=4x^2+x+1$

a.

$$g(x)-f(x)=(4x^2+x+1)-(x+2)$$

$$=4x^2+x+1-x-2$$

$$=4x^2-1$$

b.

$$g(2)-f(2)=\left[4(2)^2+(2)+1\right]-[(2)+2]$$

$$=(19)-(4)$$

$$=15$$

c.
$$f(x)g(x) = (x+2)(4x^2+x+1)$$
$$= 4x^3 + x^2 + x + 8x^2 + 2x + 2$$
$$= 4x^3 + 9x^2 + 3x + 2$$

d.
$$f(2)g(2) = \left[(2)+2\right]\left[4(2)^2+(2)+1\right]$$
$$= (4)(19)$$
$$= 76$$

Section 3.3

1. $f(x) = 2x + 4$ and $g(x) = 6x + 9$

a.
$$f(g(x)) = 2(6x+9)+4$$
$$= 12x + 18 + 4$$
$$= 12x + 22$$

b.
$$g(f(x)) = 6(2x+4)+9$$
$$= 12x + 24 + 9$$
$$= 12x + 33$$

3. $f(x) = 3x + 8$ and $g(x) = -6x + 9$

a.
$$(f \circ g)(x) = 3(-6x+9)+8$$
$$= -18x + 27 + 8$$
$$= -18x + 35$$

b.
$$(g \circ f)(x) = -6(3x+8)+9$$
$$= -18x - 48 + 9$$
$$= -18x - 39$$

5.

a. Let $K(w)$ be the total weekly cost in dollars for week w of production.
$$K(w) = C(T(w))$$
$$= 1.75(500w + 3000) + 5000$$
$$= 875w + 5250 + 5000$$
$$= 875w + 10,250$$

b. $T(5) = 500(5) + 3000 = 5500$. There were 5500 toys produced during week five.

c. $C(5000) = 1.75(5000) + 5000 = 13,750$. The total weekly cost from the production of 5000 toys per week, is $13,750.

d. $K(7) = 875(7) + 10,250 = 16,375$. The total weekly cost of production for week 7 was $16,375.

e.
$$18,500 = 875w + 10,250$$
$$\underline{-10,250 \qquad -10,250}$$
$$8250 = 875w$$
$$\frac{8250}{875} = \frac{875w}{875}$$
$$9.4 \approx w$$

The weekly cost will reach, and surpass, $18,500 in week ten of production.

7.

a. $v(10) = 0.25(10) + 1.5 = 4$. A West Tech employee will get four weeks of vacation per year after working with the company for ten years.

b. $C(4) = 1500(4) + 575 = 6575$. West Tech's cost for a ten-year employee's vacation is $6575.

c. Let $K(y)$ be the cost for vacation taken by an employee who has been with the company y years.
$$K(y) = C(v(y))$$
$$= 1500(0.25y + 1.5) + 575$$
$$= 375y + 2250 + 575$$
$$= 375y + 2825$$

d. $K(20) = 375(20) + 2825 = 10,325$. West Tech's cost for a 20-year employee's vacation is $10,325.

e. $K(30) = 375(30) + 2825 = 14,075$. West Tech's cost for a 30-year employee's vacation is $14,075.

9.

a. $A(4) = -400(4)^2 + 3500(4) - 1000 = 6600$. On the 4th day of the Renaissance fair, there were 6,600 people in attendance.

b. $P(6600) = 2(6600) - 2500 = 10,700$. On the 4th day of the Renaissance fair, with 6600 in attendance, a profit of $10700 was made.

Chapter 3 Exponents, Polynomials, and Functions

c. Let $T(d)$ be the profit made at the Renaissance fair d days after the fair opens.

$$T(d) = P(A(d))$$
$$= 2(-400d^2 + 3500d - 1000) - 2500$$
$$= -800d^2 + 7000d - 2000 - 2500$$
$$= -800d^2 + 7000d - 4500$$

d. $T(3) = -800(3)^2 + 7000(3) - 4500 = 9300$. The Renaissance fair made $9300 profit on the 3rd day of the fair.

e.

d	Profit
1	1700
2	6300
3	9300
4	10700
5	10500
6	8700
7	5300

$T(1) + T(2) + T(3) + T(4) + T(5) + T(6) + T(7) = 52,500$

If the Renaissance fair lasts a total of seven days, the total profit would be $52,500.

11. $P(B(t))$

$P =$ profit in thousands of $
$b =$ number of bikes $\Big\}$ same
$B =$ number of bikes
$t =$ year

Use $B(t)$ in place of b in $P(b)$ to get $P(B(t))$.

Input units: the year

Output units: profit in thousands of dollars

13. $H(T(d))$

$H =$ number of homeless who seek space
$t =$ low night-time temperature in °F $\Big\}$ same
$T =$ low night-time temperature in °F
$d =$ day of the week

Use $T(d)$ in place of t in $H(t)$ to get $H(T(d))$.

Input units: day of the week

Output units: number of homeless who seek space

15. $f(x) = 4x + 5$ and $g(x) = 2x + 6$

a.
$$f(g(x)) = f(2x + 6)$$
$$= 4(2x + 6) + 5$$
$$= 8x + 24 + 5$$
$$= 8x + 29$$

b.
$$g(f(x)) = g(4x + 5)$$
$$= 2(4x + 5) + 6$$
$$= 8x + 10 + 6$$
$$= 8x + 16$$

17. $f(x) = 7x + 11$ and $g(x) = -8x + 15$

a.
$$(f \circ g)(x) = f(g(x))$$
$$= f(-8x + 15)$$
$$= 7(-8x + 15) + 11$$
$$= -56x + 105 + 11$$
$$= -56x + 116$$

b.
$$(g \circ f)(x) = g(f(x))$$
$$= g(7x + 11)$$
$$= -8(7x + 11) + 15$$
$$= -56x - 88 + 15$$
$$= -56x - 73$$

19. $f(x) = \frac{1}{2}x + \frac{1}{5}$ and $g(x) = 2x - \frac{3}{5}$

a.
$$f(g(x)) = f\left(2x - \frac{3}{5}\right)$$
$$= \frac{1}{2}\left(2x - \frac{3}{5}\right) + \frac{1}{5}$$
$$= x - \frac{3}{10} + \frac{1}{5}$$
$$= x - \frac{1}{10}$$

b.

$$g(f(x)) = g\left(\frac{1}{2}x + \frac{1}{5}\right)$$
$$= 2\left(\frac{1}{2}x + \frac{1}{5}\right) - \frac{3}{5}$$
$$= x + \frac{2}{5} - \frac{3}{5}$$
$$= x - \frac{1}{5}$$

21. $f(x) = 0.68x + 2.36$ and $g(x) = 3.57x + 6.49$

a.

$$f(g(x)) = f(3.57x + 6.49)$$
$$= 0.68(3.57x + 6.49) + 2.36$$
$$= 2.4276x + 4.4132 + 2.36$$
$$= 2.4276x + 6.7732$$

b.

$$g(f(x)) = g(0.68x + 2.36)$$
$$= 3.57(0.68x + 2.36) + 6.49$$
$$= 2.4276x + 8.4252 + 6.49$$
$$= 2.4276x + 14.9152$$

23. $f(x) = 3x + 5$ and $g(x) = x^2 + 4x + 10$

a.

$$f(g(x)) = f(x^2 + 4x + 10)$$
$$= 3(x^2 + 4x + 10) + 5$$
$$= 3x^2 + 12x + 30 + 5$$
$$= 3x^2 + 12x + 35$$

b.

$$g(f(x)) = g(3x + 5)$$
$$= (3x + 5)^2 + 4(3x + 5) + 10$$
$$= (3x + 5)(3x + 5) + 12x + 20 + 10$$
$$= 9x^2 + 15x + 15x + 25 + 12x + 30$$
$$= 9x^2 + 42x + 55$$

25. $f(x) = x + 2$ and $g(x) = 4x^2 + x + 1$

a.

$$(f \circ g)(x) = f(g(x))$$
$$= f(4x^2 + x + 1)$$
$$= (4x^2 + x + 1) + 2$$
$$= 4x^2 + x + 1 + 2$$
$$= 4x^2 + x + 3$$

b.

$$(g \circ f)(x) = g(f(x))$$
$$= g(x + 2)$$
$$= 4(x + 2)^2 + (x + 2) + 1$$
$$= 4(x + 2)(x + 2) + x + 2 + 1$$
$$= 4(x^2 + 2x + 2x + 4) + x + 3$$
$$= 4(x^2 + 4x + 4) + x + 3$$
$$= 4x^2 + 16x + 16 + x + 3$$
$$= 4x^2 + 17x + 19$$

27. Lilybell multiplied $f(x)$ times $g(x)$.

$f(x) = 3x - 10$ and $g(x) = 7x + 8$

$$f(g(x)) = f(7x + 8)$$
$$= 3(7x + 8) - 10$$
$$= 21x + 24 - 10$$
$$= 21x + 14$$

29. Colter reversed the composition and found $g(f(x))$.

$f(x) = 3x - 10$ and $g(x) = 7x + 8$

$$f(g(x)) = f(7x + 8)$$
$$= 3(7x + 8) - 10$$
$$= 21x + 24 - 10$$
$$= 21x + 14$$

31. $f(x) = 3x + 8$ and $g(x) = -6x + 9$

a.

$$f(g(5)) = f(-21) \quad \text{*Since } g(5) = -6(5) + 9 = -21$$
$$= 3(-21) + 8$$
$$= -63 + 8$$
$$= -55$$

Chapter 3 Exponents, Polynomials, and Functions

b.
$g(f(5)) = g(23)$ *Since $f(5) = 3(5) + 8 = 23$
$= -6(23) + 9$
$= -138 + 9$
$= -129$

33. $f(x) = \frac{1}{2}x + \frac{1}{5}$ and $g(x) = 2x - \frac{3}{5}$

a.
$f(g(2)) = f\left(\frac{17}{5}\right)$ *Since $g(2) = 2(2) - \frac{3}{5} = \frac{17}{5}$
$= \frac{1}{2}\left(\frac{17}{5}\right) + \frac{1}{5}$
$= \frac{17}{10} + \frac{1}{5}$
$= \frac{19}{10} = 1.9$

b.
$g(f(2)) = g\left(\frac{6}{5}\right)$ *Since $f(2) = \frac{1}{2}(2) + \frac{1}{5} = \frac{6}{5}$
$= 2\left(\frac{6}{5}\right) - \frac{3}{5}$
$= \frac{12}{5} - \frac{3}{5}$
$= \frac{9}{5} = 1.8$

35. $f(x) = 0.68x + 2.36$ and $g(x) = 3.57x + 6.49$

a.
$(f \circ g)(-3) = f(g(-3))$
$= f(-4.22)$ *Since $g(-3) = 3.57(-3) + 6.49 = -4.22$
$= 0.68(-4.22) + 2.36$
$= -2.8696 + 2.36$
$= -0.5096$

b.
$(g \circ f)(-3) = g(f(-3))$
$= g(0.32)$ *Since $f(-3) = 0.68(-3) + 2.36 = 0.32$
$= 3.57(0.32) + 6.49$
$= 1.1424 + 6.49$
$= 7.6324$

37. $f(x) = 3x + 5$ and $g(x) = x^2 + 4x + 10$

a.
$f(g(5)) = f(55)$ *Since $g(5) = (5)^2 + 4(5) + 10 = 55$
$= 3(55) + 5$
$= 165 + 5$
$= 170$

b.
$g(f(5)) = g(20)$ *Since $f(5) = 3(5) + 5 = 20$
$= (20)^2 + 4(20) + 10$
$= 400 + 80 + 10$
$= 490$

39. $f(x) = x + 2$ and $g(x) = 4x^2 + x + 1$

a.
$f(g(2)) = f(19)$ *Since $g(2) = 4(2)^2 + (2) + 1 = 19$
$= (19) + 2$
$= 21$

b.
$g(f(2)) = g(4)$ *Since $f(2) = (2) + 2 = 4$
$= 4(4)^2 + (4) + 1$
$= 64 + 4 + 1$
$= 69$

41. $T(t) - K(t)$

Input units: the year

Output units: number of people over 12-years-old who go on the tour

43. $A(T(t) - K(t))$

Input units: the year

Output units: cost for all people over 12-years-old who go on the tour

45. $C(K(t)) - A(T(t) - K(t))$

Input units: the year

Output units: difference between the cost for all children under 12-years-old and the cost for all people over 12-years-old who traveled on the tour

47. $f(x) = (9x+2)$ and $g(x) = (3x+7)$

a.
$$f(x)g(x) = (9x+2)(3x+7)$$
$$= 27x^2 + 63x + 6x + 14$$
$$= 27x^2 + 69x + 14$$

b.
$$f(x) + g(x) = (9x+2) + (3x+7)$$
$$= 12x + 9$$

c.
$$f(g(x)) = 9(3x+7) + 2$$
$$= 27x + 63 + 2$$
$$= 27x + 65$$

49. $f(x) = -9x + 1$ and $g(x) = x + 8$

a.
$$f(3) + g(3) = (-9(3)+1) + ((3)+8)$$
$$= -27 + 1 + 3 + 8$$
$$= -15$$

b.
$$f(4)g(4) = (-9(4)+1)(4+8)$$
$$= (-35)(12)$$
$$= -420$$

c.
$$g(7) = 7 + 8 = 15$$
$$f(g(7)) = f(15) = -9(15) + 1$$
$$f(g(7)) = -134$$

51. $f(x) = \frac{2}{3}x + \frac{1}{3}$ and $g(x) = \frac{1}{4}x + \frac{3}{4}$

a.
$$f(g(x)) = \frac{2}{3}\left(\frac{1}{4}x + \frac{3}{4}\right) + \frac{1}{3}$$
$$= \frac{2}{12}x + \frac{6}{12} + \frac{1}{3}$$
$$= \frac{1}{6}x + \frac{1}{2} + \frac{1}{3}$$
$$= \frac{1}{6}x + \frac{3}{6} + \frac{2}{6}$$
$$= \frac{1}{6}x + \frac{5}{6}$$

b.
$$f(42)g(42) = \left(\frac{2}{3}(42) + \frac{1}{3}\right)\left(\frac{1}{4}(42) + \frac{3}{4}\right)$$
$$= \left(\frac{85}{3}\right)\left(\frac{45}{4}\right)$$
$$= 318.75$$

c.
$$f(x) + g(x) = \left(\frac{2}{3}x + \frac{1}{3}\right) + \left(\frac{1}{4}x + \frac{3}{4}\right)$$
$$= \frac{8}{12}x + \frac{4}{12} + \frac{3}{12}x + \frac{9}{12}$$
$$= \frac{11}{12}x + \frac{13}{12}$$

53. $f(x) = 1.5x + 4.5$ and $g(x) = 3.5x + 9$

a.
$$f(g(x)) = 1.5(3.5x + 9) + 4.5$$
$$= 5.25x + 13.5 + 4.5$$
$$= 5.25x + 18$$

b.
$$f(6) = 1.5(6) + 4.5 = 13.5$$
$$g(f(6)) = g(13.5) = 3.5(13.5) + 9$$
$$g(f(6)) = 56.25$$

c.
$$g(x) - f(x) = (3.5x + 9) - (1.5x + 4.5)$$
$$= 3.5x + 9 - 1.5x - 4.5$$
$$= 2x + 4.5$$

Chapter 3 Exponents, Polynomials, and Functions
Section 3.4

1.
$6x + 8$
$2(3x + 4)$

3.
$4h^2 + 7h$
$h(4h + 7)$

5.
$25x^3 + x^2 - 2x$
$x(25x^2 + x - 2)$

7.
$4a^2b + 6ab$
$2ab(2a + 3)$

9.
$4x^2 + 6x - 2$
$2(2x^2 + 3x - 1)$

11.
$15x^2yz + 6xyz^2 - 3xy^2z$
$3xyz(5x + 2z - y)$

13.
$3x(x+5) + 2(x+5)$
$(x+5)(3x+2)$

15.
$7(5w+4) - 2w(5w+4)$
$(5w+4)(7-2w)$

17.
$x^2 - 8x + 7x - 56$
$x(x-8) + 7(x-8)$
$(x-8)(x+7)$

19.
$7r^2 + 28r + 2r + 8$
$7r(r+4) + 2(r+4)$
$(r+4)(7r+2)$

21.
$16x^2 + 18x + 24xy + 27y$
$2x(8x+9) + 3y(8x+9)$
$(8x+9)(2x+3y)$

23.
$8m^2 - 28m - 10mn + 35n$
$4m(2m-7) - 5n(2m-7)$
$(2m-7)(4m-5n)$

25.
$x^2 - 4x - 21$
$ac = 1 \times -21 = -21$
Find the factors of -21 that sum to -4:
$-7 \times 3 = -21$
$-7 + 3 = -4$
We get:
$x^2 - 4x - 21 = (x-7)(x+3)$

27.
$x^2 + 6x + 5$
$ac = 1 \times 5 = 5$
Find the factors of 5 that sum to 6:
$1 \times 5 = 5$
$1 + 5 = 6$
We get:
$x^2 + 6x + 5 = (x+5)(x+1)$

29.
$w^2 - 7w - 18$
$ac = 1 \times -18 = -18$
Find the factors of -18 that sum to -7:
$-9 \times 2 = -18$
$-9 + 2 = -7$
We get:
$w^2 - 7w - 18 = (w-9)(w+2)$

31.
$t^2 - 11t + 28$
$ac = 1 \times 28 = 28$
Find the factors of 28 that sum to -11:
$-7 \times -4 = 28$
$-7 + (-4) = -11$
We get:
$t^2 - 11t + 28 = (t-7)(t-4)$

33.

$2x^2 + 13x + 15$

$ac = 2 \times 15 = 30$

Find the factors of 30 that sum to 13:

$10 \times 3 = 30$

$10 + 3 = 13$

We get:

$2x^2 + 13x + 15$

$2x^2 + 3x + 10x + 15$

$x(2x+3) + 5(2x+3)$

$(2x+3)(x+5)$

35.

$7m^2 - 25m + 12$

$ac = 7 \times 12 = 84$

Find the factors of 84 that sum to -25:

$-4 \times -21 = 84$

$-4 + (-21) = -25$

We get:

$7m^2 - 4m + 12$

$7m^2 - 4m - 21m + 12$

$m(7m-4) - 3(7m-4)$

$(7m-4)(m-3)$

37.

$4x^2 - 31x - 45$

$ac = 4 \times -45 = -180$

Find the factors of -180 that sum to -31:

$-36 \times 5 = -180$

$-36 + 5 = -31$

We get:

$4x^2 - 31x - 45$

$4x^2 + 5 - 36x - 45$

$x(4x+5) - 9(4x+5)$

$(4x+5)(x-9)$

39.

$x^2 + 5x + 3$

$ac = 1 \times 3 = 3$

There are no factors of 3 that sum to 5, so this is a prime polynomial.

41. Dusty grouped the last two terms without first rewriting with addition. This caused a positive 7 to be factored out of the 21 when in fact a -7 needed to be factored out. Then, he continued to factor as if there were common factors, and when none were available, wrote all the binomial groupings as factors. Here is the correct factorization:

$2x^2 - 13x + 21$

$2x^2 - 6x - 7x + 21$

$(2x^2 - 6x) + (-7x + 21)$

$2x(x-3) - 7(x-3)$

$(x-3)(2x-7)$

43.

$2x^2 + 16x + 2 = 2(x^2 + 8x + 1)$

Use the ac method on the remaining quadratic: $x^2 + 8x + 1$

$ac = 1 \times 1 = 1$

Since there are no factors of 1 that sum to 8, the remaining quadratic is prime.

$2x^2 + 16x + 2 = 2(x^2 + 8x + 1)$

45.

$6w^2 + 29w + 28$

$ac = 6 \times 28 = 168$

Find the factors of 168 that sum to 29:

$21 \times 8 = 168$

$21 + 8 = 29$

We get:

$6w^2 + 29w + 28 = 6w^2 + 21w + 8w + 28$

$= 3w(2w+7) + 4(2w+7)$

$= (2w+7)(3w+4)$

47.

$10t^2 - 41t - 18$

$ac = 10 \times (-18) = -180$

Find the factors of -180 that sum to -41:

$4 \times (-45) = -180$

$4 + (-45) = -41$

We get:

$10t^2 - 41t - 18 = 10t^2 + 4t - 45t - 18$

$= 2t(5t+2) - 9(5t+2)$

$= (5t+2)(2t-9)$

Chapter 3 Exponents, Polynomials, and Functions

49.

$6x^2 + 2x + 5$

$ac = 6 \times 5 = 30$

Since there are no factors of 30 that sum to 2, the quadratic is prime.

51.

$12p^2 - 4p - 9$

$ac = 12 \times (-9) = -108$

Since there are no factors of -108 that sum to -4, the quadratic is prime.

53.

$12x^2 - 43x + 56$

$ac = 12 \times 56 = 672$

Since there are no factors of 672 that sum to -43, the quadratic is prime.

55.

$2m^2 + 14m + 24 = 2(m^2 + 7m + 12)$

$ac = 1 \times 12 = 12$

Find the factors of 12 that sum to 7:

$3 \times 4 = 12$

$3 + 4 = 7$

We get:

$$2m^2 + 14m + 24 = 2(m^2 + 7m + 12)$$
$$= 2(m^2 + 3m + 4m + 12)$$
$$= 2[m(m+3) + 4(m+3)]$$
$$= 2(m+3)(m+4)$$

57.

$6x^2 - 66x + 168 = 6(x^2 - 11x + 28)$

$ac = 1 \times 28 = 28$

Find the factors of 28 that sum to -11:

$-4 \times (-7) = 28$

$-4 + (-7) = -11$

We get:

$$6x^2 - 66x + 168 = 6(x^2 - 11x + 28)$$
$$= 6(x^2 - 4x - 7x + 28)$$
$$= 6[x(x-4) - 7(x-4)]$$
$$= 6(x-4)(x-7)$$

59.

$40x^2 + 30x - 45 = 5(8x^2 + 6x - 9)$

$ac = 8 \times (-9) = -72$

Find the factors of -72 that sum to 6:

$-6 \times 12 = -18$

$-6 + 12 = 6$

We get:

$$40x^2 + 30x - 45 = 5(8x^2 - 6x + 12x - 9)$$
$$= 5[2x(4x-3) + 3(4x-3)]$$
$$= 5(4x-3)(2x+3)$$

61.

$x^3 - 4x^2 - 21x = x(x^2 - 4x - 21)$

$ac = 1 \times (-21) = -21$

Find the factors of -21 that sum to -4:

$-7 \times 3 = -21$

$-7 + 3 = -4$

We get:

$$x^3 - 4x^2 - 21x = x(x^2 - 4x - 21)$$
$$= x(x^2 - 7x + 3x - 21)$$
$$= x[x(x-7) + 3(x-7)]$$
$$= x(x-7)(x+3)$$

63.

$10x^3 + 35x^2 + 30x = 5x(2x^2 + 7x + 6)$

$ac = 2 \times 6 = 12$

Find the factors of 12 that sum to 7:

$3 \times 4 = 12$

$3 + 4 = 7$

We get:

$$10x^3 + 35x^2 + 30x = 5x(2x^2 + 7x + 6)$$
$$= 5x(2x^2 + 3x + 4x + 6)$$
$$= 5x[x(2x+3) + 2(2x+3)]$$
$$= 5x(2x+3)(x+2)$$

65.

$x^2y + 13xy + 36y = y(x^2 + 13x + 36)$

$ac = 1 \times 36 = 36$

Find the factors of 36 that sum to 13:

$9 \times 4 = 36$

$9 + 4 = 13$

We get:

$$x^2y + 13xy + 36y = y(x^2 + 13x + 36)$$
$$= y(\underline{x^2 + 9x} + \underline{4x + 36})$$
$$= y[x(x+9) + 4(x+9)]$$
$$= y(x+9)(x+4)$$

67.

$2x^2 + 11xy + 15y^2$

$ac = 2 \times 15 = 30$

Find the factors of 30 that sum to 11:

$5 \times 6 = 30$

$5 + 6 = 11$

We get:

$$2x^2 + 11xy + 15y^2 = \underline{2x^2 + 5xy} + \underline{6xy + 15y^2}$$
$$= x(2x+5y) + 3y(2x+5y)$$
$$= (2x+5y)(x+3y)$$

69.

$6a^2 - 13ab - 5b^2$

$ac = 6 \times (-5) = -30$

Find the factors of -30 that sum to -13:

$-15 \times 2 = -30$

$-15 + 2 = -13$

We get:

$$6a^2 - 13ab - 5b^2 = \underline{6a^2 - 15ab} + \underline{2ab - 5b^2}$$
$$= 3a(2a-5b) + b(2a-5b)$$
$$= (2a-5b)(3a+b)$$

71.

$-6mn^2 - 20mn - 16m = -2m(3n^2 + 10n + 8)$

$ac = 3 \times 8 = 24$

Find the factors of 24 that sum to 10:

$4 \times 6 = 24$

$4 + 6 = 10$

We get:

$$-6mn^2 - 20mn - 16m = -2m(3n^2 + 10n + 8)$$
$$= -2m(\underline{3n^2 + 4n} + \underline{6n + 8})$$
$$= -2m[n(3n+4) + 2(3n+4)]$$
$$= -2m(3n+4)(n+2)$$

73.

$24x^2 - 38x + 15$

$ac = 24 \times 15 = 360$

Find the factors of 360 that sum to -38:

$-20 \times (-18) = 360$

$-20 + (-18) = -38$

We get:

$$24x^2 - 38x + 15 = \underline{24x^2 - 20x} - \underline{18x + 15}$$
$$= 4x(6x-5) - 3(6x-5)$$
$$= (6x-5)(4x-3)$$

75.

$24x^2 - 44x - 40 = 4(6x^2 - 11x - 10)$

$ac = 6 \times (-10) = -60$

Find the factors of -60 that sum to -11:

$-15 \times 4 = -60$

$-15 + 4 = -11$

We get:

$$24x^2 - 44x - 40 = 4(6x^2 - 11x - 10)$$
$$= 4(\underline{6x^2 - 15x} + \underline{4x - 10})$$
$$= 4[3x(2x-5) + 2(2x-5)]$$
$$= 4(2x-5)(3x+2)$$

77.

$20n^2 - 32n - 15mn + 24m$

$= \underline{20n^2 - 32n} - \underline{15mn + 24m}$

$= 4n(5n-8) - 3m(5n-8)$

$= (5n-8)(4n-3m)$

79.

$12gf + 28g = 4g(3f + 7)$

81.

$a^2 + 3a - 1$

$ac = 1 \times (-1) = -1$

Since there are no factors of -1 that sum to 3, the quadratic is prime.

83.

$7d^2 + 21d - 70 = 7(d^2 + 3d - 10)$

$ac = 1 \times (-10) = -10$

Find the factors of -10 that sum to 3:

$-2 \times 5 = -10$

$-2 + 5 = 3$

We get:

$$7d^2 + 21d - 70 = 7(d^2 + 3d - 10)$$
$$= 7(d^2 - 2d + 5d - 10)$$
$$= 7[d(d-2) + 5(d-2)]$$
$$= 7(d-2)(d+5)$$

85.

$10x^2 - 20x + 15 = 5(2x^2 - 4x + 3)$

Use the *ac* method on the remaining quadratic: $2x^2 - 4x + 3$

$ac = 2 \times 3 = 6$

Since there are no factors of 6 that sum to -4, the remaining quadratic is prime.

$10x^2 - 20x + 15 = 5(2x^2 - 4x + 3)$

Section 3.5

1. Since the middle term is $8x = 2(4 \cdot x)$:

$$x^2 + 8x + 16 = x^2 + 8x + 4^2$$
$$= (x+4)^2$$

3. Since the middle term is $12g = 2(2 \cdot 3g)$:

$$9g^2 + 12g + 4 = (3g)^2 + 12g + 2^2$$
$$= (3g+2)^2$$

5. Since the middle term is $28t = 2(7 \cdot 2t)$:

$$4t^2 - 28t + 49 = (2t)^2 - 28t + 7^2$$
$$= (2t-7)^2$$

7. Since the middle term is $30xy = 2(5x \cdot 3y)$:

$$25x^2 + 30xy + 9y^2 = (5x)^2 + 30xy + (3y)^2$$
$$= (5x+3y)^2$$

9.

$$x^2 - 36 = x^2 - 6^2$$
$$= (x+6)(x-6)$$

11.

$$9k^2 - 16 = (3k)^2 - 4^2$$
$$= (3k+4)(3k-4)$$

13.

$$m^3 - 64 = m^3 - 4^3$$
$$= (m-4)(m^2 + 4 \cdot m + 4^2)$$
$$= (m-4)(m^2 + 4m + 16)$$

15.

$$x^3 + 125 = x^3 + 5^3$$
$$= (x+5)(x^2 - 5 \cdot x + 5^2)$$
$$= (x+5)(x^2 - 5x + 25)$$

17.

$$8x^3 + 27 = (2x)^3 + 3^3$$
$$= (2x+3)((2x)^2 - 3 \cdot 2x + 3^2)$$
$$= (2x+3)(4x^2 - 6x + 9)$$

19.

$$3g^3 - 24 = 3(g^3 - 8)$$
$$= 3(g^3 - 2^3)$$
$$= 3(g-2)(g^2 + 2 \cdot g + 2^2)$$
$$= 3(g-2)(g^2 + 2g + 4)$$

21.

$$50x^2 - 18 = 2(25x^2 - 9)$$
$$= 2((5x)^2 - 3^2)$$
$$= 2(5x+3)(5x-3)$$

23. Prime. The sum of two squares will not factor with real numbers.

25. Prime. The polynomial is not a perfect square trinomial, because the middle term should be $8m$.

27. Prime. The sum of two squares will not factor with real numbers.

29.

$$r^6 - 64 = (r^3)^2 - 8^2$$
$$= (r^3 + 8)(r^3 - 8)$$
$$= (r+2)(r^2 - 2r + 4) \cdot (r-2)(r^2 + 2r + 4)$$
$$= (r+2)(r-2)(r^2 - 2r + 4)(r^2 + 2r + 4)$$

31.

$$16b^4 - 625c^4 = (4b^2)^2 - (25c^2)^2$$
$$= (4b^2 + 25c^2)(4b^2 - 25c^2)$$
$$= (4b^2 + 25c^2)((2b)^2 - (5c)^2)$$
$$= (4b^2 + 25c^2)(2b + 5c)(2b - 5c)$$

33. Let $u = h^2$.

$$h^4 + 3h^2 - 10 = (h^2)^2 + 3h^2 - 10$$
$$= u^2 + 3u - 10$$
$$= (u+5)(u-2)$$

Replace u with h^2.

$$h^4 + 3h^2 - 10 = (h^2 + 5)(h^2 - 2)$$

Chapter 3 Exponents, Polynomials, and Functions

35. Let $u = g^4$.

$$g^8 + 6g^4 + 9 = u^2 + 6u + 9$$
$$= (u+3)(u+3)$$

Replace u with g^4

$$g^8 + 6g^4 + 9 = (g^4 + 3)^2$$

37. Let $u = t^6$.

$$2t^{12} + 13t^6 + 15 = 2u^2 + 13u + 15$$
$$= (2u^2 + 3u) + (10u + 15)$$
$$= u(2u+3) + 5(2u+3)$$
$$= (2u+3)(u+5)$$

Replace u with t^6.

$$2t^{12} + 13t^6 + 15 = (2t^6 + 3)(t^6 + 5)$$

39.

$$8x^8 + 12x^4 + 18$$
$$2(4x^8 + 6x^4 + 9)$$

41. Let $u = \sqrt{H}$.

$$H + 6\sqrt{H} + 9 = u^2 + 6u + 9$$
$$= (u+3)(u+3)$$

Replace u with $\sqrt{H}$.

$$H + 6\sqrt{H} + 9 = (\sqrt{H} + 3)^2$$

43. Let $u = t^{\frac{1}{2}}$.

$$4t - 20t^{\frac{1}{2}} + 25 = 4u^2 - 20u + 25$$
$$= (2u - 5)(2u - 5)$$

Replace u with $t^{\frac{1}{2}}$.

$$4t - 20t^{\frac{1}{2}} + 25 = \left(2t^{\frac{1}{2}} - 5\right)^2$$

45.

$$7g^4 - 567h^4 = 7(g^4 - 81h^4)$$
$$= 7(g^2 + 9h^2)(g^2 - 9h^2)$$
$$= 7(g^2 + 9h^2)(g + 3h)(g - 3h)$$

47.

$$5w^5x^3z + 25w^3x^2z^2 - 120wxz^3$$
$$5wxz(w^4x^2 + 5w^2xz - 24z^2)$$
$$5wxz(w^2x + 8z)(w^2x - 3z)$$

49.

$$24a^3b^2 + 11a^2b^3 - 35ab^4 =$$
$$= ab^2(24a^2 - 11ab - 35b^2)$$
$$= ab^2(24a^2 + 24ab - 35ab - 35b^2)$$
$$= ab^2\left[(24a^2 + 24ab) + (-35ab - 35b^2)\right]$$
$$= ab^2\left[24a(a+b) - 35(a+b)\right]$$
$$= ab^2(a+b)(24a - 35b)$$

51.

$a^{12} - b^{36}$ difference of 2 squares

$= (a^6 + b^{18})(a^6 - b^{18})$ sum of 2 cubes and difference of 2 squares

$= (a^2 + b^6)(a^4 - a^2b^6 + b^{12})(a^3 + b^9)(a^3 - b^9)$ sum and difference of 2 cubes

$= (a^2 + b^6)(a^4 - a^2b^6 + b^{12})(a + b^3)(a^2 - ab^3 + b^6)(a - b^3)(a^2 + ab^3 + b^6)$

53. This is repeatedly a difference of two squares.

$$x^{16} - 1$$
$$(x^8 + 1)(x^8 - 1)$$
$$(x^8 + 1)(x^4 + 1)(x^4 - 1)$$
$$(x^8 + 1)(x^4 + 1)(x^2 + 1)(x^2 - 1)$$
$$(x^8 + 1)(x^4 + 1)(x^2 + 1)(x + 1)(x - 1)$$

55.

$$20x^2 + 23x - 21$$
$$= 20x^2 - 12x + 35x - 21$$
$$= (20x^2 - 12x) + (35x - 21)$$
$$= 4x(5x - 3) + 7(5x - 3)$$
$$= (5x - 3)(4x + 7)$$

57.

perfect square trinomial

$$9m^2 - 24m + 16$$
$$(3m - 4)^2$$

59. $4t^2 - 3t + 15$ is prime

61.

difference of two cubes

$$125x^3 - 64 = (5x)^3 - 4^3$$
$$= (5x-4)(25x^2 + 20x + 16)$$

63. Let $u = r^{\frac{1}{2}}$.

$$6r + 7r^{\frac{1}{2}} - 20 = 6u^2 + 7u - 20$$
$$= 6u^2 - 8u + 15u - 20$$
$$= (6u^2 - 8u) + (15u - 20)$$
$$= 2u(3u-4) + 5(3u-4)$$
$$= (3u-4)(2u+5)$$

Replace u with $r^{\frac{1}{2}}$

$$6r + 7r^{\frac{1}{2}} - 20 = \left(3r^{\frac{1}{2}} - 4\right)\left(2r^{\frac{1}{2}} + 5\right)$$

65.

$$-30x^3 + 58x^2 + 28x =$$
$$= -2x \ (15x^2 - 29x - 14)$$
$$= -2x \ (15x^2 + 6x - 35x - 14)$$
$$= -2x \ (15x^2 + 6x) + (-35x - 14)$$
$$= -2x \ \ 3x(5x+2) + -7(5x+2)$$
$$= -2x(5x+2)(3x-7)$$

Chapter 3 Exponents, Polynomials, and Functions
Chapter 3 Review Exercises

1.
$$(2x^3y^4z)^3(5x^{-3}y^2z)^{-2} = (2^3x^9y^{12}z^3)(5^{-2}x^6y^{-4}z^{-2})$$
$$= \frac{8x^{9+6}y^{12-4}z^{3-2}}{5^2}$$
$$= \frac{8x^{15}y^8z}{25}$$

2.
$$(3x^2y^5)(5x^2y^2) = 15x^{2+2}y^{5+2}$$
$$= 15x^4y^7$$

3.
$$\frac{225a^5b^3c}{15ab^6c^{24}} = \frac{15a^{5-1}b^{3-6}c^{1-24}}{1}$$
$$= \frac{15a^4b^{-3}c^{-23}}{1}$$
$$= \frac{15a^4}{b^3c^{23}}$$

4.
$$\left(\frac{31m^5n^4}{a^5c^7}\right)^0 = 1$$
for all $a, c, m, n \neq 0$

5.
$$(32x^{10}y^{20})^{\frac{1}{5}} = 32^{\frac{1}{5}}x^{\frac{10}{5}}y^{\frac{20}{5}}$$
$$= 2x^2y^4$$

6.
$$(a^3b^{-5}c^{-2})^{-2} = a^{-6}b^{10}c^4$$
$$= \frac{b^{10}c^4}{a^6}$$

7.
$$(3m^{-1}n^3p^{-2})^2(5m^2n^{-6}p^4) =$$
$$= (3^2m^{-2}n^6p^{-4})(5m^2n^{-6}p^4)$$
$$= \frac{9n^6}{m^2p^4} \cdot \frac{5m^2p^4}{n^6}$$
$$= 45$$

8.
$$\frac{a^2b^{-3}c^5}{a^3b^5c^7} = \frac{a^{2-3}b^{-3-5}c^{5-7}}{1}$$
$$= \frac{a^{-1}b^{-8}c^{-2}}{1}$$
$$= \frac{1}{ab^8c^2}$$

9.
$$\left(\frac{3a^{-1}b^3c^5}{ab^{-2}c^3}\right)^{-2} = \left(\frac{3a^{-1-1}b^{3-(-2)}c^{5-3}}{1}\right)^{-2}$$
$$= \left(\frac{3a^{-2}b^5c^2}{1}\right)^{-2} = \left(\frac{1}{3a^{-2}b^5c^2}\right)^2$$
$$= \left(\frac{a^2}{3b^5c^2}\right)^2 = \frac{a^4}{9b^{10}c^4}$$

10.
$$\left(\frac{3x^2y^5z}{5xy^3z}\right)\left(\frac{15x^3yz^4}{10xz}\right)$$
$$= \left(\frac{3x^{2-1}y^{5-3}z}{z}\right)\left(\frac{(3)\cancel{15}x^{3-1}yz^{4-1}}{10}\right)$$
$$= \left(\frac{3xy^2 z}{z}\right)\left(\frac{3x^2yz^3}{10}\right)$$
$$= \frac{9x^{1+2}y^{2+1}z^3}{10}$$
$$= \frac{9x^3y^3z^3}{10}$$

11.

a. $x^{\frac{1}{4}}y^{\frac{3}{4}} = (xy^3)^{\frac{1}{4}} = \sqrt[4]{xy^3}$

b. $3^{\frac{1}{2}}a^{\frac{1}{2}} = (3a)^{\frac{1}{2}} = \sqrt{3a}$

12.

a. $\sqrt[5]{4x^2} = (4x^2)^{\frac{1}{5}} = 4^{\frac{1}{5}}x^{\frac{2}{5}}$

b. $\sqrt{7ab} = (7ab)^{\frac{1}{2}} = 7^{\frac{1}{2}}a^{\frac{1}{2}}b^{\frac{1}{2}}$

13. Let $P(v)$ be the monthly profit in dollars for Hope's Pottery, when v vases are produced and sold in the month.

$$P(v) = R(v) - C(v)$$
$$P(v) = 155v - (5000 + 65v)$$
$$P(v) = 155v - 5000 - 65v$$
$$P(v) = 90v - 5000$$

14.

a. Let $P(t)$ be the total prescription drug sales in the United States in billions of dollars t years since 2000.

Using two points: $(2, 182.7)$ and $(6, 249.8)$:

find the slope: $m = \dfrac{249.8 - 182.7}{6 - 2} \approx 16.8$

$P(t) = 16.8t + b$

$182.7 = 16.8(2) + b$

$182.7 = 33.6 + b$

$182.7 - 33.6 = 33.6 - 33.6 + b$

$149.1 = b$

$P(t) = 16.8t + 149.1$

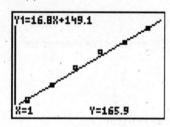

b. Let $M(t)$ be the amount of mail-order prescription drug sales in the United States in billions of dollars t years since 2000.

Using two points: $(5, 45.5)$ and $(6, 50.4)$:

find the slope: $m = \dfrac{50.4 - 45.5}{6 - 5} = 4.9$

$M(t) = 4.9t + b$

$45.5 = 4.9(5) + b$

$45.5 = 24.5 + b$

$45.5 - 24.5 = 24.5 - 24.5 + b$

$21 = b$

$M(t) = 4.9t + 21$

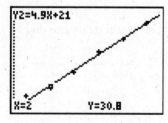

c. $P(10) = 16.8(10) + 149.1 = 317.1$. In 2010, the total prescription drug sales in the United States were about $317.1 billion.

d. $M(10) = 4.9(10) + 21 = 70$. In 2010, the mail-order prescription drug sales in the United States were about $70 billion.

e. Let $N(t)$ be the non-mail-order prescription drug sales in the United States, in billions of dollars, t years since 2000.

$N(t) = P(t) - M(t)$

$ = (16.8t + 149.1) - (4.9t + 21)$

$ = 16.8t + 149.1 - 4.9t - 21$

$N(t) = 11.9t + 128.1$

f. $N(9) = 11.9(9) + 128.1 = 235.2$. In 2009, the non-mail-order prescription drug sales in the United States were about $235.2 billion.

$N(15) = 11.9(15) + 128.1 = 306.6$. In 2015, the non-mail-order prescription drug sales in the United States will be about $306.6 billion.

15.

a.

$J(9) = 16.27(9)^3 - 321.0(9)^2 + 2089.32(9) - 3885.40$

$J(9) = 778.31$

In 1999, the federal funding for research and development at Johns Hopkins University was about $778.31 million.

$W(9) = 2.89(9)^3 - 53.94(9)^2 + 339.66(9) - 411.82 = 382.79$

$W(9) = 382.79$

In 1999, the federal funding for research and development at the University of Washington was about $382.79 million.

b. Let $D(t)$ be the difference between the federal funding at John Hopkins University and the University of Washington, in millions of dollars, t years since 1990.

$D(t) = J(t) - W(t)$

$D(t) = 16.27t^3 - 321.0t^2 + 2089.32t - 3885.40$

$ \underline{-2.89t^3 + 53.94t^2 - 339.66t + 411.82}$

$D(t) = 13.38t^3 - 267.06t^2 + 1749.66t - 3473.58$

c.

$D(10) = 13.38(10)^3 - 267.06(10)^2$

$ + 1749.66(10) - 3473.58$

$D(10) = 697.02$

In 2000, John Hopkins University had about $697.02 million more federal funding for research and development than the University of Washington.

Chapter 3 Exponents, Polynomials, and Functions

16.
$(9x-8)+(-4x+2)$
$5x-6$

17.
$(2x^2y+5xy-4y^2)-(7xy+9y^2)$
$2x^2y+5xy-4y^2-7xy-9y^2$
$2x^2y-2xy-13y^2$

18.
$(3x+7)(2x-9)$
$6x^2-27x+14x-63$
$6x^2-13x-63$

19.
$(3x+4y)(5x-7y)$
$15x^2-21xy+20xy-28y^2$
$15x^2-xy-28y^2$

20.
$(5x-3)^2$
$25x^2-30x-9$

21. The polynomial $8a^3b^2-7a^2b+19ab$ has 3 terms.

$8a^3b^2$: variable term, coefficient 8
$-7a^2b$: variable term, coefficient -7
$19ab$: variable term, coefficient 19

22. The polynomial $4m^7np^3+24m^5n^4p^3+14$ has 3 terms.

$4m^7np^3$: variable term, coefficient 4
$24m^5n^4p^3$: variable term, coefficient 24
14: constant term

23. The polynomial $5t+8$ has 2 terms.

$5t$: variable term, coefficient 5
8: constant term

24. The polynomial $8a^3b^2-7a^2b+19ab$ has degree 5.

$8a^3b^2$ has degree 5
$-7a^2b$ has degree 3
$19ab$ has degree 2

25. The polynomial $4m^7np^3+24m^5n^4p^3+14$ has degree 12.

$4m^7np^3$ has degree 11
$24m^5n^4p^3$ has degree 12
14 has degree 0

26. The polynomial $5t+8$ has degree 1.

$5t$ has degree 1
8 has degree 0

27. $7x^2y-4x+10$ is a polynomial.

28. $3\sqrt{x}-5x+2 = 3x^{\frac{1}{2}}-5x+2$ is not a polynomial because the variable x is raised to the $\frac{1}{2}$ power.

29. The constant 204 is a polynomial.

30. $2x+5x^{-1}$ is not a polynomial because the variable x is raised to a negative exponent.

31. $U(t)-C(t)=(U-C)(t)$

Input units: years since 1900

Output units: number of adults 20-years-old or older in the United States.

32. $R(l)-C(l)=(R-C)(l)$

Input units: number of limousines per year

Output units: profit in thousands of dollars

33. $E(t)S(t)=ES(t)$

Input units: the year

Output units: total number of sick days taken by Disneyland employees

34. $W(E(t))$

Input units: the year

Output units: annual worker's compensation insurance cost at Disneyland

35. $M(t)+R(t)=(M+R)(t)$

Input units: the year

Output units: total amount spent on cancer treatments and research in the U.S.

36. $f(x)=6x+3$ and $g(x)=-4x+8$

a.

$f(x)+g(x)=(6x+3)+(-4x+8)$
$=6x+3+(-4x)+8$
$=2x+11$

b.

$$f(x) - g(x) = (6x+3) - (-4x+8)$$
$$= 6x + 3 + 4x - 8$$
$$= 10x - 5$$

c.

$$f(x)g(x) = (6x+3)(-4x+8)$$
$$= -24x^2 + 48x - 12x + 24$$
$$= -24x^2 + 36x + 24$$

37. $f(x) = 15x + 34$ and $g(x) = -17x + 34$

a.

$$f(x) + g(x) = (15x + 34) + (-17x + 34)$$
$$= 15x + 34 + (-17x) + 34$$
$$= -2x + 68$$

b.

$$g(x) - f(x) = (-17x + 34) - (15x + 34)$$
$$= -17x + 34 - 15x - 34$$
$$= -32x$$

c.

$$f(x)g(x) = (15x + 34)(-17x + 34)$$
$$= -255x^2 + 510x - 578x + 1156$$
$$= -255x^2 - 68x + 1156$$

d.

$$\frac{f(x)}{g(x)} = \frac{15x + 34}{-17x + 34}$$

38. $f(x) = \dfrac{2}{5}x + \dfrac{3}{5}$ and $g(x) = 4x - 7$

a.

$$f(x) + g(x) = \left(\frac{2}{5}x + \frac{3}{5}\right) + (4x - 7)$$
$$= \frac{2}{5}x + \frac{3}{5} + 4x - 7$$
$$= 4.4x - 6.4$$

b.

$$f(x) - g(x) = \left(\frac{2}{5}x + \frac{3}{5}\right) - (4x - 7)$$
$$= \frac{2}{5}x + \frac{3}{5} - 4x + 7$$
$$= -3.6x + 7.6$$

c.

$$f(x)g(x) = \left(\frac{2}{5}x + \frac{3}{5}\right)(4x - 7)$$
$$= 1.6x^2 - 2.8x + 2.4x - 4.2$$
$$= 1.6x^2 - 0.4x - 4.2$$

39. $f(x) = \dfrac{1}{7}x + \dfrac{5}{7}$ and $g(x) = 2x - 7$

a.

$$f(4) + g(4) = \left[\frac{1}{7}(4) + \frac{5}{7}\right] + [2(4) - 7]$$
$$= \left(\frac{4}{7} + \frac{5}{7}\right) + (8 - 7)$$
$$= \left(\frac{9}{7}\right) + (1)$$
$$= \frac{16}{7}$$

b.

$$f(4) - g(4) = \left[\frac{1}{7}(4) + \frac{5}{7}\right] - [2(4) - 7]$$
$$= \left(\frac{4}{7} + \frac{5}{7}\right) - (8 - 7)$$
$$= \left(\frac{9}{7}\right) - (1)$$
$$= \frac{2}{7}$$

c.

$$f(4)g(4) = \left[\frac{1}{7}(4) + \frac{5}{7}\right][2(4) - 7]$$
$$= \left(\frac{4}{7} + \frac{5}{7}\right)(8 - 7)$$
$$= \left(\frac{9}{7}\right)(1)$$
$$= \frac{9}{7}$$

40. Adjust the data to fit the definitions in parts (a) and (b).

L1	L2	L3	1
6	15.4	1.17	
7	15.75	1.06	
8	16.22	.98	
9	16.5	.93	
10	16.86	.84	
11	17.25	.81	

L1(7)=

a. Let $L(t)$ be the number of people in California's labor force, in millions, t years since 1990.

Using two points: $(6, 15.4)$ and $(11, 17.25)$

find the slope: $m = \dfrac{17.25 - 15.4}{11 - 6} = 0.37$

$L(t) = 0.37t + b$
$15.4 = 0.37(6) + b$
$15.4 = 2.22 + b$
$15.4 - 2.22 = 2.22 - 2.22 + b$
$13.2 \approx b$
$L(t) = 0.37t + 13.2$

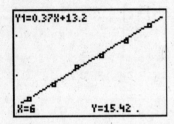

b. Let $U(t)$ be the number of people in California who are considered unemployed, in millions, t years since 1990.

Using two points: $(7, 1.06)$ and $(9, 0.93)$

Find the slope: $m = \dfrac{0.93 - 1.06}{9 - 7} \approx -0.07$

$U(t) = -0.07t + b$
$0.93 = -0.07(9) + b$
$0.93 = -0.63 + b$
$0.93 + 0.63 = -0.63 + 0.63 + b$
$1.56 = b$
$U(t) = -0.07t + 1.56$

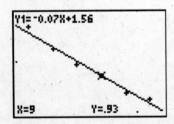

c. $L(15) = 0.37(15) + 13.2 = 18.75$. In 2005, there were about 18.75 million people in California's labor force.

d. $U(15) = -0.07(15) + 1.56 = 0.51$. In 2005, there were about 0.51 million people (510,000 people) in California who are considered unemployed. This may be too low and therefore be considered model breakdown.

e. Let $P(t)$ be the percent of California's labor force that is considered unemployed, t years since 1990.

$P(t) = \dfrac{U(t)}{L(t)} = \dfrac{-0.07t + 1.56}{0.37t + 13.2}$

f.

$P(8) = \dfrac{-0.07(8) + 1.56}{0.37(8) + 13.2} \approx 0.062$. In 1998, 6.2% of California's labor force was considered unemployed.

$P(9) = \dfrac{-0.07(9) + 1.56}{0.37(9) + 13.2} \approx 0.056$. In 1999, 5.6% of California's labor force was considered unemployed.

$P(10) = \dfrac{-0.07(10) + 1.56}{0.37(10) + 13.2} \approx 0.051$. In 2000, 5.1% of California's labor force was considered unemployed.

$P(15) = \dfrac{-0.07(15) + 1.56}{0.37(15) + 13.2} \approx 0.027$. In 2005, 2.7% of California's labor force was considered unemployed. This may be too low and therefore be considered model breakdown.

41. $f(x) = 6x + 3$ and $g(x) = -4x + 8$

a.
$f(g(x)) = f(-4x + 8)$
$ = 6(-4x + 8) + 3$
$ = -24x + 48 + 3$
$ = -24x + 51$

b.
$g(f(x)) = g(6x + 3)$
$ = -4(6x + 3) + 8$
$ = -24x - 12 + 8$
$ = -24x - 4$

42. $f(x) = 15x + 34$ and $g(x) = -17x + 34$

a.
$f(g(x)) = f(-17x + 34)$
$ = 15(-17x + 34) + 34$
$ = -255x + 510 + 34$
$ = -255x + 544$

b.
$g(f(x)) = g(15x + 34)$
$ = -17(15x + 34) + 34$
$ = -255x - 578 + 34$
$ = -255x - 544$

43. $f(x) = \frac{2}{5}x + \frac{3}{5}$ and $g(x) = 4x - 7$

a.
$$(f \circ g)(x) = f(g(x))$$
$$= \frac{2}{5}(4x - 7) + \frac{3}{5}$$
$$= \frac{8}{5}x - \frac{14}{5} + \frac{3}{5}$$
$$= \frac{8}{5}x - \frac{11}{5}$$

b.
$$(g \circ f)(x) = g(f(x))$$
$$= 4\left(\frac{2}{5}x + \frac{3}{5}\right) - 7$$
$$= \frac{8}{5}x + \frac{12}{5} - 7$$
$$= \frac{8}{5}x - \frac{23}{5}$$

44. $f(x) = \frac{1}{7}x + \frac{5}{7}$ and $g(x) = 2x - 7$

a.
$$(f \circ g)(x) = f(g(4))$$
$$= f(1) \quad \text{*Since } g(4) = 2(4) - 7 = 1$$
$$= \frac{1}{7}(1) + \frac{5}{7}$$
$$= \frac{1}{7} + \frac{5}{7}$$
$$= \frac{6}{7}$$

b.
$$(g \circ f)(4) = g(f(4))$$
$$= g\left(\frac{9}{7}\right) \quad \text{*Since } f(4) = \frac{1}{7}(4) + \frac{5}{7} = \frac{9}{7}$$
$$= 2\left(\frac{9}{7}\right) - 7$$
$$= \frac{18}{7} - 7$$
$$= -\frac{31}{7}$$

45. $f(x) = 0.6x + 2.5$ and $g(x) = 3.5x + 3.7$

a.
$$f(-3) + g(-3) = [0.6(-3) + 2.5] + [3.5(-3) + 3.7]$$
$$= (-1.8 + 2.5) + (-10.5 + 3.7)$$
$$= (0.7) + (-6.8)$$
$$= -6.1$$

b.
$$g(-3) - f(-3) = [3.5(-3) + 3.7] - [0.6(-3) + 2.5]$$
$$= (-10.5 + 3.7) - (-1.8 + 2.5)$$
$$= (-6.8) - (0.7)$$
$$= -7.5$$

c.
$$f(-3)g(-3) = [0.6(-3) + 2.5][3.5(-3) + 3.7]$$
$$= (-1.8 + 2.5)(-10.5 + 3.7)$$
$$= (0.7)(-6.8)$$
$$= -4.76$$

d.
$$f(g(-3)) = f(-6.8) \quad \text{*Since } g(-3) = 3.5(-3) + 3.7 = -6.8$$
$$= 0.6(-6.8) + 2.5$$
$$= -4.08 + 2.5$$
$$= -1.58$$

e.
$$g(f(-3)) = g(0.7) \quad \text{*Since } f(-3) = 0.6(-3) + 2.5 = 0.7$$
$$= 3.5(0.7) + 3.7$$
$$= 2.45 + 3.7$$
$$= 6.15$$

46. $f(x) = 0.35x - 2.78$ and $g(x) = 2.4x - 6.3$

a.
$$f(2) + g(2) = [0.35(2) - 2.78] + [2.4(2) - 6.3]$$
$$= (0.7 - 2.78) + (4.8 - 6.3)$$
$$= (-2.08) + (-1.5)$$
$$= -3.58$$

b.
$$g(2) - f(2) = [2.4(2) - 6.3] - [0.35(2) - 2.78]$$
$$= (4.8 - 6.3) - (0.7 - 2.78)$$
$$= (-1.5) - (-2.08)$$
$$= 0.58$$

Chapter 3 Exponents, Polynomials, and Functions

c.

$$f(2)g(2) = [0.35(2)-2.78][2.4(2)-6.3]$$
$$= (0.7-2.78)(4.8-6.3)$$
$$= (-2.08)(-1.5)$$
$$= 3.12$$

d.

$$f(g(2)) = f(-1.5) \quad \text{*Since } g(2) = 2.4(2)-6.3 = -1.5$$
$$= 0.35(-1.5)-2.78$$
$$= -0.525-2.78$$
$$= -3.305$$

47.

$x^2 + 2x - 35$
$ac = 1 \times (-35) = -35$
Find the factors of -35 that sum to 2:
$-5 \times 7 = -35$
$-5 + 7 = 2$
We get:

$$x^2 + 2x - 35 = \underline{x^2 - 5x} + \underline{7x - 35}$$
$$= x(x-5) + 7(x-5)$$
$$= (x-5)(x+7)$$

48.

$x^2 + 6x + 9 = x^2 + 6x + 3^2$ Since the middle term is $6x = 2(3 \cdot x)$
$\qquad = (x+3)^2$

49.

$$6x^2 - 16x + 21x - 56 = \underline{6x^2 - 16x} + \underline{21x - 56}$$
$$= 2x(3x-8) + 7(3x-8)$$
$$= (3x-8)(2x+7)$$

50.

$5b^2 - 14b - 3$
$ac = 5 \times (-3) = -15$
Find the factors of -15 that sum to -14:
$-15 \times 1 = -15$
$-15 + 1 = -14$
We get:

$$5b^2 - 14b - 3 = \underline{5b^2 - 15b} + \underline{b - 3}$$
$$= 5b(b-3) + 1(b-3)$$
$$= (b-3)(5b+1)$$

51.

$6p^2 + 23p + 20$
$ac = 6 \times 20 = 120$
Find the factors of 120 that sum to 23:
$15 \times 8 = 120$
$15 + 8 = 23$
We get:

$$6p^2 + 23p + 20 = \underline{6p^2 + 15p} + \underline{8p + 20}$$
$$= 3p(2p+5) + 4(2p+5)$$
$$= (2p+5)(3p+4)$$

52.

$14k^2 - 21k + 7 = 7(2k^2 - 3k + 1)$
$ac = 2 \times 1 = 2$
Find the factors of 2 that sum to -3:
$(-1) \times (-2) = 2$
$(-1) + (-2) = -3$
We get:

$$14k^2 - 21k + 7 = 7(2k^2 - 3k + 1)$$
$$= 7(\underline{2k^2 - k} - \underline{2k + 1})$$
$$= 7[k(2k-1) - 1(2k-1)]$$
$$= 7(2k-1)(k-1)$$

53.

$20x^3 - 52x^2 - 24x = 4x(5x^2 - 13x - 6)$
$ac = 5 \times (-6) = -30$
Find the factors of -30 that sum to -13:
$2 \times (-15) = -30$
$2 + (-15) = -13$
We get:

$$20x^3 - 52x^2 - 24x = 4x(5x^2 - 13x - 6)$$
$$= 4x(\underline{5x^2 + 2x} - \underline{15x - 6})$$
$$= 4x[x(5x+2) - 3(5x+2)]$$
$$= 4x(5x+2)(x-3)$$

54.

$$a^2 - a + 6ab - 6b = \underline{a^2 - a} + \underline{6ab - 6b}$$
$$= a(a-1) + 6b(a-1)$$
$$= (a-1)(a+6b)$$

55.
$$9m^2 - 100 = (3m)^2 - 10^2$$
$$= (3m+10)(3m-10)$$

56. Prime. The sum of two squares does not factor with real numbers.

57.
$$9x^2 + 30x + 25 = (3x)^2 + 30x + 5^2$$
Since the middle term is $30x = 2(5 \cdot 3x)$
$$= (3x+5)^2$$

58.
$$27x^3 - 1 = (3x)^3 - 1^3$$
$$= (3x-1)\left((3x)^2 + 1 \cdot 3x + 1^2\right)$$
$$= (3x-1)(9x^2 + 3x + 1)$$

59.
$$125w^3 + 8x^3 = (5w)^3 + (2x)^3$$
$$= (5w+2x)\left((5w)^2 - 5w \cdot 2x + (2x)^2\right)$$
$$= (5w+2x)(25w^2 - 10wx + 4x^2)$$

60.
$$t^6 - 64 = (t^3)^2 - 8^2$$
$$= (t^3+8)(t^3-8)$$
$$= (t+2)(t^2-2t+4) \cdot (t-2)(t^2+2t+4)$$
$$= (t+2)(t-2)(t^2-2t+4)(t^2+2t+4)$$

61. Let $u = m^3$.
$$10m^6 - 29m^3 + 21 = 10(m^3)^2 - 29m^3 + 21$$
$$= 10u^2 - 29u + 21$$
$$= (5u-7)(2u-3)$$
Replace u with m^3
$$= (5m^3-7)(2m^3-3)$$

62.
$$80h^2 - 125 = 5(16h^2 - 25)$$
$$= 5\left((4h)^2 - 5^2\right)$$
$$= 5(4h+5)(4h-5)$$

63.
$$18a^2b + 84ab + 98b = 2b(9a^2 + 42a + 49)$$
$$= 2b\left((3a)^2 + 42a + 7^2\right)$$
Since the middle term is $42a = 2(7 \cdot 3a)$
$$= 2b(3a+7)^2$$

64.
$$3r^4 - 48 = 3(r^4 - 16)$$
$$= 3r\left((r^2)^2 - 4^2\right)$$
$$= 3r(r^2+4)(r^2-4)$$
$$= 3r(r^2+4)(r^2-2^2)$$
$$= 3(r^2+4)(r+2)(r-2)$$

Chapter 3 Exponents, Polynomials, and Functions
Chapter 3 Test

1.

$$\left(2b^4c^{-2}\right)^5\left(3b^{-3}c^{-4}\right)^{-2} = 2^5 b^{20} c^{-10} \cdot 3^{-2} b^6 c^8$$
$$= \frac{32 b^{20+6} c^{-10+8}}{9}$$
$$= \frac{32 b^{26} c^{-2}}{9}$$
$$= \frac{32 b^{26}}{9c^2}$$

2.

$$\left(\frac{16b^{12}c^2}{2b^{-3}c^{-4}}\right)^{-\frac{1}{3}} = \left(8b^{12-(-3)}c^{2-(-4)}\right)^{-\frac{1}{3}}$$
$$= \left(8b^{15}c^6\right)^{-\frac{1}{3}}$$
$$= \frac{1}{\left(8b^{15}c^6\right)^{\frac{1}{3}}}$$
$$= \frac{1}{8^{\frac{1}{3}} b^{\frac{15}{3}} c^{\frac{6}{3}}}$$
$$= \frac{1}{2b^5 c^2}$$

3.

$$\frac{25x^{-9} y^{-8}}{35x^{-10} y^{-3}} = \frac{5 x^{-9-(-10)} y^{-8-(-3)}}{7}$$
$$= \frac{5xy^{-5}}{7}$$
$$= \frac{5x}{7y^5}$$

4.

a. $T(9) = 1.1(9) + 6.84 = 16.74$. In 1999, New Zealand spent about 16.74 million New Zealand dollars on the treatment of diabetes.

b.
$$10 = 1.3t + 0.52$$
$$10 - 0.52 = 1.3t + 0.52 - 0.52$$
$$9.48 = 1.3t$$
$$\frac{9.48}{1.3} = \frac{1.3t}{1.3}$$
$$7.3 \approx t$$

New Zealand spent about 10 million New Zealand dollars on diabetes research in 1997.

c. Let $D(t)$ be the total New Zealand spent on diabetes research and treatment, in millions of New Zealand dollars, t years since 1990.

$$D(t) = T(t) + R(t)$$
$$D(t) = (1.1t + 6.84) + (1.3t + 0.52)$$
$$D(t) = 2.4t + 7.36$$

d. $D(10) = 2.4(10) + 7.36 = 31.36$. New Zealand spent about 31.36 million New Zealand dollars on research and treatment of diabetes in 2000.

5. $f(x) = 4x + 17 \quad g(x) = 2x - 7$

a.
$$f(x) - g(x) = (4x + 17) - (2x - 7)$$
$$= 4x + 17 - 2x + 7$$
$$= 2x + 24$$

b.
$$f(x)g(x) = (4x + 17)(2x - 7)$$
$$= 8x^2 - 28x + 34x - 119$$
$$= 8x^2 + 6x - 119$$

c.
$$f(g(x)) = 4(2x - 7) + 17$$
$$= 8x - 28 + 17$$
$$= 8x - 11$$

6. $f(x) = 2.35x + 1.45 \quad g(x) = 2.4x - 6.3$

a.
$$(f+g)(4) = (2.35(4) + 1.45) + (2.4(4) - 6.3)$$
$$= 9.4 + 1.45 + 9.6 - 6.3$$
$$= 14.15$$

b.
$$fg(-2) = (2.35(-2) + 1.45)(2.4(-2) - 6.3)$$
$$= (-3.25)(-11.1)$$
$$= 36.075$$

c.
$$(f \circ g)(6) = 2.35(2.4(6) - 6.3) + 1.45$$
$$= 2.35(8.1) + 1.45$$
$$= 20.485$$

d.

$g(f(0)) = 2.4(2.35(0) + 1.45) - 6.3$
$= 2.4(1.45) - 6.3$
$= -2.82$

7.

a.

$S(10) = 1.2(10)^2 + 4.8(10) + 376.6$
$S(10) = 544.6$

In 2000, there were approximately 544.6 thousand science/engineering graduate students in doctoral programs.

b. Let $M(t)$ be the number of male science/engineering graduate students in doctoral programs, in thousands, t years since 2000.

$M(t) = S(t) - F(t)$
$= 1.2t^2 + 4.8t + 376.6$
$\underline{-0.4t^2 - 4.1t - 150.9}$
$M(t) = 0.8t^2 + 0.7t + 225.7$

c.

$M(10) = 0.8(10)^2 + 0.7(10) + 225.7$
$M(10) = 312.7$

In 2010, there were approximately 312.7 thousand male science/engineering graduate students in doctoral programs.

8.

a.

$P(6) = -0.3(6)^3 + 3.55(6)^2 - 10.41(6) + 51.23$
$P(6) = 51.77$

In 2006, about 51.8% of murders were committed by using a handgun.

b. Let $H(t)$ be the number of murders in thousands, that were committed using a handgun t years since 2000.

$H(t) = M(t)\dfrac{P(t)}{100}$

$= (-0.02t^3 + 0.28t^2 - 11t + 13.23)\dfrac{-0.3t^3 + 3.55t^2 - 10.41t + 51.23}{100}$

$= (-0.02t^3 + 0.28t^2 - 11t + 13.23)(-0.003t^3 + 0.0355t^2 - 0.1041t + 0.5123)$

$= 0.00006t^6 - 0.0013t^5 + 0.0095t^4 - 0.075t^3 + 0.58t^2 - 1.43t + 6.78$

c.

$H(7) = 0.00006(7)^6 - 0.0013(7)^5 + 0.0095(7)^4 - 0.075(7)^3$
$+ 0.58(7)^2 - 1.43(7) + 6.78 \approx 7.5$

In 2007, about 7.5 thousand murders were committed using a handgun.

9. $(T - B)(t)$ is the number of girls attending the tennis camp in year t.

10. $M(B(t))$ is the number of boys' matches at the camp in year t.

11. $C(T(t))$ is the total cost for the camp in year t.

12. $A(t) \cdot M(T(t))$ is the time it takes for all the matches to be played in year t.

13.

$(4x^2 + 2x - 7) - (3x^2 + 9)$
$4x^2 + 2x - 7 - 3x^2 - 9$
$x^2 + 2x - 16$

14.

$(4x^2y + 3xy - 2y^2) + (5x^2y - 7xy + 8)$
$9x^2y - 4xy - 2y^2 + 8$

15.

$(2x + 8)(3x - 7)$
$6x^2 - 14x + 24x - 56$
$6x^2 + 10x - 56$

16.

$(3a - 4b)(2a - 5b)$
$6a^2 - 15ab - 8ab + 20b^2$
$6a^2 - 23ab + 20b^2$

17. The polynomial $7x^2 + 8x - 10$ has three terms.

$7x^2$: variable term, coefficient 7
$8x$: variable term, coefficient 8
-10 : constant term

18. The polynomial $14m^3n^6 - 8m^2n + 205$ has three terms.

$14m^3n^6$: variable term, coefficient 14
$-8m^2n$: variable term, coefficient -8
205 : constant term

Chapter 3 Exponents, Polynomials, and Functions

19. The polynomial $7x^2 + 8x - 10$ has degree two.

$7x^2$ has degree 2
$8x$ has degree 1
-10 has degree 0

20. The polynomial $14m^3n^6 - 8m^2n + 205$ has degree nine.

$14m^3n^6$ has degree 9
$-8m^2n$ has degree 3
205 has degree 0

21. $5x^2 + \dfrac{6}{x} - 12 = 5x^2 + 6x^{-1} - 12$; No, it is not a polynomial, because the second term has a negative exponent on the variable.

22. $11a^3b^2 - 17ab + 12b$; Yes, it is a polynomial.

23.

$x^2 - 14x + 45$
$ac = 1 \times 45 = 45$
Find the factors of 45 that sum to -14:
$-9 \times (-5) = 45$
$-9 + (-5) = -14$
We get:
$x^2 - 14x + 45 = \underline{x^2 - 9x - 5x + 45}$
$ = x(x-9) - 5(x-9)$
$ = (x-9)(x-5)$

24.

$2p^3 + 8p^2 - 42p = 2p(p^2 + 4p - 21)$
$ac = 1 \times (-21) = -21$
Find the factors of -21 that sum to 4:
$7 \times (-3) = -21$
$7 + (-3) = 4$
We get:
$2p^3 + 8p^2 - 42p = 2p(p^2 + 4p - 21)$
$ = 2p\underline{(p^2 + 7p - 3p - 21)}$
$ = 2p[p(p+7) - 3(p+7)]$
$ = 2p(p+7)(p-3)$

25.

$12m^2 - 54m + 22mn - 99n = (12m^2 - 54m) + (22mn - 99n)$
$ = 6m(2m-9) + 11n(2m-9)$
$ = (2m-9)(6m+11n)$

26.

$36b^2 - 49 = (6b)^2 - 7^2$
$ = (6b+7)(6b-7)$

27.

$t^2 - 20t + 100 = t^2 - 20t + 10^2$
Since the middle term $20t = 2(10 \cdot t)$
this is a perfect square trinomial.
$ = (t-10)^2$

28.

$7x^3 - 28x = 7x(x^2 - 4)$
$ = 7x(x^2 - 2^2)$
$ = 7x(x+2)(x-2)$

29.

$8m^3 + 125n^3 = (2m)^3 + (5n)^3$
$ = (2m+5n)\left((2m)^2 - 2m \cdot 5n + (5n)^2\right)$
$ = (2m+5n)(4m^2 - 10mn + 25n^2)$

30.

$3x^2 + x + 10$
$ac = 3 \times 10 = 30$
Since there are no factors of 30 that sum to 1, the quadratic is prime.

31. Let $u = m^4$.

$6m^8 - 29m^4 + 28 = 6(m^4)^2 - 29m^4 + 28$
$ = 6u^2 - 29u + 28$
$ = (3u-4)(2u-7)$
Replace u with m^4.
$ = (3m^4 - 4)(2m^4 - 7)$

32.

$a^3 - 125b^3 = a^3 - (5b)^3$
$ = (a-5b)(a^2 + a \cdot 5b + (5b)^2)$
$ = (a-5b)(a^2 + 5ab + 25b^2)$

Section 4.1

1. This function is a quadratic function, because the term with the highest degree is 2.

3. This function is a linear function, because the term with the highest degree is 1.

5. This function is a quadratic function, because the term with the highest degree is 2.

7. Other, because the variable is in the exponent.

9. This function is a quadratic function, because the term with the highest degree is 2 once the polynomial is simplified.

11. This function is a quadratic function, because the term with the highest degree is 2.

13. This function is a quadratic function, because the term with the highest degree is 2 once the polynomial is simplified.

15.

a. $(-5,1)$

b. $x < -5$

c. $x > -5$

d. The graph crosses the horizontal x-axis at $(-7.5, 0)$ and $(-2.5, 0)$.

e. The graph crosses the vertical y-axis at $(0, -3)$.

17.

a. $(4, 25)$

b. $x < 4$

c. $x > 4$

d. The graph crosses the x-axis at $(-1, 0)$ and $(9, 0)$.

e. The graph crosses the y-axis at $(0, 9)$.

f. $f(2) = 21$, because when $x = 2$, $y = 21$.

19.

a. $(-1, 4)$

b. $x > -1$

c. $x < -1$

d. None, because the horizontal intercept is the point at which the graph crosses the x-axis and the graph does not cross the x-axis.

e. The graph crosses the vertical axis at $(0, 5)$.

f. $f(-4) = 9.5$, because when $x = -4$, $y = 9.5$.

g. $f(x) = 14$ for $x = 3$ and $x = -5$.

21. Horizontal intercepts: $(-4, 0)$ and $(3, 0)$, vertical intercept: $(0, 6)$

23. Horizontal intercepts: $(-10, 0)$ and $(15, 0)$, vertical intercept: $(0, -15)$

25.

a. Let $D(m)$ be the average number of days above $70°F$ in San Diego, California, during month m (i.e., $m = 1$ represents January).

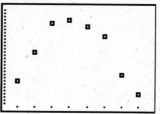

b. A quadratic model would fit best. The data looks like a downward facing parabola.

c. The vertex is $(8, 31)$, which is a maximum point. Thus, August has the most number of days, on average, above $70°F$ in San Diego, California.

d. Using the shape of the distribution, there would be about seven days above $70°F$ in San Diego, California, during the month of April.

27.

a.

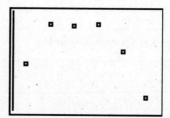

b.
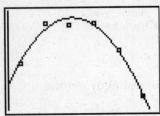

c. This model fits the data reasonably well; the point at the top is low for a vertex, but otherwise the data is shaped like a quadratic.

d. The vertex for the model is at about $(4, 11{,}238)$ and is a maximum point for this function.

e. This vertex represents that in 1994, the number of cable television systems in the United States reached a maximum of 11,238.

29.

a.
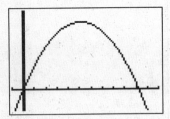

b. The vertex of this parabola is at about (5, 128.5), which means that the tennis ball will reach its highest point in the air at 128.5 meters after about 5 seconds.

c. $H(2) = 81.4$. About 2 seconds after being hit, the tennis ball will reach a height of 81.4 meters.

d. According to the graph, the tennis ball will hit the ground at about 10.224 seconds.

31.

a. According to the graph, the revenue from selling T-shirts for $10 each would be about $1000.

b. According to the graph, the maximum monthly revenue would be about $1100.

c. The bookstore should charge $15 for each T-shirt to maximize the monthly revenue.

d. The revenue may go down after the vertex in this situation because people are less likely to purchase these T-shirts for more than $15 each, according to this graph.

33. A quadratic model would best fit this data.

The vertex is $(-5, 8)$, which is a maximum point.

35. A quadratic model would best fit these data.

The vertex is $(5, -5)$, which is a minimum point.

37. A linear model would best fit these data.

39. A linear model would best fit these data.

41.
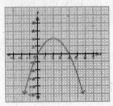

a. There are two horizontal intercepts.

b. There is one vertical intercept.

43.
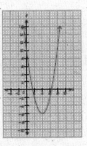

a. There are two horizontal intercepts.

b. There is one vertical intercept.

45.
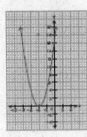

The vertex must be on the horizontal axis for the parabola to face upward and have one horizontal intercept.

47.

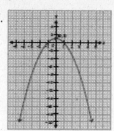

a. The vertex is above the horizontal axis.

b. The vertex must be below the horizontal axis for the parabola to face down and have no horizontal intercepts.

Chapter 4 Quadratic Functions
Section 4.2

1. h; positive.

3. k; negative.

5. $|a| > 1$

7. a; positive

9. $h = 0, k = 0$

11.
a. $(-2.5, -3)$ b. $x = -2.5$
c. $(-5.5, 4)$

13.
a. $(5.5, -6)$ b. $x = 5.5$
c. $(2, 5)$

15.
a. $(-4, -8)$ b. $x = -4$
c. positive   d. $h = -4, k = -8$
e. $(2, 15)$

17.
a. $(2, 10)$ b. $x = 2$
c. negative   d. $h = 2, k = 10$
e. $(5, -6)$

19.
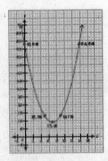

21.
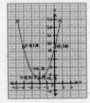

23.
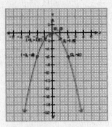

25.
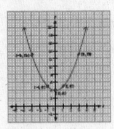

27.
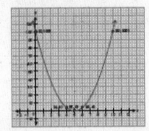

29.
a.
$P(90) = 4.95(90-57)^2 + 1406$
$P(90) = 6796.55$

In 1990, the poverty threshold was about $6,796.55 for individuals under 65 years old in the United States.

b.
$P(80) = 4.95(80-57)^2 + 1406$
$P(80) = 4024.55$

In 1980, the poverty threshold was about $4,024.55 for individuals under 65 years old in the United States.

c.
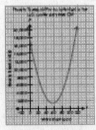

d. According to this model, the poverty threshold for individuals under 65 years old in the United States reached a minimum of about $1,406 in 1957, which seems to make sense.

e. According to the graph, the poverty threshold for individuals under 65 years old in the United States was $3000 in 1975 and 1939; 1939 shows model breakdown.

f. Domain: $[57,90]$. Within this domain, the minimum value of 1406 comes from the vertex. The maximum value comes from $t = 90$.

$P(57) = 4.95(57-57)^2 + 1406 = 1406$
$P(90) = 4.95(90-57)^2 + 1406 = 6796.55$
Range: $[1406, 6796.55]$

31.
a.

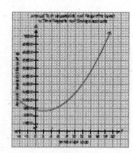

b.
$D(6) = 17(6-3)^2 + 2300$
$D(6) = 2453$

In 1996, households and nonprofit organizations invested about $2453 billion in time and savings accounts.

c. In about 1993, these time and savings accounts were at their lowest levels with about $2300 billion invested.

d. According to the graph, households and nonprofit organizations invested $3000 billion in about 1987 and again in 1999.

e. Domain: $[0,18]$. Within this domain, the minimum value of 2300 comes from the vertex. The maximum value comes from $t = 18$.

$D(18) = 17(18-3)^2 + 2300$
$D(18) = 6125$
Range: $[2300, 6125]$

33.
a. $h(2) = -16(2)^2 + 256 = 192$

Two seconds after a ball is dropped from the roof of the building, the ball will be at a height of about 192 feet.

b.

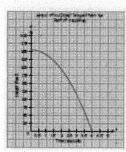

c. According to the graph, the ball will hit the ground about 4 seconds after being dropped from the roof of the building.

d. Domain: $[0,4]$. Within this domain, the maximum value of 256 comes from the vertex at time $t = 0$. The minimum value comes from $t = 4$ when the object hits the ground
Range: $[0, 256]$

35.
a.
$P(100) = -0.000025(100-120)^2 + 2.75$
$P(100) = 2.74$

The monthly profit made from selling round trip airline tickets from New York City to Orlando, Florida, when the tickets are sold for $100 each is $2.74 million.

b. From the vertex we see that selling round trip airline tickets from New York City to Orlando, Florida for $120 each will produce a maximum profit of $2.75 million.

c.

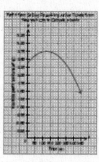

Chapter 4 Quadratic Functions

d. According to the graph, the company will have a monthly profit of about $2 million when selling round trip airline tickets from New York City to Orlando, Florida, for $295 each and –$50; –$50 is model breakdown.

e. Domain: $[60, 300]$. Within this domain, the maximum value of 2.75 comes from the vertex. The minimum value comes from $t = 300$.

$P(300) = -0.000025(300 - 120)^2 + 2.75$
$P(300) = 1.95$
Range: $[1.95, 2.75]$

37.
a.
$E(4) = -0.006(4 - 10)^2 + 8$
$E(4) = 7.784$

The epidemic threshold for the fourth week of 2006 was 7.784%

b.

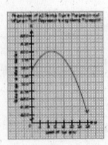

c. According to the graph, in the 23rd week of 2006, there was not a flu epidemic; the epidemic threshold was about 7%.

d. The vertex for this model is (10, 8), which means that in week 10 of 2006, the epidemic threshold reached a maximum of 8%.

e. Domain: $[1, 26]$. Within this domain, the maximum value of 8% comes from the vertex. The minimum value comes from $n = 0$.

$E(26) = -0.006(26 - 10)^2 + 8$
$E(26) = 6.464$
Range: $[6.464, 8]$

39. Domain: $(-\infty, \infty)$, range: $(-\infty, 15]$

41. Domain: $(-\infty, \infty)$, range: $[-58, \infty)$

43. Domain: $(-\infty, \infty)$, range: $[0, \infty)$

45. Domain: $(-\infty, \infty)$, range: $(-\infty, 4]$

47. Domain: $(-\infty, \infty)$, range: $y \geq -9$

The domain should be all real numbers. The range is using a greater than sign, but it should use a greater than or equal to sign.

49. Domain: $(-\infty, \infty)$, range: $(-\infty, 7]$

The domain is correct, but the range is assuming that the parabola is facing upward and there is a parenthesis on the 7 instead of a bracket.

51. Domain: $(-\infty, \infty)$, range: $[-4, \infty)$

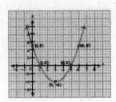

53. Domain: $(-\infty, \infty)$, range: $[0, \infty)$

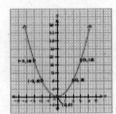

55. Domain: $(-\infty, \infty)$, range: $[2, \infty)$

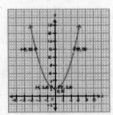

57. Domain: $(-\infty, \infty)$, range: $(-\infty, 0]$

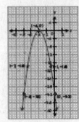

59. Domain: $(-\infty,\infty)$, range: $[2,\infty)$

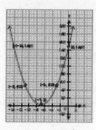

61. Domain: $(-\infty,\infty)$, range: $[-8,\infty)$

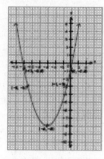

63. Domain: $(-\infty,\infty)$, range: $(-\infty,-15]$

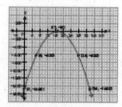

65. Domain: $(-\infty,\infty)$, range: $(-\infty,25]$

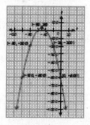

67. Domain: $(-\infty,\infty)$, range: $[-2500,\infty)$

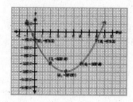

69. $f(x) = 5x^2 + 100$

a. Vertex: $(0,100)$

b. The parabola is narrow.

c. The parabola faces upward.

d. X Min: −10   X Max: 10

    Y Min: 0   Y Max: 600

71. $f(x) = (x+30)^2 - 50$

a. Vertex: $(-30,-50)$

b. The parabola is neither wide nor narrow.

c. The parabola faces upward.

d. X Min: −45   X Max: −10

    Y Min: −60   Y Max: 70

73. $f(x) = 0.002(x+20)^2 + 50$

a. Vertex: $(-20,50)$

b. The parabola is wide.

c. The parabola faces upward.

d. X Min: −100   X Max: 60

    Y Min: 45   Y Max: 60

75. $f(x) = 0.0005(x-1000)^2 + 1000$

a. Vertex: $(1000,1000)$

b. The parabola is wide.

c. The parabola faces upward.

d. X Min: 0   X Max: 2000

    Y Min: 900   Y Max: 1500

77. $f(x) = -10(x+25,000)^2 - 10,000$

a. Vertex: $(-25,000, -10,000)$

b. The parabola is narrow.

c. The parabola faces downward.

d. X Min: −25,010   X Max: −24,990

    Y Min: −11,000   Y Max: −9900

Chapter 4 Quadratic Functions
Section 4.3

1. $f(x) = 2.0(x+5)^2 - 15$

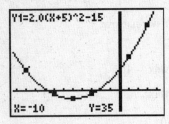

3. $f(x) = 0.4(x-4)^2 + 3.2$

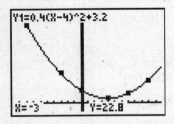

5. $f(x) = 2(x+1.5)^2 - 20$

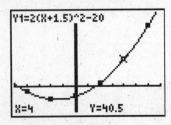

7. Adjusted to: $f(x) = 2.7(x-5)^2 - 8$

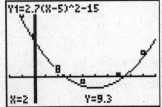 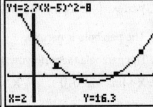

9. Adjusted to: $f(x) = -2(x-7)^2 + 15$

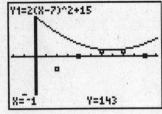 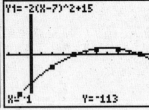

11. Adjusted to: $f(x) = 4(x+2)^2 - 10$

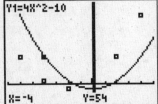 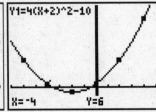

13. Increase k to shift the graph up.

15. Increase h to shift the graph right.

17. Decrease a to make the graph wider.

19. Decrease h to shift the graph left and up.

21.

a. Let $R(p)$ be the revenue in dollars from selling gloves at a price of p dollars.

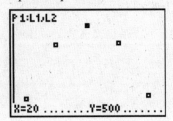

Choose $(20, 500)$ as the vertex, and $(10, 400)$ as the additional point used to solve for "a".

$$R(p) = a(p-20)^2 + 500$$
$$400 = a(10-20)^2 + 500$$
$$400 = a(-10)^2 + 500$$
$$400 = 100a + 500$$
$$\underline{-500 \qquad -500}$$
$$-100 = 100a$$
$$a = -1$$
$$R(p) = -(p-20)^2 + 500$$

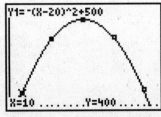

Note: Your answer may vary and yet still represent the trend of the data.

b. The total revenue reaches a maximum of $500 from selling the gloves for $20 a pair.

c.

$$R(22) = -(22-20)^2 + 500$$
$$R(22) = 496$$

This model predicts the revenue will be about $496 if the gloves sell for $22 a pair.

d. Domain: $[6, 35]$. Within this domain, the maximum value of $500 comes from the vertex. The minimum value comes from $p = 35$.

$R(35) = -(35-20)^2 + 500 = 275$
Range: $[275, 500]$

23.

a. Let $T(m)$ be the average monthly low temperatures (in degrees Fahrenheit) in Anchorage, Alaska, during month m (i.e., $m = 1$ represents January).

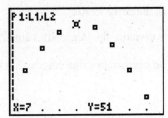

Choose $(7, 51)$ as the vertex, and $(10, 28)$ as the additional point used to solve for "a".

$T(m) = a(m-7)^2 + 51$
$28 = a(10-7)^2 + 51$
$28 = a(3)^2 + 51$
$28 = 9a + 51$
$\underline{-51 \quad\quad -51}$
$-23 = 9a$
$\dfrac{-23}{9} = \dfrac{9a}{9} \bigg\} \; a \approx -2.56$

$T(m) = -2.56(m-7)^2 + 51$

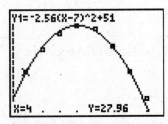

Note: Your answer may vary and yet still represent the trend of the data.

b. The highest average low temperature for Anchorage, Alaska, is about $51°F$ in July.

c.

$T(3) = -2.56(3-7)^2 + 51$
$T(3) = 10.04$

This model predicts the average low temperature in March to be $10°F$.

d. Domain: $[3, 11]$. Within this domain, the maximum value of $51°F$ comes from the vertex. The minimum value comes from $m = 3$.

$T(3) = -2.56(3-7)^2 + 51 \approx 10$
Range: $[10, 51]$

25.

a. Let $F(t)$ be the number of Hispanic families, in thousands, in the United States below the poverty level t years since 1990.

Choose $(9.5, 1520)$ as the vertex, and $(12, 1792)$ as the additional point used to solve for "a".

$F(t) = a(t-9.5)^2 + 1520$
$1792 = a(12-9.5)^2 + 1520$
$1792 = a(2.5)^2 + 1520$
$1792 = 6.25a + 1520$
$\underline{-1520 \quad\quad -1520}$
$272 = 6.25a$
$\dfrac{272}{6.25} = \dfrac{6.25a}{6.25} \bigg\} \; a = 43.52$

$F(t) = 43.52(t-9.5)^2 + 1520$

For a better fit, allow $a = 46$.

$F(t) = 46(t-9.5)^2 + 1520$

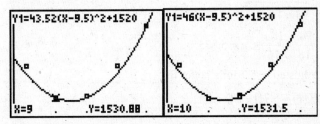

Note: Your answer may vary and yet still represent the trend of the data.

Chapter 4 Quadratic Functions

b.

$F(15) = 46(15-9.5)^2 + 1520$

$F(15) = 2911.5$

In 2005, there were approximately 2,911,500 Hispanic families in the United States under the poverty level.

c. In about 1999 or 2000, the number of Hispanic families in the United States under the poverty level reached a minimum of about 1,520,000. This is probably true only for this close time period.

d. Domain: $[6, 15]$. Within this domain, the minimum value of 1520 comes from the vertex. The maximum value comes from $t = 15$.

$F(15) = 46(15-9.5)^2 + 1520 = 2911.5$

Range: $[1520, 2911.5]$

27.

a. Let $H(t)$ be the number of hours per year the average person spent using the Internet t years since 1990.

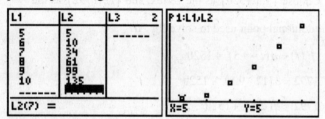

Choose $(5, 5)$ as the vertex, and $(9, 99)$ as the additional point used to solve for "a".

$H(t) = a(t-5)^2 + 5$

$99 = a(9-5)^2 + 5$

$99 = a(4)^2 + 5$

$99 = 16a + 5$

$\underline{-5 \quad\quad -5}$

$94 = 16a$

$\dfrac{94}{16} = \dfrac{16a}{16}\bigg\} \; a = 5.875$

Adjusted to 5.7 for a better fit.

$H(t) = 5.7(t-5)^2 + 5$

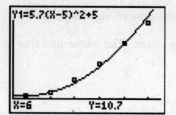

Note: Your answer may vary and yet still represent the trend of the data.

b. If you use this model for years before 1995, the model shows more and more hours spent on the Internet as you go back in time. This is model breakdown because the Internet was just beginning to take off in the early 1990s.

c. Domain: $[5, 12]$. Within this domain, the minimum value of 5 comes from the vertex. The maximum value comes from $t = 12$.

$H(12) = 5.7(12-5)^2 + 5 \approx 284$

Range: $[5, 284]$

d.

$H(11) = 5.7(11-5)^2 + 5$

$H(11) = 210.2$

In 2001, the average person spent 210 hours using the Internet.

e.

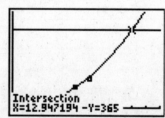

According to this model, people will spend 365 hours per year on the Internet in 2003.

29.

a. Let $H(d)$ be the height of the baseball (in feet) from the ground, and let d be the horizontal distance of the ball (in feet) from home plate.

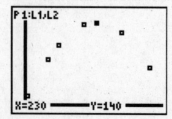

Choose $(230,140)$ as the vertex, and $(400,65)$ as the additional point used to solve for "a".

$H(d) = a(d-230)^2 + 140$
$65 = a(400-230)^2 + 140$
$65 = a(170)^2 + 140$
$65 = 28{,}900a + 140$
$\underline{-140 \qquad\qquad -140}$
$-75 = 28{,}900a$
$\dfrac{-75}{28{,}900} = \dfrac{28{,}900a}{28{,}900} \Big\}\ a \approx -0.0026$
$H(d) = -0.0026(d-230)^2 + 140$

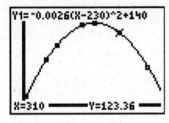

Note: Your answer may vary and yet still represent the trend of the data.

b. Domain: $[0, 400]$. Within this domain, the maximum value of 140 comes from the vertex. The minimum value comes from $h = 0$.

$H(0) = -0.0026(0-230)^2 + 140$
$H(0) = 2.46$
Range: $[2.46, 140]$

c. This vertex of this model is (230, 140), which means that at a distance of 230 feet after being hit, the baseball will be at a maximum height of about 140 feet from the ground.

d.
$H(450) = -0.0026(450-230)^2 + 140$
$H(450) = 14.16$

If the center wall of the stadium is 450 feet from home plate and is 10 feet tall, the ball will make it over the wall.

31. Domain: $(-\infty, \infty)$, range: $(-\infty, 8]$

Vertex $= (-5, 8)$
Other point $= (-3, -8)$
$-8 = a(-3+5)^2 + 8$
$-8 = a(2)^2 + 8$
$-8 = 4a + 8$
$-8 - 8 = 4a + 8 - 8$
$-16 = 4a$
$\dfrac{-16}{4} = \dfrac{4a}{4}$
$a = -4$
Since $h = -5, k = 8$, and $a = -4$
$y = -4(x+5)^2 + 8$

33. Domain: $(-\infty, \infty)$, range: $[-4, \infty)$

Vertex $= (5, -4)$
Other point $= (2, 5)$
$5 = a(2-5)^2 - 4$
$5 = a(-3)^2 - 4$
$5 = 9a - 4$
$5 + 4 = 9a - 4 + 4$
$9 = 9a$
$\dfrac{9}{9} = \dfrac{9a}{9}$
$a = 1$
Since $h = 5, k = -4$, and $a = 1$
$y = (x-5)^2 - 4$

35. Domain: $(-\infty, \infty)$, range: $[-9000, \infty)$

Vertex $= (-55, -9000)$
Other point $= (-30, -1500)$
$-1500 = a(-30+55)^2 - 9000$
$-1500 = a(25)^2 - 9000$
$-1500 = 625a - 9000$
$-1500 + 9000 = 625a - 9000 + 9000$
$7500 = 625a$
$\dfrac{7500}{625} = \dfrac{625a}{625}$
$a = 12$
Since $h = -55, k = -9000$, and $a = 12$
$y = 12(x+55)^2 - 9000$

Chapter 4 Quadratic Functions

37. Domain: $(-\infty,\infty)$, range: $(-\infty,14]$

Vertex $= (6,14)$

Other point $= (4,10)$

$10 = a(4-6)^2 + 14$

$10 = a(-2)^2 + 14$

$10 = 4a + 14$

$10 - 14 = 4a + 14 - 14$

$-4 = 4a$

$\dfrac{-4}{4} = \dfrac{4a}{4}$

$a = -1$

Since $h = 6, k = 14,$ and $a = -1$

$y = -(x-6)^2 + 14$

39. Domain: $(-\infty,\infty)$, range: $(-\infty,\infty)$

This is a linear finction

First point $= (0,-15)$

Other point $= (6, 4.2)$

$y - y_1 = m(x - x_1)$

$4.2 + 15 = m(6 - 0)$

$19.2 = 6m$

$\dfrac{19.2}{6} = \dfrac{6m}{6}$

$m = 3.2$

Since $m = 3.2$ and $b = -15$

$y = 3.2x - 15$

41. Domain: $(-\infty,\infty)$, range: $(-\infty,8]$

Vertex $= (0,8)$

Other point $= (2,2)$

$2 = a(2-0)^2 + 8$

$2 = a(2)^2 + 8$

$2 = 4a + 8$

$2 - 8 = 4a + 8 - 8$

$-6 = 4a$

$\dfrac{-6}{4} = \dfrac{4a}{4}$

$a = -1.5$

Since $h = 0, k = 8,$ and $a = -1.5$

$y = -1.5x^2 + 8$

43. Domain: $(-\infty,\infty)$, range: $(-\infty,0]$

Vertex $= (-5,0)$

Other point $= (0,-10)$

$-10 = a(0+5)^2 + 0$

$-10 = a(5)^2$

$-10 = 25a$

$\dfrac{-10}{25} = \dfrac{25a}{25}$

$a = -0.4$

Since $h = -5, k = 0,$ and $a = -0.4$

$y = -0.4(x+5)^2$

Section 4.4

1.
$x^2 = 100$
$x = \pm\sqrt{100}$
$x = 10$, $x = -10$

3.
$2x^2 = 162$
$\dfrac{2x^2}{2} = \dfrac{162}{2} = 81$
$x^2 = 81$
$x = \pm\sqrt{81}$
$x = 9$, $x = -9$

5.
$x^2 + 12 = 181$
$ -12 \ \ -12$
$x^2 = 169$
$x = \pm\sqrt{169}$
$x = 13$, $x = -13$

7.
$(x-5)^2 = 49$
$x - 5 = \pm\sqrt{49} = \pm 7$
$x = 5 + 7 = 12$
$x = 5 - 7 = -2$
$x = 12$, $x = -2$

9.
$4(x+7)^2 = 400$
$\dfrac{4(x+7)^2}{4} = \dfrac{400}{4}$
$(x+7)^2 = 100$
$x + 7 = \pm\sqrt{100} = \pm 10$
$x = -7 + 10 = 3$
$x = -7 - 10 = -17$
$x = 3$, $x = -17$

11.
$3(x-7)^2 + 8 = 647.48$
$ -8 \ \ -8$
$3(x-7)^2 = 639.48$
$\dfrac{3(x-7)^2}{3} = \dfrac{639.48}{3}$
$(x-7)^2 = 213.16$
$x - 7 = \pm\sqrt{213.16} = \pm 14.6$
$x = 7 + 14.6 = 21.6$
$x = 7 - 14.6 = -7.6$
$x = 21.6$, $x = -7.6$

13.
$-5(x-6)^2 - 15 = -30$
$ +15 \ \ +15$
$-5(x-6)^2 = -15$
$\dfrac{-5(x-6)^2}{-5} = \dfrac{-15}{-5}$
$(x-6)^2 = 3$
$x - 6 = \pm\sqrt{3} \approx \pm 1.732$
$x \approx 6 + 1.732 = 7.732$
$x \approx 6 - 1.732 = 4.268$
$x \approx 4.268$, $x \approx 7.732$

15.
$-0.5(x+4)^2 - 9 = -20$
$ +9 \ \ +9$
$-0.5(x+4)^2 = -11$
$\dfrac{-0.5(x+4)^2}{-0.5} = \dfrac{-11}{-0.5}$
$(x+4)^2 = 22$
$x + 4 = \pm\sqrt{22} \approx \pm 4.690$
$x \approx -4 + 4.690 = 0.690$
$x \approx -4 - 4.690 = -8.690$
$x \approx 0.690$, $x \approx -8.690$

17. The correct answers are $x = 2$, $x = -10$. When applying the square root property, the student did not include the negative square root, so one answer was missed.

$$27 = (x+4)^2 - 9$$
$$+9 \qquad\qquad +9$$
$$36 = (x+4)^2$$
$$\pm\sqrt{36} = x + 4$$
$$\pm 6 = x + 4$$

The two solutions are:
$x = -4 + 6 = 2$
$x = -4 - 6 = -10$
$x = 2$, $x = -10$

19.

a.
$$C(18) = -2.336(18-13)^2 + 85.6$$
$$= -2.336(5)^2 + 85.6$$
$$= 27.2$$

In 1998, 27.2 million cassette singles were shipped by major recording media manufacturers.

b.
$$-2.336(t-13)^2 + 85.6 = 50$$
$$\qquad\qquad -85.6 \quad -85.6$$
$$-2.336(t-13)^2 = -35.6$$
$$\frac{-2.336(t-13)^2}{-2.336} = \frac{-35.6}{-2.336}$$
$$(t-13)^2 \approx 15.240$$
$$t - 13 \approx \pm\sqrt{15.240} \approx \pm 3.9$$
$$t \approx 13 - 3.9 = 9.1 \to 1989$$
$$t \approx 13 + 3.9 = 16.9 \to 1997$$

In about 1989 and 1997, about 50 million cassette singles were shipped by major recording media manufacturers.

c. Vertex $= (13, 85.6)$. In 1993, the number of cassette singles shipped by major recording media manufacturers reached a maximum of 85.6 million.

d. Anytime before about 1988 is probably model breakdown, and anytime after 1999 will also be model breakdown because the model gives a negative number of cassette singles being shipped.

21.

a.
$$D(10) = 17(10-3)^2 + 2300$$
$$= 17(7)^2 + 2300$$
$$= 3133$$

In 2000, personal households and nonprofit organizations invested about $3133 billion in time and savings accounts.

b.
$$17(t-3)^2 + 2300 = 7500$$
$$\qquad\qquad -2300 \quad -2300$$
$$17(t-3)^2 = 5200$$
$$\frac{17(t-3)^2}{17} = \frac{5200}{17}$$
$$(t-3)^2 \approx 305.882$$
$$t - 3 \approx \pm\sqrt{305.882} \approx 17.5$$
$$t \approx 3 - 17.5 = -14.5 \to 1976$$
$$t \approx 3 + 17.5 = 20.5 \to 2010$$

According to this model, personal households and nonprofit organizations invested about $7500 billion in time and savings deposits in the years 1976 and 2010. 1976 might be model breakdown.

c. Vertex $= (3, 2300)$. In about 1993, personal households and nonprofit organizations invested about $2300 billion in time and savings deposits. This was the minimum amount invested around this time.

23.

a. Let $P(t)$ be the poverty threshold (in dollars) for a family of four t years since 1900.

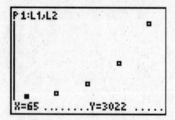

Choose $(65, 3022)$ as the vertex, and $(80, 5500)$ as the additional point used to solve for "a".

$P(t) = a(t-65)^2 + 3022$

$5500 = a(80-65)^2 + 3022$

$5500 = a(15)^2 + 3022$

$\underline{-3022 \qquad\qquad -3022}$

$2478 = 225a$

$\left.\dfrac{2478}{225} = \dfrac{225a}{225}\right\} \; a \approx 11$

~~$P(t) = 11(t-65)^2 + 3022$~~

This model needs adjusting.

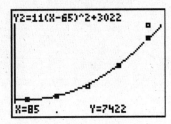

If we adjust the vertex to $(67, 2920)$, and make $a = 17$ slightly larger, the model has a much better fit.

$P(t) = 17(t-67)^2 + 2920$

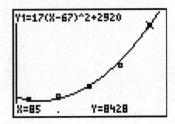

Note: Your answer may vary and yet still represent the trend of the data.

b.

$P(90) = 17(90-67)^2 + 2920$

$\qquad = 17(23)^2 + 2920$

$\qquad = 11,913$

The poverty threshold for a family of four in 1990 was about $11,913.

c. Domain: $[67, 95]$. Within this domain, the minimum value of 2920 comes from the vertex. The maximum value comes from $t = 95$.

$P(95) = 17(95-67)^2 + 2920$

$P(95) = 16,248$

Range: $[2920, 16248]$

d.

$17(t-67)^2 + 2920 = 5000$

$\underline{\qquad\qquad -2920 \; -2920}$

$17(t-67)^2 = 2080$

$\dfrac{17(t-67)^2}{17} = \dfrac{2080}{17}$

$(t-67)^2 \approx 122.353$

$t - 67 \approx \pm\sqrt{122.353} \approx 11.1$

$t \approx 67 - 11.1 = 55.9 \;\rightarrow\; 1956$

This is model breakdown.

$t \approx 67 + 11.1 = 78.1 \;\rightarrow\; 1978$

The poverty threshold for a family of four was $5000 in about 1978.

25.

a.

$V(5) = 2000\left(1 - \dfrac{5}{20}\right)^2$

$V(5) = 1125$

After 5 minutes, there would be about 1,125 gallons of water remaining in the tank.

b.

$2000\left(1 - \dfrac{t}{20}\right)^2 = 500$

$\dfrac{2000\left(1 - \dfrac{t}{20}\right)^2}{2000} = \dfrac{500}{2000}$

$\left(1 - \dfrac{t}{20}\right)^2 = 0.25$

$1 - \dfrac{t}{20} = \pm\sqrt{0.25} = \pm 0.5$

The first solution is:

$-\dfrac{t}{20} = -1 - 0.5 = -1.5$

$(-20)\left(-\dfrac{t}{20}\right) = (-1.5)(-20)$

$t = 30 \;\rightarrow\;$ This is model breakdown since the tank is empty after 20 minutes.

Chapter 4 Quadratic Functions

The second solution is:

$$-\frac{t}{20} = -1 + 0.5 = -0.5$$
$$(-20)\left(-\frac{t}{20}\right) = (-0.5)(-20)$$
$$t = 10 \rightarrow \text{This is within the domain of the function.}$$

After about 10 minutes, there would be only 500 gallons of water remaining in the tank.

c.

$$2000\left(1 - \frac{t}{20}\right)^2 = 20$$

$$\frac{2000\left(1 - \frac{t}{20}\right)^2}{2000} = \frac{20}{2000}$$

$$\left(1 - \frac{t}{20}\right)^2 = 0.01$$

$$1 - \frac{t}{20} = \pm\sqrt{0.01} = \pm 0.1$$

The first solution is:

$$-\frac{t}{20} = -1 - 0.1 = -1.1$$
$$(-20)\left(-\frac{t}{20}\right) = (-1.1)(-20)$$
$$t = 22 \rightarrow \text{This is model breakdown since the tank is empty after 20 minutes.}$$

The second solution is:

$$-\frac{t}{20} = -1 + 0.1 = -0.9$$
$$(-20)\left(-\frac{t}{20}\right) = (-0.9)(-20)$$
$$t = 18 \rightarrow \text{This is within the domain of the function.}$$

After about 18 minutes, there would be only 20 gallons of water remaining in the tank.

27.

$$x^2 + 6x = 7$$
$$x^2 + 6x + \square = 7 + \square$$

Completing the square: $\left(\frac{6}{2}\right)^2 = 9$

$$x^2 + 6x + \boxed{9} = 7 + \boxed{9}$$
$$(x + 3)^2 = 16$$
$$x + 3 = \pm\sqrt{16} = \pm 4$$

The two solutions are:
$$x + 3 = 4, \; x + 3 = -4$$
$$x = 1, \; x = -7$$

29.

$$k^2 - 16k = 3$$
$$k^2 - 16k + \square = 3 + \square$$

Completing the square: $\left(\frac{-16}{2}\right)^2 = 64$

$$k^2 - 16k + \boxed{64} = 3 + \boxed{64}$$
$$(k - 8)^2 = 67$$
$$k - 8 = \pm\sqrt{67} \approx \pm 8.185$$

The two solutions are:
$$k - 8 \approx 8.185, \; k - 8 \approx -8.185$$
$$k \approx 16.185, \; k \approx -0.185$$

31.

$$x^2 + 4x + 7 = 0$$
$$x^2 + 4x = -7$$
$$x^2 + 4x + \square = -7 + \square$$

Completing the square: $\left(\frac{4}{2}\right)^2 = 4$

$$x^2 + 4x + \boxed{4} = -7 + \boxed{4}$$
$$(x + 2)^2 = -3$$
$$x + 2 = \pm\sqrt{-3}$$

There are no real solutions.

4.4 Exercises

33.

$t^2 + 11t = 4$

$t^2 + 11t + \square = 4 + \square$

Completing the square: $\left(\dfrac{11}{2}\right)^2 = 30.25$

$t^2 + 11t + \boxed{30.25} = 4 + \boxed{30.25}$

$(t+5.5)^2 = 34.25$

$t + 5.5 = \pm\sqrt{34.25} \approx \pm 5.852$

The two solutions are:

$t + 5.5 \approx 5.852$, $t + 5.5 \approx -5.852$

$t \approx 0.352$, $t \approx -11.352$

35.

$x^2 - 9x + 14 = 0$

$x^2 - 9x = -14$

$x^2 - 9x + \square = -14 + \square$

Completing the square: $\left(\dfrac{-9}{2}\right)^2 = 20.25$

$x^2 - 9x + \boxed{20.25} = -14 + \boxed{20.25}$

$(x - 4.5)^2 = 6.25$

$x - 4.5 = \pm\sqrt{6.25} = \pm 2.5$

The two solutions are:

$x - 4.5 = 2.5$, $x - 4.5 = -2.5$

$x = 7$, $x = 2$

37.

$m^2 + 8m + 20 = 0$

$m^2 + 8m = -20$

$m^2 + 8m + \square = -20 + \square$

Completing the square: $\left(\dfrac{8}{2}\right)^2 = 16$

$m^2 + 8m + \boxed{16} = -20 + \boxed{16}$

$(m+4)^2 = -4$

$m + 4 = \pm\sqrt{-4}$

There are no real solutions.

39.

$3x^2 + 12x - 15 = 0$

$3x^2 + 12x = 15$

$\dfrac{3x^2 + 12x}{3} = \dfrac{15}{3}$

$x^2 + 4x = 5$

$x^2 + 4x + \square = 5 + \square$

Completing the square: $\left(\dfrac{4}{2}\right)^2 = 4$

$x^2 + 4x + \boxed{4} = 5 + \boxed{4}$

$(x+2)^2 = 9$

$x + 2 = \pm\sqrt{9} = \pm 3$

The two solutions are:

$x + 2 = 3$, $x + 2 = -3$

$x = 1$, $x = -5$

41.

$5x^2 + 7x = 0$

$\dfrac{5x^2 + 7x}{5} = \dfrac{0}{5}$

$x^2 + \dfrac{7}{5}x = 0$

$x^2 + \dfrac{7}{5}x + \square = \square$

Completing the square: $\left(\dfrac{1}{2} \cdot \dfrac{7}{5}\right)^2 = \dfrac{49}{100}$

$x^2 + \dfrac{7}{5}x + \boxed{\dfrac{49}{100}} = \boxed{\dfrac{49}{100}}$

$\left(x + \dfrac{7}{10}\right)^2 = \dfrac{49}{100}$

$x + \dfrac{7}{10} = \pm\sqrt{\dfrac{49}{100}} = \pm\dfrac{7}{10}$

The two solutions are:

$x + \dfrac{7}{10} = \dfrac{7}{10}$, $x + \dfrac{7}{10} = -\dfrac{7}{10}$

$x = 0$, $x = -\dfrac{14}{10} = -\dfrac{7}{5}$

43.

$-7x^2 + 4x + 20 = 0$

$-7x^2 + 4x = -20$

$\dfrac{-7x^2 + 4x}{-7} = \dfrac{-20}{-7}$

$x^2 - \dfrac{4}{7}x = \dfrac{20}{7}$

$x^2 - \dfrac{4}{7}x + \square = \dfrac{20}{7} + \square$

Completing the square: $\left(\dfrac{1}{2} \cdot \dfrac{-4}{7}\right)^2 = \left(\dfrac{-2}{7}\right)^2 = \dfrac{4}{49}$

$x^2 - \dfrac{4}{7}x + \boxed{\dfrac{4}{49}} = \dfrac{20}{7} + \boxed{\dfrac{4}{49}} = \dfrac{140}{49} + \dfrac{4}{49} = \dfrac{144}{49}$

$\left(x - \dfrac{2}{7}\right)^2 = \dfrac{144}{49}$

$x - \dfrac{2}{7} = \pm\sqrt{\dfrac{144}{49}} = \pm\dfrac{12}{7}$

The two solutions are:

$x - \dfrac{2}{7} = \dfrac{12}{7}$, $x - \dfrac{2}{7} = -\dfrac{12}{7}$

$x = 2$, $x = -\dfrac{10}{7}$

45. Use the formula $A = P(1+r)^t$:

$900 = 800(1+r)^2$

$\dfrac{900}{800} = \dfrac{800(1+r)^2}{800}$

$1.125 = (1+r)^2$

$1+r = \pm\sqrt{1.125}$

$r = -1 + \sqrt{1.125} \approx 0.0607$

$r = -1 - \sqrt{1.125} \approx \cancel{-2.0607}$

A 6.07% interest rate would turn a deposit of $800 into $900 in 2 years.

47. Use the formula $A = P(1+r)^t$:

$2150 = 2000(1+r)^2$

$\dfrac{2150}{2000} = \dfrac{2000(1+r)^2}{2000}$

$1.075 = (1+r)^2$

$1+r = \pm\sqrt{1.075}$

$r = -1 + \sqrt{1.075} \approx 0.0368$

$r = -1 - \sqrt{1.075} \approx \cancel{-2.0368}$

A 3.68% interest rate would turn a deposit of $2000 into $2150 in 2 years.

49. Use the Pythagorean Theorem.

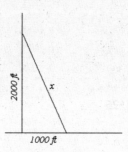

$x^2 = 2000^2 + 1000^2$

$x^2 = 5,000,000$

$x = \sqrt{5,000,000}$

$x \approx 2236.07$

The guy wire is approximately 2236.068 feet long.

51. Use the Pythagorean Theorem.

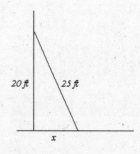

$25^2 = x^2 + 20^2$

$625 = x^2 + 400$

$\underline{-400 \quad\quad -400}$

$225 = x^2$

$x = \sqrt{225} = 15$

Abida should put the base of the ladder 15 feet away from the house.

53.

$f(x) = x^2 + 6x + 8$

$f(x) = (x^2 + 6x + \underline{}) + 8 - \underline{}$ Completing the square: $\left(\dfrac{6}{2}\right)^2 = 9$

$f(x) = (x^2 + 6x + 9) + 8 - 9$

$f(x) = (x + 3)(x + 3) - 1$

$f(x) = (x + 3)^2 - 1$

55.

$g(t) = t^2 - 8t - 20$

$g(t) = (t^2 - 8t + \underline{}) - 20 - \underline{}$ Completing the square: $\left(\dfrac{-8}{2}\right)^2 = 16$

$g(t) = (t^2 - 8t + 16) - 20 - 16$

$g(t) = (t - 4)(t - 4) - 36$

$g(t) = (t - 4)^2 - 36$

57.

$f(x) = x^2 - 7x + 10$

$f(x) = (x^2 - 7x + \underline{}) + 10 - \underline{}$ Completing the square: $\left(\dfrac{-7}{2}\right)^2 = \dfrac{49}{4}$

$f(x) = \left(x^2 - 7x + \dfrac{49}{4}\right) + 10 - \dfrac{49}{4}$

$f(x) = \left(x - \dfrac{7}{2}\right)\left(x - \dfrac{7}{2}\right) - \dfrac{9}{4}$

$f(x) = \left(x - \dfrac{7}{2}\right)^2 - \dfrac{9}{4}$

59.

$h(x) = 3x^2 + 12x + 24$

$h(x) = 3(x^2 + 4x) + 24$

$h(x) = 3(x^2 + 4x + \underline{} - \underline{}) + 24$ Completing the square: $\left(\dfrac{4}{2}\right)^2 = 4$

$h(x) = 3(x^2 + 4x + 4 - 4) + 24$

$h(x) = 3(x^2 + 4x + 4) - 3(4) + 24$

$h(x) = 3(x + 2)(x + 2) - 12 + 24$

$h(x) = 3(x + 2)^2 + 12$

61.

$f(t) = 2t^2 - 16t - 12$

$f(t) = 2(t^2 - 8t) - 12$

$f(t) = 2(t^2 - 8t + \underline{} - \underline{}) - 12$ Completing the square: $\left(\dfrac{-8}{2}\right)^2 = 16$

$f(t) = 2(t^2 - 8t + 16 - 16) - 12$

$f(t) = 2(t^2 - 8t + 16) - 2(16) - 12$

$f(t) = 2(t - 4)(t - 4) - 32 - 12$

$f(t) = 2(t - 4)^2 - 44$

63.

$f(x) = 4x^2 + 5x - 20$

$f(x) = 4\left(x^2 + \dfrac{5}{4}x\right) - 20$

$f(x) = 4\left(x^2 + \dfrac{5}{4}x + \underline{} - \underline{}\right) - 20$ Completing the square: $\left(\dfrac{\frac{5}{4}}{2}\right)^2 = \dfrac{25}{64}$

$f(x) = 4\left(x^2 + \dfrac{5}{4}x + \dfrac{25}{64} - \dfrac{25}{64}\right) - 20$

$f(x) = 4\left(x^2 + \dfrac{5}{4}x + \dfrac{25}{64}\right) - 4\left(\dfrac{25}{64}\right) - 20$

$f(x) = 4\left(x + \dfrac{5}{8}\right)\left(x + \dfrac{5}{8}\right) - \dfrac{25}{16} - 20$

$f(x) = 4\left(x + \dfrac{5}{8}\right)^2 - \dfrac{345}{16}$

65.

$g(x) = 0.5x^2 + 7x - 30$

$g(x) = 0.5(x^2 + 14x) - 30$

$g(x) = 0.5(x^2 + 14x + \underline{} - \underline{}) - 30$

Completing the square: $\left(\dfrac{14}{2}\right)^2 = 49$

$g(x) = 0.5(x^2 + 14x + 49 - 49) - 30$

$g(x) = 0.5(x^2 + 14x + 49) - 0.5(49) - 30$

$g(x) = 0.5(x + 7)(x + 7) - 24.5 - 30$

$g(x) = 0.5(x + 7)^2 - 54.5$

Chapter 4 Quadratic Functions

67.

$f(x) = 0.2x^2 - 7x - 10$

$f(x) = 0.2(x^2 - 35x) - 10$

$f(x) = 0.2(x^2 - 35x + \_\_ - \_\_) - 10$

Completing the square: $\left(\dfrac{-35}{2}\right)^2 = 306.25$

$f(x) = 0.2(x^2 - 35x + 306.25 - 306.25) - 10$

$f(x) = 0.2(x^2 - 35x + 306.25) - 0.2(306.25) - 10$

$f(x) = 0.2(x - 17.5)(x - 17.5) - 61.25 - 10$

$f(x) = 0.2(x - 17.5)^2 - 71.25$

69.

$c(p) = \dfrac{2}{7}p^2 - 5p - \dfrac{3}{7}$

$c(p) = \dfrac{2}{7}\left(p^2 - \dfrac{35}{2}p\right) - \dfrac{3}{7}$

$c(p) = \dfrac{2}{7}\left(p^2 - \dfrac{35}{2}p + \_\_ - \_\_\right) - \dfrac{3}{7}$

Completing the square: $\left(\dfrac{\frac{35}{2}}{2}\right)^2 = \dfrac{1225}{16}$

$c(p) = \dfrac{2}{7}\left(p^2 - \dfrac{35}{2}p + \dfrac{1225}{16} - \dfrac{1225}{16}\right) - \dfrac{3}{7}$

$c(p) = \dfrac{2}{7}\left(p^2 - \dfrac{35}{2}p + \dfrac{1225}{16}\right) - \dfrac{2}{7}\left(\dfrac{1225}{16}\right) - \dfrac{3}{7}$

$c(p) = \dfrac{2}{7}\left(p - \dfrac{35}{4}\right)\left(p - \dfrac{35}{4}\right) - \dfrac{175}{8} - \dfrac{3}{7}$

$c(p) = \dfrac{2}{7}\left(p - \dfrac{35}{4}\right)^2 - \dfrac{1249}{56}$

71. Domain: $(-\infty, \infty)$, range: $[-16, \infty)$

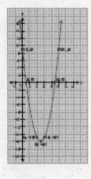

73. Domain: $(-\infty, \infty)$, range: $[-5, \infty)$

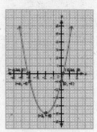

75. Domain: $(-\infty, \infty)$, range: $[10, \infty)$

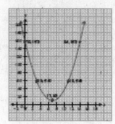

77. Domain: $(-\infty, \infty)$, range: $(-\infty, 18]$

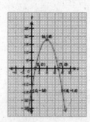

79. Domain: $(-\infty, \infty)$, range: $(-\infty, 32]$

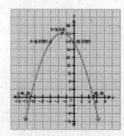

81. Domain: $(-\infty, \infty)$, range: $[-12, \infty)$

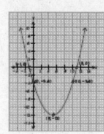

4.4 Exercises

83. Domain: $(-\infty,\infty)$, range: $[0.2,\infty)$

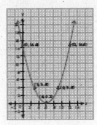

85. Domain: $(-\infty,\infty)$, range: $[-16,\infty)$

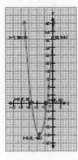

87. Domain: $(-\infty,\infty)$, range: $[-1,\infty)$

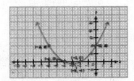

89. Domain: $(-\infty,\infty)$, range: $[-36,\infty)$

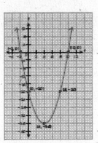

91. Domain: $(-\infty,\infty)$, range: $[-2.25,\infty)$

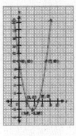

93. Domain: $(-\infty,\infty)$, range: $[12,\infty)$

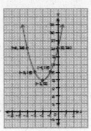

95. Domain: $(-\infty,\infty)$, range: $[-44,\infty)$

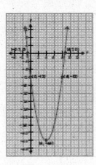

97. Domain: $(-\infty,\infty)$, range: $[-21.5625,\infty)$

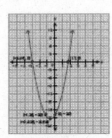

99. Domain: $(-\infty,\infty)$, range: $[-54.5,\infty)$

Chapter 4 Quadratic Functions
Section 4.5

1.

$(x+3)(x-2)=0$
$x+3=0, x-2=0$
$x=-3, x=2$

3.

$2(w+7)(3w+10)=0$
$w+7=0, 3w+10=0$
$w=-7, 3w=-10$
$w=-7, w=-\dfrac{10}{3}$

5.

$x(x-4)(x+5)=0$
$x=0, x-4=0, x+5=0$
$x=0, x=4, x=-5$

7.

$4x(x-9)(x+7)=0$
$4x=0, x-9=0, x+7=0$
$x=0, x=9, x=-7$

9.

$4.5w(3w-8)(7w+5)(w-6)=0$
$4.5w=0, 3w-8=0, 7w+5=0, w-6=0$
$w=0, 3w=8, 7w=-5, w=6$
$w=0, w=\dfrac{8}{3}, w=-\dfrac{5}{7}, w=6$

11.

a.

$H\left(\dfrac{1}{2}\right)=-16\left(\dfrac{1}{2}\right)^2+4\left(\dfrac{1}{2}\right)+20$

$H\left(\dfrac{1}{2}\right)=18$

The height of the professor $\dfrac{1}{2}$ second after jumping off the waterfall was about 18 feet.

b.

$0=-16t^2+4t+20$

$\dfrac{0}{-4}=\dfrac{-4(4t^2-t-5)}{-4}$

$0=4t^2-t-5$
$0=(4t-5)(t+1)$
$4t-5=0, t+1=0$
$4t=5, \cancel{t=-1}$
$t=\dfrac{5}{4}=1.25$

It took about 1.25 seconds for the professor to hit the pool of water below.

c.

$8=-16t^2+4t+20$
$\underline{-8 \qquad\qquad\qquad -8}$
$0=-16t^2+4t+12$

$\dfrac{0}{-4}=\dfrac{-4(4t^2-t-3)}{-4}$

$0=4t^2-t-3$
$0=(4t+3)(t-1)$
$4t+3=0, t-1=0$
$4t=-3, t=1$
$\cancel{t=-\dfrac{3}{4}}$

It took about 1 second after jumping for the professor to reach a height of 8 feet above the pool of water.

13.

a.

$144=p(80-4p)$
$144=80p-4p^2$
$\underline{-144 \qquad\qquad -144}$
$0=-4p^2+80p-144$

$\dfrac{0}{-4}=\dfrac{-4(p^2-20p+36)}{-4}$

$0=p^2-20p+36$
$0=(p-2)(p-18)$
$p=2, p=18$

4.5 Exercises

The company needs to make either 2 or 18 parts to make $144 profit per part.

b.

$$300 = p(80-4p)$$
$$300 = 80p - 4p^2$$
$$\underline{-300 \qquad\qquad -300}$$
$$0 = -4p^2 + 80p - 300$$
$$\frac{0}{-4} = \frac{-4(p^2 - 20p + 75)}{-4}$$
$$0 = p^2 - 20p + 75$$
$$0 = (p-5)(p-15)$$
$$p = 5, \; p = 15$$

The company needs to make either 5 or 15 parts to make $300 profit per part.

c.

$$400 = p(80-4p)$$
$$400 = 80p - 4p^2$$
$$\underline{-400 \qquad\qquad -400}$$
$$0 = -4p^2 + 80p - 400$$
$$\frac{0}{-4} = \frac{-4(p^2 - 20p + 100)}{-4}$$
$$0 = p^2 - 20p + 100$$
$$0 = (p-10)^2$$
$$p = 10$$

The company needs to make 10 parts to make $400 profit per part.

15.

a.

$$390 = m(108-6m)$$
$$390 = 108q - 6m^2$$
$$\underline{-390 \qquad\qquad -390}$$
$$0 = -6m^2 + 108m - 390$$
$$\frac{0}{-6} = \frac{-6(m^2 - 18m + 65)}{-6}$$
$$0 = m^2 - 18m + 65$$
$$0 = (m-5)(m-13)$$
$$m - 5 = 0, \; m - 13 = 0$$
$$m = 5, \; m = 13$$

The company needs to manufacture either 5 or 13 machines to have an average cost of $390 per machine.

b.

$$480 = m(108-6m)$$
$$480 = 108q - 6m^2$$
$$\underline{-480 \qquad\qquad -480}$$
$$0 = -6m^2 + 108m - 480$$
$$\frac{0}{-6} = \frac{-6(m^2 - 18m + 80)}{-6}$$
$$0 = m^2 - 18m + 80$$
$$0 = (m-8)(m-10)$$
$$m - 8 = 0, \; m - 10 = 0$$
$$m = 8, \; m = 10$$

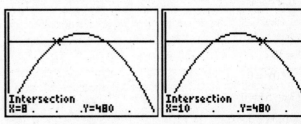

The company needs to manufacture either 8 or 10 machines to have an average cost of $480 per machine.

17.
$x^2 - 4x - 21 = 0$
$(x+3)(x-7) = 0$
$x+3 = 0, x-7 = 0$
$x = -3, x = 7$

19.
$h^2 + 12h + 27 = 0$
$(h+3)(h+9) = 0$
$h+3 = 0, h+9 = 0$
$h = -3, h = -9$

21.
$x^2 - 36 = 0$
$(x+6)(x-6) = 0$
$x+6 = 0, x-6 = 0$
$x = -6, x = 6$

23.
$5x^2 - 80 = 0$
$5(x^2 - 16) = 0$
$5(x+4)(x-4) = 0$
$x+4 = 0, x-4 = 0$
$x = -4, x = 4$

25.
$x^2 + 50 = 150$
$x^2 - 100 = 0$
$(x+10)(x-10) = 0$
$x+10 = 0, x-10 = 0$
$x = -10, x = 10$

27.
$5x^2 + 20x = 0$
$5x(x+4) = 0$
$5x = 0, x+4 = 0$
$x = 0, x = -4$

29.
$6x^2 = 10x$
$6x^2 - 10x = 0$
$2x(3x-5) = 0$
$2x = 0, 3x-5 = 0$
$x = 0, 3x = 5$
$x = 0, x = \dfrac{5}{3}$

31.
$x^2 + 9x + 20 = 0$
$(x+4)(x+5) = 0$
$x+4 = 0, x+5 = 0$
$x = -4, x = -5$

33. The two factors must be equal to zero. Set the equation equal to zero and factor.
$x^2 + 5x + 6 = 2$
$x^2 + 5x + 4 = 0$
$(x+1)(x+4) = 0$
$x+1 = 0 \rightarrow x = -1$
$x+4 = 0 \rightarrow x = -4$

35.
$t^2 - 11t + 21 = -7$
$t^2 - 11t + 28 = 0$
$(t-4)(t-7) = 0$
$t-4 = 0, t-7 = 0$
$t = 4, t = 7$

37.
$7m^2 - 25m + 18 = 6$
$7m^2 - 25m + 12 = 0$
$(7m-4)(m-3) = 0$
$7m-4 = 0, m-3 = 0$
$7m = 4, m = 3$
$m = \dfrac{4}{7}, m = 3$

39.
$4x^2 - 31x - 45 = 0$
$(4x+5)(x-9) = 0$
$4x+5 = 0, x-9 = 0$
$4x = -5, x = 9$
$x = -\dfrac{5}{4}, x = 9$

41.
$-41t + 10t^2 = 18$
$10t^2 - 41t - 18 = 0$
$(5t+2)(2t-9) = 0$
$5t+2 = 0, 2t-9 = 0$
$5t = -2, 2t = 9$
$t = -\dfrac{2}{5}, t = \dfrac{9}{2}$

43.

$x^2 + 7x - 9 = 7x + 16$
$x^2 - 25 = 0$
$(x+5)(x-5) = 0$
$x + 5 = 0, x - 5 = 0$
$x = -5, x = 5$

45.

$28p^2 + 3p + 60 = 100$
$28p^2 + 3p - 40 = 0$
$(4p+5)(7p-8) = 0$
$4p + 5 = 0, 7p - 8 = 0$
$4p = -5, 7p = 8$
$p = -\dfrac{5}{4}, p = \dfrac{8}{7}$

47.

$55 + 55h = 10h^2 - 50$
$0 = 10h^2 - 55h - 105$
$0 = (10h + 15)(h - 7)$
$0 = 10h + 15, 0 = h - 7$
$10h = -15, h = 7$
$h = -1.5, h = 7$

49.

$x^3 - 4x^2 - 21x = 0$
$x(x^2 - 4x - 21) = 0$
$x(x+3)(x-7) = 0$
$x = 0, x + 3 = 0, x - 7 = 0$
$x = 0, x = -3, x = 7$

51.

$w^3 - 7w^2 + 15w = 5w$
$w^3 - 7w^2 + 10w = 0$
$w(w^2 - 7w + 10) = 0$
$w(w-5)(w-2) = 0$
$w = 0, w - 5 = 0, w - 2 = 0$
$w = 0, w = 5, w = 2$

53.

$10x^3 + 30x^2 + 40x + 15 = -5x^2 + 10x + 15$
$10x^3 + 35x^2 + 30x = 0$
$5x(2x^2 + 7x + 6) = 0$
$5x(2x+3)(x+2) = 0$
$5x = 0, 2x + 3 = 0, x + 2 = 0$
$x = 0, 2x = -3, x = -2$
$x = 0, x = -1.5, x = -2$

55.

a. $(\$0.75)(5500) = \$4125$

The profit for this school fundraiser is $4125 total, if the profit per candy bar is $0.75.

b. Let $P(n)$ represent the total profit made at the school's sports program fundraiser, in dollars, and let n represent the number of each $0.50 increase on the total price of the candy bar.

Profit = (price in $)(number of candy bars)
$P(n) = (0.75 + 0.50n)(5500 - 1000n)$

c.

$P(2) = (0.75 + 0.50(2))(5500 - 1000(2))$
$P(2) = 6125$

If the price per candy bar is increased by $1, the total profit would be $6125.

d.

$P(4) = (0.75 + 0.50(4))(5500 - 1000(4))$
$P(4) = 4125$

If the price per candy bar is increased by $2, the total profit would be $4125.

e. The highest profit occurs at the vertex, (h, k).

$P(n) = (0.75 + 0.50n)(5500 - 1000n)$
$P(n) = 4125 - 750n + 2750n - 500n^2$
$P(n) = -500n^2 + 2000n + 4125$
$h = \dfrac{-2000}{2(-500)} = 2$
$k = P(2) = 6125$

Increasing the price of each candy bar by $1 would bring the school the highest profit.

57.

a. $(\$15)(100) = \$1500$

The store's weekly revenue from selling boxes of golf balls for $15 would be $1500.

b. For each $1 increase in the price of the boxes golf balls, the number of boxes sold will decrease by 20 boxes.

$15 → 100$ boxes sold
$16 → 80$ boxes sold
$17 → 60$ boxes sold
Revenue $= (\$17)(60) = \$1020$

The store's weekly revenue from selling boxes of golf balls for $17 would be about $1020.

c. Let $R(x)$ represent the store's weekly revenue from selling boxes of golf balls if the store lowers the price by x dollars.

$R(x) = (15-x)(100+20x)$

d. If the boxes are selling for $8 per box, then the price has decreased by $7.

$R(7) = (15-7)(100+20(7))$
$R(7) = 1920$

The stores weekly revenue from selling boxes of golf balls would be $1920 if each box was sold for $7.

e. To maximize the revenue, we are looking for the vertex of $R(x) = (15-x)(100+20x)$.

$R(x) = (15-x)(100+20x)$
$R(x) = 1500 + 300x - 100x - 20x^2$
$R(x) = -20x^2 + 200x + 1500$

$h = \dfrac{-200}{2(-20)} = 5$

$k = R(5) = 2000$

Selling the boxes of golf balls for $10 each would maximize the store's weekly revenue.

59.

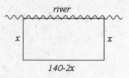

$x(140-2x) = 2400$
$140x - 2x^2 = 2400$
$-2x^2 - 140x - 2400 = 0$
$\dfrac{-2x^2 - 140x - 2400}{-2} = \dfrac{0}{-2}$
$x^2 - 70x + 1200 = 0$
$(x-30)(x-40) = 0$
$x - 30 = 0, \; x - 40 = 0$
$x = 30, \; x = 40$

The enclosure was either 30 feet by 80 feet or 40 feet by 60 feet.

61. Let x be the width of the frame, then $(32-2x)$ represents the width of the picture and $(29-2x)$ represents the length of the picture.

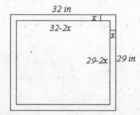

$(32-2x)(29-2x) = 550$
$928 - 64x - 58x + 4x^2 = 550$
$4x^2 - 122x + 928 = 550$
$ \underline{-550 \quad -550}$
$4x^2 - 122x + 378 = 0$
$\dfrac{4x^2 - 122x + 378}{2} = \dfrac{0}{2}$
$2x^2 - 61x + 189 = 0$
$(2x-7)(x-27) = 0$
$2x - 7 = 0 , \ x - 27 = 0$
$2x = 7, \ \cancel{x = 27}$
$x = \dfrac{7}{2} = 3.5$

The width of the frame is 3.5 inches.

63.

Use: $x = -3, \ x = 2$
with $f(0) = -6$.
$x + 3 = 0 , \ x - 2 = 0$
$f(x) = a(x+3)(x-2)$
$f(x) = a(x^2 + x - 6)$
$f(0) = a(-6) = -6$
$-6a = -6 \ \rightarrow \ a = 1$
$f(x) = x^2 + x - 6$

65.

Use: $x = -3, \ x = 3$
with $f(0) = 9$.
$x + 3 = 0 , \ x - 3 = 0$
$f(x) = a(x+3)(x-3)$
$f(x) = a(x^2 - 9)$
$f(0) = a(-9) = 9$
$-9a = 9 \ \rightarrow \ a = -1$
$f(x) = -(x^2 - 9)$
$f(x) = -x^2 + 9$

67.

Use: $x = -2, \ x = 2$
with $f(0) = 8$.
$x + 2 = 0 , \ x - 2 = 0$
$f(x) = a(x+2)(x-2)$
$f(x) = a(x^2 - 4)$
$f(0) = a(-4) = 8$
$-4a = 8 \ \rightarrow \ -2 = a$
$f(x) = -2(x^2 - 4)$
$f(x) = -2x^2 + 8$

69.

$f(x) = a(x-h)^2 + k$
Vertex $= (-2, 5)$
Second point $= (1, 15)$
$f(x) = a(x+2)^2 + 5$
$15 = a(1+2)^2 + 5$
$15 = a(3)^2 + 5$
$\underline{-5 -5}$
$10 = 9a$
$\dfrac{10}{9} = a$
$f(x) = \dfrac{10}{9}(x+2)^2 + 5$
$f(x) = \dfrac{10}{9}(x^2 + 4x + 4) + 5$
$f(x) = \dfrac{10}{9}x^2 + \dfrac{40}{9}x + \dfrac{40}{9} + 5$
$f(x) = \dfrac{10}{9}x^2 + \dfrac{40}{9}x + \dfrac{85}{9}$

71.

Given: $x = 3, \ x = -7$.
$x - 3 = 0 , \ x + 7 = 0$
$f(x) = (x-3)(x+7)$
$f(x) = x^2 + 4x - 21$

73.

Given: $x = 2, \ x = 4$.
$x - 2 = 0 , \ x - 4 = 0$
$f(x) = (x-2)(x-4)$
$f(x) = x^2 - 6x + 8$

75.

Given: $x = \dfrac{2}{3}$, $x = 4$.

$3x = 2$

$3x - 2 = 0$, $x - 4 = 0$

$f(x) = (3x - 2)(x - 4)$

$f(x) = 3x^2 - 14x + 8$

77.

Given: $x = 4$, $x = 2$ with $f(0) = 40$.

$x - 4 = 0$, $x - 2 = 0$

$f(x) = a(x - 4)(x - 2)$

$f(x) = a(x^2 - 6x + 8)$

$f(0) = a(8) = 40$

$8a = 40$

$a = 5$

$f(x) = 5(x^2 - 6x + 8)$

$f(x) = 5x^2 - 30x + 40$

79.

Given: $x = \dfrac{1}{4}$, $x = -2$ with $f(2) = -21$.

$4x = 1$

$4x - 1 = 0$, $x + 2 = 0$

$f(x) = a(4x - 1)(x + 2)$

$f(x) = a(4x^2 + 7x - 2)$

$f(2) = a(4(2)^2 + 7(2) - 2) = -21$

$28a = -21$

$\dfrac{28a}{28} = \dfrac{-21}{28}$

$a = -\dfrac{3}{4}$

$f(x) = -\dfrac{3}{4}(4x^2 + 7x - 2)$

$f(x) = -3x^2 - 5.25x + 1.5$

81.

Given: $x = -\dfrac{1}{3}$, $x = \dfrac{9}{2}$ with $f(3) = 20$.

$3x = -1$, $2x = 9$

$3x + 1 = 0$, $2x - 9 = 0$

$f(x) = a(3x + 1)(2x - 9)$

$f(x) = a(6x^2 - 25x - 9)$

$f(3) = a(6(3)^2 - 25(3) - 9) = 20$

$-30a = 20$

$\dfrac{-30a}{-30} = \dfrac{20}{-30}$

$a = -\dfrac{2}{3}$

$f(x) = -\dfrac{2}{3}(6x^2 - 25x - 9)$

$f(x) = -4x^2 + \dfrac{50}{3}x + 6$

83.

$4x^2 - 100 = 0$

$4x^2 = 100$

$x^2 = 25$

$x = \pm\sqrt{25}$

$x = \pm 5$

85.

$2n^3 + 5n^2 - 7n = 0$

$n(2n^2 + 5n - 7) = 0$

$n(2n + 7)(n - 1) = 0$

$n = 0, 2n + 7 = 0, n - 1 = 0$

$n = 0, 2n = -7, n = 1$

$n = 0, n = -3.5, n = 1$

87.

$3(h + 5) = 22$

$3h + 15 = 22$

$3h = 7$

$h = \dfrac{7}{3}$

89.

$p^2 - 7p + 10 = 0$

$(p - 5)(p - 2) = 0$

$p - 5 = 0, p - 2 = 0$

$p = 5, p = 2$

91.

$6b^2 + 18b = 0$

$6b(b+3) = 0$

$6b = 0, b+3 = 0$

$b = 0, b = -3$

93.

$4h^2 + h + 10 = 0$

$4\left(h^2 + \dfrac{1}{4}h\right) = -10$

$4\left(h^2 + \dfrac{1}{4}h + \underline{}\right) = -10 + \underline{}$

Completing the square: $\left(\dfrac{1}{2} \cdot \dfrac{1}{4}\right)^2 = \dfrac{1}{64}$

$4\left(h^2 + \dfrac{1}{4}h + \dfrac{1}{64} - \dfrac{1}{64}\right) = -10 + \dfrac{1}{64}$

$4\left(h^2 + \dfrac{1}{4}h + \dfrac{1}{64}\right) - 4\left(\dfrac{1}{64}\right) = -\dfrac{639}{64}$

$4\left(h + \dfrac{1}{8}\right)^2 - \dfrac{1}{16} = -\dfrac{639}{64}$

$4\left(h + \dfrac{1}{8}\right)^2 - \dfrac{1}{16} + \dfrac{1}{16} = -\dfrac{639}{64} + \dfrac{1}{16}$

$4\left(h + \dfrac{1}{8}\right)^2 = -\dfrac{635}{64}$

$\left(\dfrac{1}{4}\right)(4)\left(h + \dfrac{1}{8}\right)^2 = -\dfrac{635}{64}\left(\dfrac{1}{4}\right)$

$\left(h + \dfrac{1}{8}\right)^2 = -\dfrac{635}{256}$

$h + \dfrac{1}{8} = \pm\sqrt{-\dfrac{635}{256}}$

No real solution.

95.

$4m^2 + 5m = 4m^2 + 10$

$5m = 10$

$m = 2$

97.

$-3(y+5)^2 + 21 = -9$

$-3(y+5)^2 = -30$

$(y+5)^2 = 10$

$y + 5 = \pm\sqrt{10}$

$y = -5 \pm \sqrt{10}$

$y = -5 + \sqrt{10} \approx -1.84$

$y = -5 - \sqrt{10} \approx -8.16$

Chapter 4 Quadratic Functions
Section 4.6

1.

$x^2 + 8x + 15 = 0$

$a = 1, b = 8, c = 15$

$x = \dfrac{-8 \pm \sqrt{8^2 - 4(1)(15)}}{2(1)}$

$x = \dfrac{-8 \pm \sqrt{4}}{2}$

$x = \dfrac{-8 \pm 2}{2}$

$x = \dfrac{-8 + 2}{2}, x = \dfrac{-8 - 2}{2}$

$x = \dfrac{-6}{2}, x = \dfrac{-10}{2}$

$x = -3, x = -5$

3.

$x^2 - 7x + 10 = 0$

$a = 1, b = -7, c = 10$

$x = \dfrac{-(-7) \pm \sqrt{(-7)^2 - 4(1)(10)}}{2(1)}$

$x = \dfrac{7 \pm \sqrt{9}}{2}$

$x = \dfrac{7 \pm 3}{2}$

$x = \dfrac{7 + 3}{2}, x = \dfrac{7 - 3}{2}$

$x = \dfrac{10}{2}, x = \dfrac{4}{2}$

$x = 5, x = 2$

5.

$3x^2 + 9x - 12 = 0$

$a = 3, b = 9, c = -12$

$x = \dfrac{-9 \pm \sqrt{9^2 - 4(3)(-12)}}{2(3)}$

$x = \dfrac{-9 \pm \sqrt{225}}{6}$

$x = \dfrac{-9 \pm 15}{6}$

$x = \dfrac{-9 + 15}{6}, x = \dfrac{-9 - 15}{6}$

$x = \dfrac{6}{6}, x = \dfrac{-24}{6}$

$x = 1, x = -4$

7.

$2x^2 - 9x = 5$

$2x^2 - 9x - 5 = 0$

$a = 2, b = -9, c = -5$

$x = \dfrac{-(-9) \pm \sqrt{(-9)^2 - 4(2)(-5)}}{2(2)}$

$x = \dfrac{9 \pm \sqrt{121}}{4}$

$x = \dfrac{9 \pm 11}{4}$

$x = \dfrac{9 + 11}{4}, x = \dfrac{9 - 11}{4}$

$x = \dfrac{20}{4}, x = \dfrac{-2}{4}$

$x = 5, x = -0.5$

9.

$-4x^2 + 7x - 8 = -20$

$-4x^2 + 7x + 12 = 0$

$a = -4, b = 7, c = 12$

$x = \dfrac{-7 \pm \sqrt{7^2 - 4(-4)(12)}}{2(-4)}$

$x = \dfrac{-7 \pm \sqrt{241}}{-8}$

$x = \dfrac{-7 + \sqrt{241}}{-8}, x = \dfrac{-7 - \sqrt{241}}{-8}$

$x \approx -1.07, x \approx 2.82$

11. $L(t) = 0.0572t^2 - 8.5587t + 320.3243$

a.

$L(55) = 0.0572(55)^2 - 8.5587(55) + 320.3243$

$L(55) = 22.63$

In 1955, about 22.63 thousand tons of lead were used in paint in the United States

b.

$5.5 = 0.0572t^2 - 8.5587t + 320.3243$
$-5.5 \qquad\qquad -5.5$
$0 = 0.0572t^2 - 8.5587t + 314.8243$

$t = \dfrac{8.5587 \pm \sqrt{(-8.5587)^2 - 4(0.0572)(314.8243)}}{2(0.0572)}$

$t = \dfrac{8.5587 + \sqrt{1.21954585}}{0.1144} \approx \cancel{84.5}$

$t = \dfrac{8.5587 - \sqrt{1.21954585}}{0.1144} \approx 65.2$

In about 1965, 5500 tons of lead were used in paints in the United States. 1984 would be considered model breakdown.

c.

$51 = 0.0572t^2 - 8.5587t + 320.3243$
$-51 \qquad\qquad -51$
$0 = 0.0572t^2 - 8.5587t +$

$t = \dfrac{8.5587 \pm \sqrt{(-8.5587)^2 - 4(0.0572)(269.3243)}}{2(0.0572)}$

$t = \dfrac{8.5587 + \sqrt{11.62994585}}{0.1144} \approx \cancel{104.6}$

$t = \dfrac{8.5587 - \sqrt{11.62994585}}{.01144} \approx 45.0$

In 1945, about 51 thousand tons of lead was used in paints in the United States. 2004 would be model breakdown.

13. $I(t) = 3t^2 - 4t + 5$

a.

$I(0.5) = 3(0.5)^2 - 4(0.5) + 5$
$I(0.5) = 3.75$

The current in the wire after 0.5 second would be about 3.75 amperes.

b.

$20 = 3t^2 - 4t + 5$
$-20 \qquad -20$
$0 = 3t^2 - 4t - 15$

$t = \dfrac{4 \pm \sqrt{(4)^2 - 4(3)(-15)}}{2(3)}$

$t = \dfrac{4 + \sqrt{196}}{6} = 3$

$t = \dfrac{4 - \sqrt{196}}{6} = \cancel{\dfrac{5}{3}}$

After about 3 seconds, the current would reach 20 amperes.

c.

$60 = 3t^2 - 4t + 5$
$-60 \qquad -60$
$0 = 3t^2 - 4t - 55$

$t = \dfrac{4 \pm \sqrt{(4)^2 - 4(3)(-55)}}{2(3)}$

$t = \dfrac{4 + \sqrt{676}}{6} = 5$

$t = \dfrac{4 - \sqrt{676}}{6} = \cancel{-\dfrac{11}{3}}$

After about 5 seconds, the current would reach 60 amperes.

15. $M(n) = 0.00045n^2 + 0.02n + 5$

a.

$M(200) = 0.00045(200)^2 + 0.02(200) + 5$
$M(200) = 27$

The marginal cost of producing the 201st pair of shoes would be $27.

b.

$20 = 0.00045n^2 + 0.02n + 5$
$-20 \qquad\qquad\qquad -20$
$0 = 0.00045n^2 + 0.02n - 15$

$n = \dfrac{-0.02 \pm \sqrt{0.02^2 - 4(0.02)(-15)}}{2(0.02)}$

$n = \dfrac{-0.02 + \sqrt{0.0274}}{0.04} \approx 161.7 \approx 162$

$n = \dfrac{-0.02 - \sqrt{0.0274}}{0.04} \approx \cancel{206}$

For the marginal cost to be $20 per pair of shoes you would be producing the 163rd pair of shoes.

17. $M(n) = 0.0005n^2 + 0.07n + 50$

a.

$M(400) = 0.0005(400)^2 + 0.07(400) + 50$
$M(400) = 158$

The marginal cost of producing the 401st bike would be $158.

b.

$700 = 0.0005n^2 + 0.07n + 50$
$\underline{-700 \qquad\qquad\qquad -700}$
$0 = 0.0005n^2 + 0.07n - 650$

$n = \dfrac{-0.07 \pm \sqrt{0.07^2 - 4(0.0005)(-650)}}{2(0.0005)}$

$n = \dfrac{-0.07 + \sqrt{1.3049}}{0.001} \approx 1072$

$n = \dfrac{-0.07 - \sqrt{1.3049}}{0.001} \approx \cancel{1212}$

For the marginal cost to be $700 per bike, you would be producing the 1073rd bike.

19. $h(t) = -16t^2 + 200t + 2$

a.

$h(1) = -16(1)^2 + 200(1) + 2$
$h(1) = 186$

One second after launch, the rocket was 186 feet high.

b.

$450 = -16t^2 + 200t + 2$
$\underline{-450 \qquad\qquad\qquad -450}$
$0 = -16t^2 + 200t - 448$

$t = \dfrac{-200 \pm \sqrt{200^2 - 4(-16)(-448)}}{2(-16)}$

$t = \dfrac{-200 + \sqrt{11{,}328}}{-32} \approx 2.9$

$t = \dfrac{-200 - \sqrt{11{,}328}}{-32} \approx 9.6$

The rocket first reached 450 feet 2.9 seconds into the flight.

c.

$600 = -16t^2 + 200t + 2$
$\underline{-600 \qquad\qquad\qquad -600}$
$0 = -16t^2 + 200t - 598$

$t = \dfrac{-200 \pm \sqrt{200^2 - 4(-16)(-598)}}{2(-16)}$

$t = \dfrac{-200 + \sqrt{1728}}{-32} \approx 4.95$

$t = \dfrac{-200 - \sqrt{1728}}{-32} \approx 7.55$

The rocket will first reach a height of 600 feet about 4.95 seconds into the flight.

d.

$700 = -16t^2 + 200t + 2$
$\underline{-700 \qquad\qquad\qquad -700}$
$0 = -16t^2 + 200t - 698$

$t = \dfrac{-200 \pm \sqrt{200^2 - 4(-16)(-698)}}{2(-16)}$

$t = \dfrac{-200 \pm \sqrt{-4672}}{-32}$

Nonreal answers.

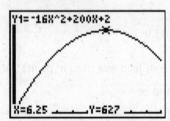

The rocket never reached a height of 700 feet. The rocket reached a maximum of 627 feet.

21. $I(t) = 3.8t^2 - 29.5t + 56.0$

a.

$I(10) = 3.8(10)^2 - 29.5(10) + 56.0$
$I(10) = 141$

In 2000, an average person spent 141 hours on the Internet.

b.

$100 = 3.8t^2 - 29.5t + 56.0$
$\underline{-100 \qquad\qquad -100}$
$0 = 3.8t^2 - 29.5t - 44$

$t = \dfrac{29.5 \pm \sqrt{(-29.5)^2 - 4(3.8)(-44)}}{2(3.8)}$

$t = \dfrac{29.5 + \sqrt{1539.05}}{7.6} \approx 9.04$

$t = \dfrac{29.5 - \sqrt{1539.05}}{7.6} \approx \cancel{-1.28}$

In 1999, an average person spent 100 hours on the Internet. 1989 would be model breakdown.

c.

$365 = 3.8t^2 - 29.5t + 56.0$
$\underline{-365 \qquad\qquad -365}$
$0 = 3.8t^2 - 29.5t - 309$

$t = \dfrac{29.5 \pm \sqrt{(-29.5)^2 - 4(3.8)(-309)}}{2(3.8)}$

$t = \dfrac{29.5 + \sqrt{5567.05}}{7.6} \approx 14$

$t = \dfrac{29.5 - \sqrt{5567.05}}{7.6} \approx \cancel{-6}$

In 2004, an average person will spend about 1 hour per day on the Internet. 1984 would be model breakdown.

23.

$a^2 + 2a = 15$
$a^2 + 2a - 15 = 0$
$(a+5)(a-3) = 0$
$a + 5 = 0, a - 3 = 0$
$a = -5, a = 3$

25.

$5t^2 - 14 = 0$
$5t^2 = 14$
$t^2 = \dfrac{14}{5}$
$\sqrt{t^2} = \pm\sqrt{\dfrac{14}{5}}$
$t = \pm\sqrt{\dfrac{14}{5}}$
$t \approx \pm 1.67$

27.

$\dfrac{1}{7}x^2 - \dfrac{5}{7}x = 0$
$7\left(\dfrac{1}{7}x^2\right) - 7\left(\dfrac{5}{7}x\right) = 7(0)$
$x^2 - 5x = 0$
$x(x-5) = 0$
$x = 0, x - 5 = 0$
$x = 0, x = 5$

29.

$4.7x^2 - 2.6x = 0$
$x(4.7x - 2.6) = 0$
$x = 0, 4.7x - 2.6 = 0$
$x = 0, 4.7x = 2.6$
$x = 0, x = \dfrac{2.6}{4.7}$
$x = 0, x \approx 0.55$

31.

$5x - 12 = 80$
$5x = 92$
$x = \dfrac{92}{5}$
$x = 18.4$

33.

$5x^2 + 3x = 0$
$x(5x + 3) = 0$
$x = 0, 5x + 3 = 0$
$x = 0, 5x = -3$
$x = 0, x = -\dfrac{3}{5}$

35.

$(x-9)^2 + 8 = 24$
$(x-9)^2 = 16$
$\sqrt{(x-9)^2} = \pm\sqrt{16}$
$x - 9 = \pm 4$
$x - 9 = 4, x - 9 = -4$
$x = 13, x = 5$

Chapter 4 Quadratic Functions

37.

$3(4x-5)^2 + 20 = 47$

$3(4x-5)^2 = 27$

$(4x-5)^2 = 9$

$\sqrt{(4x-5)^2} = \pm\sqrt{9}$

$4x - 5 = \pm 3$

$4x - 5 = 3, 4x - 5 = -3$

$4x = 8, 4x = 2$

$x = 2, x = 0.5$

39.

$33 = -7(4-w)^2 + 59$

$-26 = -7(4-w)^2$

$\dfrac{-26}{-7} = (4-w)^2$

$\pm\sqrt{\dfrac{26}{7}} = \sqrt{(4-w)^2}$

$\pm\sqrt{\dfrac{26}{7}} = 4-w$

$\pm\sqrt{\dfrac{26}{7}} - 4 = 4-w-4$

$\pm\sqrt{\dfrac{26}{7}} - 4 = -w$

$\dfrac{\pm\sqrt{\dfrac{26}{7}} - 4}{-1} = w$

$w = 4 \pm \sqrt{\dfrac{26}{7}}$

$w = 4 + \sqrt{\dfrac{26}{7}}, w = 4 - \sqrt{\dfrac{26}{7}}$

$w \approx 5.93, w \approx 2.07$

41.

$r^2 + 1.4r - 14.9 = 0$

$a = 1, b = 1.4, c = -14.9$

$r = \dfrac{-1.4 \pm \sqrt{1.4^2 - 4(1)(-14.9)}}{2(1)}$

$r = \dfrac{-1.4 \pm \sqrt{61.56}}{2}$

$r = \dfrac{-1.4 + \sqrt{61.56}}{2}, r = \dfrac{-1.4 - \sqrt{61.56}}{2}$

$r \approx 3.22, r \approx -4.62$

43.

$d^2 + 2d - 35 = 0$

$(d+7)(d-5) = 0$

$d + 7 = 0, d - 5 = 0$

$d = -7, d = 5$

45.

$b^2 - 3b = 28$

$b^2 - 3b - 28 = 0$

$(b+4)(b-7) = 0$

$b + 4 = 0, b - 7 = 0$

$b = -4, b = 7$

47.

$25 = -2(s+9)^2 + 5$

$20 = -2(s+9)^2$

$-10 = (s+9)^2$

$(s+9)^2 = -10$

$\sqrt{(s+9)^2} = \pm\sqrt{-10}$

This is not a real number.

No real solutions.

49.

$\dfrac{2}{3}c^2 - \dfrac{5}{6} = \dfrac{1}{6}c - 2$

$6\left(\dfrac{2}{3}c^2 - \dfrac{5}{6}\right) = 6\left(\dfrac{1}{6}c - 2\right)$

$4c^2 - 5 = c - 12$

$4c^2 - c + 7 = 0$

$a = 4, b = -1, c = 7$

$c = \dfrac{-(-1) \pm \sqrt{(-1)^2 - 4(4)(7)}}{2(4)}$

$c = \dfrac{1 \pm \sqrt{-111}}{8}$

This is not a real number.

No real solutions.

4.6 Exercises

51.

$120 = -28f + 7f^2 - 939$

$0 = 7f^2 - 28f - 1059$

$a = 7, b = -28, c = -1059$

$f = \dfrac{-(-28) \pm \sqrt{(-28)^2 - 4(7)(-1059)}}{2(7)}$

$f = \dfrac{28 \pm \sqrt{30436}}{14}$

$f = \dfrac{28 + \sqrt{30436}}{14}, f = \dfrac{28 - \sqrt{30436}}{14}$

$f \approx 14.46, f \approx -10.46$

53.

$(3p - 4)(p + 3) = 15$

$3p^2 + 5p - 12 = 15$

$3p^2 + 5p - 27 = 0$

$a = 3, b = 5, c = -27$

$p = \dfrac{-5 \pm \sqrt{5^2 - 4(3)(-27)}}{2(3)}$

$p = \dfrac{-5 \pm \sqrt{349}}{6}$

$p = \dfrac{-5 + \sqrt{349}}{6}, p = \dfrac{-5 - \sqrt{349}}{6}$

$p \approx 2.28, p \approx -3.95$

55.

$3x^2 + 4x + 20 = 0$

$a = 3, b = 4, c = 20$

$x = \dfrac{-4 \pm \sqrt{4^2 - 4(3)(20)}}{2(3)}$

$x = \dfrac{-4 \pm \sqrt{-224}}{6}$

This is not a real number.
No real solutions.

57.

$\dfrac{3}{2}(x - 8)^2 + 10 = 1$

$\dfrac{3}{2}(x - 8)^2 = -9$

$\dfrac{2}{3}\left(\dfrac{3}{2}(x - 8)^2\right) = \dfrac{2}{3}(-9)$

$(x - 8)^2 = -6$

$\sqrt{(x - 8)^2} = \pm\sqrt{-6}$

This is not a real number.
No real solutions.

59.

$3x^3 - 15x^2 = 252x$

$3x^3 - 15x^2 - 252x = 0$

$3x(x^2 - 5x - 84) = 0$

$3x(x - 12)(x + 7) = 0$

$3x = 0, x - 12 = 0, x + 7 = 0$

$x = 0, x = 12, x = -7$

61.

$7(x + 3) - 8 = 20$

$7x + 21 - 8 = 20$

$7x + 13 = 20$

$7x = 7$

$x = 1$

63.

$-1.5(d + 5) - 7 = 2d + 8$

$-1.5d - 7.5 - 7 = 2d + 8$

$-1.5d - 14.5 = 2d + 8$

$-1.5d = 2d + 22.5$

$-3.5d = 22.5$

$d \approx -6.43$

65.

$x^2 + 6x + 25 = 0$

$a = 1, b = 6, c = 25$

$x = \dfrac{-6 \pm \sqrt{6^2 - 4(1)(25)}}{2(1)}$

$x = \dfrac{-6 \pm \sqrt{-64}}{2}$

This is not a real number.
No real solutions.

Chapter 4 Quadratic Functions

67.
$\frac{1}{4}x^2 - \frac{3}{4}x + 7 = 13$

$4\left(\frac{1}{4}x^2 - \frac{3}{4}x + 7\right) = 4(13)$

$x^2 - 3x + 28 = 52$

$x^2 - 3x - 24 = 0$

$a = 1, b = -3, c = -24$

$x = \frac{-(-3) \pm \sqrt{(-3)^2 - 4(1)(-24)}}{2(1)}$

$x = \frac{3 \pm \sqrt{105}}{2}$

$x = \frac{3 + \sqrt{105}}{2}, x = \frac{3 - \sqrt{105}}{2}$

$x \approx 6.62, x \approx -3.62$

69. $(3,10),(-4,10)$

71. No real solution.

73. $(-5,-1)$

75. $(-3,-25)$

77. $(-4,12),(3,12)$

79. $\approx (-3.71, 0.295),(2.71, 6.71)$

81.
$5x - 8 = x^2 + 3x - 9$
$\underline{-5x + 8 \quad\quad -5x + 8}$
$0 = x^2 - 2x - 1$

$x = \frac{2 \pm \sqrt{(-2)^2 - 4(1)(-1)}}{2(1)}$

$x = \frac{2 + \sqrt{8}}{2} \approx 2.41, x = \frac{2 - \sqrt{8}}{2} \approx -0.41$

$y \approx 5(2.41) - 8, y \approx 5(-0.41) - 8$

$y \approx 4.05, y \approx -10.05$

$(2.41, 4.05), (-0.41, -10.05)$

83.
$2x - 10 = x^2 + 5x - 3$
$\underline{-2x + 10 \quad\quad -2x + 10}$
$0 = x^2 + 3x + 7$

$x = \frac{-3 \pm \sqrt{3^2 - 4(1)(7)}}{2(1)}$

$x = \frac{-3 \pm \sqrt{-19}}{2}$

No real solution.

85.
$-0.5x^2 - 3x + 15 = 3x^2 + 5x - 9$
$\underline{+0.5x^2 + 3x - 15 \quad +0.5x^2 + 3x - 15}$
$0 = 3.5x^2 + 8x - 24$

$x = \frac{-8 \pm \sqrt{(8)^2 - 4(3.5)(-24)}}{2(3.5)}$

$x = \frac{-8 + \sqrt{400}}{7} \approx 1.71$

$x = \frac{-8 - \sqrt{400}}{7} = -4$

$y = 3(1.71)^2 + 5(1.71) - 9 \approx 8.32$

$y = 3(-4)^2 + 5(-4) - 9 = 19$

$(1.71, 8.32), (-4, 19)$

87.
$-x^2 + 7x - 4 = x^2 - 4x + 11$
$\underline{+x^2 - 7x + 4 \quad +x^2 - 7x + 4}$
$0 = 2x^2 - 11x + 15$

$0 = (2x - 5)(x - 3)$

$2x - 5 = 0 \rightarrow x = \frac{5}{2} = 2.5$

$y = (2.5)^2 - 4(2.5) + 11 = -3.75$

$x - 3 = 0 \rightarrow x = 3$

$y = (3)^2 - 4(3) + 11 = 8$

$(2.5, 7.25), (3, 8)$

89.

$$2.5x^2 + 3.4x - 8.5 = -1.8x^2 - 2.3x + 4.7$$
$$\underline{-2.5x^2 - 3.4x + 8.5} \quad \underline{-2.5x^2 - 3.4x + 8.5}$$
$$0 = -4.3x^2 - 5.7x + 13.2$$

$$x = \frac{5.7 \pm \sqrt{(-5.7)^2 - 4(-4.3)(13.2)}}{2(-4.3)}$$

$$x = \frac{5.7 + \sqrt{259.53}}{-8.6} \approx 1.2105 \approx 1.21$$

$$x = \frac{5.7 - \sqrt{259.53}}{-8.6} \approx -2.536 \approx -2.54$$

$$y \approx 2.5(1.2105)^2 + 3.4(1.2105) - 8.5 \approx -0.72$$
$$y \approx 2.5(-2.536)^2 + 3.4(-2.536) - 8.5 \approx -1.04$$
$$(1.21, -0.72), (-2.54, -1.04)$$

91.

$$-x^2 - 6x - 38 = x^2 + 6x - 20$$
$$\underline{+x^2 + 6x + 38} \quad \underline{+x^2 + 6x + 38}$$
$$0 = 2x^2 + 12x + 18$$
$$\frac{0}{2} = \frac{2x^2 + 12x + 18}{2}$$
$$0 = x^2 + 6x + 9$$
$$0 = (x+3)^2$$
$$x + 3 = 0 \rightarrow x = -3$$
$$y = (-3)^2 + 6(-3) - 20 = -29$$
$$(-3, -29)$$

93.

$$-2x^2 + 2x - 18 = x^2 + 2x - 8$$
$$\underline{+2x^2 - 2x + 18} \quad \underline{+2x^2 - 2x + 18}$$
$$0 = 3x^2 + 10$$
$$-3x^2 = 10$$
$$x^2 = \frac{10}{-3}$$
$$x = \sqrt{\frac{10}{-3}}$$

No real solution.

95.

$$-x^2 + 6x - 15 = x^2 - 10x + 30$$
$$\underline{+x^2 - 6x + 15} \quad \underline{+x^2 - 6x + 15}$$
$$0 = 2x^2 - 16x + 45$$

$$x = \frac{16 \pm \sqrt{(-16)^2 - 4(2)(45)}}{2(2)}$$

$$x = \frac{16 \pm \sqrt{-104}}{4}$$

No real solution.

97.

$$9x^2 + 2x - 15 = 6x^2 + 2x - 9$$
$$\underline{-9x^2 - 2x + 15} \quad \underline{-9x^2 - 2x + 15}$$
$$0 = -3x^2 + 6$$
$$3x^2 = 6$$
$$x^2 = 2 \rightarrow x = \pm\sqrt{2} \approx \pm 1.41$$
$$y = 6(-1.41)^2 + 2(-1.41) - 9 \approx 0.11$$
$$y = 6(1.41)^2 + 2(1.41) - 9 \approx 5.75$$
$$(-1.41, 0.11), (1.41, 5.75)$$

Chapter 4 Quadratic Functions
Section 4.7

1.

$y = -2x^2 - 6x - 7$: C, downward, y-int: $(0, -7)$.

$y = -0.5x^2 - x + 7$: A, downward, y-int: $(0, 7)$.

$y = x^2 + 4x + 4$: B, upward, y-int: $(0, 4)$.

$y = x^2 - 5x - 6$: None, upward, y-int: $(0, -6)$.

3.

$y = -2(x-3)^2 - 9$: None, downward.

$y = 2(x-3)^2 - 9$: A, upward, vertex: $(3, -9)$.

$y = 2(x+4)^2 + 1$: C, upward, vertex: $(-4, 1)$.

$y = 2(x+4)^2 - 9$: B, upward, vertex: $(-4, -9)$.

5.

$f(x) = x^2 + 6x + 8$

Vertex: (h, k)

$h = \dfrac{-6}{2} = -3$

$k = f(-3) = (-3)^2 + 6(-3) + 8 = -1$

$(-3, -1)$

Vertical intercept:

$f(0) = 8$

$(0, 8)$

Horizontal intercepts:

$x^2 + 6x + 8 = 0$

$(x+4)(x+2) = 0$

$x + 4 = 0$, $x + 2 = 0$

$x = -4$, $x = -2$

$(-4, 0), (-2, 0)$

7. $h(x) = 3x^2 - 18x + 15$

Vertex: (h, k)

$h = \dfrac{18}{2(3)} = 3$

$k = h(3) = 3(3)^2 - 18(3) + 15 = -12$

$(3, -12)$

Vertical intercept:

$h(0) = 15$

$(0, 15)$

Horizontal intercepts:

$3x^2 - 18x + 15 = 0$

$3(x^2 - 6x + 5) = 0$

$3(x-5)(x-1) = 0$

$x - 5 = 0$, $x - 1 = 0$

$x = 5$, $x = 1$

$(5, 0), (1, 0)$

9. $h(x) = 4(x-8)^2 - 20$

Vertex: $(8, -20)$

Vertical intercept:

$h(0) = 4(-8)^2 - 20$

$h(0) = 236$

$(0, 236)$

Horizontal intercepts:

$4(x-8)^2 - 20 = 0$

$4(x-8)^2 = 20$

$\dfrac{4(x-8)^2}{4} = \dfrac{20}{4}$

$(x-8)^2 = 5$

$x - 8 = \pm\sqrt{5}$

$x = 8 + \sqrt{5} \approx 10.236$

$x = 8 - \sqrt{5} \approx 5.764$

$(10.236, 0), (5.764, 0)$

11. $g(x) = 5x^2 + 12x + 10$

Vertex: (h, k)

$h = \dfrac{-12}{2(5)} = -1.2$

$k = g(-1.2) = 5(-1.2)^2 + 12(-1.2) + 10 = 2.8$

$(-1.2, 2.8)$

Vertical intercept:

$g(0) = 10$

$(0, 10)$

Horizontal intercepts:

$5x^2 + 12x + 10 = 0$

$x = \dfrac{-12 \pm \sqrt{(12)^2 - 4(5)(10)}}{2(5)}$

$x = \dfrac{-12 + \sqrt{-56}}{10}$

There are no horizontal intercepts.

13. $f(x) = 1.5x^2 - 6x + 4$

Vertex: (h, k)

$h = \dfrac{6}{2(1.5)} = \dfrac{6}{3} = 2$

$k = f(2) = 1.5(2)^2 - 6(2) + 4 = -2$

$(2, -2)$

Vertical intercept:

$f(0) = 4$

$(0, 4)$

Horizontal intercepts:

$1.5x^2 - 6x + 4 = 0$

$x = \dfrac{6 \pm \sqrt{(-6)^2 - 4(1.5)(4)}}{2(1.5)}$

$x = \dfrac{6 + \sqrt{12}}{3} \approx 3.155$

$x = \dfrac{6 - \sqrt{12}}{3} \approx 0.845$

$(3.155, 0), (0.845, 0)$

15. $C(u) = 0.25u^2 - 25u + 3500$

a.

$C(30) = 0.25(30)^2 - 25(30) + 3500$

$C(30) = 2975$

The cost to produce 30 uniforms would be about $2975.

b. Vertex: (h, k)

$h = \dfrac{25}{2(0.25)} = \dfrac{25}{0.5} = 50$

$k = C(50) = 0.25(50)^2 - 25(50) + 3500$

$C(50) = 2875$

The vertex is $(50, 2875)$, which means the minimum cost would be $2875, producing 50 uniforms.

c.

$1600 = 0.25u^2 - 25u + 3500$

$\underline{-1600 \qquad\qquad\qquad -1600}$

$0 = 0.25u^2 - 25u + 1900$

$u = \dfrac{25 \pm \sqrt{(25)^2 - 4(0.25)(1900)}}{2(0.25)}$

$u = \dfrac{25 \pm \sqrt{-2875}}{0.5}$

According to the model, the school can never produce uniforms for $1600, which would be model breakdown.

17. $R(c) = -3c^2 + 90c$

a.

$R(5) = -3(5)^2 + 90(5)$

$R(5) = 375$

The revenue from selling 5 thousand digital cameras would be $375 thousand.

b.

$600 = -3c^2 + 90c$

$3c^2 - 90c + 600 = 0$

$\dfrac{3c^2 - 90c + 600}{3} = \dfrac{0}{3}$

$c^2 - 30c + 200 = 0$

$(c - 10)(c - 20) = 0$

$c = 10$ or $c = 20$

The company must sell 10 thousand cameras or 20 thousand cameras to generate revenue of $600,000.

c. The maximum would occur at the input value h for the vertex of this model.

(h, k)

$h = \dfrac{-90}{2(-3)} = \dfrac{-90}{-6} = 15$

The company must sell 15 thousand cameras to maximize its revenue.

Chapter 4 Quadratic Functions

19. $N(t) = 376.5t^2 + 548.1t + 2318.4$

a.

$N(5) = 376.5(5)^2 + 548.1(5) + 2318.4$

$N(5) = 14,471.4$

According to this model, the annual net sales for Home Depot in 1995, was about $14,471 million.

b.

$30,000 = 376.5t^2 + 548.1t + 2318.4$

$\underline{-30,000 \qquad\qquad\qquad -30,000}$

$0 = 376.5t^2 + 548.1t - 27,681.6$

$t = \dfrac{-548.1 \pm \sqrt{548.1^2 - 4(376.5)(-27,681.6)}}{2(376.5)}$

$t = \dfrac{-548.1 + \sqrt{41,988,903.21}}{753} \approx 7.878$

$t = \dfrac{-548.1 - \sqrt{41,988,903.21}}{753} \approx \cancel{-9.333}$

Home Depot's net sales were $30,000 million in about 1998.

c.

Vertex: (h, k)

$h = \dfrac{-548.1}{2(376.5)} = \dfrac{-10}{1.5} \approx -0.728$

$k \approx N(-0.728) = 376.5(-0.728)^2 + 548.1(-0.728) + 2318.4$

$N(-0.728) \approx 2118.9$

The vertex is $(-0.728, 2118.9)$, which means that in 1989, Home Depot's net sales were at a low point of $2118.90 million.

21. $I(t) = -1.5t^2 + 32.3t - 138.8$

a.

$I(9) = -1.5(9)^2 + 32.3(9) - 138.8$

$I(9) = 30.4$

According to this model, the net income for Quicksilver in 1999 was about $30.4 million.

b.

$18 = -1.5t^2 + 32.3t - 138.8$

$\underline{-18 \qquad\qquad\qquad -18}$

$0 = -1.5t^2 + 32.3t - 156.8$

$t = \dfrac{-32.3 \pm \sqrt{32.3^2 - 4(-1.5)(-156.8)}}{2(-1.5)}$

$t = \dfrac{-32.3 + \sqrt{102.49}}{-3} \approx 7.39$

$t = \dfrac{-32.3 - \sqrt{102.49}}{-3} \approx 14.14$

According to this model, Quicksilver's net income reached $18 million in 1997 and again in 2004.

c. Vertex: (h, k)

$h = \dfrac{-32.3}{2(-1.5)} \approx 10.77$

$k \approx I(10.77) = -1.5(10.77)^2 + 32.3(10.77) - 138.8$

$I(10.77) \approx 35.08$

The vertex is $(10.77, 35.08)$, which means that in about 2001, Quicksilver reached a maximum net sales of $35.08 million.

23. $h(t) = -16t^2 + 40t + 4$

a.

$h(0) = -16(0)^2 + 40(0) + 4$

$h(0) = 4$

The ball is at a height of 4 feet when it is hit.

b.

$20 = -16t^2 + 40t + 4$

$\underline{-20 \qquad\qquad\qquad -20}$

$0 = -16t^2 + 40t - 16$

$\dfrac{0}{-8} = \dfrac{-16t^2 + 40t - 16}{-8}$

$0 = 2t^2 - 5t + 2$

$0 = (2t - 1)(t - 2)$

$2t - 1 = 0 \quad , \quad t - 2 = 0$

$2t = 1$

$t = 0.5 \quad , \quad t = 2$

The ball reached a height of 20 feet about 0.5 second and again 2 seconds after being hit.

c. Find the input value for the vertex of this model.

$$\frac{-b}{2a} = \frac{-40}{2(-16)} = 1.25$$

The ball reached its maximum height 1.25 seconds after being hit.

d. Find the output value for the vertex of this model when the input value is 1.25.

$$h(1.25) = -16(1.25)^2 + 40(1.25) + 4$$
$$h(1.25) = 29$$

The ball reached a maximum height of 29 feet.

e.

$$0 = -16t^2 + 40t + 4$$
$$t = \frac{-40 \pm \sqrt{40^2 - 4(-16)(4)}}{2(-16)}$$
$$t = \frac{-40 + \sqrt{1856}}{-32} \approx \cancel{-0.096}$$
$$t = \frac{-40 - \sqrt{1856}}{-32} \approx 2.596$$

If the ball does not get caught, the ball will hit the ground 2.596 seconds after being hit.

25. $h(t) = -16t^2 + 60t + 4.2$

a.

$$h(0) = -16(0)^2 + 60(0) + 4.2$$
$$h(0) = 4.2$$

The ball is at a height of 4.2 feet when it is hit.

b.

$$40 = -16t^2 + 60t + 4.2$$
$$\underline{-40 \qquad\qquad -40}$$
$$0 = -16t^2 + 60t - 35.8$$
$$t = \frac{-60 \pm \sqrt{60^2 - 4(-16)(4.2)}}{2(-16)}$$
$$t = \frac{-60 + \sqrt{1308.8}}{-32} \approx 0.744$$
$$t = \frac{-60 - \sqrt{1308.8}}{-32} \approx 3.0055$$

The ball reached a height of 40 feet at 0.744 second and again 3.0055 seconds after being hit.

c. Find the vertex for this model.

$$\frac{-b}{2a} = \frac{-60}{2(-16)} = 1.875$$
$$h(1.875) = -16(1.875)^2 + 60(1.875) + 4.2$$
$$h(1.875) = 60.45$$

The ball's maximum height is 60.45 feet.

d.

$$0 = -16t^2 + 60t + 4.2$$
$$t = \frac{-60 \pm \sqrt{60^2 - 4(-16)(4.2)}}{2(-16)}$$
$$t = \frac{-60 + \sqrt{3868.8}}{-32} \approx \cancel{-0.069}$$
$$t = \frac{-60 - \sqrt{3868.8}}{-32} \approx 3.819$$

If the ball does not get caught, the ball will hit the ground 3.819 seconds after being hit.

27. $H(m) = 0.9m^2 - 13m + 104$

a.

$$H(6) = 0.9(6)^2 - 13(6) + 104$$
$$H(6) = 58.4$$

In June, the average high temperature in Melbourne, Australia is 58.4°F

b.

$$\frac{-b}{2a} = \frac{13}{2(0.9)} \approx 7.22$$
$$H(7.22) = 0.9(7.22)^2 - 13(7.22) + 104$$
$$H(7.22) \approx 57.056$$

The vertex is $(7.22, 57.056)$, which means that in July, the average high temperature in Melbourne, Australia, reaches its minimum of 57.056°F.

Chapter 4 Quadratic Functions

c.

$60 = 0.9m^2 - 13m + 104$
$\underline{-60 \qquad\qquad -60}$
$0 = 0.9m^2 - 13m + 44$

$m = \dfrac{13 \pm \sqrt{(13)^2 - 4(0.9)(44)}}{2(.9)}$

$m = \dfrac{13 + \sqrt{10.6}}{1.8} \approx 9.03$

$m = \dfrac{13 - \sqrt{10.6}}{1.8} \approx 5.41$

During September and May, the average temperature of Melbourne, Australia, is about 60°F.

29. $f(x) = x^2 + 2x - 15$

To find the vertex use $\left(-\dfrac{b}{2a}, f\left(-\dfrac{b}{2a}\right)\right)$.

$-\dfrac{b}{2a} = -\dfrac{2}{2(1)} = -1$

$f\left(-\dfrac{b}{2a}\right) = f(-1) = (-1)^2 + 2(-1) - 15 = -16$

Vertex $= (-1, -16)$

To find the vertical intercept set $x = 0$.

$f(0) = (0)^2 + 2(0) - 15 = -15$

Vertical intercept $= (0, -15)$

To find the horizontal intercepts set $f(x) = 0$.

$x^2 + 2x - 15 = 0$
$(x+5)(x-3) = 0$
$x + 5 = 0, x - 3 = 0$
$x = -5, x = 3$

Horizontal intercepts $= (-5, 0)$ and $(3, 0)$.

Domain: $(-\infty, \infty)$, range: $[-16, \infty)$

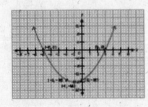

31. $m(b) = -b^2 + 11b - 24$

To find the vertex use $\left(-\dfrac{b}{2a}, m\left(-\dfrac{b}{2a}\right)\right)$.

$-\dfrac{b}{2a} = -\dfrac{11}{2(-1)} = 5.5$

$m\left(-\dfrac{b}{2a}\right) = m(5.5) = -(5.5)^2 + 11(5.5) - 24 = 6.25$

Vertex $= (5.5, 6.25)$

To find the vertical intercept set $b = 0$.

$m(0) = -(0)^2 + 11(0) - 24 = -24$

Vertical intercept $= (0, -24)$

To find the horizontal intercepts set $m(b) = 0$.

$-b^2 + 11b - 24 = 0$
$(-b + 8)(b - 3) = 0$
$-b + 8 = 0, b - 3 = 0$
$b = 8, b = 3$

Horizontal intercepts $= (8, 0)$ and $(3, 0)$

Domain: $(-\infty, \infty)$, range: $(-\infty, 6.25]$

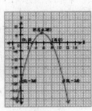

33. $g(s) = 2s^2 - 62s + 216$

To find the vertex use $\left(-\dfrac{b}{2a}, g\left(-\dfrac{b}{2a}\right)\right)$.

$-\dfrac{b}{2a} = -\dfrac{-62}{2(2)} = 15.5$

$g\left(-\dfrac{b}{2a}\right) = g(15.5) = 2(15.5)^2 - 62(15.5) + 216 = -264.5$

Vertex $= (15.5, -264.5)$

To find the vertical intercept set $s = 0$.

$g(0) = 2(0)^2 - 62(0) + 216 = 216$

Vertical intercept $= (0, 216)$

To find the horizontal intercepts set $g(s) = 0$.
$2s^2 - 62s + 216 = 0$
$(2s - 54)(s - 4) = 0$
$2s - 54 = 0, s - 4 = 0$
$s = 27, s = 4$
Horizontal intercepts $= (27, 0)$ and $(4, 0)$

Domain: $(-\infty, \infty)$, range: $[-264.5, \infty)$

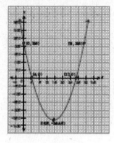

35. $f(x) = 2x^2 + 5$

To find the vertex use $\left(-\dfrac{b}{2a}, f\left(-\dfrac{b}{2a}\right)\right)$.

$-\dfrac{b}{2a} = -\dfrac{0}{2(2)} = 0$

$f\left(-\dfrac{b}{2a}\right) = f(0) = 2(0)^2 + 5 = 5$

Vertex $= (0, 5)$

To find the vertical intercept set $x = 0$.
$f(0) = 2(0)^2 + 5 = 5$
Vertical intercept $= (0, 5)$

To find the horizontal intercepts set $f(x) = 0$.
$2x^2 + 5 = 0$
$a = 2, b = 0, c = 5$
$x = \dfrac{-0 \pm \sqrt{0^2 - 4(2)(5)}}{2(2)}$
$x = \dfrac{\pm\sqrt{-40}}{4}$ This is not a real number.
There are no horizontal intercepts.

Domain: $(-\infty, \infty)$, range: $[5, \infty)$

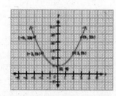

37. $d(p) = -1.5p^2 - 3$

To find the vertex use $\left(-\dfrac{b}{2a}, d\left(-\dfrac{b}{2a}\right)\right)$.

$-\dfrac{b}{2a} = -\dfrac{0}{2(-1.5)} = 0$

$d\left(-\dfrac{b}{2a}\right) = d(0) = -1.5(0)^2 - 3 = -3$

Vertex $= (0, -3)$

To find the vertical intercept set $p = 0$.
$d(0) = -1.5(0)^2 - 3 = -3$
Vertical intercept $= (0, -3)$

To find the horizontal intercepts set $d(p) = 0$.
$-1.5p^2 - 3 = 0$
$a = -1.5, b = 0, c = -3$
$p = \dfrac{-0 \pm \sqrt{0^2 - 4(-1.5)(-3)}}{2(-1.5)}$
$p = \dfrac{\pm\sqrt{-18}}{-3}$ This is not a real number.
There are no horizontal intercepts.

Domain: $(-\infty, \infty)$, range: $(-\infty, -3]$

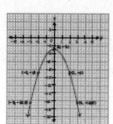

39. $h(x) = -\dfrac{1}{4}x^2 - 5$

To find the vertex use $\left(-\dfrac{b}{2a}, h\left(-\dfrac{b}{2a}\right)\right)$.

$-\dfrac{b}{2a} = -\dfrac{0}{2\left(-\dfrac{1}{4}\right)} = 0$

$h\left(-\dfrac{b}{2a}\right) = -\dfrac{1}{4}(0)^2 - 5 = -5$

Vertex $= (0, -5)$

To find the vertical intercept set $x = 0$.
$h(0) = -\dfrac{1}{4}(0)^2 - 5 = -5$
Vertical intercept $= (0, -5)$

Chapter 4 Quadratic Functions

To find the horizontal intercepts set $h(x) = 0$.

$-\dfrac{1}{4}x^2 - 5 = 0$

$a = -\dfrac{1}{4}, b = 0, c = -5$

$x = \dfrac{-0 \pm \sqrt{0^2 - 4\left(-\dfrac{1}{4}\right)(-5)}}{2\left(-\dfrac{1}{4}\right)}$

$x = \dfrac{\pm\sqrt{-5}}{-0.5}$ This is not a real number.

There are no horizontal intercepts.

Domain: $(-\infty, \infty)$, range: $(-\infty, -5]$

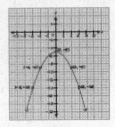

41. $p(k) = -5k^2 - 17.5k - 12.5$

To find the vertex use $\left(-\dfrac{b}{2a}, p\left(-\dfrac{b}{2a}\right)\right)$.

$-\dfrac{b}{2a} = -\dfrac{-17.5}{2(-5)} = -1.75$

$p\left(-\dfrac{b}{2a}\right) = p(-1.75) = -5(-1.75)^2 - 17.5(-1.75) - 12.5 = 2.8125$

Vertex $= (-1.75, 2.8125)$

To find the vertical intercept set $k = 0$.

$p(0) = -5(0)^2 - 17.5(0) - 12.5$

Vertical intercept $= (0, -12.5)$

To find the horizontal intercepts set $p(k) = 0$.

$-5k^2 - 17.5k - 12.5 = 0$

$a = -5, b = -17.5, c = -12.5$

$k = \dfrac{-(-17.5) \pm \sqrt{(-17.5)^2 - 4(-5)(-12.5)}}{2(-5)}$

$k = \dfrac{17.5 \pm \sqrt{56.25}}{-10}$

$k = \dfrac{17.5 \pm 7.5}{-10}$

$k = \dfrac{17.5 + 7.5}{-10}, k = \dfrac{17.5 - 7.5}{-10}$

$k \approx -2.5, k \approx -1$

Horizontal intercepts $= (-2.5, 0)$ and $(-1, 0)$

Domain: $(-\infty, \infty)$, range: $(-\infty, 2.8125]$

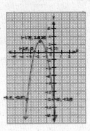

43. $h(w) = 0.4w^2 - 3.6w - 44.8$

To find the vertex use $\left(-\dfrac{b}{2a}, h\left(-\dfrac{b}{2a}\right)\right)$.

$-\dfrac{b}{2a} = -\dfrac{(-3.6)}{2(0.4)} = 4.5$

$h\left(-\dfrac{b}{2a}\right) = h(4.5) = 0.4(4.5)^2 - 3.6(4.5) - 44.8 = -52.9$

Vertex $= (4.5, -52.9)$

To find the vertical intercept set $w = 0$.

$h(0) = 0.4(0)^2 - 3.6(0) - 44.8 = -44.8$

Vertical intercept $= (0, -44.8)$

To find the horizontal intercepts set $h(w) = 0$.
$0.4w^2 - 3.6w - 44.8 = 0$
$a = 0.4, b = -3.6, c = -44.8$

$w = \dfrac{-(-3.6) \pm \sqrt{(-3.6)^2 - 4(0.4)(-44.8)}}{2(0.4)}$

$w = \dfrac{3.6 \pm \sqrt{84.64}}{0.8}$

$w = \dfrac{3.6 \pm 9.2}{0.8}$

$w = \dfrac{3.6 - 9.2}{0.8}, w = \dfrac{3.6 + 9.2}{0.8}$

$w = -7, w = 16$

Horizontal intercepts $= (-7, 0)$ and $(16, 0)$

Domain: $(-\infty, \infty)$, range: $[-52.8, \infty)$

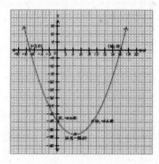

45. $p(x) = \dfrac{2}{5}x^2 - 2x - \dfrac{3}{5}$

To find the vertex use $\left(-\dfrac{b}{2a}, p\left(-\dfrac{b}{2a}\right)\right)$.

$-\dfrac{b}{2a} = -\dfrac{(-2)}{2\left(\dfrac{2}{5}\right)} = 2.5$

$p\left(-\dfrac{b}{2a}\right) = p(2.5) = \dfrac{2}{5}(2.5)^2 - 2(2.5) - \dfrac{3}{5} = -3.1$

Vertex $= (2.5, -3.1)$

To find the vertical intercept set $x = 0$.

$p(0) = \dfrac{2}{5}(0)^2 - 2(0) - \dfrac{3}{5} = -0.6$

Vertical intercept $= (0, -0.6)$

To find the horizontal intercepts set $p(x) = 0$.
$\dfrac{2}{5}x^2 - 2x - \dfrac{3}{5} = 0$

$a = \dfrac{2}{5}, b = -2, c = -\dfrac{3}{5}$

$x = \dfrac{-(-2) \pm \sqrt{(-2)^2 - 4\left(\dfrac{2}{5}\right)\left(-\dfrac{3}{5}\right)}}{2\left(\dfrac{2}{5}\right)}$

$x = \dfrac{2 \pm \sqrt{4.96}}{0.8}$

$x = \dfrac{2 - \sqrt{4.96}}{0.8}, x = \dfrac{2 + \sqrt{4.96}}{0.8}$

$x = -0.3, x = 5.3$

Horizontal intercepts $= (-0.3, 0)$ and $(5.3, 0)$

Domain: $(-\infty, \infty)$, range: $[-3.1, \infty)$

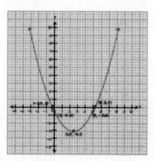

47. $Q(p) = -0.3p^2 - 2.4p + 82$

To find the vertex use $\left(-\dfrac{b}{2a}, Q\left(-\dfrac{b}{2a}\right)\right)$.

$-\dfrac{b}{2a} = -\dfrac{(-2.4)}{2(-0.3)} = -4$

$Q\left(-\dfrac{b}{2a}\right) = Q(-4) = -0.3(-4)^2 - 2.4(-4) + 82 = 86.8$

Vertex $= (-4, 86.8)$

To find the vertical intercept set $p = 0$.

$Q(0) = -0.3(0)^2 - 2.4(0) + 82 = 82$

Vertical intercept $= (0, 82)$

Chapter 4 Quadratic Functions

To find the horizontal intercepts set $Q(p) = 0$.

$-0.3p^2 - 2.4p + 82 = 0$

$a = -0.3, b = -2.4, c = 82$

$p = \dfrac{-(-2.4) \pm \sqrt{(-2.4)^2 - 4(-0.3)(82)}}{2(-0.3)}$

$p = \dfrac{2.4 \pm \sqrt{104.16}}{-0.6}$

$p = \dfrac{2.4 + \sqrt{104.16}}{-0.6}, p = \dfrac{2.4 - \sqrt{104.16}}{-0.6}$

$p \approx -21.0, p \approx 13.0$

Horizontal intercepts $= (-21.0, 0)$ and $(13, 0)$

Domain: $(-\infty, \infty)$, range: $(-\infty, 86.8]$

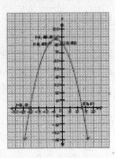

49. $W(g) = -0.3(g+2)^2 + 17$

Vertex $= (-2, 17)$

To find the vertical intercept set $g = 0$.

$W(0) = -0.3(0+2)^2 + 17 = 15.8$

Vertical intercept $= (0, 15.8)$

To find the horizontal intercepts set $W(g) = 0$.

$-0.3(g+2)^2 + 17 = 0$

$-0.3(g+2)^2 = -17$

$(g+2)^2 = \dfrac{17}{0.3}$

$\sqrt{(g+2)^2} = \pm\sqrt{\dfrac{17}{0.3}}$

$g + 2 = \pm\sqrt{\dfrac{17}{0.3}}$

$g = -2 \pm \sqrt{\dfrac{17}{0.3}}$

$g = -2 - \sqrt{\dfrac{17}{0.3}}, g = -2 + \sqrt{\dfrac{17}{0.3}}$

$g \approx -9.5, g \approx 5.5$

Horizontal intercepts $= (-9.5, 0)$ and $(5.5, 0)$

Domain: $(-\infty, \infty)$, range: $(-\infty, 17]$

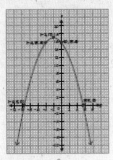

51.

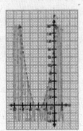

53.

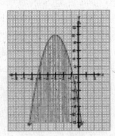

55.

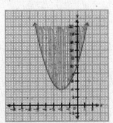

57.

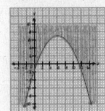

59.

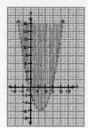

61.

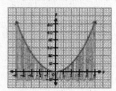

63.

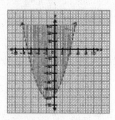

65. Note: Dashed boundary curve, shaded below, so use "$y <$" in the answer.

Use: $x = -4$, $x = 3$
with $f(0) = -12$.
$x + 4 = 0$, $x - 3 = 0$
$f(x) = a(x+4)(x-3)$
$f(x) = a(x^2 + x - 12)$
$f(0) = a(-12) = -12$
$a = 1$
$f(x) = x^2 + x - 12$
$y < x^2 + x - 12$

67. Note: Solid boundary curve, shaded above so use "$y \geq$" in the answer.

Use: $x = -5$, $x = -1$
with $f(-3) = 2$.
$x + 5 = 0$, $x + 1 = 0$
$f(x) = a(x+5)(x+1)$
$f(-3) = a(-3+5)(-3+1) = 2$
$a(2)(-2) = 2$
$-4a = 2 \rightarrow a = -0.5$
$f(x) = -0.5(x+5)(x+1)$
$y \geq -0.5(x+5)(x+1)$
or
$y \geq -0.5x^2 - 3x - 2.5$

Chapter 4 Quadratic Functions
Chapter 4 Review Exercises

1.
a. Vertex: $(-1, 9)$
b. $x < -1$
c. $x > -1$
d. $(-4, 0), (2, 0)$
e. $(0, 8)$

2.
a. $x = -1$
b. a would be negative, since the graph faces downward.
c. $h = -1$, $k = 9$
d. $f(1) = 5$
e. $x = -5.5$, $x = 3.5$

3.
a. $x = 1$
b. a would be positive, since the graph faces upward.
c. $h = 1$, $k = -4$
d. $f(-1.5) = 2.25$
e. $x = -1.6$, $x = 3.6$

4.
a. Vertex: $(1, -4)$
b. $x > 1$
c. $x < 1$
d. $(-1, 0), (3, 0)$
e. $(0, -3)$

5. $P(t) = -0.89(t - 5.6)^2 + 643$

a.
$P(8) = -0.89(8 - 5.6)^2 + 643$
$P(8) = -0.89(2.4)^2 + 643$
$P(8) = 637.87$

In 1998, the population of North Dakota was approximately 638 thousand.

b.

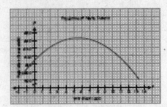

c. Vertex: $(5.6, 643)$. In about 1996, the population of North Dakota reached a maximum of about 643 thousand people.

d.
$-0.89(t - 5.6)^2 + 643 = 640$
$\underline{\quad -643 \, -643 \quad}$
$-0.89(t - 5.6)^2 = -3$
$\dfrac{-0.89(t - 5.6)^2}{-0.89} = \dfrac{-3}{-0.89}$
$(t - 5.6)^2 \approx 3.3708$
$t - 5.6 \approx \pm\sqrt{3.3708}$
$t \approx 5.6 + \sqrt{3.3708} \approx 7.44$
$t \approx 5.6 - \sqrt{3.3708} \approx 3.76$

The population of North Dakota was about 640,000 in about 1993 and about 1997.

e. Domain: $0 \le t \le 12$. Within this domain, the maximum value of 643 comes from the vertex. The minimum value comes from $t = 12$.

$P(12) = -0.89(12 - 5.6)^2 + 643$
$P(12) \approx 606.55$
Range: $606.55 \le P(t) \le 643$

6. $M(t) = -0.42(t - 2.5)^2 + 23$

a.
$M(6) = -0.42(6 - 2.5)^2 + 23$
$M(6) = -0.42(3.5)^2 + 23$
$M(6) = 17.855$

There were approximately 17.9 thousand murders in the United States in 1996.

b. Vertex: $(2.5, 23)$. In about 1993, the most murders occurred in the United States at 23 thousand.

c.

$-0.42(t-2.5)^2 + 23 = 14.5$
$ \underline{-23 \; -23}$
$-0.42(t-2.5)^2 = -8.5$
$\dfrac{-0.42(t-2.5)^2}{-0.42} = \dfrac{-8.5}{-0.42}$
$(t-2.5)^2 \approx 20.2381$
$t - 2.5 \approx \pm\sqrt{20.2381}$
$t \approx 2.5 + \sqrt{20.2381} \approx 7.00$
$t \approx 2.5 - \sqrt{20.2381} \approx -2.00$

In 1997 and 1988, there were about 14,500 murders in the United States.

d. Domain: $[0,8]$. Within this domain, the maximum value of 23 comes from the vertex. The minimum value comes from $t = 8$.

$M(8) = -0.42(8-2.5)^2 + 23$
$M(8) = 10.295$
Range: $[10.295, 23]$

7. $h(x) = (x+3)^2 - 16$

Domain: $(-\infty, \infty)$, range: $[-16, \infty)$

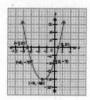

8. $f(x) = 1.25(x-4)^2 - 20$

Domain: $(-\infty, \infty)$; range: $[-20, \infty)$

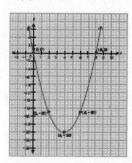

9. $g(x) = -\dfrac{1}{4}(x-8)^2 + 9$

Domain: $(-\infty, \infty)$; range: $(-\infty, 9]$

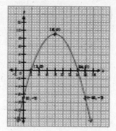

10. $h(x) = -2(x+4.5)^2 + 12.5$

Domain: $(-\infty, \infty)$; range: $(-\infty, 12.5]$

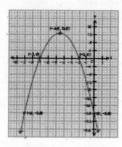

11. $g(x) = -0.5(x+7)^2 - 3$

Domain: $(-\infty, \infty)$; range: $(-\infty, -3]$

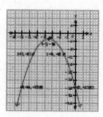

12. $f(x) = 3(x-8)^2 + 4$

Domain: $(-\infty, \infty)$; range: $[4, \infty)$

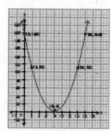

13. Let $J(t)$ be the number of juveniles in thousands arrested for possession of drugs t years since 1990.

a.

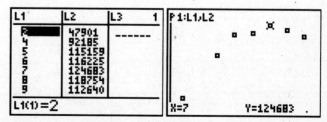

Chapter 4 Quadratic Functions

Choose $(7, 124683)$ as the vertex, and $(2, 47901)$ as the additional point used to solve for "a".

$$J(t) = a(t-7)^2 + 124,683$$
$$47,901 = a(2-7)^2 + 124,683$$
$$47,901 = a(-5)^2 + 124,683$$
$$47,901 = 25a + 124,683$$
$$-124,683 \qquad -124,683$$
$$-76,782 = 25a$$
$$\left.\frac{-76,782}{25} = \frac{25a}{25}\right\} \quad a = -3071.28$$
$$J(t) = -3071.28(t-7)^2 + 124,683$$

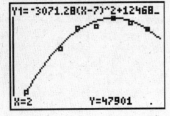

Note: Your answer may vary and yet still represent the trend of the data.

b. Domain: $[1, 11]$. Within this domain, the maximum value of 124,683 comes from the vertex. The minimum value comes from $t = 1$.

$$J(1) = -3071.28(1-7)^2 + 124,683$$
$$J(1) \approx 14,117$$

Range: $[14117, 124683]$

c.

$$J(10) = -3071.28(10-7)^2 + 124,683$$
$$J(10) = -3071.28(3)^2 + 124,683$$
$$J(10) \approx 97,041$$

In 2000, there were approximately 97,041,000 juveniles arrested for drug possession.

d.

$$-3071.28(t-7)^2 + 124,683 = 50,000$$
$$\qquad -124,683 \quad -124,683$$
$$-3071.28(t-7)^2 = -74,683$$
$$\frac{-3071.28(t-7)^2}{-3071.28} = \frac{-74,683}{-3071.28}$$
$$(t-7)^2 \approx 24.3166$$
$$t - 7 \approx \pm\sqrt{24.3166}$$
$$t \approx 7 + \sqrt{24.3166} \approx 11.9$$
$$t \approx 7 - \sqrt{24.3166} \approx 2.1$$

The juvenile arrests were down to 50,000,000 again in about 2002.

14. Let $R(t)$ be the obligations of the U.S. Department of Commerce for research and development in millions of dollars t years since 1990.

a.

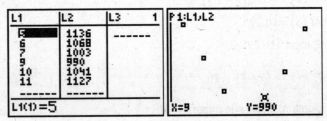

Note here that the lowest point does not look like a good choice for the vertex.

Choose $(8.2, 982)$ (for example) as the vertex, and $(6, 1068)$ as the additional point used to solve for "a".

$$R(t) = a(t - 8.2)^2 + 982$$
$$1068 = a(6 - 8.2)^2 + 982$$
$$1068 = a(-2.2)^2 + 982$$
$$1068 = 4.42a + 982$$
$$-982 \qquad -982$$
$$86 = 4.42a$$
$$\left.\frac{86}{4.42} = \frac{4.42a}{4.42}\right\} \quad a \approx 17.8$$

Adjust "$a = 17$" for a better fit.

$$R(t) = 17(t - 8.2)^2 + 982$$

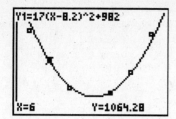

Note: Your answer may vary and yet still represent the trend of the data.

b. Domain: $[4,12]$. Within this domain, the minimum value of 982 comes from the vertex. The maximum value comes from $t = 4$.

$R(4) = 17(4-8.2)^2 + 982$
$R(4) \approx 1282$
Range: $[982, 1282]$

c.

$R(8) = 17(8-8.2)^2 + 982$
$R(8) = 17(-0.2)^2 + 982$
$R(8) = 982.68$

In 1998, the research and development obligations of the U.S. Department of Commerce, was approximately $983 million.

d.

$17(t-8.2)^2 + 982 = 1500$
$\underline{\ -982\ -982}$
$17(t-8.2)^2 = 518$
$\dfrac{17(t-8.2)^2}{17} = \dfrac{518}{17}$
$(t-8.2)^2 \approx 30.4706$
$t - 8.2 \approx \pm\sqrt{30.4706}$
$t \approx 8.2 + \sqrt{30.4706} \approx 13.72$
$t \approx 8.2 - \sqrt{30.4706} \approx 2.68$

The research and development obligations for the U.S. Department of Commerce will have reached $1.5 billion in 1993 and again in 2004.

15. $P(t) = 1.073t^2 - 5.84t + 144.68$

a.

$P(5) = 1.073(5)^2 - 5.84(5) + 144.68$
$P(5) = 142.31$

In 1995, the median sales price of a new home in the western United States was $142,310.

b.

$250 = 1.073t^2 - 5.84t + 144.68$
$\underline{-250 -250}$
$0 = 1.073t^2 - 5.84t - 105.32$

$t = \dfrac{5.84 \pm \sqrt{(-5.84)^2 - 4(1.073)(-105.32)}}{2(1.073)}$

$t = \dfrac{5.84 + \sqrt{486.13904}}{2.146} \approx 12.996$

$t = \dfrac{5.84 - \sqrt{486.13904}}{2.146} \approx \cancel{-7.553}$

In 2003, the median sales price of a new home in the western United States reached $250,000.

c. Vertex: (h, k).

$h = \dfrac{5.84}{2(1.073)} \approx 2.72$

$k = P(2.72) = 1.073(2.72)^2 - 5.84(2.72) + 144.68$
$k \approx 136.73$

Vertex: $(2.72, 136.73)$. In about 2003, the median sales price of a new home in the western United States reached a low of $136,730.

16. $P(t) = -1.43t^2 + 19.89t + 101.68$

a.

$P(4) = -1.43(4)^2 + 19.89(4) + 101.68$
$P(4) = 158.36$

In 2004, the median asking price of houses in Memphis, Tennessee, was $158,360.

b.

$165 = -1.43t^2 + 19.89t + 101.68$
$\underline{-165 -165}$
$0 = -1.43t^2 + 19.89t - 63.32$

$t = \dfrac{-19.89 \pm \sqrt{19.89^2 - 4(-1.43)(-63.32)}}{2(-1.43)}$

$t = \dfrac{-19.89 + \sqrt{33.4217}}{-2.86} \approx 4.933$

$t = \dfrac{-19.89 - \sqrt{33.4217}}{-2.86} \approx 8.976$

In 2005 and again in 2009, the median asking price of houses in Memphis, Tennessee, reached $165,000.

Chapter 4 Quadratic Functions

c. Vertex: (h,k).

$h = \dfrac{-19.89}{2(-1.43)} \approx 6.95$

$k = P(6.95) = -1.43(6.95)^2 + 19.89(6.95) + 101.68$

$k \approx 170.84$

Vertex: (6.95, 170.84). In about 2007, the median asking price of houses in Memphis, Tennessee, reached a high of $170,840.

17.

$t^2 = 169$

$\sqrt{t^2} = \pm\sqrt{169}$

$t = 13,\ t = -13$

18.

$-6m^2 + 294 = 0$

$-6m^2 = -294$

$m^2 = 49$

$\sqrt{m^2} = \pm\sqrt{49}$

$m = 7,\ m = -7$

19.

$4(x+7)^2 - 36 = 0$

$4(x+7)^2 = 36$

$(x+7)^2 = 9$

$\sqrt{(x+7)^2} = \pm\sqrt{9}$

$x + 7 = \pm 3$

$x + 7 = 3,\ x + 7 = -3$

$x = -4,\ x = -10$

20.

$-0.25(x-6)^2 + 8 = 0$

$-0.25(x-6)^2 = -8$

$(x-6)^2 = \dfrac{-8}{-0.25}$

$(x-6)^2 = 32$

$\sqrt{(x-6)^2} = \pm\sqrt{32}$

$x - 6 = \pm\sqrt{32}$

$x - 6 = \sqrt{32},\ x - 6 = -\sqrt{32}$

$x = 6 + \sqrt{32},\ x = 6 - \sqrt{32}$

$x \approx 11.66,\ x \approx 0.34$

21.

$3x^2 + 75 = 0$

$3x^2 = -75$

$x^2 = -25$

$\sqrt{x^2} = \pm\sqrt{-25}$

$x = \pm\sqrt{-25}$

This is not a real number.

No real solutions.

22.

$-4(t+3)^2 - 100 = 0$

$-4(t+3)^2 = 100$

$(t+3)^2 = \dfrac{100}{-4}$

$(t+3)^2 = -25$

$\sqrt{(t+3)^2} = \pm\sqrt{-25}$

$t + 3 = \pm\sqrt{-25}$

This is not a real number.

No real solutions.

23.

$\dfrac{1}{2}(c+6)^2 - \dfrac{5}{2} = 0$

$\dfrac{1}{2}(c+6)^2 = \dfrac{5}{2}$

$2\left(\dfrac{1}{2}(c+6)^2\right) = 2\left(\dfrac{5}{2}\right)$

$(c+6)^2 = 5$

$\sqrt{(c+6)^2} = \pm\sqrt{5}$

$c + 6 = \pm\sqrt{5}$

$c + 6 = \sqrt{5},\ c + 6 = -\sqrt{5}$

$c = -6 + \sqrt{5},\ c = -6 - \sqrt{5}$

$c \approx -3.76,\ c \approx -8.23$

24.

$\dfrac{2}{7}(p-4)^2 - \dfrac{3}{14} = 0$

$\dfrac{2}{7}(p-4)^2 = \dfrac{3}{14}$

$\dfrac{7}{2}\left(\dfrac{2}{7}(p-4)^2\right) = \dfrac{7}{2}\left(\dfrac{3}{14}\right)$

$(p-4)^2 = \dfrac{3}{4}$

$\sqrt{(p-4)^2} = \pm\sqrt{\dfrac{3}{4}}$

$p - 4 = \pm\sqrt{\dfrac{3}{4}}$

$p - 4 = \sqrt{\dfrac{3}{4}}, p - 4 = -\sqrt{\dfrac{3}{4}}$

$p = 4 + \sqrt{\dfrac{3}{4}}, p = 4 - \sqrt{\dfrac{3}{4}}$

$p \approx 4.87, p \approx 3.13$

25.

$x^2 + 26x = 30$

$x^2 + 26x + \underline{} = 30 + \underline{}$

Completing the square: $\left(\dfrac{26}{2}\right)^2 = 169$

$x^2 + 26x + 169 = 30 + 169$

$(x+13)(x+13) = 199$

$(x+13)^2 = 199$

$\sqrt{(x+13)^2} = \pm\sqrt{199}$

$x + 13 = \pm\sqrt{199}$

$x + 13 = \sqrt{199}, x + 13 = -\sqrt{199}$

$x = -13 + \sqrt{199}, x = -13 - \sqrt{199}$

$x \approx 1.11, x \approx -27.11$

26.

$x^2 - 12x - 13 = 0$

$x^2 - 12x = 13$

$x^2 - 12x + \underline{} = 13 + \underline{}$

Completing the square: $\left(\dfrac{-12}{2}\right)^2 = 36$

$x^2 - 12x + 36 = 13 + 36$

$(x-6)(x-6) = 49$

$(x-6)^2 = 49$

$\sqrt{(x-6)^2} = \pm\sqrt{49}$

$x - 6 = \pm 7$

$x - 6 = 7, x - 6 = -7$

$x = 13, x = -1$

27.

$3x^2 + 15x = 198$

$\dfrac{3x^2 + 15x}{3} = \dfrac{198}{3}$

$x^2 + 5x = 66$

$x^2 + 5x + \underline{} = 66 + \underline{}$

Completing the square: $\left(\dfrac{5}{2}\right)^2 = \dfrac{25}{4}$

$x^2 + 5x + \dfrac{25}{4} = 66 + \dfrac{25}{4}$

$\left(x + \dfrac{5}{2}\right)\left(x + \dfrac{5}{2}\right) = \dfrac{289}{4}$

$\left(x + \dfrac{5}{2}\right)^2 = \dfrac{289}{4}$

$\sqrt{\left(x + \dfrac{5}{2}\right)^2} = \pm\sqrt{\dfrac{289}{4}}$

$x + \dfrac{5}{2} = \pm\dfrac{17}{2}$

$x + \dfrac{5}{2} = \dfrac{17}{2}, x + \dfrac{5}{2} = -\dfrac{17}{2}$

$x = 6, x = -11$

28.

$4x^2 - 6x + 20 = 0$

$4x^2 - 6x = -20$

$\dfrac{4x^2 - 6x}{4} = \dfrac{-20}{4}$

$x^2 - \dfrac{3}{2}x = -5$

$x^2 - \dfrac{3}{2}x + \underline{} = -5 + \underline{}$

Completing the square: $\left(\dfrac{-\dfrac{3}{2}}{2}\right)^2 = \dfrac{9}{16}$

$x^2 - \dfrac{3}{2}x + \dfrac{9}{16} = -5 + \dfrac{9}{16}$

$\left(x - \dfrac{3}{4}\right)\left(x - \dfrac{3}{4}\right) = -\dfrac{71}{16}$

$\left(x - \dfrac{3}{4}\right)^2 = -\dfrac{71}{16}$

$\sqrt{\left(x-\frac{3}{4}\right)^2} = \pm\sqrt{-\frac{71}{16}}$

This is not a real number.
No real solutions.

29.

$f(x) = x^2 + 8x + 11$

$f(x) = \left(x^2 + 8x + \underline{}\right) + 11 - \underline{}$

Completing the square: $\left(\frac{8}{2}\right)^2 = 16$

$f(x) = \left(x^2 + 8x + 16\right) + 11 - 16$

$f(x) = (x+4)(x+4) - 5$

$f(x) = (x+4)^2 - 5$

30.

$g(x) = -6x^2 + 20x - 18$

$g(x) = -6\left(x^2 - \frac{10}{3}x\right) - 18$

$g(x) = -6\left(x^2 - \frac{10}{3}x + \underline{} - \underline{}\right) - 18$

Completing the square: $\left(\frac{-\frac{10}{3}}{2}\right)^2 = \frac{25}{9}$

$g(x) = -6\left(x^2 - \frac{10}{3}x + \frac{25}{9} - \frac{25}{9}\right) - 18$

$g(x) = -6\left(x^2 - \frac{10}{3}x + \frac{25}{9}\right) - (-6)\left(\frac{25}{9}\right) - 18$

$g(x) = -6\left(x - \frac{5}{3}\right)\left(x - \frac{5}{3}\right) + \frac{50}{3} - 18$

$g(x) = -6\left(x - \frac{5}{3}\right)^2 - \frac{4}{3}$

31.

$t^2 - 12t + 20 = 0$

$(t-10)(t-2) = 0$

$t - 10 = 0, t - 2 = 0$

$t = 10, t = 2$

32.

$p^2 + 6p = 27$

$p^2 + 6p - 27 = 0$

$(p+9)(p-3) = 0$

$p + 9 = 0, p - 3 = 0$

$p = -9, p = 3$

33.

$6x^2 - 8x = 0$

$2x(3x - 4) = 0$

$2x = 0, 3x - 4 = 0$

$x = 0, 3x = 4$

$x = 0, x = \frac{4}{3}$

34.

$3x^2 - x = 2$

$3x^2 - x - 2 = 0$

$(3x+2)(x-1) = 0$

$3x + 2 = 0, x - 1 = 0$

$3x = -2, x = 1$

$x = -\frac{2}{3}, x = 1$

35.

$m^2 - 64 = 0$

$(m+8)(m-8) = 0$

$m + 8 = 0, m - 8 = 0$

$m = -8, m = 8$

$m = \pm 8$

36.

$8m^2 - 50 = 0$

$2(4m^2 - 25) = 0$

$2(2m+5)(2m-5) = 0$

$2m + 5 = 0, 2m - 5 = 0$

$2m = -5, 2m = 5$

$m = -\frac{5}{2}, m = \frac{5}{2}$

$m = \pm\frac{5}{2}$

37.

$9x^2 - 24x + 5 = -11$

$9x^2 - 24x + 16 = 0$

$(3x-4)(3x-4) = 0$

$3x - 4 = 0$

$3x = 4$

$x = \dfrac{4}{3}$

38.

$2x^2 - 8x - 120 = 0$

$\dfrac{2x^2 - 8x - 120}{2} = \dfrac{0}{2}$

$x^2 - 4x - 60 = 0$

$(x+6)(x-10) = 0$

$x + 6 = 0, x - 10 = 0$

$x = -6, x = 10$

39.

$x^3 + 7x^2 + 10x = 0$

$x(x^2 + 7x + 10) = 0$

$x(x+5)(x+2) = 0$

$x = 0, x + 5 = 0, x + 2 = 0$

$x = 0, x = -5, x = -2$

40.

$12h^3 - 60h^2 = 168h$

$12h^3 - 60h^2 - 168h = 0$

$12h(h^2 - 5h - 14) = 0$

$12h(h+2)(h-7) = 0$

$12h = 0, h + 2 = 0, h - 7 = 0$

$h = 0, h = -2, h = 7$

41.

Vertex $= (-1, -9)$

Other point $= (2, 0)$

$y = a(x-h)^2 + k$

where $h = -1, k = -9, x = 2, y = 0$

$0 = a(2-(-1))^2 + (-9)$

$0 = a(9) - 9$

$9 = 9a$

$a = 1$

Since $a = 1, h = -1, k = -9$

$y = 1(x-(-1))^2 + (-9)$

$y = (x+1)^2 - 9$

42.

Vertex $= (-1, 12)$

Other point $= (1, 0)$

$y = a(x-h)^2 + k$

where $h = -1, k = 12, x = 1, y = 0$

$0 = a(1-(-1))^2 + 12$

$0 = a(4) + 12$

$-12 = 4a$

$a = -3$

Since $a = -3, h = -1, k = 12$

$y = -3(x-(-1))^2 + 12$

$y = -3(x+1)^2 + 12$

43.

$x^2 - 4x - 12 = 0$

$a = 1, b = -4, c = -12$

$x = \dfrac{-(-4) \pm \sqrt{(-4)^2 - 4(1)(-12)}}{2(1)}$

$x = \dfrac{4 \pm \sqrt{64}}{2}$

$x = \dfrac{4 \pm 8}{2}$

$x = \dfrac{4-8}{2}, x = \dfrac{4+8}{2}$

$x = \dfrac{-4}{2}, x = \dfrac{12}{2}$

$x = -2, x = 6$

44.

$2t^2 + 11t - 63 = 0$

$a = 2, b = 11, c = -63$

$t = \dfrac{-11 \pm \sqrt{11^2 - 4(2)(-63)}}{2(2)}$

$t = \dfrac{-11 \pm \sqrt{625}}{4}$

$t = \dfrac{-11 \pm 25}{4}$

$t = \dfrac{-11 - 25}{4}, t = \dfrac{-11 + 25}{4}$

$t = -9, t = \dfrac{7}{2}$

45.

$x^2 + 6x + 18 = 0$

$a = 1, b = 6, c = 18$

$x = \dfrac{-6 \pm \sqrt{6^2 - 4(1)(18)}}{2(1)}$

$x = \dfrac{-6 \pm \sqrt{-36}}{2}$

This is not a real number.
No real solutions.

46.

$3x^2 + 9x - 20 = 14$

$3x^2 + 9x - 34 = 0$

$a = 3, b = 9, c = -34$

$x = \dfrac{-9 \pm \sqrt{9^2 - 4(3)(-34)}}{2(3)}$

$x = \dfrac{-9 \pm \sqrt{489}}{6}$

$x = \dfrac{-9 - \sqrt{489}}{6}, x = \dfrac{-9 + \sqrt{489}}{6}$

$x \approx -5.19, x \approx 2.19$

47.

$-4.5x^2 + 3.5x + 12.5 = 0$

$a = -4.5, b = 3.5, c = 12.5$

$x = \dfrac{-3.5 \pm \sqrt{3.5^2 - 4(-4.5)(12.5)}}{2(-4.5)}$

$x = \dfrac{-3.5 \pm \sqrt{237.25}}{-9}$

$x = \dfrac{-3.5 + \sqrt{237.25}}{-9}, x = \dfrac{-3.5 - \sqrt{237.25}}{-9}$

$x \approx -1.32, x \approx 2.10$

48.

$3.25n^2 - 4.5n - 42.75 = 0$

$a = 3.25, b = -4.5, c = -42.75$

$n = \dfrac{-(-4.5) \pm \sqrt{(-4.5)^2 - 4(3.25)(-42.75)}}{2(3.25)}$

$n = \dfrac{4.5 \pm \sqrt{576}}{6.5}$

$n = \dfrac{4.5 \pm 24}{6.5}$

$n = \dfrac{4.5 - 24}{6.5}, n = \dfrac{4.5 + 24}{6.5}$

$n = \dfrac{-19.5}{6.5}, n = \dfrac{28.5}{6.5} = \dfrac{285}{65}$

$n = -3, n = \dfrac{57}{13}$

49.

$3x^2 + 8x = -2$

$3x^2 + 8x + 2 = 0$

$a = 3, b = 8, c = 2$

$x = \dfrac{-8 \pm \sqrt{8^2 - 4(3)(2)}}{2(3)}$

$x = \dfrac{-8 \pm \sqrt{40}}{6}$

$x = \dfrac{-8 - \sqrt{40}}{6}, x = \dfrac{-8 + \sqrt{40}}{6}$

$x \approx -2.39, x \approx -0.28$

50.

$2a^2 + 6a - 10 = 6a + 20$

$2a^2 - 30 = 0$

$a = 2, b = 0, c = -30$

$a = \dfrac{0 \pm \sqrt{0^2 - 4(2)(-30)}}{2(2)}$

$a = \dfrac{\pm\sqrt{240}}{4}$

$a \approx 3.87, a \approx -3.87$

51. Domain: $(-\infty, \infty)$, range: $(-\infty, 2.5]$

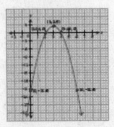

52. Domain: $(-\infty, \infty)$, range: $[7, \infty)$

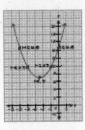

53. Domain: $(-\infty, \infty)$, range: $[-15, \infty)$

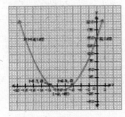

54. Domain: $(-\infty, \infty)$, range: $(-\infty, 28]$

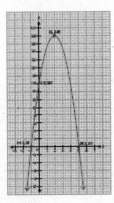

55.

$5x - 9 = -2x^2 + 7x + 15$
$\underline{-5x + 9 \qquad\quad -5x + 9}$
$0 = -2x^2 + 2x + 24$
$\dfrac{0}{-2} = \dfrac{-2x^2 + 2x + 24}{-2}$
$0 = x^2 - x - 12$
$0 = (x + 3)(x - 4)$
$x + 3 = 0, \ x - 4 = 0$
$x = -3 \quad, \ x = 4$
$y = 5(-3) - 9 = -24$
$y = 5(4) - 9 = 11$
$(-3, -24), (4, 11)$

56.

$-4x + 3 = x^2 - 8x - 9$
$\underline{+4x - 3 \qquad\quad +4x - 3}$
$0 = x^2 - 4x - 12$
$0 = (x + 2)(x - 6)$
$x + 2 = 0, \ x - 6 = 0$
$x = -2, \ x = 6$
$y = -4(-2) + 3 = 11$
$y = -4(6) + 3 = -21$
$(-2, 11), (6, -21)$

57.

$6x^2 - 3x + 10 = 4x^2 + 8x + 1$
$\underline{-4x^2 - 8x - 1 \quad -4x^2 - 8x - 1}$
$2x^2 - 11x + 9 = 0$
$(2x - 9)(x - 1) = 0$
$2x - 9 = 0 \to x = \dfrac{9}{2} = 4.5$
$y = 4(4.5)^2 + 8(4.5) + 1 = 118$
$x - 1 = 0 \to x = 1$
$y = 4(1)^2 + 8(1) + 1 = 13$
$(1, 13), (4.5, 118)$

58.

$0.25x^2 + 4x - 7 = 0.5x^2 + 3x - 9$
$\underline{-0.25x^2 - 4x + 7 \quad -0.25x^2 - 4x + 7}$
$0 = 0.25x^2 - x - 2$
$x = \dfrac{1 \pm \sqrt{(-1)^2 - 4(0.25)(-2)}}{2(0.25)}$
$x = \dfrac{1 + \sqrt{3}}{0.5} \approx 5.46, \ x = \dfrac{1 - \sqrt{3}}{0.5} \approx -1.46$
$y \approx 0.5(5.46)^2 + 3(5.46) - 9 \approx 22.32$
$y \approx 0.5(-1.46)^2 + 3(-1.46) - 9 \approx -12.32$
$(-1.46, -12.32), (5.46, 22.32)$

59. $(3, 9), (5, 9)$

Chapter 4 Quadratic Functions

60.

a.

$2w + l = 100 \rightarrow l = 100 - 2w$

$A = lw \rightarrow A = (100 - 2w)w$

Use function notation.

$A(w) = -2w^2 + 100w$

Find the vertex of this function.

$\dfrac{-b}{2a} = \dfrac{-100}{4} = -25$

$A(-25) = (-25)^2 + 100(-25)$

$A(-25) = 1250$

The largest possible rectangular area is 1250 ft^2.

b. The largest possible area requires a length of 50 feet and a width of 25 feet.

61. $y < 2x^2 + 24x - 10$

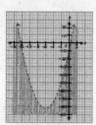

62. $y \geq 0.5x^2 - 4x - 6$

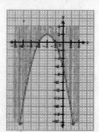

63. $y \leq 2.5(x-4)^2 - 18$

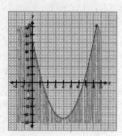

64. $y > -0.8(x+1)^2 - 3$

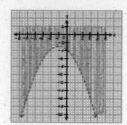

65. $(-4, 3), (2, 6)$

66. $(-4, 0), (2, 9)$

Chapter 4 Test

1. $V(t) = 13(t-4)^2 + 1370$

a.

$2008 \to t = 8$

$V(8) = 13(8-4)^2 + 1370$

$V(8) = 13(4)^2 + 1370$

$V(8) = 1578$

In 2008, there were approximately 1578 thousand violent crimes in the United States.

b.

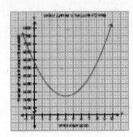

c. Vertex: $(4, 1370)$. In 2004, the number of violent crimes in the United States was at a minimum of 1370 thousand.

d. According to this graph, there were 1.5 million violent crimes in 2001 and again in 2007.

2. $(-6, 4)$ and $(4, 8)$

3.

$3x^2 - 5x + 5 = 33$

$3x^2 - 5x - 28 = 0$

$(3x+7)(x-4) = 0$

$3x + 7 = 0, x - 4 = 0$

$3x = -7, x = 4$

$x = -\dfrac{7}{3}, x = 4$

4.

$8x^2 - 34x = -35$

$8x^2 - 34x + 35 = 0$

$(4x-7)(2x-5) = 0$

$4x - 7 = 0, 2x - 5 = 0$

$4x = 7, 2x = 5$

$x = \dfrac{7}{4}, x = \dfrac{5}{2}$

5. Let $O(t)$ be the total outlays in billions of dollars for national defense and veterans benefits by the United States t years since 1990.

a.

Choose $(6, 308)$ as the vertex, and $(10, 337.4)$ as the additional point used to solve for "a".

$O(t) = a(t-6)^2 + 308$

$337.4 = a(10-6)^2 + 308$

$337.4 = a(4)^2 + 308$

$337.4 = 16a + 308$

$\underline{-308 \qquad\quad -308}$

$29.4 = 16a$

$\dfrac{29.4}{16} = \dfrac{16a}{16}$

$a = 1.8375$

Adjusted to $a = 1.9$.

$O(t) = 1.9(t-6)^2 + 308$

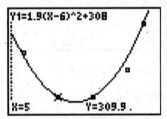

Note: Your answer may vary and yet still represent the trend of the data.

b. Domain: $[1, 11]$. Within this domain, the minimum value of 308 comes from the vertex. The maximum value comes from $t = 1$.

$O(1) = 1.9(1-6)^2 + 308$

$O(1) = 355.5$

Range: $[308, 355.5]$

c.

$O(8) = 1.9(8-6)^2 + 308$

$O(8) = 315.6$

Chapter 4 Quadratic Functions

In 1998, the total outlays for national defense and veterans benefits by the United States were about $315.6 billion.

d.

$1.9(t-6)^2 + 308 = 500$
$\ -308\ -308$
$1.9(t-6)^2 = 192$
$\dfrac{1.9(t-6)^2}{1.9} = \dfrac{192}{1.9}$
$(t-6)^2 \approx 101.0526$
$t - 6 \approx \pm\sqrt{101.0526}$
$t \approx 6 + \sqrt{101.0526} \approx 16.05$
$t \approx 6 - \sqrt{101.0526} \approx \cancel{-4.05}$

Total outlays for national defense and veterans benefits by the United States will reach half a trillion dollars in about 2006.

6. Let $R(t)$ be the revenue for the U.S. commercial space industry from satellite manufacturing in billions of dollars t years since 1990.

a.

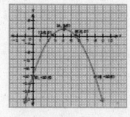

Choose $(8, 11.8)$ as the vertex, and $(6, 7.3)$ as the additional point used to solve for "a".

$R(t) = a(t-8)^2 + 11.8$
$7.3 = a(6-8)^2 + 11.8$
$7.3 = a(-2)^2 + 11.8$
$7.3 = 4a + 11.8$
$-11.8\ \ \ \ -11.8$
$-4.5 = 4a$
$\dfrac{-4.5}{4} = \dfrac{4a}{4}\Big\}\ a = -1.125$
$R(t) = -1.125(t-8)^2 + 11.8$

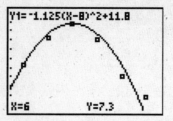

Note: Your answer may vary and yet still represent the trend of the data.

b. Domain: $[6, 10]$. Within this domain, the maximum value of 11.8 comes from the vertex. The minimum value comes from $t = 10$.

$R(10) = -1.125(10-8)^2 + 11.8$
$R(10) = 7.3$
Range: $[7.3, 11.8]$

In this case, we did not go back from the data because it resulted in a very small revenue in 1995, and that seemed like model breakdown.

c.

$R(10) = -1.125(10-8)^2 + 11.8$
$R(10) = -1.125(2)^2 + 11.8$
$R(10) = 7.3$

In 2000, the revenue from satellite manufacturing was about $7.3 billion.

d. Vertex: $(8, 11.8)$. In 1998, the revenue from satellite manufacturing reached a maximum of $11.8 billion.

7. Domain: $(-\infty, \infty)$, range: $(-\infty, 3.5]$

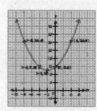

8. Domain: $(-\infty, \infty)$, range: $[10, \infty)$

9. Domain: $(-\infty, \infty)$, range: $(-\infty, -17]$

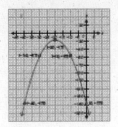

10.
$\begin{cases} y = 2x^2 + 5x - 9 \\ y = -7x^2 + 3x + 2 \end{cases}$

$2x^2 + 5x - 9 = -7x^2 + 3x + 2$

$9x^2 + 2x - 11 = 0$

$(x-1)(9x+11) = 0$

$x - 1 = 0, 9x + 11 = 0$

$x = 1, x = -\dfrac{11}{9}$

Substitute x back into either equation solve for y:

$y = 2(1)^2 + 5(1) - 9 = -2$

$y = 2\left(-\dfrac{11}{9}\right)^2 + 5\left(-\dfrac{11}{9}\right) - 9 = -\dfrac{982}{81}$

The solutions to the system are $(1, -2)$ and $\left(-\dfrac{11}{9}, -\dfrac{982}{81}\right)$.

11. $H(t) = -52t^2 + 916.2t - 2731.2$

a.

$H(10) = -52(10)^2 + 916.2(10) - 2731.2$

$H(10) = 1230.8$

In 2000, there were about 1230.8 thousand privately owned single-unit houses started.

b.

$\begin{array}{r} 1000 = -52t^2 + 916.2t - 2731.2 \\ -1000 -1000 \\ \hline 0 = -52t^2 + 916.2t - 3731.2 \end{array}$

$t = \dfrac{-916.2 \pm \sqrt{916.2^2 - 4(-52)(-3731.2)}}{2(-52)}$

$t = \dfrac{-916.2 + \sqrt{63{,}332.84}}{-104} \approx 6.390$

$t = \dfrac{-916.2 - \sqrt{63{,}332.84}}{-104} \approx 11.229$

There were 1000 thousand privately owned single-unit houses started in about 2001 and in about 1996.

12.

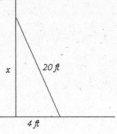

$20^2 = x^2 + 4^2$

$400 = x^2 + 16$

$\begin{array}{r} -16 -16 \\ \hline 384 = x^2 \end{array}$

$x = \sqrt{384} \approx 19.6$

The ladder will reach about 19.6 feet up the house.

13. $(-7, 4), (-4, 8)$

14.

a. $(-3, -4)$

b. $x > -3$

c. $x < -3$

d. $x = -3$

e. $(0, 5)$

f. $(-5, 0), (-1, 0)$

g.

$f(x) = a(x+3)^2 - 4$

$f(0) = a(3)^2 - 4 = 5$

$9a - 4 = 5$

$9a = 9 \rightarrow a = 1$

$f(x) = (x+3)^2 - 4$

or

$f(x) = x^2 + 6x + 5$

15.

$x^2 + 12x + 30 = 0$

$x^2 + 12x = -30$

$x^2 + 12x + \square = -30 + \square$

Chapter 4 Quadratic Functions

Completing the square: $\left(\dfrac{12}{2}\right)^2 = 36$

$x^2 + 12x + \boxed{36} = -30 + \boxed{36}$

$(x+6)^2 = 6$

$x + 6 = \pm\sqrt{6} \approx \pm 2.449$

The two solutions are:

$x \approx -6 + 2.449$, $x \approx -6 - 2.449$

$x \approx -3.551$, $x \approx -8.449$

19.

16.

$4(x-9)^2 - 20 = 124$

$4(x-9)^2 = 144$

$(x-9)^2 = 36$

$\sqrt{(x-9)^2} = \pm\sqrt{36}$

$x - 9 = \pm 6$

$x - 9 = -6, x - 9 = 6$

$x = 3, x = 15$

17.

$8x^2 + 15 = 65$

$8x^2 = 50$

$x^2 = \dfrac{50}{8}$

$x^2 = \dfrac{25}{4}$

$\sqrt{x^2} = \pm\sqrt{\dfrac{25}{4}}$

$x = \pm\dfrac{5}{2}$

$x = \pm 2.5$

18.

$f(x) = 3x^2 + 24x - 30$

$f(x) = 3(x^2 + 8x) - 30$

$f(x) = 3(x^2 + 8x + \_\_ - \_\_) - 30$

Completing the square: $\left(\dfrac{8}{2}\right)^2 = 16$

$f(x) = 3(x^2 + 8x + 16 - 16) - 30$

$f(x) = 3(x^2 + 8x + 16) - 3(16) - 30$

$f(x) = 3(x+4)(x+4) - 48 - 30$

$f(x) = 3(x+4)^2 - 78$

Cumulative Review of Chapters 1-4

1.
$10m + 4 = 3m + 32$
$10m = 3m + 28$
$7m = 28$
$m = 4$

2.
$t^2 - 11t + 10 = 0$
$(t-1)(t-10) = 0$
$t - 1 = 0, t - 10 = 0$
$t = 1, t = 10$

3.
$4(x-7)^2 + 30 = 46$
$4(x-7)^2 = 16$
$\dfrac{4(x-7)^2}{4} = \dfrac{16}{4}$
$(x-7)^2 = 4$
$\sqrt{(x-7)^2} = \pm\sqrt{4}$
$x - 7 = \pm 2$
$x - 7 = 2, x - 7 = -2$
$x = 9, x = 5$

4.
$\dfrac{7}{15}a + \dfrac{3}{5} = \dfrac{2}{3}(a-4) - \dfrac{5}{3}$
$\dfrac{7}{15}a + \dfrac{3}{5} = \dfrac{2}{3}a - \dfrac{8}{3} - \dfrac{5}{3}$
$\dfrac{7}{15}a + \dfrac{3}{5} = \dfrac{2}{3}a - \dfrac{13}{3}$
$15\left(\dfrac{7}{15}a + \dfrac{3}{5}\right) = 15\left(\dfrac{2}{3}a - \dfrac{13}{3}\right)$
$7a + 9 = 10a - 65$
$7a = 10a - 74$
$-3a = -74$
$\dfrac{-3a}{-3} = \dfrac{-74}{-3}$
$a = \dfrac{74}{3}$

5.
$3x^2 - 8x = 0$
$x(3x - 8) = 0$
$x = 0, 3x - 8 = 0$
$x = 0, 3x = 8$
$x = 0, x = \dfrac{8}{3}$

6.
$3.4m^2 - 4.6m = 108$
$3.4m^2 - 4.6m - 108 = 0$
$a = 3.4, b = -4.6, c = -108$
$m = \dfrac{-(-4.6) \pm \sqrt{(-4.6)^2 - 4(3.4)(-108)}}{2(3.4)}$
$m = \dfrac{4.6 \pm \sqrt{1489.96}}{6.8}$
$m = \dfrac{4.6 \pm 38.6}{6.8}$
$m = \dfrac{4.6 - 38.6}{6.8}, m = \dfrac{4.6 + 38.6}{6.8}$
$m = \dfrac{-34}{6.8}, m = \dfrac{43.2}{6.8}$
$m = -5, m \approx 6.35$

7.
$\dfrac{1}{3}(c-9) - 6 = \dfrac{5}{6}c - 15$
$\dfrac{1}{3}c - 3 - 6 = \dfrac{5}{6}c - 15$
$\dfrac{1}{3}c - 9 = \dfrac{5}{6}c - 15$
$6\left(\dfrac{1}{3}c - 9\right) = 6\left(\dfrac{5}{6}c - 15\right)$
$2c - 54 = 5c - 90$
$2c = 5c - 36$
$-3c = -36$
$c = 12$

8.
$2n^3 + 12n^2 - 432n = 0$
$2n(n^2 + 6n - 216) = 0$
$2n(n+18)(n-12) = 0$
$2n = 0, n + 18 = 0, n - 12 = 0$
$n = 0, n = -18, n = 12$

Chapter 4 Quadratic Functions

9.
$(x+3)(x-7) = -16$
$x^2 - 4x - 21 = -16$
$x^2 - 4x - 5 = 0$
$(x+1)(x-5) = 0$
$x+1 = 0, x-5 = 0$
$x = -1, x = 5$

10.
$3.8h - 4.2 = 7.5h + 27.9$
$3.8h = 7.5h + 32.1$
$-3.7h = 32.1$
$\dfrac{-3.7h}{-3.7} = \dfrac{32.1}{-3.7}$
$h \approx -8.68$

11.
$4mn - 5 = n$
$4mn = n + 5$
$\dfrac{4mn}{4n} = \dfrac{n+5}{4n}$
$m = \dfrac{n+5}{4n}$

12.
$\dfrac{2}{5}a^2b + 7a = c$
$\dfrac{2}{5}a^2b = c - 7a$
$\dfrac{5}{2}\left(\dfrac{2}{5}a^2b\right) = \dfrac{5}{2}(c - 7a)$
$a^2b = \dfrac{5}{2}(c - 7a)$

$\dfrac{a^2b}{a^2} = \dfrac{\frac{5}{2}(c-7a)}{a^2}$
$b = \dfrac{\frac{5}{2}(c-7a)}{a^2}$
$b = \dfrac{\frac{5}{2}c - \frac{35}{2}a}{a^2}$
$b = \dfrac{\frac{5}{2}c}{a^2} - \dfrac{\frac{35}{2}a}{a^2}$
$b = \dfrac{5c}{2a^2} - \dfrac{35}{2a}$

13.
$\begin{cases} 4x - 7y = -8 \\ 3x + 10y = 55 \end{cases}$
$\begin{cases} 10(4x - 7y) = 10(-8) \\ 7(3x + 10y) = 7(55) \end{cases}$
$\begin{cases} 40x - 70y = -80 \\ 21x + 70y = 385 \end{cases}$
$61x = 305$
$x = 5$

Substitute x back into either equation solve for y:
$3(5) + 10y = 55$
$15 + 10y = 55$
$10y = 40$
$y = 4$

The solution to the system is $(5, 4)$.

14.
$\begin{cases} y = 3x - 7 \\ y = x^2 + 6x - 17 \end{cases}$
$3x - 7 = x^2 + 6x - 17$
$0 = x^2 + 3x - 10$
$0 = (x+5)(x-2)$
$x + 5 = 0, x - 2 = 0$
$x = -5, x = 2$

Substitute x back into either equation solve for y:
$y = 3(-5) - 7 = -22$
$y = 3(2) - 7 = -1$

The solutions to the system are $(-5, -22)$ and $(2, -1)$.

15.
$\begin{cases} y = 2.5x - 9 \\ -10x + 4y = 20 \end{cases}$
$-10x + 4(2.5x - 9) = 20$
$-10x + 10x - 36 = 20$
$-36 = 20$

Inconsistent system.
No solution.

16.

$\begin{cases} m = 7n+8 \\ 3m+2n = 1 \end{cases}$

$3(7n+8)+2n = 1$
$21n+24+2n = 1$
$23n+24 = 1$
$23n = -23$
$n = -1$

Substitute n back into either equation solve for m:
$m = 7(-1)+8 = 1$

17.

$\begin{cases} b = 3a^2+3a-18 \\ b = -8a^2+4a+30 \end{cases}$

$3a^2+3a-18 = -8a^2+4a+30$
$11a^2-a-48 = 0$
$a = 11, b = -1, c = -48$

$a = \dfrac{-(-1) \pm \sqrt{(-1)^2 - 4(11)(-48)}}{2(11)}$

$a = \dfrac{1 \pm \sqrt{2113}}{22}$

$a = \dfrac{1+\sqrt{2113}}{22}, a = \dfrac{1-\sqrt{2113}}{22}$

$a \approx 2.13, a \approx -2.04$

Substitute a back into either equation solve for b:
$b = 3(2.13)^2 + 3(2.13) - 18 \approx 2.00$
$b = 3(-2.04)^2 + 3(-2.04) - 18 \approx -11.64$

The solutions to the system are $(2.13, 2.00)$ and $(-2.04, -11.64)$.

18.

$\begin{cases} h = 1.4g+2.6 \\ 7g-5h = -13 \end{cases}$

$7g-5(1.4g+2.6) = -13$
$7g-7g-13 = -13$
$-13 = -13$

Consistent system with dependent lines.
Infinite number of solutions.

19.

Using two points: $(4,8)$ and $(10,35)$

find the slope: $m = \dfrac{35-8}{10-4} = \dfrac{27}{6} = \dfrac{9}{2}$.

Use $m = \dfrac{9}{2}$ and $(4,8)$.

$8 = \dfrac{9}{2}(4)+b$
$8 = 18+b$
$b = -10$

Since $m = \dfrac{9}{2}$, and $b = -10$:

$y = \dfrac{9}{2}x - 10$.

20.

Since the two lines are perpendicular, $m_1 = -\dfrac{1}{m_2}$.

Find the slope: $m_1 = 8$, therefore $m_2 = -\dfrac{1}{8}$.

Use $m = -\dfrac{1}{8}$, and $(12, 4.5)$.

$4.5 = 12\left(-\dfrac{1}{8}\right) + b$
$4.5 = -1.5+b$
$b = 6$

Since $m = -\dfrac{1}{8}$, and $b = 6$:

$y = -\dfrac{1}{8}x + 6$.

21.

Since the two lines are perpendicular, $m_1 = m_2$.
$2x+5y = 11 \rightarrow$ Solve for y to find the slope of this equation.
$2x+5y-2x = 11-2x$
$5y = -2x+11$
$\dfrac{5y}{5} = \dfrac{-2x}{5} + \dfrac{11}{5}$
$y = -0.4x+2.2$
Find the slope: $m_1 = -0.4$ therefore $m_2 = -0.4$

Use $m = -0.4$, and $(-3, 10)$.
$10 = -0.4(-3)+b$
$10 = 1.2+b$
$b = 8.8$

Since $m = -0.4$, and $b = 8.8$:
$y = -0.4x+8.8$.

22. Let $C(n)$ be the cost to produce custom printed tote bag, and let n be the number of bags.

a.

$(25, 138.75)$ & $(100, 405)$

Chapter 4 Quadratic Functions

Slope: $m = \dfrac{405 - 138.75}{100 - 25} = 3.55$

Use the point slope form of the equation:

$C - 405 = 3.55(n - 100)$

$C - 405 = 3.55n - 355$

$ +405 +405$

$C = 3.55n + 50$

Using function notation:

$C(n) = 3.55n + 50$

b.

$C(200) = 3.55(200) + 50$

$C(200) = 760$

200 bags will cost $760.

c. Slope = 3.55. The cost to produce custom printed tote bags is increased by $3.55 per bag.

23. Let $P(t)$ be the population of Pittsburgh, Pennsylvania, in millions t years since 2000.

a.

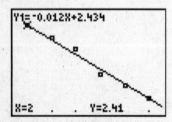

$P(t) = mt + b$

Using two points: $(2, 2.41)$ and $(7, 2.35)$,

find the slope: $m = \dfrac{2.35 - 2.41}{7 - 2} = -0.012$.

$P(t) = -0.012t + b$

$b = 2.41 - (-0.012)2$

$b = 2.434$

$P(t) = -0.012t + 2.434$

b. Domain: $[-2, 12]$

$P(-2) = -0.012(-2) + 2.434 = 2.458$

$P(12) = -0.012(12) + 2.434 = 2.29$

Range: $[2.29, 2.458]$

c. $P(12) = -0.012(12) + 2.434 = 2.29$

The population of Pittsburgh, Pennsylvania, will be about 2.29 million in 2012.

d. The slope is -0.012. The population of Pittsburgh, Pennsylvania, is decreasing by 12,000 people per year.

e.

$2.25 = -0.012t + 2.434$

$-2.434 -2.434$

$-0.184 = -0.012t$

$\dfrac{-0.184}{-0.012} = \dfrac{-0.012t}{-0.012}$

$15.33 \approx t$

In 2015, the population of Pittsburgh, Pennsylvania, will be 2.25 million.

24. Let $R(p)$ be the revenue from selling four-person tents (in dollars) if they are priced at p dollars each.

a.

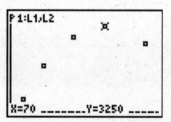

Choose $(70, 3250)$ as the vertex, and $(40, 2360)$ as the additional point used to solve for "a".

$R(p) = a(p - 70)^2 + 3250$

$2360 = a(40 - 70)^2 + 3250$

$2360 = a(-30)^2 + 3250$

$2360 = 900a + 3250$

$-3250 -3250$

$-890 = 900a$

$\left. \dfrac{-890}{900} = \dfrac{900a}{900} \right\} \; a \approx -0.989$

$R(p) = -0.989(p - 70)^2 + 3250$

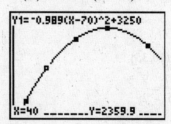

b. The vertex is (70, 3250). When the price is $70 per four-person tent, the revenue generated is at a maximum of $3250.

c.
$R(60) = -0.989(60-70)^2 + 3250$
$R(60) = 3151.1$

When the price is $60 per four-person tent, the revenue generated is $3151.10.

d. Domain: $[20, 100]$. Within this domain, the maximum value of $3250 comes from the vertex. The minimum value comes from $p = 20$.

$R(20) = -0.989(20-70)^2 + 3250 = 777.5$
Range: $[777.50, 3250]$

25. Let $E(t)$ be the net summer electricity capacity in the United States (in millions of kilowatts) t years since 2000.

a.

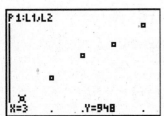

$E(t) = mt + b$
Using two points: $(4, 963)$ and $(7, 999)$,
find the slope: $m = \dfrac{999-963}{7-4} = 12$.

$E(t) = 12t + b$
$b = 963 - (12)4$
$b = 915$
$E(t) = 12t + 915$

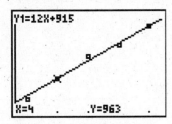

b. Domain: $[-2, 12]$

$E(-2) = 12(-2) + 915 = 891$
$E(12) = 12(12) + 915 = 1059$

Range: $[891, 1059]$

c.
$E(15) = 12(15) + 915$
$E(15) = 1095$

In 2015, the net summer electricity capacity in the United States will be about 1095 million kilowatts.

d. The slope is 12. The net summer electricity capacity in the United States is increasing by 12 million kilowatts per year.

e. The vertical intercept is (0, 915), which means in the year 2000, the net summer electricity capacity in the United States was 915 million kilowatts.

26.

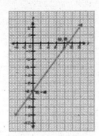

27.

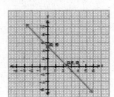

28.

29.

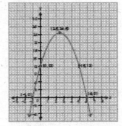

30.

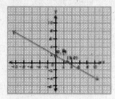

31.

32.

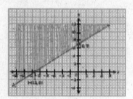

33.

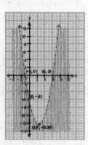

34. $f(x) = 2x+8$ and $g(x) = 7x-18$

a.

$f(x) - g(x) = (2x+8) - (7x-18)$
$= 2x + 8 - 7x + 18$
$= -5x + 26$

b.

$f(x)g(x) = (2x+8)(7x-18)$
$= 14x^2 + 20x - 144$

c.

$f(g(x)) = 2(7x-18) + 8$
$= 14x - 36 + 8$
$= 14x - 28$

35. $f(x) = 3x-4$ and $g(x) = 2x^2 + 5x - 9$

a.

$f(x) - g(x) = (3x-4) - (2x^2 + 5x - 9)$
$= 3x - 4 - 2x^2 - 5x + 9$
$= -2x^2 - 2x + 5$

b.

$f(x)g(x) = (3x-4)(2x^2 + 5x - 9)$
$= 6x^3 + 15x^2 - 27x - 8x^2 - 20x + 36$
$= 6x^3 + 7x^2 - 47x + 36$

c.

$f(g(x)) = 3(2x^2 + 5x - 9) - 4$
$= 6x^2 + 15x - 27 - 4$
$= 6x^2 + 15x - 31$

d.

$g(f(x)) = 2(3x-4)^2 + 5(3x-4) - 9$
$= 2(3x-4)(3x-4) + 15x - 20 - 9$
$= 2(9x^2 - 24x + 16) + 15x - 29$
$= 18x^2 - 48x + 32 + 15x - 29$
$= 18x^2 - 33x + 3$

36. $f(x) = 4.8x + 1.4$ and $g(x) = -1.6x + 3.2$

a.

$(f+g)(5) = f(5) + g(5)$
$= [4.8(5) + 1.4] + [-1.6(5) + 3.2]$
$= [25.4] + [-4.8]$
$= 20.6$

b.

$fg(-3) = f(-3)g(-3)$
$= [4.8(-3) + 1.4][-1.6(-3) + 3.2]$
$= [-13][8]$
$= -104$

c.

$f(g(2)) = f(-1.6(2) + 3.2)$
$= f(0)$
$= 4.8(0) + 1.4$
$= 1.4$

d.

$g(f(1)) = g(4.8(1)+1.4)$
$= g(6.2)$
$= -1.6(6.2)+3.2$
$= -6.72$

37. $f(x) = x - 8$ and $g(x) = x^2 - 7x - 20$

a.

$(f+g)(8) = f(8) + g(8)$
$= [(8)-8] + [(8)^2 - 7(8) - 20]$
$= [0] + [-12]$
$= -12$

b.

$fg(5) = f(5)g(5)$
$= [(5)-8][(5)^2 - 7(5) - 20]$
$= [-3][-30]$
$= 90$

c.

$f(g(12)) = f((12)^2 - 7(12) - 20)$
$= f(40)$
$= (40) - 8$
$= 32$

d.

$g(f(6)) = g((6)-8)$
$= g(-2)$
$= (-2)^2 - 7(-2) - 20$
$= -2$

38. $f(x) = -\dfrac{4}{3}x + 5$

a.

$f(21) = -\dfrac{4}{3}(21) + 5$
$= -28 + 5$
$= -23$

b.

$-\dfrac{4}{3}x + 5 = \dfrac{41}{3}$

$3\left(-\dfrac{4}{3}x + 5\right) = 3\left(\dfrac{41}{3}\right)$

$-4x + 15 = 41$
$-4x = 26$
$x = -6.5$

c.

Domain: $(-\infty, \infty)$, range: $(-\infty, \infty)$

39. $g(x) = 20$

a.

$g(9) = 20$

b.

Domain: $(-\infty, \infty)$, range: $\{20\}$

40. $h(x) = 4x + 20$

a.

$h(-8) = 4(-8) + 20$
$= -32 + 20$
$= -12$

b.

$4x + 20 = 75$
$4x = 55$
$x = 13.75$

c.

Domain: $(-\infty, \infty)$, range: $(-\infty, \infty)$

41. During the first 8 minutes of a jog-a-thon, an eighth grader has run 3.5 laps.

42.

a.

When $x = 0$ and $y = 9$
$(0, 9)$

b.

When $x = 15$ and $y = 0$
$(15, 0)$

Chapter 4 Quadratic Functions

c.

Using two points: $(0, 9.0)$ and $(15, 0)$, find the slope:

$$m = \frac{9.0 - 0}{0 - 15} = -\frac{9}{15} = -\frac{3}{5}$$

d.

$$m = -\frac{3}{5}$$

y-intercept $= (0, 9)$, therefore $b = 9$.

$$y = -\frac{3}{5}x + 9$$

43.

a. $f(8) = 4$

b. $x = -4$

c.

Domain: $(-\infty, \infty)$, range: $(-\infty, \infty)$

44.

a.

Using two points: $(0, 2)$ and $(9, 0)$, find the slope:

$$m = \frac{2 - 0}{0 - 9} = -\frac{2}{9}$$

b. $(0, 2)$

c. $(9, 0)$

d.

$$m = -\frac{2}{9}$$

y-intercept $= (0, 2)$, therefore $b = 2$.

$$y = -\frac{2}{9}x + 2$$

45. Let x be the amount in dollars Greg is investing at 3% and y be the amount in dollars Greg is investing at 2.5%.

a.

$x + y = 900{,}000 \to$ Total amount invested.
$0.03x + 0.025y = 25{,}800 \to$ Total interest earned.

b. Substitute $x = 900{,}000 - y$ in the second equation and solve.

$$0.03(900{,}000 - y) + 0.025y = 25{,}800$$
$$27{,}000 - 0.03y + 0.025y = 25{,}800$$
$$27{,}000 - 0.005y = 25{,}800$$
$$\underline{-27{,}000 \qquad\qquad -27{,}000}$$
$$-0.005y = 1200$$

$$\frac{-0.005y}{-0.005} = \frac{1200}{-0.005}$$
$$y = 240{,}000$$
$$x = 900{,}000 - 240{,}000$$
$$x = 660{,}000$$

$240{,}000 should be invested at 2.5% and $660{,}000 should be invested at 3%.

46. Let x be the number of pounds of bold roast coffee, and let y be the number of pounds of mild roast coffee.

$$x + y = 200$$
$$8.99x + 11.99y = (10.19)(200) = 2038$$

Substitute $x = 200 - y$ in the second equation and solve.

$$8.99(200 - y) + 11.99y = 2038$$
$$1798 - 8.99y + 11.99y = 2038$$
$$1798 + 3y = 2038$$
$$\underline{-1798 \qquad -1798}$$
$$3y = 240$$
$$\frac{3y}{3} = \frac{240}{3}$$
$$y = 80$$
$$x = 200 - 80 = 120$$

They need to mix 80 pounds of the $11.99 per pound mild roast coffee with 120 pounds of the $8.99 per pound bold roast coffee to make 200 pounds of a $10.19 per pound blend.

47. Let P be the salary Deeva earns if she has s dollars in sales.

$$P = 300 + 0.04s$$
$$410 = 300 + 0.04s$$
$$110 = 0.04s$$
$$\frac{110}{0.04} = \frac{0.04s}{0.04}$$
$$2750 = s$$

Deeva needs to make at least $2750 per week in sales to earn at least $410 per week.

48. $24x^3y - 7x^2y^2 + 12y^3 - 10$

There are four terms.

$24x^3y$: variable term, coefficient 24

$-7x^2y^2$: variable term, coefficient -7

$12y^3$: variable term, coefficient 12

-10: constant term

49. $-6x^5 + 4x^2 - 2x + 4$

There are four terms

$-6x^5$: variable term, coefficient -6

$4x^2$: variable term, coefficient 4

$-2x$: variable term, coefficient -2

4: constant term

50. $24x^3y - 7x^2y^2 + 12y^3 - 10$

$24x^3y$ has degree 4

$-7x^2y^2$ has degree 4

$12y^3$ has degree 3

-10 has degree 0

The degree of the polynomial is 4.

51. $-6x^5 + 4x^2 - 2x + 4$

$-6x^5$ has degree 5

$4x^2$ has degree 2

$-2x$ has degree 1

4 has degree 0

The degree of the polynomial is 5.

52. Polynomial

53. Not a polynomial. The variable y has a negative exponent.

54.

$4x \cancel{-11} > \cancel{10x} + 1$
$\underline{-10x \cancel{+11} \quad \cancel{-10x} + 11}$
$-6x > 12$
$\dfrac{-6x}{-6} < \dfrac{12}{-6}$
$x < -2$

55.

$\dfrac{1}{4}(x+3) - \dfrac{5}{12} \leq \dfrac{1}{2}x + \dfrac{5}{6}$

$\dfrac{x}{4} + \dfrac{3}{4} - \dfrac{5}{12} \leq \dfrac{1}{2}x + \dfrac{5}{6}$

$(12)\left(\dfrac{x}{4} + \dfrac{3}{4} - \dfrac{5}{12}\right) \leq \left(\dfrac{1}{2}x + \dfrac{5}{6}\right)(12)$

$3x + 9 - 5 \leq 6x + 10$

$3x \cancel{+4} \leq \cancel{6x} + 10$
$\underline{-6x \cancel{-4} \quad \cancel{-6x} - 4}$
$-3x \leq 6$

$\dfrac{-3x}{-3} \geq \dfrac{6}{-3}$

$x \geq -2$

56.

$|x - 12| < 9$
$-9 < x - 12 < 9$
$\underline{+12 \quad +12 \quad +12}$
$3 < x < 21$

57.

$|-3x + 5| \leq 11$
$-11 \leq -3x \cancel{+5} \leq 11$
$\underline{-5 \qquad \cancel{-5} \quad -5}$
$-16 \leq -3x \leq 6$

$\dfrac{-16}{-3} \geq \dfrac{-3x}{-3} \geq \dfrac{6}{-3}$

$\dfrac{16}{3} \geq x \geq -2$

$-2 \leq x \leq \dfrac{16}{3}$

58.

$|x - 9| \geq 3$

$x - 9 \leq -3$ or $x - 9 \geq 3$

$x - 9 + 9 \leq -3 + 9$ or $x - 9 + 9 \geq 3 + 9$

$x \leq 6$ or $x \geq 12$

59.

$-5|x-4| + 20 < -15$
$ -20 \phantom{<} -20$
$-5|x-4| < -35$
$\dfrac{-5|x-4|}{-5} > \dfrac{-35}{-5}$
$|x-4| > 7$
$x - 4 < -7$ or $x - 4 > 7$
$x < -3$ or $x > 11$

60. $y_1 < y_2$ for $x < 8$

61.

$|2m - 5| + 7 = 22$
$ -7 -7$
$|2m - 5| = 15$
$2m - 5 = -15$ or $2m - 5 = 15$
$2m = -10$ or $2m = 20$
$\dfrac{2m}{2} = \dfrac{-10}{2}$ or $\dfrac{2m}{2} = \dfrac{20}{2}$
$m = -5$ or $m = 10$

62.

$|a - 9| - 4 = -2$
$ +4 +4$
$|a - 9| = 2$
$a - 9 = -2$ or $a - 9 = 2$
$a = 7$ or $a = 11$

63.

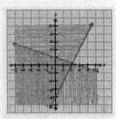

64.

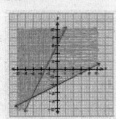

65.

$\left(\dfrac{3}{5} m^4 n^5\right)^4 = \dfrac{3^4}{5^4} m^{4 \times 4} n^{5 \times 4}$
$= \dfrac{81 m^{16} n^{20}}{625}$

66.

$(10 x^3 y^7)(-4 x y^2) = -40 x^{3+1} y^{7+2}$
$= -40 x^4 y^9$

67.

$\dfrac{24 g^3 h^6}{16 g h^2} = \dfrac{\cancel{8} \cdot 3 g^{3-1} h^{6-2}}{\cancel{8} \cdot 2}$
$= \dfrac{3 g^2 h^4}{2}$

68.

$\dfrac{-14 x^4 y z^{-3}}{20 x y^{-5} z^{-3}} = \dfrac{-7 \cdot \cancel{2} x^{4-1} y^{1-(-5)} z^{-3-(-3)}}{\cancel{2} \cdot 10}$
$= -\dfrac{7 x^3 y^6}{10}$

69.

$\left(-27 a^6 b^{-3}\right)^{\frac{1}{3}} \left(16 a^{12} b^{-20}\right)^{-\frac{1}{4}}$
$= (-27)^{\frac{1}{3}} a^{\frac{6}{3}} b^{-\frac{3}{3}} \cdot 16^{-\frac{1}{4}} a^{-\frac{12}{4}} b^{\frac{-20}{-4}}$
$= -3 a^2 b^{-1} \cdot \dfrac{1}{2} a^{-3} b^5$
$= -\dfrac{3}{2} a^{2-3} b^{-1+5}$
$= -\dfrac{3 a^{-1} b^4}{2}$
$= -\dfrac{3 b^4}{2a}$

70.

$\left(\dfrac{72 m^3 n^{12}}{50 m^{-1} n^8}\right)^{\frac{1}{2}} = \left(\dfrac{\cancel{2} \cdot 36 m^{3-(-1)} n^{12-8}}{\cancel{2} \cdot 25}\right)^{\frac{1}{2}}$
$= \dfrac{36^{\frac{1}{2}} m^{\frac{4}{2}} n^{\frac{4}{2}}}{25^{\frac{1}{2}}}$
$= \dfrac{6 m^2 n^2}{5}$

71. a.

$S(6) = 1.88(6) + 54.28$
$S(6) = 65.56$

In 2006, Sweden's GDP per employed person was $65,560.

b.

$$S(t) > N(t)$$
$$1.88t + 54.28 > 1.01t + 61.3$$
$$-1.01t - 54.28 \quad -1.01t - 54.28$$
$$0.87t > 7.02$$
$$\frac{0.87t}{0.87} > \frac{7.02}{0.87}$$
$$t > 8.07$$

Sweden's GDP per employed person will be greater than Netherlands in 2008.

72. a.

$$A(7) = 1.26(7) - 0.04$$
$$A(7) = 8.78$$

In 2007, there were about 8780 loans to African American owned small businesses.

b. Let $T(t)$ be the total number of loans (in thousands) given to African American and Hispanic American owned small businesses t years past 2000.

$$T(t) = A(t) + H(t)$$
$$T(t) = (1.26t - 0.04) + (1.33t + 2.31)$$
$$T(t) = 2.59t + 2.27$$

c.

$$T(10) = 2.59(10) + 2.27$$
$$T(10) = 28.17$$

There will be about 28,170 loans given to African American and Hispanic American owned small businesses in 2010.

73.

$$a^2 + 11a + 24 = (a+3)(a+8)$$

74.

$$3n^3 - 27n^2 + 42n = 3n(n^2 - 9n + 14)$$
$$= 3n(n-7)(n-2)$$

75.

$$16x^2 - 40xy - 14x + 35y = \underline{16x^2 - 40xy} - \underline{14x + 35y}$$
$$= 8x(2x - 5y) - 7(2x - 5y)$$
$$= (2x - 5y)(8x - 7)$$

76.

$$25k^2 - 49 = (5k+7)(5k-7)$$

77.

$$12t^2 - 19t - 21 = (4t+3)(3t-7)$$
$$ac = 12 \times -21 = -252$$

Find the factors of -252 that sum to -19:

$$-28 \times 9 = -252$$
$$-28 + 9 = -19$$

We get:

$$12t^2 - 19t - 21 = 12t^2 - 28t + 9t - 21$$
$$= 4t(3t-7) + 3(3t-7)$$
$$= (4t+3)(3t-7)$$

78.

$$5z^3 - 40 = 5(z^3 - 8)$$
$$= 5(z^3 - 2^3)$$
$$= 5(z-2)(z^2 + 2z + 2^2)$$
$$= 5(z-2)(z^2 + 2z + 4)$$

79.

a. $P(15) = 2(15) + 48 = 78$

The population of this city will be about 78 thousand in 2015.

b. $B(78) = 0.6(78) + 0.2 = 47$

There are about 47 thousand burglaries in this city in 2015.

c. Let $K(t)$ be the number of burglaries in this city t years since 2000:

$$K(t) = B(P(t))$$
$$K(t) = 0.6(2t + 48) + 0.2$$
$$K(t) = 1.2t + 28.8 + 0.2$$
$$K(t) = 1.2t + 29$$

d.

$$K(20) = 1.2(20) + 29$$
$$K(20) = 53$$

In 2020, there will be about 53 thousand burglaries in this city.

Chapter 4 Quadratic Functions

80.

$f(x) = x^2 + 10x - 14$

$f(x) = \left(x^2 + 10x + \Box - \Box\right) - 14$

Completing the square: $\left(\dfrac{10}{2}\right)^2 = 25$

$f(x) = \left(x^2 + 10x + \boxed{25} - \boxed{25}\right) - 14$

$f(x) = \left(x^2 + 10x + 25\right) - 25 - 14$

$f(x) = (x+5)^2 - 39$

Section 5.1

1. This function is an exponential function because the variable is in the exponent.

3. This function is an exponential function because the variable is in the exponent.

5. This function is a quadratic function because the degree of the polynomial is 2.

7. This function is a quadratic function because the degree of the polynomial is 2.

9. Other, because there is a variable in the numerator and in the denominator.

11. Let $L(h)$ be the number of Lactobacillus acidophilus bacteria present after h hours have passed.

a. The bacteria double every hour:

h	L(h)
0	30
1	$30(2)^1$
2	$30(2)^2$
3	$30(2)^3$
h	$30(2)^h$

$L(h) = 30(2)^h$

b. $L(12) = 30(2)^{12} = 122{,}880$. After 12 hours, there are 122,880 Lactobacillus acidophilus bacteria present.

c. $L(24) = 30(2)^{24} = 503{,}316{,}480$. After 24 hours, there are 503,316,480 Lactobacillus acidophilus bacteria present.

d.

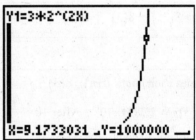

After just 15 hours, there are 1 million bacteria present.

13. Let $E(h)$ be the number of E coli bacteria present after h hours have passed.

a.

h	E(h)
0	3
1	$3(2)(2) = 3(2)^2$
2	$3(2^2)(2^2) = 3(2^2)^2 = 3(2)^{2 \times 2} = 3(2)^4$
3	$3(2^2)(2^2)(2^2) = 3(2^2)^3 = 3(2)^{2 \times 3} = 3(2)^6$
h	$3(2)^{2h}$

The bacteria double 2 times each hour: $E(h) = 3(2)^{2h}$

b.

$B(5) = 3(2)^{2 \times 5} = 3(2)^{10}$

$B(5) = 3072$

After 5 hours, there are approximately 3072 bacteria present.

c.

After about 9.17 hours, there are 1 million bacteria present.

15. Let $S(n)$ be the number of Streptococcus lactis bacteria present after n 26-minute time intervals have passed.

a.

n	S(n)
0	8
1	$8(2)$
2	$8(2)(2) = 8(2)^2$
3	$8(2)(2)(2) = 8(2)^3$
n	$8(2)^n$

The bacteria double every 26-minute time period:

$S(n) = 8(2)^n$

Chapter 5 Exponential Functions

b.

$$6.5 \text{ hours} = (6.5 \text{ hours})\left(\frac{60 \text{ min}}{1 \text{ hour}}\right) = 390 \text{ min}$$

$$n = \frac{390 \text{ min}}{26 \text{ min}} = 15$$

There are 15 of the 26-minute time periods in 390 minutes.

$S(15) = 8(2)^{15} = 262{,}144$. After 6.5 hours, there are 262,144 bacteria present.

17. Let $B(h)$ be the number of bacteria present after h hours have passed.

a.

h	$B(h)$
0	8
1	$8(3)(3) = 8(3)^2$
2	$8(3^2)(3^2) = 8(3^2)^2 = 8(3)^{2\times 2} = 8(3)^4$
3	$8(3^2)(3^2)(3^2) = 8(3^2)^3 = 8(3)^{2\times 3} = 8(3)^6$
h	$8(3)^{2h}$

The bacteria triple 2 times each hour: $B(h) = 8(3)^{2h}$

b. $B(10) = 8(3)^{2\times 10} = 8(3)^{20} \approx 2.7894 \times 10^{10}$. After 10 hours, there are approximately 2.7894×10^{10} bacteria present.

19. Let $T(h)$ be the number of Treponema pallidum bacteria present after h hours have passed. Let n be the number of 33 hour time intervals.

a. The bacteria double every 33 hours:

h	$T(h)$
0	12
33	$12(2)$
$2(33)$	$12(2)^2$
$3(33)$	$12(2)^3$
$n(33)$	$12(2)^n$

$n(33) = h$

$\frac{n(33)}{33} = \frac{h}{33}$

$n = \frac{h}{33}$

$T(h) = 12(2)^{\frac{h}{33}}$

b.

$$7 \text{ days} = (7 \text{ days})\left(\frac{24 \text{ hours}}{1 \text{ day}}\right) = 168 \text{ hours}$$

$T(168) = 12(2)^{\frac{168}{33}}$

$T(168) \approx 408.98$

$T(168) \approx 409$. After 1 week, there are approximately 409 bacteria present.

21. Let $F(n)$ be the number of square meters burnt after n 6-minute time intervals have passed.

a.

n	$F(n)$
0	2
1	$2(2)$
2	$2(2)^2 = 2^3$
3	$2(2)^3 = 2^4$
n	$2(2)^n = 2^{n+1}$

$F(n) = 2^{n+1}$

b.

$\frac{1}{2} \text{ hour} = 30 \text{ min}$

$n = \frac{30 \text{ min}}{6 \text{ min}} = 5$

There are 5 of the 6-minute time intervals in $\frac{1}{2}$ hour.

$F(5) = 2^{5+1} = 2^6$

$F(5) = 64$

After $\frac{1}{2}$ hour, there are 64 square meters burnt.

5.1 Exercises

23. Let $H(t)$ be the number of Hispanic centenarians in the year t years since 1990. Let n be the number of 7.5 year time intervals.

a. The number of centenarians double every 7.5 years:

t	$H(t)$
0	2072
7.5	$2072(2)$
$2(7.5)$	$2072(2)^2$
$3(7.5)$	$2072(2)^3$
$n(7.5)$	$2072(2)^n$

$n(7.5) = t$

$\dfrac{n(7.5)}{7.5} = \dfrac{t}{7.5}$

$n = \dfrac{t}{7.5}$

$H(t) = 2072(2)^{\frac{t}{7.5}}$

b.

$H(60) = 2072(2)^{\frac{60}{7.5}}$

$H(60) = 530{,}432$

In 2050, there will be approximately 530,432 Hispanic centenarians in the U.S.

25. Let $I(t)$ be the population of India (in millions) in the year t years since 1971. Let n be the number of 33 year time intervals.

a. The population of India doubles every 33 years:

t	$I(t)$
0	560
33	$560(2)$
$2(33)$	$560(2)^2$
$3(33)$	$560(2)^3$
$n(33)$	$560(2)^n$

$n(33) = t$

$\dfrac{n(33)}{33} = \dfrac{t}{33}$

$n = \dfrac{t}{33}$

$I(t) = 560(2)^{\frac{t}{33}}$

b.

$2015 \to t = 44$

$I(44) = 560(2)^{\frac{44}{33}}$.

$I(44) \approx 1411$

In 2015, there will be approximately 1411 million people in India.

c.

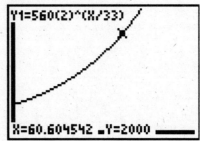

Midway through the year 2031, there will be approximately 2 billion people in India.

27. Internet search: approximately 65 years.

29.

a.

Option 1:

$2000

$\times$ 25

$50,000

b. Your opinion.

c. Option 2:

Let $S(w)$ be salary option 2 (in dollars) for week w of the job:

w	$S(w)$
1	0.01
2	$0.01(2)$
3	$0.01(2)^2$
4	$0.01(2)^3$
w	$0.01(2)^{w-1}$

$S(w) = 0.01(2)^{w-1}$

Chapter 5 Exponential Functions

d.

Week	Option 2
1	0.01
2	0.02
3	0.04
4	0.08
5	0.16
6	0.32
7	0.64
8	1.28
9	2.56
10	5.12
11	10.24
12	20.48
13	40.96
14	81.92
15	163.84
16	327.68
17	655.36
18	1310.72
19	2621.44
20	5242.88
21	10485.76
22	20971.52
23	41943.04
24	83886.08
25	167772.16
Total:	335544.31

The total salary from option 2 for 25 weeks of work is $335,544.31.

e. Option 2 is the best deal for 25 weeks of work.

f.

Week	Option1	Option2
1	2000	0.01
2	2000	0.02
3	2000	0.04
4	2000	0.08
5	2000	0.16
6	2000	0.32
7	2000	0.64
8	2000	1.28
9	2000	2.56
10	2000	5.12
11	2000	10.24
12	2000	20.48
13	2000	40.96
14	2000	81.92
15	2000	163.84
16	2000	327.68
17	2000	655.36
18	2000	1310.72
19	2000	2621.44
20	2000	5242.88
Totals:	40000	10485.75

Option 1 offers $40,000 for 20 weeks of work and option 2 offers $10,485.75 for 20 weeks of work therefore option 1 is the best deal for 20 weeks of work.

31. Let $P(t)$ be the percent of lead-210 left in a body t years after a person has died. Let n be the number of 22-year time intervals.

a. The percent of lead-210 is cut in half every 22 years:

t	$P(t)$
0	100
22	$100\left(\frac{1}{2}\right)$
2(22)	$100\left(\frac{1}{2}\right)^2$
3(22)	$100\left(\frac{1}{2}\right)^3$
$n(22)$	$100\left(\frac{1}{2}\right)^n$

$$n(22) = t$$
$$\frac{n(22)}{22} = \frac{t}{22}$$
$$n = \frac{t}{22}$$
$$P(t) = 100\left(\frac{1}{2}\right)^{t/22}$$

b. $P(60) = 100\left(\frac{1}{2}\right)^{60/22} \approx 15.10$. Approximately 60 years after death, there will be about 15.10% of lead-210 left in the body.

c.

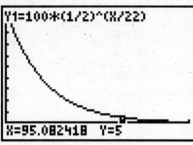

Approximately 95 years after death, there will be about 5% of lead-210 left in the body.

33. Let $P(d)$ be the percent of radon-222 left after d days. Let n be the number of 3.825-day time intervals.

a. The percent of radon-222 is cut in half every 3.825 days:

d	$P(d)$
0	100
3.825	$100\left(\frac{1}{2}\right)$
2(3.825)	$100\left(\frac{1}{2}\right)^2$
3(3.825)	$100\left(\frac{1}{2}\right)^3$
$n(3.825)$	$100\left(\frac{1}{2}\right)^n$

$n(3.825) = d$

$\dfrac{n(3.825)}{3.825} = \dfrac{d}{3.825}$

$n = \dfrac{d}{3.825}$

$P(d) = 100\left(\frac{1}{2}\right)^{d/3.825}$

b. $P(30) = 100\left(\frac{1}{2}\right)^{30/3.825} \approx 0.44$. After approximately 30 days, there will be about 0.44% of radon-222 left.

c.

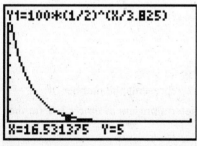

After approximately 16.5 days, there will be about 5% of radon-222 left.

35.
a. The graph is decreasing.
b. This is an example of exponential decay.
c. $f(10) = 22$
d. $x = 6$ for $f(x) = 40$
e. $f(0) = 100$

37.
a. The graph is increasing.
b. This is an example of exponential growth.
c. $h(5) = 20$
d. $x = 15$ for $h(x) = 100$
e. $h(0) = 10$

39.
a. The graph is decreasing.
b. This is an example of exponential decay.
c. $g(7.5) = 20$
d. $x = 4$ for $g(x) = 30$
e. $g(0) = 60$

41.

x	$f(x)$	base
0	25	$\dfrac{100}{25} = 4$
1	100	$\dfrac{400}{100} = 4$
2	400	$\dfrac{1600}{400} = 4$
3	1600	$\dfrac{6400}{1600} = 4$
4	6400	

$a = 25$ because the initial value is 25 when $x = 0$.

$b = 4$ because the common base is 4 from the above table.

Therefore the exponential model is $f(x) = 25(4)^x$.

Chapter 5 Exponential Functions

43.

x	$f(x)$	base
0	-35	$\dfrac{-245}{-35}=7$
1	-245	$\dfrac{-1715}{-245}=7$
2	-1715	$\dfrac{-12005}{-1715}=7$
3	-12005	$\dfrac{-84035}{-12005}=7$
4	-84035	

$a = -35$ because the initial value is -35 when $x = 0$.

$b = 7$ because the common base is 7 from the above table.

Therefore the exponential model is $f(x) = -35(7)^x$.

45.

x	$f(x)$	base
0	2000	$\dfrac{400}{2000}=0.2$
1	400	$\dfrac{80}{400}=0.2$
2	80	$\dfrac{16}{80}=0.2$
3	16	$\dfrac{3.2}{16}=0.2$
4	3.2	

$a = 2000$ because the initial value is 2000 when $x = 0$.

$b = 0.2$ because the common base is 0.2 from the above table.

Therefore the exponential model is $f(x) = 2000(0.2)^x$.

47.

x	$f(x)$	base
0	32	$\dfrac{48}{32}=1.5$
1	48	$\dfrac{72}{48}=1.5$
2	72	$\dfrac{108}{72}=1.5$
3	108	$\dfrac{162}{108}=1.5$
4	162	

$a = 32$ because the initial value is 32 when $x = 0$.

$b = 1.5$ because the common base is 1.5 from the above table.

Therefore the exponential model is $f(x) = 32(1.5)^x$.

49.

x	$f(x)$	base
0	6400	$\dfrac{800}{6400}=0.125$
1	800	$\dfrac{100}{800}=0.125$
2	100	$\dfrac{12.5}{100}=0.125$
3	12.5	$\dfrac{1.5625}{12.5}=0.125$
4	1.5625	

$a = 6400$ because the initial value is 6400 when $x = 0$.

$b = 0.125$ because the common base is 0.125 from the above table. Therefore the exponential model is $f(x) = 6400(0.125)^x$.

51. For $f(x) = 5(3^x)$: The base $b = 3$ and the coefficient $a = 5$.

53. For $f(x) = 200\left(\dfrac{1}{5}\right)^x$: The base $b = \dfrac{1}{5}$ and the coefficient $a = 200$.

55. For $f(x) = -4(2)^x$: The base $b = 2$ and the coefficient $a = -4$.

57. Let $E(h)$ be the number of E coli bacteria present after h hours have passed: $E(h) = 3(2)^{2h}$.

a. The base $b = 2$.

b. The coefficient $a = 3$.

c. The coefficient $a = 3$ represents the number of E coli bacteria present at time $t = 0$ (at the start).

59. Let $T(h)$ be the number of Treponema pallidum bacteria present after h hours have passed:

$T(h) = 12(2)^{\frac{h}{33}}$

a. The base $b = 2$.

b. The coefficient $a = 12$.

c. The coefficient $a = 12$ represents the number of Treponema pallidum bacteria present at time $t = 0$ (at the start).

5.1 Exercises

61.

x	f(x)	base	pattern
0	3		3
5	12	$\frac{12}{3}=4$	$3(4)^1=12$
10	48	$\frac{48}{12}=4$	$3(4)^2=48$
15	192	$\frac{192}{48}=4$	$3(4)^3=192$
20	768	$\frac{768}{192}=4$	$3(4)^4=768$

$a = 3$ because the initial value is 3 when $x = 0$.
$b = 4$ because the common base is 4 from the above table.
Since the exponents are the input values divided by 5, the exponent in the model is $\frac{x}{5}$. Therefore the exponential model is $f(x) = 3(4)^{\frac{x}{5}}$.

63.

x	f(x)	base	pattern
0	-7		-7
10	-21	$\frac{-21}{-7}=3$	$-7(3)^1=-21$
20	-63	$\frac{-63}{-21}=3$	$-7(3)^2=-63$
30	-189	$\frac{-189}{-63}=3$	$-7(3)^3=-189$
40	-567	$\frac{-567}{-189}=3$	$-7(3)^4=-567$

$a = -7$ because the initial value is -7 when $x = 0$.
$b = 3$ because the common base is 3 from the above table.
Since the exponents are the input values divided by 10, the exponent in the model is $\frac{x}{10}$. Therefore the exponential model is $f(x) = -7(3)^{\frac{x}{10}}$.

65.

x	f(x)	base	pattern
0	1701		1701
6	567	$\frac{567}{1701}=\frac{1}{3}$	$1701\left(\frac{1}{3}\right)^1=567$
12	189	$\frac{189}{567}=\frac{1}{3}$	$1701\left(\frac{1}{3}\right)^2=189$
18	63	$\frac{63}{189}=\frac{1}{3}$	$1701\left(\frac{1}{3}\right)^3=63$
24	21	$\frac{21}{63}=\frac{1}{3}$	$1701\left(\frac{1}{3}\right)^4=21$

$a = 1701$ because the initial value is 1701 when $x = 0$.
$b = \frac{1}{3}$ because the common base is $\frac{1}{3}$ from the above table.
Since the exponents are the input values divided by 6, the exponent in the model is $\frac{x}{6}$. Therefore the exponential model is $f(x) = 1701\left(\frac{1}{3}\right)^{\frac{x}{6}}$.

67.

x	f(x)	base	pattern
2	80		80
3	320	$\frac{320}{80}=4$	$80(4)^1=320$
4	1280	$\frac{1280}{320}=4$	$80(4)^2=1280$
5	5120	$\frac{5120}{1280}=4$	$80(4)^3=5120$
6	20480	$\frac{20480}{5120}=4$	$80(4)^4=20480$

$b = 4$ because the common base is 4 from the above table. The input values do not include $x = 0$ so we do not know the initial value of the function. The pattern in the above table does show that we have 80 times the common base 4 raised to an exponent. Since the exponents are the input values minus 2, the exponent in the model is $x - 2$. Therefore the exponential model is $f(x) = 80(4)^{x-2}$.

Chapter 5 Exponential Functions

69.

x	f(x)	base	pattern
5	2		2
6	12	$\frac{12}{2} = 6$	$2(6)^1 = 12$
7	72	$\frac{72}{12} = 6$	$2(6)^2 = 72$
8	432	$\frac{432}{72} = 6$	$2(6)^3 = 432$
9	2592	$\frac{2592}{432} = 6$	$2(6)^4 = 2592$

$b = 6$ because the common base is 6 from the above table. The input values do not include $x = 0$ so we do not know the initial value of the function. The pattern in the above table does show that we have 2 times the common base 6 raised to an exponent. Since the exponents are the input values minus 5, the exponent in the model is $x - 5$. Therefore the exponential model is $f(x) = 2(6)^{x-5}$.

71.

x	f(x)	base	pattern
4	3584		3584
5	896	$\frac{896}{3584} = 0.25$	$3584(0.25)^1 = 896$
6	224	$\frac{224}{896} = 0.25$	$3584(0.25)^2 = 224$
7	56	$\frac{56}{224} = 0.25$	$3584(0.25)^3 = 56$
8	14	$\frac{14}{56} = 0.25$	$3584(0.25)^4 = 14$

$b = 0.25$ because the common base is 0.25 from the above table. The input values do not include $x = 0$ so we do not know the initial value of the function. The pattern in the above table does show that we have 3584 times the common base 0.25 raised to an exponent. Since the exponents are the input values minus 4, the exponent in the model is $x - 4$. Therefore the exponential model is $f(x) = 3584(0.25)^{x-4}$.

73. The function is increasing because the base is greater than 1.

75. The function is decreasing because the base is less than 1.

77. The function is increasing because the base is greater than 1.

79. The function is decreasing because the base is less than 1.

Section 5.2

1. This is an exponential equation because the variable is in the exponent.

3. This is a power equation because it has a constant (a number) as an exponent.

5. This is a power equation because it has a constant (a number) as an exponent.

7. This is an exponential equation because the variable is in the exponent.

9. This is a power equation because it has a constant (a number) as an exponent.

11.
$2^x = 8$
$2^x = 2^3$
$x = 3$

13.
$5^c = 3125$
$5^c = 5^5$
$c = 5$

15.
$\frac{1}{9} = 3^t$
$\frac{1}{9} = \frac{1}{3^2} = 3^{-2}$
$3^t = 3^{-2}$
$t = -2$

17.
$3^x = \frac{1}{81}$
$\frac{1}{81} = \frac{1}{3^4} = 3^{-4}$
$3^x = 3^{-4}$
$x = -4$

19.
$(-2)^d = -32$
$(-2)^d = (-2)^5$
$d = 5$

21.
$(-6)^w = 36$
$(-6)^w = (-6)^2$
$w = 2$

23.
$\left(\frac{1}{2}\right)^x = \frac{1}{16}$
$\left(\frac{1}{2}\right)^x = \left(\frac{1}{2}\right)^4$
$x = 4$

25.
$\left(\frac{1}{2}\right)^x = 16$
$(2^{-1})^x = 2^4$
$2^{-x} = 2^4$
$-x = 4$
$x = -4$

27.
$\left(\frac{1}{2}\right)^x = 32$
$(2^{-1})^x = 2^5$
$2^{-x} = 2^5$
$-x = 5$
$x = -5$

29.
$10^x = 1000$
$10^x = 10^3$
$x = 3$

31.
$10^x = 1$
$x = 0$

33.
$(-5)^m = 1$
$m = 0$

35.
$\left(\frac{2}{3}\right)^t = 1$
$t = 0$

37. Tom evaluated $\frac{1}{16} = 4^2$ as instead of $\frac{1}{16} = 4^{-2}$. So he missed the negative sign in the exponent.

$4^x = \frac{1}{16}$
$4^x = \frac{1}{4^2} = 4^{-2}$
$x = -2$

Chapter 5 Exponential Functions

39. $B(h) = 100(2^h)$

a. $B(8) = 100(2^8) = 25,600$ There are 25,600 Lactobacillus acidophilus bacteria present after 8 hours have passed.

b.
$800 = 100(2^h)$
$\dfrac{800}{100} = \dfrac{100(2^h)}{100}$
$8 = (2^h)$
$2^3 = 2^h$
$h = 3$

There are 800 Lactobacillus acidophilus bacteria present after 3 hours have passed.

41. $R(h) = 5(3^h)$

a. $R(8) = 5(3^8) = 32,805$ After 8 hours have passed, 32,805 people have heard the rumor.

b.
$405 = 5(3^h)$
$\dfrac{405}{5} = \dfrac{5(3^h)}{5}$
$81 = 3^h$
$3^4 = 3^h$
$h = 4$

It takes 4 hours for the rumor to have spread to 405 people.

43.
$5(2^x) = 40$
$\dfrac{5(2^x)}{5} = \dfrac{40}{5}$
$2^x = 8$
$2^x = 2^3$
$x = 3$

45.
$-2(5^c) = -250$
$\dfrac{-2(5^c)}{-2} = \dfrac{-250}{-2}$
$5^c = 125$
$5^c = 5^3$
$c = 3$

47.
$3^c + 5 = 32$
$3^c + 5 - 5 = 32 - 5$
$3^c = 27$
$3^c = 3^3$
$c = 3$

49.
$7^t + 8 = 57$
$7^t + 8 - 8 = 57 - 8$
$7^t = 49$
$7^t = 7^2$
$t = 2$

51.
$-4(5^m) - 9 = -109$
$-4(5^m) - 9 + 9 = -109 + 9$
$-4(5^m) = -100$
$\dfrac{-4(5^m)}{-4} = \dfrac{-100}{-4}$
$5^m = 25 = 5^2$
$m = 2$

53.
$3(6^x) + 2(6^x) = 180$
$5(6^x) = 180$
$\dfrac{5(6^x)}{5} = \dfrac{180}{5}$
$6^x = 36$
$6^x = 6^2$
$x = 2$

55.
$10(2^x) - 7(2^x) = 48$
$3(2^x) = 48$
$\dfrac{3(2^x)}{3} = \dfrac{48}{3}$
$2^x = 16$
$2^x = 2^4$
$x = 4$

5.2 Exercises

57.
$7(2^x) = 48 + 4(2^x)$
$7(2^x) - 4(2^x) = 48 + 4(2^x) - 4(2^x)$
$3(2^x) = 48$
$\dfrac{3(2^x)}{3} = \dfrac{48}{3}$
$2^x = 16$
$2^x = 2^4$
$x = 4$

59.
$13(7^w) + 20 = 5(7^w) + 2764$
$13(7^w) + 20 - 5(7^w) = 5(7^w) + 2764 - 5(7^w)$
$8(7^w) + 20 = 2764$
$8(7^w) + 20 - 20 = 2764 - 20$
$8(7^w) = 2744$
$\dfrac{8(7^w)}{8} = \dfrac{2744}{8}$
$7^w = 343$
$7^w = 7^3$
$w = 3$

61.
$x^5 = 32$
$(x^5)^{\frac{1}{5}} = (32)^{\frac{1}{5}}$
$x = 2$

63.
$x^6 = 15625$
$(x^6)^{\frac{1}{6}} = \pm(15625)^{\frac{1}{6}}$
$x = \pm 5$

65.
$12w^4 = 7500$
$\dfrac{12w^4}{12} = \dfrac{7500}{12}$
$w^4 = 625$
$(w^4)^{\frac{1}{4}} = \pm(625)^{\frac{1}{4}}$
$w = \pm 5$

67.
$-7x^3 = 1512$
$\dfrac{-7x^3}{-7} = \dfrac{1512}{-7}$
$x^3 = -216$
$(x^3)^{\frac{1}{3}} = (-216)^{\frac{1}{3}}$
$x = -6$

69. Warrick forgot to put the $\pm$ symbol on the final answer. Both positive 6 and negative 6 are correct answers. The correct answer is $x = \pm 6$.

71. Frank divided 216 by 3 instead of raising 216 to the power $\dfrac{1}{3}$. The correct answer is $r = 6$.

75.
The volume of the cube is given by the formula:
$V = s^3$
Since the volume is given as 91.125 in^3 substituting this value into the formula we get:
$91.125 = s^3$
$(91.125)^{\frac{1}{3}} = (s^3)^{\frac{1}{3}}$
$s = 4.5$
The length of the side is 4.5 *in*.

77.
$3x^5 + 94 = 190$
$3x^5 + 94 - 94 = 190 - 94$
$3x^5 = 96$
$x^5 = 32$
$(x^5)^{\frac{1}{5}} = (32)^{\frac{1}{5}}$
$x = 2$

79.
$5x^6 - 30 = 1475$
$5x^6 - 30 + 30 = 1475 + 30$
$5x^6 = 1505$
$x^6 = 301$
$(x^6)^{\frac{1}{6}} = \pm(301)^{\frac{1}{6}}$
$x \approx \pm 2.6$

81.
$$96 - 24c^4 = 40$$
$$96 - 24c^4 - 96 = 40 - 96$$
$$-24c^4 = -56$$
$$\frac{-24c^4}{-24} = \frac{-56}{-24}$$
$$c^4 = \frac{7}{3}$$
$$\left(c^4\right)^{\frac{1}{4}} = \pm\left(\frac{7}{3}\right)^{\frac{1}{4}}$$
$$c \approx \pm 1.2$$

83.
$$\frac{4x^3 + 5}{8} = 63.125$$
$$8\left(\frac{4x^3 + 5}{8}\right) = 8(63.125)$$
$$4x^3 + 5 = 505$$
$$4x^3 + 5 - 5 = 505 - 5$$
$$4x^3 = 500$$
$$\frac{4x^3}{4} = \frac{500}{4}$$
$$x^3 = 125$$
$$\left(x^3\right)^{\frac{1}{3}} = (125)^{\frac{1}{3}}$$
$$x = 5$$

85.
$$\frac{7b^4 + 58}{8} = 1141.25$$
$$8\left(\frac{7b^4 + 58}{8}\right) = 8(1141.25)$$
$$7b^4 + 58 = 9130$$
$$7b^4 + 58 - 58 = 9130 - 58$$
$$7b^4 = 9072$$
$$\frac{7b^4}{7} = \frac{9072}{7}$$
$$b^4 = 1296$$
$$\left(b^4\right)^{\frac{1}{4}} = \pm(1296)^{\frac{1}{4}}$$
$$b = \pm 6$$

87. Exponential
$$3^k = 81$$
$$3^k = 3^4$$
$$k = 4$$

89. Power
$$150r^5 = 1725$$
$$\frac{150r^5}{150} = \frac{1725}{150}$$
$$r^5 = 11.5$$
$$\left(r^5\right)^{\frac{1}{5}} = (11.5)^{\frac{1}{5}}$$
$$r \approx 1.6$$

91. Exponential
$$3(4^x) + 20 = 212$$
$$3(4^x) + 20 - 20 = 212 - 20$$
$$3(4^x) = 192$$
$$\frac{3(4^x)}{3} = \frac{192}{3}$$
$$4^x = 64$$
$$4^x = 4^3$$
$$x = 3$$

93. Power
$$45 - 3.5g^5 = -67$$
$$45 - 3.5g^5 - 45 = -67 - 45$$
$$-3.5g^5 = -112$$
$$\frac{-3.5g^5}{-3.5} = \frac{-112}{-3.5}$$
$$g^5 = 32$$
$$\left(g^5\right)^{\frac{1}{5}} = (32)^{\frac{1}{5}}$$
$$g = 2$$

95. Power
$$\frac{3.6h^8 - 56}{33} = 58.168$$
$$33\left(\frac{3.6h^8 - 56}{33}\right) = 33(58.168)$$
$$3.6h^8 - 56 = 1919.544$$
$$3.6h^8 - 56 + 56 = 1919.544 + 56$$
$$3.6h^8 = 1975.544$$
$$\frac{3.6h^8}{3.6} = \frac{1975.544}{3.6}$$
$$h^8 \approx 548.76222$$
$$\left(h^8\right)^{\frac{1}{8}} \approx \pm(548.76222)^{\frac{1}{8}}$$
$$h \approx \pm 2.2$$

97. Exponential

$$\left(\frac{1}{5}\right)^x = \frac{1}{125}$$

$$\left(\frac{1}{5}\right)^x = \left(\frac{1}{5}\right)^3$$

$$x = 3$$

99. Power

$$5x^6 + 20 = 3x^6 + 1478$$

$$5x^6 + 20 - 3x^6 = 3x^6 + 1478 - 3x^6$$

$$2x^6 + 20 = 1478$$

$$2x^6 + 20 - 20 = 1478 - 20$$

$$2x^6 = 1458$$

$$\frac{2x^6}{2} = \frac{1458}{2}$$

$$x^6 = 729$$

$$\left(x^6\right)^{\frac{1}{6}} = \pm(729)^{\frac{1}{6}}$$

$$x = \pm 3$$

101. Exponential

$$5\left(\frac{1}{2}\right)^x - \frac{2}{32} = 3\left(\frac{1}{2}\right)^x - \frac{3}{64}$$

$$5\left(\frac{1}{2}\right)^x - \frac{2}{32} - 3\left(\frac{1}{2}\right)^x = 3\left(\frac{1}{2}\right)^x - \frac{3}{64} - 3\left(\frac{1}{2}\right)^x$$

$$2\left(\frac{1}{2}\right)^x - \frac{2}{32} = -\frac{3}{64}$$

$$2\left(\frac{1}{2}\right)^x - \frac{2}{32} + \frac{2}{32} = -\frac{3}{64} + \frac{2}{32}$$

$$2\left(\frac{1}{2}\right)^x = -\frac{3}{64} + \frac{2}{32}$$

$$2\left(\frac{1}{2}\right)^x = \frac{1}{64}$$

$$\frac{2\left(\frac{1}{2}\right)^x}{2} = \frac{\frac{1}{64}}{2}$$

$$\left(\frac{1}{2}\right)^x = \frac{1}{128}$$

$$\left(\frac{1}{2}\right)^x = \left(\frac{1}{2}\right)^7$$

$$x = 7$$

Chapter 5 Exponential Functions
Section 5.3

1.
a. a is positive because the graph is above the x-axis.
b. $b > 1$ because we have exponential growth with $a > 0$.
c. The graph is increasing.
d. Domain: All real numbers $(-\infty, \infty)$
e. Range: $(0, \infty)$

3.
a. a is positive because the graph is above the x-axis.
b. $b < 1$ because we have exponential decay with $a > 0$.
c. The graph is decreasing.
d. Domain: All real numbers
e. Range: $(0, \infty)$

5.
a. a is negative because the graph is below the x-axis.
b. $b < 1$ because we have exponential growth with $a < 0$.
c. The graph is increasing.
d. Domain: All real numbers $(-\infty, \infty)$
e. Range: $(-\infty, 0)$

7.
a. a is negative because the graph is below the x-axis.
b. $b > 1$ because we have exponential decay with $a < 0$.
c. The graph is decreasing.
d. Domain: All real numbers
e. Range: $(-\infty, 0)$

9.
a. a is negative because the graph is below the x-axis.
b. $b > 1$ because we have exponential decay with $a < 0$.
c. The graph is decreasing.
d. Domain: All real numbers $(-\infty, \infty)$
e. Range: $(-\infty, 0)$

11.
$f(x) = 7(2)^x$
$a = 7$ so the y-intercept is $(0, 7)$.
$b = 2$, $b > 1$ and $a > 0$ so the graph is increasing.
Domain: $(-\infty, \infty)$
Range: $(0, \infty)$

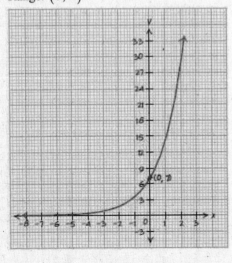

13.
$g(x) = 3(1.2)^x$
$a = 3$ so the y-intercept is $(0, 3)$.
$b = 1.2$, $b > 1$ and $a > 0$ so the graph is increasing.
Domain: $(-\infty, \infty)$
Range: $(0, \infty)$

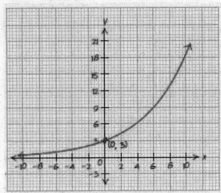

15.
$h(x) = 12(1.4)^x$
$a = 12$ so the y-intercept is $(0, 12)$.
$b = 1.4$, $b > 1$ and $a > 0$ so the graph is increasing.
Domain: $(-\infty, \infty)$
Range: $(0, \infty)$

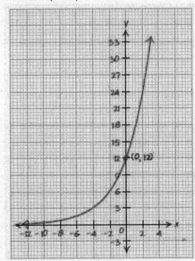

17.
$p(t) = 140\left(\dfrac{1}{2}\right)^t$
$a = 140$ so the t-intercept is $(0, 140)$.
$b = \dfrac{1}{2}$, $b < 1$ and $a > 0$ so the graph is decreasing.
Domain: $(-\infty, \infty)$
Range: $(0, \infty)$

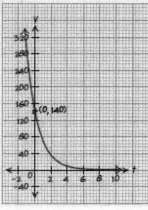

19.

$f(x) = 250\left(\dfrac{1}{4}\right)^x$

$a = 250$ so the x-intercept is $(0, 250)$.

$b = \dfrac{1}{4}$, $b < 1$ and $a > 0$ so the graph is decreasing.

Domain: $(-\infty, \infty)$

Range: $(0, \infty)$

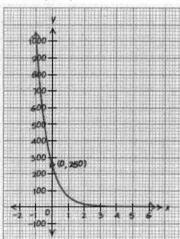

21.

a. $B(n) = 4(2)^n$

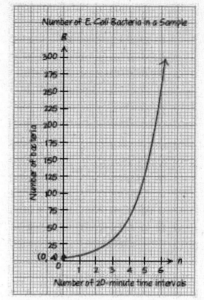

b.

$B(0) = 4(2)^0 = 4$

The B-intercept is $(0, 4)$.

At the start, there were 4 E coli bacteria present in the sample.

23.

a. $N(t) = 5000\left(\dfrac{1}{2}\right)^t$

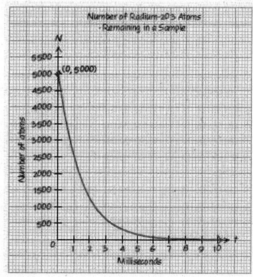

b.

$N(0) = 5000\left(\dfrac{1}{2}\right)^0 = 5000$

The N-intercept is $(0, 5000)$.

At the start, there were 5000 Radium-203 atoms present in the sample.

Chapter 5 Exponential Functions

25.

$f(x) = -2(1.4)^x$

$a = -2$ so the y-intercept is $(0, -2)$.

$b = 1.4$, $b > 1$ and $a < 0$ so the graph is decreasing.

Domain: $(-\infty, \infty)$

Range: $(-\infty, 0)$

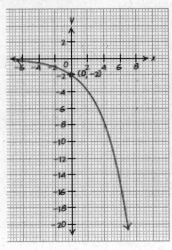

27.

$g(t) = -3(0.7)^t$

$a = -3$ so the vertical intercept is $(0, -3)$.

$b = 0.7$, $b < 1$ and $a < 0$ so the graph is increasing.

Domain: $(-\infty, \infty)$

Range: $(-\infty, 0)$

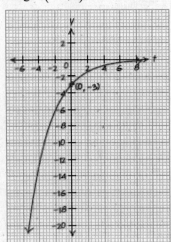

29.

$h(m) = 0.5(2.5)^m$

$a = 0.5$ so the vertical intercept is $(0, 0.5)$.

$b = 2.5$, $b > 1$ and $a > 0$ so the graph is increasing.

Domain: $(-\infty, \infty)$

Range: $(0, \infty)$

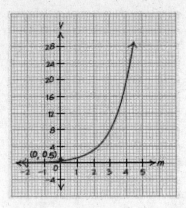

31.

$j(w) = -0.5(4)^w$

$a = -0.5$ so the vertical intercept is $(0, -0.5)$.

$b = 4$, $b > 1$ and $a < 0$ so the graph is decreasing.

Domain: $(-\infty, \infty)$

Range: $(-\infty, 0)$

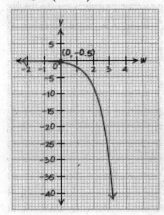

33.

$h(t) = -0.4(1.5)^t$

$a = -0.4$ so the vertical intercept is $(0, -0.4)$.

$b = 1.5$, $b > 1$ and $a < 0$ so the graph is decreasing.

Domain: $(-\infty, \infty)$

Range: $(-\infty, 0)$

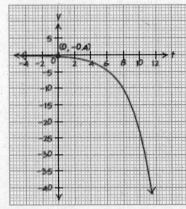

35.

$c(n) = 550\left(\dfrac{3}{4}\right)^n$

$a = 550$ so the vertical intercept is $(0, 550)$.

$b = \dfrac{3}{4}$, $b < 1$ and $a > 0$ so the graph is decreasing.

Domain: $(-\infty, \infty)$

Range: $(0, \infty)$

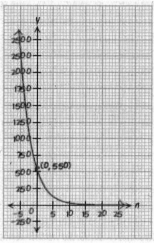

37.

$f(x) = -500\left(\dfrac{1}{5}\right)^x$

$a = -500$ so the vertical intercept is $(0, -500)$.

$b = \dfrac{1}{5}$, $b < 1$ and $a < 0$ so the graph is increasing.

Domain: $(-\infty, \infty)$

Range: $(-\infty, 0)$

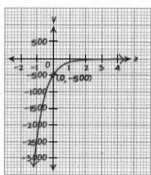

39.

$f(x) = -700(0.95)^x$

$a = -700$ so the vertical intercept is $(0, -700)$.

$b = 0.95$, $b < 1$ and $a < 0$ so the graph is increasing.

Domain: $(-\infty, \infty)$

Range: $(-\infty, 0)$

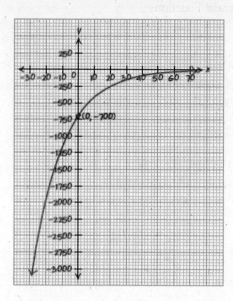

41. The student wrote the domain started at zero and the range started at 7. Domain: $(-\infty, \infty)$ Range: $(0, \infty)$

43. The student wrote the domain correctly but the range should be from zero to infinity.

Domain: $(-\infty, \infty)$ Range: $(0, \infty)$

45.

a. $y = 10$

b. Domain: $(-\infty, \infty)$

c. Range: $(10, \infty)$

47.

a. $y = -5$

b. Domain: $(-\infty, \infty)$

c. Range: $(-\infty, -5)$

49. $f(x) = 4(5)^x + 3$

a.

The horizontal asymptote is $y = 3$.

The graph is shifted up 3 units so the horizontal asymptote is also shifted up 3 units.

b.

Domain: $(-\infty, \infty)$

The domain of an exponential function is all real numbers.

c.

Range: $(3, \infty)$

The function is an exponential growth function because the base is greater than 1.

It will also be above the x-axis since a is positive.

51. $f(x) = 7(4)^x$

a. The horizontal asymptote is $y = 0$.

b.

Domain: $(-\infty, \infty)$

The domain of an exponential function is all real numbers.

c.
Range: $(0, \infty)$

The function is an exponential growth function because the base is greater than 1.
It will also be above the x-axis since a is positive.

53. $f(x) = -40(1.25)^x - 6$

a.
The horizontal asymptote is $y = -6$.
The graph is shifted down 6 units so the horizontal asymptote is also shifted down 6 units.

b.
Domain: $(-\infty, \infty)$
The domain of an exponential function is all real numbers.

c.
Range: $(-\infty, -6)$
The function is an exponential growth function because the base is greater than 1,
but a is negative so the function is flipped below the x-axis.

55. Horizontal asymptote: $y = 30$
Domain: $(-\infty, \infty)$ Range: $(30, \infty)$

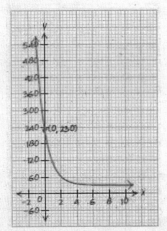

57. Horizontal asymptote: $y = 3$
Domain: $(-\infty, \infty)$ Range: $(3, \infty)$

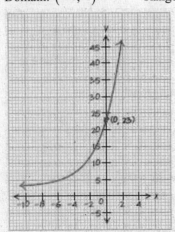

59. Horizontal asymptote: $y = -12$
Domain: $(-\infty, \infty)$ Range: $(-\infty, -12)$

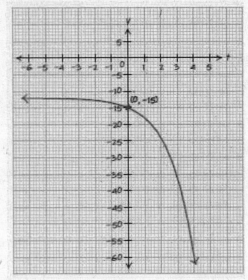

61. Horizontal asymptote: $y = -50$
Domain: $(-\infty, \infty)$ Range: $(-\infty, -50)$

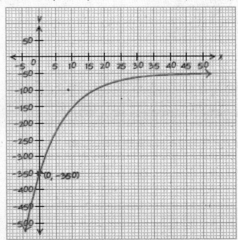

63. Horizontal asymptote: $y = 25$
Domain: $(-\infty, \infty)$ Range: $(25, \infty)$

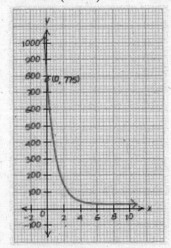

65. Domain: $(-\infty, \infty)$ Range: $(-\infty, \infty)$

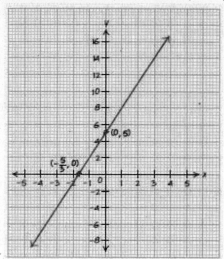

67. Domain: $(-\infty, \infty)$ Range: $[3, \infty)$

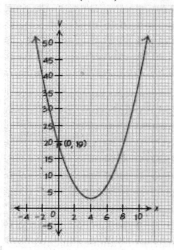

69. Domain: $(-\infty, \infty)$ Range: $(0, \infty)$

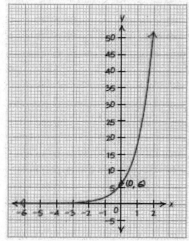

71. Domain: $(-\infty, \infty)$ Range: $(-\infty, \infty)$

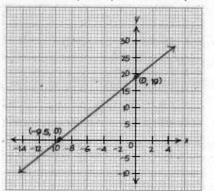

73. Domain: $(-\infty, \infty)$ Range: $[-75, \infty)$

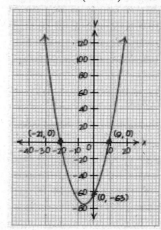

Chapter 5 Exponential Functions
Section 5.4

1. $f(x) = 25(1.5)^x$

3. $f(x) = 900(0.9)^x$

5. b, because the function is increasing too rapidly.

7. a, because the y-intercept is too low.

9. Adjust $a = 3.5$, because the value of the functions equals 3.5 when $x = 0$.
The new equation is $f(x) = 3.5(1.2)^x$.

11. Adjust $a = 78$, because the value of the functions equals 78 when $x = 0$.
The new equation is $f(x) = 78(0.45)^x$.

13. Adjust $b = 1.3$, because the original graph is increasing too fast.
The new equation is $f(x) = 8(1.3)^x$.

15. Adjust $b = 0.5$, because the original graph is increasing too slowly.
The new equation is $f(x) = 120(0.5)^x$.

17. Adjust $a = 5$, because the value of the functions equals 5 when $x = 0$.
Adjust $b = 1.2$, because the original graph is increasing too fast.
The new equation is $f(x) = 5(1.2)^x$.

19.
a. Let $I(t)$ be the number of internet hosts (in millions) in the year t years since 1990.

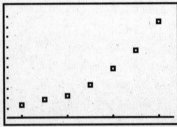

An exponential model is appropriate so choose a point from the flatter end and one from the steeper end.
Note: Your model may differ, but it must follow the trend of the trend of the data.

$I(t) = ab^t$

To find a and b choose two points $(3, 1.3)$ and $(5.5, 6.6)$.

$1.3 = ab^3$ and $6.6 = ab^{5.5}$

$$\frac{6.6}{1.3} = \frac{ab^{5.5}}{ab^3}$$

$$\frac{6.6}{1.3} = b^{2.5}$$

$$\left(\frac{6.6}{1.3}\right)^{\frac{1}{2.5}} = \left(b^{2.5}\right)^{\frac{1}{2.5}}$$

$b \approx 1.92$

Use b and one of the above equations to find a.

$1.3 = a(1.92)^3$

$$\frac{1.3}{(1.92)^3} = \frac{a(1.92)^3}{(1.92)^3}$$

$a \approx 0.184$

Then $I(t) = 0.184(1.92)^t$

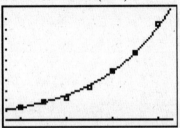

b.
$I(7) = 0.184(1.92)^7$
$I(7) \approx 17.7$

In 1997 there were approximately 17.7 million internet hosts.

c. Domain: $[0, 8]$ Range: $[0.18, 33.98]$

d.
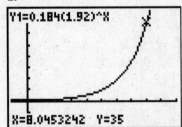

Partway through the year 1998 the number of internet hosts reached 35 million.

21.

a. Let $N(t)$ be the number of nuclear warheads in the U.S. arsenal in the year t years since 1940.

t	$N(t)$
5	6
7	32
9	235
11	640
13	1436
15	3057
17	6444
19	15,468

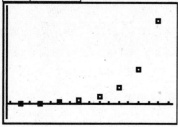

An exponential model is appropriate so choose a point from the flatter end and one from the steeper end.

Note: Your model may differ, but it must follow the trend of the data.

$N(t) = ab^t$

To find a and b choose two points $(7, 32)$ and $(17, 6444)$.

$32 = ab^7$ and $6444 = ab^{17}$

$\dfrac{6444}{32} = \dfrac{ab^{17}}{ab^7}$

$201.375 = b^{10}$

$(201.375)^{\frac{1}{10}} = (b^{10})^{\frac{1}{10}}$

NOTE: We do not need the $\pm$ on the answer because $b > 0$.

$b \approx 1.70$

Use b and one of the above equations to find a.

$32 = a(1.70)^7$

$\dfrac{32}{(1.70)^7} = \dfrac{a(1.70)^7}{(1.70)^7}$

$a \approx 0.78$

Then $N(t) = 0.78(1.70)^t$

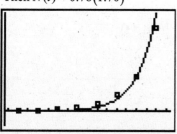

b.
$N(10) = 0.78(1.70)^{10}$
$N(10) \approx 157.25$

In 1970 there were approximately 157 nuclear warheads stockpiled in the U.S. arsenal.

c. Domain: $[5, 20]$ Range: $[11, 31701]$

d.

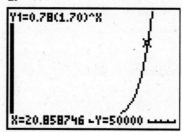

By the end of 1960 the number of stockpiled nuclear warheads in the U.S. arsenal surpassed 50,000.

23.

a. Let $V(t)$ be the volume of water left in the cylinder t seconds after the start of the experiment.

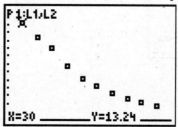

An exponential model is appropriate so choose a point from the flatter end and one from the steeper end.

Note: Your model may differ, but it must follow the trend of the data.

$V(t) = ab^t$

To find a and b choose two points $(30, 13.24)$ and $(300, 2.66)$.

$13.24 = ab^{30}$ and $2.66 = ab^{300}$

$\dfrac{2.66}{13.24} = \dfrac{ab^{300}}{ab^{30}}$

$\dfrac{2.66}{13.24} = b^{270}$

$\left(\dfrac{2.66}{13.24}\right)^{\frac{1}{270}} = \left(b^{270}\right)^{\frac{1}{270}}$

$b \approx 0.994$

Use b and one of the above equations to find a.

$13.24 = a(0.994)^{30}$

$\dfrac{13.24}{(0.994)^{30}} = \dfrac{a(0.994)^{30}}{(0.994)^{30}}$

$a \approx 15.86$

Then $V(t) = 15.86(0.994)^t$

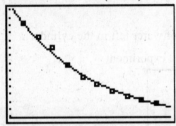

b.

6 minutes → $t = 360$

$V(360) = 15.86(0.994)^{360}$

$V(360) \approx 1.82$

Six minutes after the start of the experiment, approximately 1.82 liters of water remain in the cylinder.

c. Domain: $[0, 500]$ Range: $[0.78, 15.86]$

d.

X	Y1
458.5	1.0045
458.75	1.003
459	1.0015
459.25	1
459.5	.9985
459.75	.997
460	.9955

X=459.25

It took approximately 459 seconds for there to be only 1 liter of water remaining in the cylinder.

25.

a. Let $R(d)$ be the river gage height, in feet above normal, d days after the rainfall event.

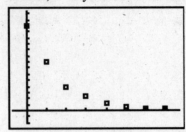

An exponential model is appropriate so choose the vertical intercept and a point from the flatter end.

Note: Your model may differ, but must it follow the trend of the data.

$R(d) = ab^d$

From $(0, 14)$ we get $a = 14$.

Giving us: $R(d) = 14b^d$

To find b use a second point: $(6, 0.4)$

$0.4 = 14b^6$

$\dfrac{0.4}{14} = \dfrac{14b^6}{14}$

$\dfrac{0.4}{14} \approx b^6$

$\left(\dfrac{0.4}{14}\right)^{\frac{1}{6}} \approx \left(b^6\right)^{\frac{1}{6}}$

NOTE: We do not need the $\pm$ on the answer because $b > 0$.

$b \approx 0.553$

Then $R(d) = 14(0.553)^d$

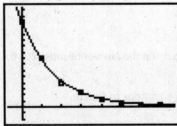

b.

$R(10) = 14(0.553)^{10}$

$R(10) \approx 0.037$

Ten days after the rainfall event the gage height will read 0.037 feet above normal.

c. Domain: $[0, 12]$ Range: $[0.01, 14]$

5.4 Exercises

27.

To find a use the point $(0,4)$

$4 = ab^0$

$4 = a(1)$

$a = 4$

Giving us: $f(x) = 4b^x$

To find b use a point that falls along the exponential path: $(5,16)$

$16 = 4b^5$

$\dfrac{16}{4} = \dfrac{4b^5}{4}$

$4 = b^5$

$(4)^{\frac{1}{5}} = (b^5)^{\frac{1}{5}}$

$b \approx 1.32$

The new equation is $f(x) = 4(1.32)^x$

Domain: $(-\infty, \infty)$ Range: $(0, \infty)$

29.

To find a use the point $(0, 94)$

$94 = ab^0$

$94 = a(1)$

$a = 94$

Giving us: $f(x) = 94b^x$

To find b use a point that falls along the exponential path: $(4, 16)$

$16 = 94b^4$

$\dfrac{16}{94} = \dfrac{94b^4}{94}$

$\dfrac{8}{47} = b^4$

$\left(\dfrac{8}{47}\right)^{\frac{1}{4}} = (b^4)^{\frac{1}{4}}$

NOTE: We do not need the $\pm$ on the answer because $b > 0$.

$b \approx 0.63$

The new equation is $f(x) = 94(0.63)^x$

Domain: $(-\infty, \infty)$ Range: $(0, \infty)$

31.

To find a and b use the points $(5, 14)$ and $(12, 76)$.

$14 = ab^5$ and $76 = ab^{12}$

$\dfrac{76}{14} = \dfrac{ab^{12}}{ab^5}$

$\dfrac{38}{7} = b^7$

$\left(\dfrac{38}{7}\right)^{\frac{1}{7}} = (b^7)^{\frac{1}{7}}$

$b \approx 1.27$

Use b and one of the above equations to find a.

$14 = a(1.27)^5$

$\dfrac{14}{(1.27)^5} = \dfrac{a(1.27)^5}{(1.27)^5}$

$\dfrac{14}{(1.27)^5} = a$

By rounding a we get:

$a \approx 4$

The new equation is $f(x) = 4(1.27)^x$

Domain: $(-\infty, \infty)$ Range: $(0, \infty)$

33.

To find a and b use the points $(-1, 855.56)$ and $(3, 73.75)$.

$855.56 = ab^{-1}$ and $73.75 = ab^3$

$\dfrac{73.75}{855.56} = \dfrac{ab^3}{ab^{-1}}$

$\dfrac{73.75}{855.56} = b^4$

$\left(\dfrac{73.75}{855.56}\right)^{\frac{1}{4}} = (b^4)^{\frac{1}{4}}$

NOTE: We do not need the $\pm$ on the answer because $b > 0$.

$b \approx 0.54$

Use b and one of the above equations to find a.

$855.56 = a(0.54)^{-1}$

$\dfrac{855.56}{(0.54)^{-1}} = \dfrac{a(0.54)^{-1}}{(0.54)^{-1}}$

$\dfrac{855.56}{(0.54)^{-1}} = a$

By rounding a we get:

$a \approx 462$

The new equation is $f(x) = 462(0.54)^x$

Domain: $(-\infty, \infty)$ Range: $(0, \infty)$

Chapter 5 Exponential Functions

35. The data plot shows an increasing curve with positive values so we know that $a > 0$ and $b > 1$. Therefore, the student's model is incorrect. The student miscalculated the value of b by multiplying both sides of $3.58 \approx b^7$ by $\frac{1}{7}$ instead of raising each side to the $\frac{1}{7}^{th}$ power. As a result the value of a is incorrect as well. The correct steps are:

$3.58 \approx b^7$

$(3.58)^{\frac{1}{7}} \approx (b^7)^{\frac{1}{7}}$

$1.2 \approx b$

$1.44 = a(1.2)^1$

$\dfrac{1.44}{1.2} = \dfrac{a(1.2)^1}{1.2}$

$a = 1.2$

$R(t) = 1.2(1.2)^t$

37.
To find a use the point $(0, -17)$

$-17 = ab^0$

$-17 = a(1)$

$a = -17$

Giving us: $f(x) = -17b^x$

To find b use a point that falls along the exponential path: $(1, -21)$

$-21 = -17b^1$

$\dfrac{-21}{-17} = \dfrac{-17b}{-17}$

$\dfrac{21}{17} = b$

$b \approx 1.22$

39.
To find a use the point $(0, -56)$

$-56 = ab^0$

$-56 = a(1)$

$a = -56$

Giving us: $f(x) = -56b^x$

To find b use a point that falls along the exponential path: $(3, -5.103)$

$-5.103 = -56b^3$

$\dfrac{-5.103}{-56} = \dfrac{-56b^3}{-56}$

$\dfrac{5.103}{56} = b^3$

$\left(\dfrac{5.103}{56}\right)^{\frac{1}{3}} = (b^3)^{\frac{1}{3}}$

$b \approx 0.45$

The new equation is $f(x) = -56(0.45)^x$

Domain: $(-\infty, \infty)$ Range: $(-\infty, 0)$

41.
To find a and b use the points $(1, -306.3)$ and $(8, -1380.61)$.

$-306.3 = ab^1$ and $-1380.61 = ab^8$

$\dfrac{-1380.61}{-306.3} = \dfrac{ab^8}{ab^1}$

$\dfrac{1380.61}{306.3} = b^7$

$\left(\dfrac{1380.61}{306.3}\right)^{\frac{1}{7}} = (b^7)^{\frac{1}{7}}$

$b \approx 1.24$

Use b and one of the above equations to find a.

$-306.3 = a(1.24)^1$

$-306.3 = 1.24a$

$\dfrac{-306.3}{1.24} = \dfrac{1.24a}{1.24}$

By rounding a we get:

$a \approx -247$

The new equation is $f(x) = -247(1.24)^x$ Domain: $(-\infty, \infty)$ Range: $(-\infty, 0)$

43.

To find a and b use the points $(-1,-282.54)$ and $(9,-2.78)$.

$-282.54 = ab^{-1}$ and $-2.78 = ab^9$

$$\frac{-2.78}{-282.54} = \frac{ab^9}{ab^{-1}}$$

$$\frac{2.78}{282.54} = b^{10}$$

$$\left(\frac{2.78}{282.54}\right)^{\frac{1}{10}} = \left(b^{10}\right)^{\frac{1}{10}}$$

$b \approx 0.63$

Use b and one of the above equations to find a.

$-282.54 = a(0.63)^{-1}$

$$\frac{-282.54}{(0.63)^{-1}} = \frac{a(0.63)^{-1}}{(0.63)^{-1}}$$

By rounding a we get:

$a \approx -178$

The new equation is $f(x) = -178(0.63)^x$

Domain: $(-\infty, \infty)$ Range: $(-\infty, 0)$

45. Exponential: Text messaging continues to grow rapidly.

47. Exponential: Music downloads continue to grow rapidly.

49. Linear: Steady growth of New Hampshire's population.

51.

a. Let $V(m)$ be the number of visitors to MySpace.com (in millions) m months after the site started.

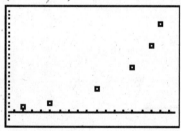

An exponential model is appropriate.

$V(m) = ab^m$

To find a and b choose two points $(10,11)$ and $(24,15)$.

$1 = ab^{10}$ and $15 = ab^{24}$

$$\frac{15}{1} = \frac{ab^{24}}{ab^{10}}$$

$15 = b^{14}$

$(15)^{\frac{1}{14}} = \left(b^{14}\right)^{\frac{1}{14}}$

NOTE: We do not need the $\pm$ on the answer because $b > 0$.

$b \approx 1.21$

Use b and one of the above equations to find a.

$1 = a(1.21)^{10}$

$$\frac{1}{(1.21)^{10}} = \frac{a(1.21)^{10}}{(1.21)^{10}}$$

$a \approx 0.15$

Then $V(m) = 0.15(1.21)^m$

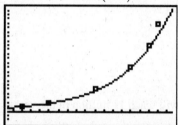

b. Domain: $[9,36]$ Range: $[0.8, 143.3]$

c.

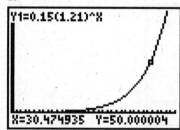

MySpace.com reached 50 million visitors after about 30 months.

53. a. Let $P(t)$ be the population of Georgia (in millions) in the year t years since 2000.

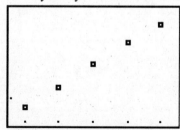

A linear model is appropriate. Choose two points: $(4, 8.91)$ and $(8, 9.69)$.

Chapter 5 Exponential Functions

$\text{Slope} = \dfrac{9.69 - 8.91}{8 - 4} = 0.195$

So $P(t) = 0.195t + b$

$8.91 = (0.195)(4) + b$

$8.91 = 0.78 + b$

$b = 8.13$

Then $P(t) = 0.195t + 8.13$

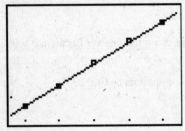

b. Domain: $[0, 15]$ Range: $[8.13, 11.06]$

c.

$11 = 0.195t + 8.13$

$11 - 8.13 = 0.195t + 8.13 - 8.13$

$2.87 = 0.195t$

$\dfrac{2.87}{0.195} = \dfrac{0.195t}{0.195}$

$14.7 \approx t$

The population of George will reach 11 million in the year 2014.

55.

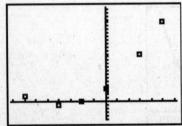

A quadratic model is appropriate. Choose the vertex and a second point.

Vertex: $(-4, -10)$

$f(x) = a(x + 4)^2 - 10$

Use a second point to solve for a: $(5, 233)$

$233 = a(5 + 4)^2 - 10$

$233 = 81a - 10$

$243 = 81a$

$\dfrac{243}{81} = \dfrac{81a}{81}$

$a = 3$

Then $f(x) = 3(x + 4)^2 - 10$

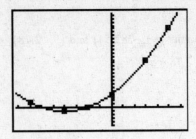

57.

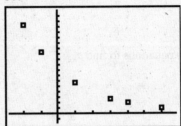

An exponential model is appropriate.

$f(x) = ab^x$

To find a and b choose two points $(-2, 163.3)$ and $(6, 9.4)$.

$163.3 = ab^{-2}$ and $9.4 = ab^6$

$\dfrac{9.4}{163.3} = \dfrac{ab^6}{ab^{-2}}$

$\dfrac{9.4}{163.3} = b^{6-(-2)} = b^8$

$\left(\dfrac{9.4}{163.3}\right)^{\frac{1}{8}} = \left(b^8\right)^{\frac{1}{8}}$

NOTE: We do not need the $\pm$ on the answer because $b > 0$.

$b \approx 0.70$

Use b and one of the above equations to find a.

$9.4 = a(0.70)^6$

$\dfrac{9.4}{(0.70)^6} = \dfrac{a(0.70)^6}{(0.70)^6}$

$a \approx 79.90$

Then $f(x) = 79.9(0.70)^x$

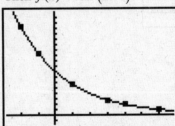

59.

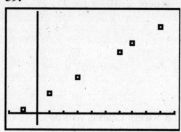

A linear model is appropriate. Choose two points
$(-1, 0.1)$ and $(9, 2.1)$

$f(x) = mx + b$

$m = \dfrac{2.1 - 0.1}{9 - (-1)} = \dfrac{2}{10} = 0.2$

$b = y - mx$

$b = 0.1 - (0.2)(-1) = 0.3$

Then $f(x) = 0.2x + 0.3$

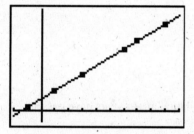

Chapter 5 Exponential Functions
Section 5.5

1.
$r = b - 1$ where $b = 1.03$
$r = 1.03 - 1$
$r = .03$
$r = 3\%$
3% growth

3.
$r = b - 1$ where $b = 1.25$
$r = 1.25 - 1$
$r = .25$
$r = 25\%$
25% growth

5.
$r = b - 1$ where $b = 3.5$
$r = 3.5 - 1$
$r = 2.5$
$r = 250\%$
250% growth

7.
$r = b - 1$ where $b = 0.95$
$r = 0.95 - 1$
$r = -0.05$
$r = -5\%$
5% decay

9.
$r = b - 1$ where $b = 0.36$
$r = 0.36 - 1$
$r = -0.64$
$r = -64\%$
64% decay

11. $W(t) = 1.19(1.08)^t$ where $1980 \to t = 0$

a.
$1990 \to t = 10$
$W(10) = 1.19(1.08)^{10}$
$W(10) \approx 2.57$

In 1990, the population of white-tailed deer was approximately 2.57 million.

b.
$\left. \begin{array}{l} b = 1.08 \\ r = b - 1 \end{array} \right\} r = 1.08 - 1 = 0.08$

The population of white-tailed deer grows at a rate of 8% per year.

c. This population might stop growing at the rate of 8% per year because of limited food or space.

13. $V(t) = 109.8(0.99939)^t$ where $2010 \to t = 0$

a.
$2015 \to t = 5$
$V(5) = 109.8(0.99939)^5$
$V(5) \approx 109.47$

In 2015, the population of the Virgin Islands will be approximately 109.47 thousand.

b.
$\left. \begin{array}{l} b = 0.99939 \\ r = b - 1 \end{array} \right\} r = 0.99939 - 1 = -0.00061$

The population of the Virgin Islands is decreasing at a rate of 0.061% per year.

15. Let $P(t)$ be the population after t years have passed.
$P(t) = ab^t$
$a = 40$ and $r = 3\% = 0.03$
$b = 1 + r$ so $b = 1.03$
Then $P(t) = 40(1.03)^t$

17. Let $P(t)$ be the population after t years have passed.
$P(t) = ab^t$
$a = 200$ and $r = 2\%$ decrease per year
$b = 1 + r$ with $r = -0.02$
$b = 1 - 0.02 = 0.98$
Then $P(t) = 200(0.98)^t$

19. Let $H(t)$ be the population of humpback whales t years after 1981.

a.
$H(t) = ab^t$
$a = 350$ and $r = 14\% = 0.14$
$b = 1 + r$ so $b = 1.14$
Then $H(t) = 350(1.14)^t$

b. Domain: $[-2, 15]$ Range: $[269, 2498]$

c. $H(9) = 350(1.14)^9 \approx 1138$ In 1990 the humpback whale population was approximately 1138.

21. Let $D(t)$ be the population of Denmark (in millions) t years after 2008

a.

$D(t) = ab^t$

$a = 5.5$ and $r = 0.295\% = 0.00295$

$b = 1 + r$ so $b = 1.00295$

Then $D(t) = 5.5(1.00295)^t$

b.

$2020 \to t = 12$

$D(12) = 5.5(1.00295)^{12}$

$D(12) \approx 5.70$

In 2020 the population of Denmark will be approximately 5.70 million.

22. Let $P(t)$ be the population of sandhill cranes (in thousands) in the U.S. t years after 1996.

a.

$P(t) = ab^t$

$a = 500$ and $r = 4.3\% = 0.043$

$b = 1 + r$ so $b = 1.043$

Then $P(t) = 500(1.043)^t$

b.

$2002 \to t = 6$

$P(6) = 500(1.043)^6$

$P(6) \approx 643.7$

23. Compounding interest formula: $A = P\left(1 + \dfrac{r}{n}\right)^{nt}$

Use this formula because the interested in compounded daily.

25. Compounding interest formula: $A = P\left(1 + \dfrac{r}{n}\right)^{nt}$

Use this formula because the interested in compounded weekly.

27. Compounding interest formula: $A = P\left(1 + \dfrac{r}{n}\right)^{nt}$

Use this formula because the interested in compounded monthly.

29. Continuously compounding interest formula: $A = Pe^{rt}$
Use this formula because the interested in compounded continuously.

31. The missing variable is A.

$A = ?$

$P = 20,000$

$r = 0.03$

$n = 365$

$t = 5$

33. The missing variable is t.

$A = 8,000$

$P = 4,000$

$r = 0.0275$

$n = 12$

$t = ?$

35. The missing variable is r.

$A = 1,000$

$P = 500$

$r = ?$

$n = 365$

$t = 10$

37.

Using the formula $A = P\left(1 + \dfrac{r}{n}\right)^{nt}$

because the interest is compounded monthly where:

$P = \$10,000$

$r = 5\% = 0.05$

$t = 10$ years

$n = 12$

$A = ?$

Solve for A:

$A = 10000\left(1 + \dfrac{0.05}{12}\right)^{12 \cdot 10}$

$A \approx 16,470.09$

The amount of money in the account after 10 years is $\$16,470.09$.

Chapter 5 Exponential Functions

39.

Using the formula $A = P\left(1+\dfrac{r}{n}\right)^{nt}$

because the interest is compounded daily where:

$P = \$10,000$

$r = 5\% = 0.05$

$t = 10$ years

$n = 365$

$A = ?$

Solve for A:

$A = 10000\left(1+\dfrac{0.05}{365}\right)^{365 \cdot 10}$

$A \approx 16,486.65$

The amount of money in the account after 10 years is $\$16,486.65$.

41.

Using the formula $A = Pe^{rt}$

because the interest is compounded continuously where:

$P = \$500,000$

$r = 9\% = 0.09$

$t = 15$ years

$A = ?$

Solve for A:

$A = 500000 e^{0.09 \cdot 15}$

$A \approx 1,928,712.77$

The amount of money in the account after 15 years is $\$1,928,712.77$.

43.

Using the formula $A = P\left(1+\dfrac{r}{n}\right)^{nt}$

because the interest is compounded hourly where:

$P = \$500,000$

$r = 9\% = 0.09$

$t = 15$ years

$n = 8760$

$A = ?$

Solve for A:

$A = 500000\left(1+\dfrac{0.09}{8760}\right)^{8760 \cdot 15}$

$A \approx 1,928,699.39$

The amount of money in the account after 15 years is $\$1,928,699.39$.

45.

Using the formula $A = P\left(1+\dfrac{r}{n}\right)^{nt}$

because the interest is compounded quarterly where:

$P = \$100,000$

$r = 7\% = 0.07$

$t = 20$ years

$n = 4$

$A = ?$

Solve for A:

$A = 100000\left(1+\dfrac{0.07}{4}\right)^{4 \cdot 20}$

$A \approx 400,639.19$

The amount of money in the account after 20 years is $\$400,639.19$.

47.

Using the formula $A = Pe^{rt}$

because the interest is compounded continuously where:

$P = \$100,000$

$r = 7\% = 0.07$

$t = 20$ years

$A = ?$

Solve for A:

$A = 100000 e^{0.07 \cdot 20}$

$A \approx 405,520.00$

The amount of money in the account after 20 years is $\$405,520.00$.

49.

For Option #1 use the formula $A = P\left(1+\dfrac{r}{n}\right)^{nt}$

$P = \$400,000$

$r = 4\% = 0.04$

$t = 5$ years

$n = 1$

Solve for A:

$A = 400,000\left(1+\dfrac{0.04}{1}\right)^{(1 \times 5)}$

$A \approx \$486,661.16$

For Option #2 use the formula $A = Pe^{rt}$

$P = \$400,000$

$r = 3.95\% = 0.0395$

$t = 5$ years

Solve for A:

$A = 400,000e^{(0.0395 \times 5)}$

$A \approx \$487,341.23$

The second option, with an interest rate of 3.95% compounded continuously, will have the larger balance after 5 years.

Chapter 5 Exponential Functions
Chapter 5 Review

1.
a. Let $P(t)$ be the population of Africa (in millions) in the year t years since 1996. Let n be the number of 28 year time intervals.

t	$P(t)$
0	731.5
28	$731.5(2)$
$2(28)$	$731.5(2)^2$
$3(28)$	$731.5(2)^3$
$n(28)$	$731.5(2)^n$

$n(28) = t$

$\dfrac{n(28)}{28} = \dfrac{t}{28}$

$n = \dfrac{t}{28}$

$P(t) = 731.5(2)^{t/28}$

b.
$2005 \to t = 9$

$P(9) = 731.5(2)^{9/28}$

$P(9) = 914.06$

In 2005, the population of Africa was approximately 914.06 million.

2. Let $F(d)$ be the number of people with flu symptoms after d days.

a. The bacteria triple every hour:

d	$F(d)$
0	1
1	$(3)^1$
2	$(3)(3) = (3)^2$
3	$(3)(3)(3) = (3)^3$
d	$(3)^d$

$F(d) = (3)^d$

b.
Two weeks $\to d = 14$

$F(14) = (3)^{14} = 4{,}782{,}969$

After 2 weeks approximately 4,782,969 people have flu symptoms. This is possibly model breakdown.

3.
a. Let $P(d)$ be the percentage of polonium-210 left in a sample after d days. Let n be the number of 138-day time intervals. The percent of polonium-210 is cut in half every 138 days:

d	$P(d)$
0	100
138	$100\left(\frac{1}{2}\right)$
$2(138)$	$100\left(\frac{1}{2}\right)^2$
$3(138)$	$100\left(\frac{1}{2}\right)^3$
$n(138)$	$100\left(\frac{1}{2}\right)^n$

$n(138) = d$

$\dfrac{n(138)}{138} = \dfrac{d}{138}$

$n = \dfrac{d}{138}$

$P(d) = 100\left(\frac{1}{2}\right)^{\frac{d}{138}}$

b.
$P(300) = 100\left(\frac{1}{2}\right)^{\frac{300}{138}}$

$P(300) = 22.16$

After 300 days, there will be about 22.16% of the polonium-210 sample left.

4.
a. Let $P(t)$ be the percentage of thorium-228 left in a sample after t years. Let n be the number of 1.9-year time intervals. The percent of thorium-228 is cut in half every 1.9 years:

t	$P(t)$
0	100
1.9	$100\left(\frac{1}{2}\right)$
$2(1.9)$	$100\left(\frac{1}{2}\right)^2$
$3(1.9)$	$100\left(\frac{1}{2}\right)^3$
$n(1.9)$	$100\left(\frac{1}{2}\right)^n$

$n(1.9) = t$

$\dfrac{n(1.9)}{1.9} = \dfrac{t}{1.9}$

$n = \dfrac{t}{1.9}$

$P(t) = 100\left(\frac{1}{2}\right)^{\frac{t}{1.9}}$

b.

$P(50) = 100\left(\frac{1}{2}\right)^{\frac{50}{1.9}}$

$P(50) = 0.000001197$

After 50 years, there will be about 0.000001197% of the thorium-228 sample left. This amount approaches zero percent.

5.

x	f(x)	base
0	5000	$\frac{4000}{5000} = 0.8$
1	4000	$\frac{3200}{4000} = 0.8$
2	3200	$\frac{2560}{3200} = 0.8$
3	2560	$\frac{2048}{2560} = 0.8$
4	2048	

$a = 5000$ because the initial value is 5000 when $x = 0$.
$b = 0.8$ because the common base is 0.8 from the above table.
Therefore the exponential model is $f(x) = 5000(0.8)^x$

6.

x	f(x)	base
0	0.2	$\frac{1.52}{0.2} = 7.6$
1	1.52	$\frac{11.53}{1.52} = 7.6$
2	11.53	$\frac{87.58}{11.53} = 7.6$
3	87.58	$\frac{665.05}{87.05} = 7.6$
4	665.05	

$a = 0.2$ because the initial value is 0.2 when $x = 0$.
$b = 7.6$ because the common base is 7.6 from the above table.
Therefore the exponential model is $f(x) = 0.2(7.6)^x$

7.

$6^x = 216$

$6^x = 6^3$

$x = 3$

8.

$4^x = 1024$

$4^x = 4^5$

$x = 5$

9.

$6(7^x) = 14406$

$\dfrac{6(7^x)}{6} = \dfrac{14406}{6}$

$7^x = 2401$

$7^x = 7^4$

$x = 4$

10.

$5(3^x) = 10935$

$\dfrac{5(3^x)}{5} = \dfrac{10935}{5}$

$3^x = 2187$

$3^x = 3^7$

$x = 7$

11.

$2x^5 + 7 = 1057.4375$

$2x^5 + 7 - 7 = 1057.4375 - 7$

$2x^5 = 1050.4375$

$\dfrac{2x^5}{2} = \dfrac{1050.4375}{2}$

$x^5 = 525.21875$

$\left(x^5\right)^{\frac{1}{5}} = (525.21875)^{\frac{1}{5}}$

$x = 3.5$

12.

$3x^4 - 600 = 1275$

$3x^4 - 600 + 600 = 1275 + 600$

$3x^4 = 1875$

$\dfrac{3x^4}{3} = \dfrac{1875}{3}$

$x^4 = 625$

$\left(x^4\right)^{\frac{1}{4}} = \pm(625)^{\frac{1}{4}}$

$x = \pm 5$

Chapter 5 Exponential Functions

13.
$$-\frac{1}{3}x^7 + 345 = -92967$$
$$-\frac{1}{3}x^7 + 345 - 345 = -92967 - 345$$
$$-\frac{1}{3}x^7 = -93312$$
$$-3\left(-\frac{1}{3}x^7\right) = -3(-93312)$$
$$x^7 = 279936$$
$$(x^7)^{\frac{1}{7}} = (279936)^{\frac{1}{7}}$$
$$x = 6$$

14.
$$-4x^8 + 2768 = -6715696$$
$$-4x^8 + 2768 - 2768 = -6715696 - 2768$$
$$-4x^8 = -6718464$$
$$\frac{-4x^8}{-4} = \frac{-6718464}{-4}$$
$$x^8 = 1679616$$
$$(x^8)^{\frac{1}{8}} = \pm(1679616)^{\frac{1}{8}}$$
$$x = \pm 6$$

15.
$$2^x = \frac{1}{32}$$
$$2^x = \frac{1}{2^5}$$
$$2^x = 2^{-5}$$
$$x = -5$$

16.
$$(-3)^x = \frac{1}{81}$$
$$(-3)^x = \frac{1}{(-3)^4}$$
$$(-3)^x = (-3)^{-4}$$
$$x = -4$$

17.
$$\left(\frac{2}{3}\right)^x = \frac{27}{8}$$
$$\left(\frac{2}{3}\right)^x = \left(\frac{3}{2}\right)^3$$
$$\left(\frac{2}{3}\right)^x = \left(\frac{2}{3}\right)^{-3}$$
$$x = -3$$

18.
$$\left(\frac{1}{4}\right)^x = 1024$$
$$\left(\frac{1}{4}\right)^x = 4^5$$
$$\left(\frac{1}{4}\right)^x = \left(\frac{1}{4}\right)^{-5}$$
$$x = -5$$

19.
a. a is positive because the y-intercept is above the origin.

b. b is greater than 1, because $a > 0$ and the function is increasing.

c. Increasing.

d. Domain: $(-\infty, \infty)$ Range: $(0, \infty)$

20.
a. a is negative because the y-intercept is below the origin.

b. b is greater than 1, because $a < 0$ and the function is decreasing.

c. Decreasing.

d. Domain: $(-\infty, \infty)$ Range: $(-\infty, 0)$

21.
a. $a > c$, because the y-intercept for $f(x)$ is greater than the y-intercept of $g(x)$.

b. $b < d$, because $f(x)$ is increasing at a slower rate than $g(x)$.

22.
a. Decreasing.

b. This is an example is exponential decay.

c. $f(20) = 5$

d. $x = 10$

e. $(0, 20)$, because when $x = 0$ $y = 20$.

23.
a. Increasing, because $b > 1$ and $a > 0$.

b. This is an example is exponential growth, because $b > 0$.

c. $(0, 2)$, because when $x = 0$ $y = 2$.

Chapter 5 Review

24. Since $a > 0$ and $b > 1$ the function is increasing.

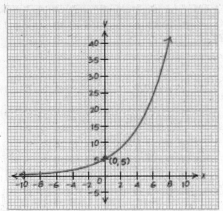

25. Since $a < 0$ and $b < 1$ the function is increasing.

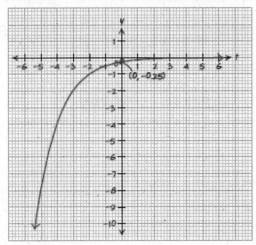

26. Since $a > 0$ and $b < 1$ the function is decreasing.

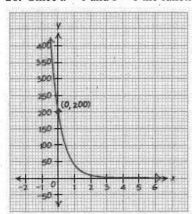

27. Since $a < 0$ and $b < 1$ the function is increasing.

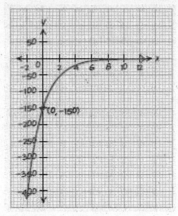

28. Since $a > 0$ and $b < 1$ the function is decreasing.

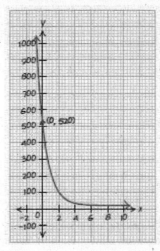

29. Since $a < 0$ and $b > 1$ the function is decreasing.

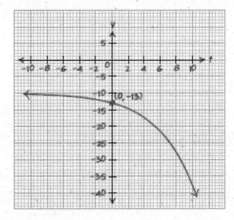

30. Domain: $(-\infty, \infty)$ Range: $(0, \infty)$

31. Domain: $(-\infty, \infty)$ Range: $(-\infty, 0)$

32. Domain: $(-\infty, \infty)$ Range: $(20, \infty)$

33. Domain: $(-\infty, \infty)$ Range: $(-\infty, -10)$

34.

a. Let $W(t)$ be the number of non-strategic warheads in the U.S. arsenal in the year t years since 1950.

t	$W(t)$
1	91
2	205
3	436
4	563
5	857
6	1,618
7	2,244
8	4,122
9	8,462
10	13,433

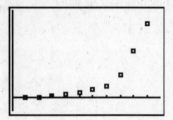

An exponential model is appropriate so choose a point from the flatter end and one from the steeper end.

Note: Your model may differ, but it must follow the trend of the data.

$W(t) = ab^t$

To find a and b choose two points $(1, 91)$ and $(10, 13433)$.

$91 = ab$ and $13{,}433 = ab^{10}$

$$\frac{13{,}433}{91} = \frac{ab^{10}}{ab}$$

$$\frac{13{,}433}{91} = b^9$$

$$\left(\frac{13{,}433}{91}\right)^{\frac{1}{9}} = \left(b^9\right)^{\frac{1}{9}}$$

$b \approx 1.74$

Use b and one of the above equations to find a.

$91 = a(1.74)$

$$\frac{91}{1.74} = \frac{a(1.74)}{1.74}$$

$a \approx 52.3$

Then $W(t) = 52.3(1.74)^t$

b. Domain: $[0, 11]$

$W(0) = 52.3(1.74)^0 \approx 52$

$W(11) = 52.3(1.74)^{11} \approx 23{,}150$

Range: $[52, 23150]$

c.

$1965 \rightarrow t = 15$

$W(15) = 52.3(1.74)^{15}$

$W(15) \approx 212{,}197.6$

The model predicts that there were approximately 212,198 non-strategic warheads in the U.S. arsenal in the year 1965. This could be model breakdown.

35.

a. Let $C(s)$ be the number of counts from a Geiger counter from the decay of barium-137 after s seconds.

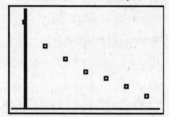

An exponential model is appropriate so choose the vertical intercept and a point from the flatter end.

Note: Your model may differ, but must it follow the trend of the data.

$C(s) = ab^s$

From $(0, 3098)$ we get $a = 3098$.

Giving us: $C(s) = 3098b^s$

To find b use a second point: $(300, 914)$

$914 = 3098b^{300}$

$$\frac{914}{3098} = \frac{3098b^{300}}{3098}$$

$$\frac{914}{3098} = b^{300}$$

$$\left(\frac{914}{3098}\right)^{\frac{1}{300}} = \left(b^{300}\right)^{\frac{1}{300}}$$

$b \approx 0.996$

Then $C(s) = 3098(0.996)^s$

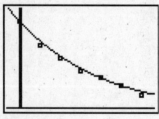

b.

Ten minutes $\rightarrow s = 600$

$C(600) = 3098(0.996)^{600}$

$C(600) \approx 279.69$

After 10 minutes the Geiger counter ticks off about 280 counts per minute.

c.

Domain: $[0, 1800]$

$C(0) = 3098(0.996)^0 = 3098$

$C(1800) = 3098(0.996)^{1800} \approx 2$

Range: $[2, 3098]$

36.

a.

$F(t) = 750(0.9405)^t$

$F(5) = 750(0.9405)^5$

$F(5) \approx 551.89$

In 2005 the population of Franklin's gulls was approximately 551.89 thousand.

b.

$r = b - 1$

$r = 0.9405 - 1$

$r = -0.0595$

The population of Franklin's gulls decays at a rate of 5.95% per year.

c. This population might stop declining at the rate of 5.95% per year due to environmental intervention.

37. Let $E(t)$ be the population of the European Union (in millions) t years after 2008.

a.

$E(t) = ab^t$

$a = 491$ and $r = 0.12\% = 0.0012$

$b = 1 + r$ so $b = 1.0012$

Then $E(t) = 491(1.0012)^t$

b.

$2020 \rightarrow t = 12$

$E(12) = 491(1.0012)^{12}$

$E(12) \approx 498.12$

In 2020 the population of the European Union will be approximately 498.12 million.

38.

Use the formula $A = P\left(1 + \dfrac{r}{n}\right)^{nt}$

because the interest is compounded monthly:

$P = \$7,000$

$r = 6\% = 0.06$

$t = 10$ years

$n = 12$

$A = ?$

Solve for A:

$A = 7000\left(1 + \dfrac{0.06}{12}\right)^{(12 \times 10)}$

$A \approx 12,735.78$

The amount of money in the account after 10 years is $12,735.78.

39.

Use the formula $A = Pe^{rt}$

because the interest is compounded continuously:

$P = \$100,000$

$r = 8.5\% = 0.085$

$t = 25$ years

$A = ?$

Solve for A:

$A = 100000e^{(0.085 \times 25)}$

$A \approx 837,289.75$

The amount of money in the account after 25 years is $837,289.75.

40.

Use the formula $A = P\left(1 + \dfrac{r}{n}\right)^{nt}$

because the interest is compounded daily:

$P = \$50,000$

$r = 4\% = 0.04$

$t = 15$ years

$n = 365$

$A = ?$

Solve for A:

$A = 50000\left(1 + \dfrac{0.04}{365}\right)^{(365 \times 15)}$

$A \approx 91,102.95$

The amount of money in the account after 15 years is $91,102.95.

Chapter 5 Exponential Functions
Chapter 5 Test

1. Let $B(h)$ be the number of bacteria present (in millions) h hours after 12 noon have passed.

a. The number of bacteria doubles every hour: $B(h) = 5(2)^h$

h	$B(h)$
0	5
1	$5(2)^1$
2	$5(2)^2$
3	$5(2)^3$
h	$5(2)^h$

b. $B(6) = 5(2)^6 = 320$ By 6:00pm there are 320 million bacteria on the bathroom door handle.

c. The bacteria double 4 times each hour: $B(h) = 5(2)^{4h}$

h	$B(h)$
0	5
1	$5(2)(2)(2)(2) = 5(2)^4$
2	$5(2)^4(2)^4 = 5(2)^{4\times 2} = 5(2)^8$
3	$5(2)^4(2)^4(2)^4 = 5(2)^{4\times 3} = 5(2)^{12}$
h	$5(2)^{4h}$

2. Let $P(m)$ be the percent of thallium-210 left in a sample after m minutes. Let n be the number of 1.32-minute time intervals.

a. The percent of thallium-210 is cut in half every 1.32 minutes: $P(m) = 100\left(\frac{1}{2}\right)^{m/1.32}$

m	$P(m)$
0	100
1.32	$100\left(\frac{1}{2}\right)$
2(1.32)	$100\left(\frac{1}{2}\right)^2$
3(1.32)	$100\left(\frac{1}{2}\right)^3$
$n(1.32)$	$100\left(\frac{1}{2}\right)^n$

$n(1.32) = m$

$\dfrac{n(1.32)}{1.32} = \dfrac{m}{1.32}$

$n = \dfrac{m}{1.32}$

$P(m) = 100\left(\frac{1}{2}\right)^{\frac{m}{1.32}}$

b.

$P(15) = 100\left(\frac{1}{2}\right)^{\frac{15}{1.32}}$

$P(15) \approx 0.038$

After 15 minutes, there is 0.038% of the thallium-210 sample left, which means there is approximately 0% left.

3.

Use the formula $A = P\left(1 + \dfrac{r}{n}\right)^{nt}$

because the interest is compounded monthly:

$P = \$1{,}000$

$r = 3.5\% = 0.035$

$t = 12$ years

$n = 12$

$A = ?$

Solve for A:

$A = 1000\left(1 + \dfrac{0.035}{12}\right)^{(12\times 12)}$

$A \approx 1521.03$

The amount of money in the account after 12 years is $1521.03.

4.

Use the formula $A = Pe^{rt}$

because the interest is compounded continuously:

$P = \$40{,}000$

$r = 6\% = 0.06$

$t = 35$ years

$A = ?$

Solve for A:

$A = 40000e^{(0.06\times 35)}$

$A \approx 326{,}646.80$

The amount of money in the account after 35 years is $326,646.80.

5.

a. a is negative because the graph is below the x-axis.

b. $b < 1$ because we have exponential growth with $a < 0$.

c. Domain: $(-\infty, \infty)$ Range: $(-\infty, 0)$

6.

a. The graph is increasing.

b. $f(1) = -4$

c. $f(x) = -36$ and $x = -2$

d. The vertical intercept is $(0, -8)$.

7.

a. Let $C(t)$ be the number of CD singles shipped (in millions) in the mid-1990's in the year t years since 1990.

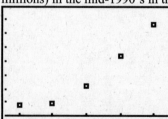

An exponential model is appropriate so choose a point from the flatter end and one from the steeper end.

Note: Your model may differ, but it must follow the trend of the data.

$C(t) = ab^t$

To find a and b choose two points $(3, 7.8)$ and $(7, 66.7)$.

$7.8 = ab^3$ and $66.7 = ab^7$

$\dfrac{66.7}{7.8} = \dfrac{ab^7}{ab^3}$

$\dfrac{66.7}{7.8} = b^4$

$\left(\dfrac{66.7}{7.8}\right)^{\frac{1}{4}} = \left(b^4\right)^{\frac{1}{4}}$

NOTE: We do not need the $\pm$ on the answer because $b > 0$.

$b \approx 1.71$

Use b and one of the above equations to find a.

$7.8 = ab^3$

$\dfrac{7.8}{(1.71)^3} = \dfrac{a(1.71)^3}{(1.71)^3}$

$a \approx 1.56$

Then $C(t) = 1.56(1.71)^t$

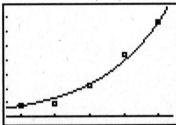

b. Domain: $[0, 10]$ Range: $[1.56, 333.49]$

c. In 2000 there were approximately 333.49 million CD singles shipped.

8. Let $P(t)$ be the number of professionals in developing countries (in millions) in the year t years since 2009.

a.

$P(t) = ab^t$

$a = 4$ and $r = 6.5\% = 0.065$

$b = 1 + r$ so $b = 1.065$

Then $P(t) = 4(1.065)^t$

b.

$2013 \rightarrow t = 4$

$P(4) = 4(1.065)^4$

$P(4) \approx 5.15$

By the year 2013 there will be approximately 5.15 million professionals in developing countries.

9. Since $a < 0$ and $b > 1$ the function is decreasing.

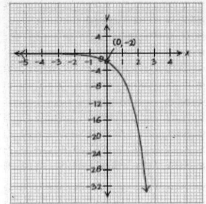

10. Since $a > 0$ and $b < 1$ the function is decreasing.

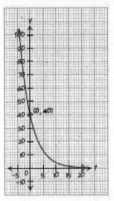

11. Since $a > 0$ and $b < 1$ the function is decreasing.

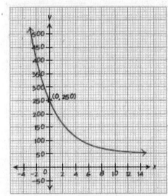

12. Since $a < 0$ and $b < 1$ the function is increasing.

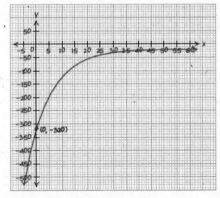

Chapter 5 Exponential Functions

13.
$12(4^x) = 12288$
$4^x = 1024$
$4^x = 4^5$
$x = 5$

14.
$2(3^x) - 557 = -503$
$2(3^x) - 557 + 557 = -503 + 557$
$2(3^x) = 54$
$3^x = 27$
$3^x = 3^3$
$x = 3$

15.
$-6(5^x) - 510 = -4260$
$-6(5^x) - 510 + 510 = -4260 + 510$
$-6(5^x) = -3750$
$5^x = 625$
$5^x = 5^4$
$x = 4$

16.
$x^3 = 216$
$(x^3)^{\frac{1}{3}} = (216)^{\frac{1}{3}}$
$x = 6$

17.
$-3x^5 + 650 = 3722$
$-3x^5 + 650 - 650 = 3722 - 650$
$-3x^5 = 3072$
$x^5 = -1024$
$(x^5)^{\frac{1}{5}} = (-1024)^{\frac{1}{5}}$
$x = -4$

18.
$4.5x^6 - 2865 = 415.5$
$4.5x^6 - 2865 + 2865 = 415.5 + 2865$
$4.5x^6 = 3280.5$
$x^6 = 729$
$(x^6)^{\frac{1}{6}} = \pm(729)^{\frac{1}{6}}$
$x = \pm 3$

19.
$300x^4 - 580 = 218768.48$
$300x^4 - 580 + 580 = 218768.48 + 580$
$300x^4 = 219348.48$
$\dfrac{300x^4}{300} = \dfrac{219348.48}{300}$
$x^4 = 731.1616$
$(x^4)^{\frac{1}{4}} = \pm(731.1616)^{\frac{1}{4}}$
$x = \pm 5.2$

Section 6.1

1. This function passes the horizontal line test because each horizontal line that you could draw across the graph would hit the function only once. Therefore, it is a one-to-one function.

3. This function passes the horizontal line test because each horizontal line that you could draw across the graph would hit the function only once. Therefore, it is a one-to-one function.

5. This function fails the horizontal line test because almost any horizontal line that you draw through this graph hits the graph more than once. Therefore, this is not a one-to-one function.

7. This function passes the horizontal line test because each horizontal line that you could draw across the graph would hit the function only once. Therefore, it is a one-to-one function.

9. This line is horizontal, so when it is tested by using the horizontal line test, it fails to be one-to-one.

11. $f(x) = \dfrac{1}{2}x + 3$

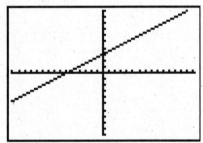

This function passes the horizontal line test because each horizontal line that you could draw across the graph would hit the function only once. Therefore, it is a one-to-one function.

13. $h(x) = 2.5x^2 + 3x - 9$

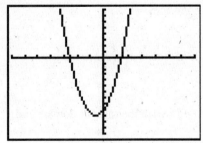

This function fails the horizontal line test because almost any horizontal line that you draw through this graph hits the graph more than once. Therefore, this is not a one-to-one function.

15. $f(x) = 2x^3 + 4$

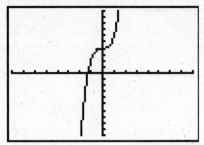

This function passes the horizontal line test because each horizontal line that you could draw across the graph would hit the function only once. Therefore, it is a one-to-one function.

17. $f(x) = 4x^3 + 2x^2 - 5x - 4$

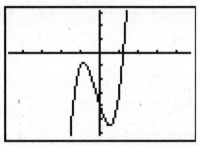

This function fails the horizontal line test because some horizontal lines that you draw through this graph hit the graph more than once. Therefore, this is not a one-to-one function.

19. $f(x) = 2x^4$

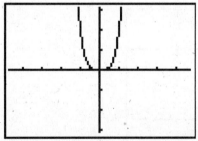

This function fails the horizontal line test because almost any horizontal line that you draw through this graph hits the graph more than once. Therefore, this is not a one-to-one function.

21. $g(x) = 3(1.2)^x$

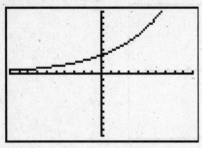

This function passes the horizontal line test because each horizontal line that you could draw across the graph would hit the function only once. Therefore, it is a one-to-one function.

23. $f(x) = 100(0.4)^x + 20$

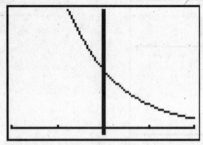

This function passes the horizontal line test because each horizontal line that you could draw across the graph would hit the function only once. Therefore, it is a one-to-one function.

25. $f(x) = 20$

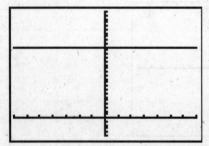

This line is horizontal, so when it is tested by using the horizontal line test, it fails to be one-to-one.

27.

a.

$N(t) = -315.9t + 4809.8$
$N = -315.9t + 4809.8$
$N - 4809.8 = -315.9t$
$\dfrac{N - 4809.8}{-315.9} = \dfrac{-315.9t}{-315.9}$
$\dfrac{N}{-315.9} - \dfrac{4809.8}{-315.9} = t$
$-0.003N + 15.23 \approx t$
$t(N) = -0.003N + 15.23$

b.

If the original model has a domain of $[-1, 12]$ and a range of $[1019, 5125.7]$, then the inverse function has a domain of $[1019, 5125.7]$ and a range of $[-1, 12]$.

c.

$t(2000) = -0.003(2000) + 15.23$
$t(2000) = 9.23$

In 1999, there were approximately 2000 homicides of 15-19 year olds in the United States.

29.

a.

If $P(b)$ has domain $0 \le b \le 5000$, then
$P(0) = 5.5(0) - 2500 = -2500$
$P(5000) = 5.5(5000) - 2500 = 25{,}000$

Therefore, the range is $-2500 \le P \le 25{,}000$.

b.

$P(b) = 5.5b - 2500$
$P = 5.5b - 2500$
$P + 2500 = 5.5b$
$\dfrac{P + 2500}{5.5} = \dfrac{5.5b}{5.5}$
$\dfrac{P}{5.5} + \dfrac{2500}{5.5} = b$
$0.18P + 454.55 \approx b$
$b(P) = 0.18P + 454.55$

c.

$P(1000) = 5.5(1000) - 2500$
$P(1000) = 3000$

There is a $3000 profit from selling 1000 books.

d.

$b(5000) = 0.18(5000) + 454.55$
$b(5000) = 1354.55$

You would need to sell 1355 books in order to make a $5000 profit.

e.

Referring to part "a" above, switch the roles of domain and range.

Domain: $-2500 \le P \le 25{,}000$
Range: $0 \le b \le 5000$

31.

a.

$P(t) = 2.57t + 249.78$

$P = 2.57t + 249.78$

$P - 249.78 = 2.57t$

$\dfrac{P - 249.78}{2.57} = \dfrac{2.57t}{2.57}$

$\dfrac{P}{2.57} - \dfrac{249.78}{2.57} = t$

$0.389P - 97.191 \approx t$

$t(P) = 0.389P - 97.191$

b.

$t(260) = 0.389(260) - 97.191$

$t(260) = 3.95$

The U.S. population reached 260 million in about 1994.

c.

For the inverse function $t(P) = 0.389P - 97.191$; the input variable P represents the U.S. population in millions, the output variable t represents the number of years since 1990.

33.

a.

$M(c) = 10,000 + 1.5c$

$M = 10,000 + 1.5c$

$M - 10,000 = 1.5c$

$\dfrac{M - 10,000}{1.5} = \dfrac{1.5c}{1.5}$

$\dfrac{M}{1.5} - \dfrac{10,000}{1.5} = c$

$0.667M - 6666.667 \approx c$

$c(M) = 0.667M - 6666.667$

b.

$c(40,000) = 0.667(40,000) - 6666.667$

$c(40,000) \approx 20013.333$

They would have to sell about 20,014 CD's in order for "Math Dude" to earn \$40,000.

c. If the domain of $M(c)$ is $10,000 \leq c \leq 50,000$, then

$M(10,000) = 10,000 + 1.5(10,000) = 25,000$

$M(50,000) = 10,000 + 1.5(50,000) = 85,000$

Therefore, the range of this original model is [25000, 85000].

Switching these roles gives the inverse domain and range.

Domain: $25,000 \leq M \leq 85,000$

Range: $10,000 \leq c \leq 50,000$

35.

$f(x) = 3x + 5$

$y = 3x + 5$

$y - 5 = 3x$

$\dfrac{1}{3}(y - 5) = \dfrac{1}{3}(3x)$

$\dfrac{1}{3}y - \dfrac{5}{3} = x$

$\dfrac{1}{3}x - \dfrac{5}{3} = y$

$f^{-1}(x) = \dfrac{1}{3}x - \dfrac{5}{3}$

37.

$g(t) = -4t + 8$

$g = -4t + 8$

$g - 8 = -4t$

$\left(-\dfrac{1}{4}\right)(g - 8) = \left(-\dfrac{1}{4}\right)(-4t)$

$-\dfrac{1}{4}g + 2 = t$

$-\dfrac{1}{4}t + 2 = g$

$g^{-1}(t) = -\dfrac{1}{4}t + 2$

39.

$h(x) = \dfrac{2}{3}x - 9$

$y = \dfrac{2}{3}x - 9$

$y + 9 = \dfrac{2}{3}x$

$\left(\dfrac{3}{2}\right)(y + 9) = \left(\dfrac{3}{2}\right)\left(\dfrac{2}{3}x\right)$

$\dfrac{3}{2}y + \dfrac{27}{2} = x$

$\dfrac{3}{2}x + \dfrac{27}{2} = y$

$h^{-1}(x) = \dfrac{3}{2}x + \dfrac{27}{2}$

41.

$f(x) = \frac{1}{5}x + \frac{3}{5}$

$y = \frac{1}{5}x + \frac{3}{5}$

$y - \frac{3}{5} = \frac{1}{5}x$

$5\left(y - \frac{3}{5}\right) = 5\left(\frac{1}{5}x\right)$

$5y - 3 = x$

$5x - 3 = y$

$f^{-1}(x) = 5x - 3$

43.

$h(x) = 0.4x - 1.6$

$y = 0.4x - 1.6$

$y + 1.6 = 0.4x$

$\frac{y + 1.6}{0.4} = \frac{0.4x}{0.4}$

$\frac{y}{0.4} + \frac{1.6}{0.4} = x$

$2.5y + 4 = x$

$2.5x + 4 = y$

$h^{-1}(x) = 2.5x + 4$

45.

$P(t) = -2.5t - 7.5$

$P = -2.5t - 7.5$

$P + 7.5 = -2.5t$

$\frac{P + 7.5}{-2.5} = \frac{-2.5t}{-2.5}$

$\frac{P}{-2.5} + \frac{7.5}{-2.5} = t$

$-0.4P - 3 = t$

$-0.4t - 3 = P$

$P^{-1}(t) = -0.4t - 3$

47. Yes, they are inverses.

$f(g(x)) = f\left(\frac{1}{3}x + 3\right)$

$f(g(x)) = 3\left(\frac{1}{3}x + 3\right) - 9$

$f(g(x)) = x + 9 - 9$

$f(g(x)) = x$

$g(f(x)) = g(3x - 9)$

$g(f(x)) = \frac{1}{3}(3x - 9) + 3$

$g(f(x)) = x - 3 + 3$

$g(f(x)) = x$

49. Yes, they are inverses.

$f(g(x)) = f(0.25x - 3)$

$f(g(x)) = 4(0.25x - 3) + 12$

$f(g(x)) = x - 12 + 12$

$f(g(x)) = x$

$g(f(x)) = g(4x + 12)$

$g(f(x)) = 0.25(4x + 12) - 3$

$g(f(x)) = x + 3 - 3$

$g(f(x)) = x$

51. No, they are not inverses.

$f(h(x)) = f(7x + 3)$

$f(h(x)) = \frac{1}{7}(7x + 3) + 21$

$f(h(x)) = x + \frac{3}{7} + 21$

$f(h(x)) = x + \frac{150}{7}$

$f(h(x)) \neq x$

53.

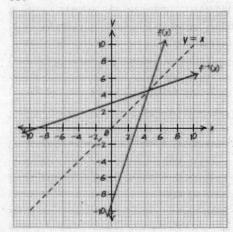

55.

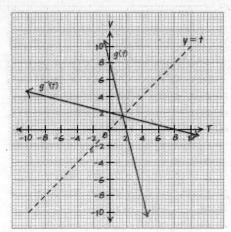

57.

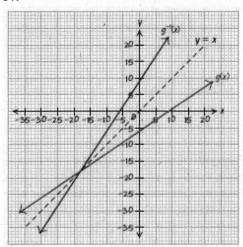

59.

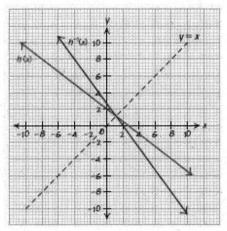

61.

x	$y = 2^x$	$f(x) = (x, y)$	$f^{-1}(x) = (y, x)$
-1	$2^{-1} = \dfrac{1}{2}$	$\left(-1, \dfrac{1}{2}\right)$	$\left(\dfrac{1}{2}, -1\right)$
0	$2^0 = 1$	$(0, 1)$	$(1, 0)$
1	$2^1 = 2$	$(1, 2)$	$(2, 1)$

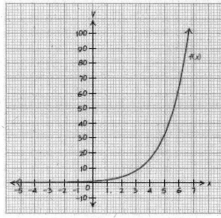

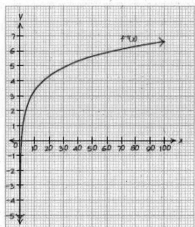

63.

x	$y = 150(0.5)^x$	$h(x) = (x, y)$	$h^{-1}(x) = (y, x)$
0	$150(0.5)^0 = 150$	$(0, 150)$	$(150, 0)$
2	$150(0.5)^2 = 37.5$	$(2, 37.5)$	$(37.5, 2)$
4	$150(0.5)^4 = 9.375$	$(4, 9.375)$	$(9.375, 4)$

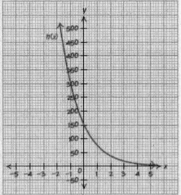

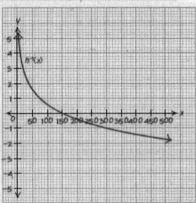

65.

x	$y = -20(1.25)^x$	(x, y)	(y, x)
−6	$-20(1.25)^{-6} \approx -5.2$	$(-6, -5.2)$	$(-5.2, -6)$
−2	$-20(1.25)^{-2} = -12.8$	$(-2, -12.8)$	$(-12.8, -2)$
0	$-20(1.25)^0 = -20$	$(0, -20)$	$(-20, 0)$
1	$-20(1.25)^1 = -25$	$(1, -25)$	$(-25, 1)$

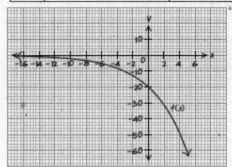

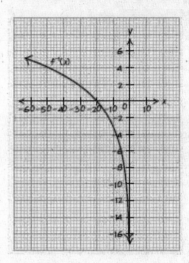

67.

x	$y=2(3)^x+5$	$h(x)=(x,y)$	$h^{-1}(x)=(y,x)$
0	$2(3)^0+5=7$	$(0,7)$	$(7,0)$
1	$2(3)^1+5=11$	$(1,11)$	$(11,1)$
2	$2(3)^2+5=23$	$(2,23)$	$(23,2)$

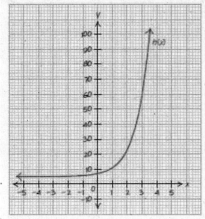

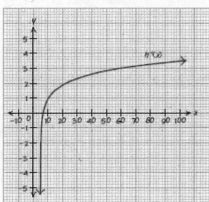

69.

x	$y=-2(2.5)^x-8$	$f(x)=(x,y)$	$f^{-1}(x)=(y,x)$
0	$-2(2.5)^0-8=-10$	$(0,-10)$	$(-10,0)$
1	$-2(2.5)^1-8=-13$	$(1,-13)$	$(-13,1)$
2	$-2(2.5)^2-8=-20.5$	$(2,-20.5)$	$(-20.5,2)$

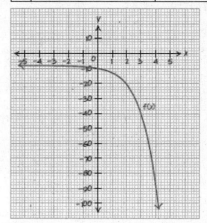

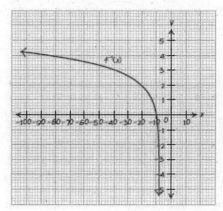

Chapter 6 Logarithmic Functions

71.

x	$y = 40\left(\dfrac{1}{10}\right)^x + 15$	(x,y)	(y,x)
0	$40\left(\dfrac{1}{10}\right)^0 + 15 = 55$	$(0,55)$	$(55,0)$
1	$40\left(\dfrac{1}{10}\right)^1 + 15 = 19$	$(1,19)$	$(19,1)$
2	$40\left(\dfrac{1}{10}\right)^2 + 15 = 15.4$	$(2,15.4)$	$(15.4,2)$

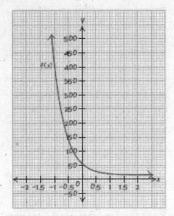

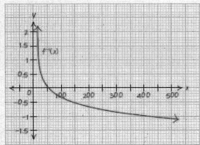

73.

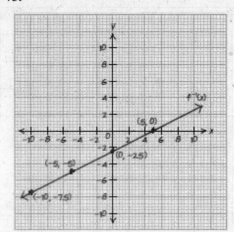

75.

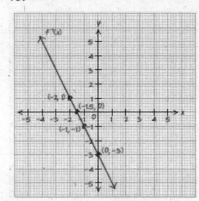

77.

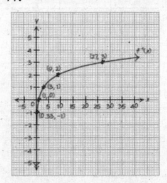

79. Yes, it is a function. Yes, it has an inverse. It passes the vertical line test so it is a function, and it passes the horizontal line test so it must have an inverse.

81. No, it is not a function. This is a vertical line and it will not pass the vertical line test.

Section 6.2

1. The base is 5.
3. The base is 10.
5. The base is e.
7. $\log_3 27 = \log_3(3^3) = 3$
9. $\log_5 125 = \log_5(5^3) = 3$
11. $\log_2 256 = \log_2(2^8) = 8$
13. $\log 1000 = \log(10^3) = 3$
15. $\log_5\left(\dfrac{1}{5}\right) = \log_5(5^{-1}) = -1$
17. $\log 0.1 = \log\left(\dfrac{1}{10}\right) = \log(10^{-1}) = -1$
19. $\log_7\left(\dfrac{1}{49}\right) = \log_7\left(\dfrac{1}{7^2}\right) = \log_7(7^{-2}) = -2$
21. $\log_5\left(\dfrac{1}{25}\right) = \log_5\left(\dfrac{1}{5^2}\right) = \log_5(5^{-2}) = -2$
23. $\log_7 1 = 0$
25. $\log_8(8^4) = 4$
27. $\ln(e^3) = 3$
29. $\ln e = 1$
31. $\log_{19} 19 = 1$
33. $\log 125 \approx 2.097$
35. $\log 275 \approx 2.439$
37. $\ln 45 \approx 3.807$
39. $\ln 120 \approx 4.787$
41.
   a) $\log_3 40 = \dfrac{\log 40}{\log 3} \approx 3.358$
   b) $\log_3 40 = \dfrac{\ln 40}{\ln 3} \approx 3.358$
43.
   a) $\log_5 63 = \dfrac{\log 63}{\log 5} \approx 2.574$
   b) $\log_5 63 = \dfrac{\ln 63}{\ln 5} \approx 2.574$
45. $\log_7 25 = \dfrac{\log 25}{\log 7} \approx 1.654$
47. $\log_2 0.473 = \dfrac{\log 0.473}{\log 2} \approx -1.080$
49. $\log_{14} 478 = \dfrac{\log 478}{\log 14} \approx 2.338$
51. $\log_{11} 0.254 = \dfrac{\log 0.254}{\log 11} \approx -0.572$
53. $\log_{12} 36478 = \dfrac{\log 36478}{\log 12} \approx 4.227$
55.
$\log 1000 = 3$
$10^3 = 1000$
57.
$\log 0.01 = -2$
$10^{-2} = 0.01$
59.
$\log_2(8) = 3$
$2^3 = 8$
61.
$\log_3(81) = 4$
$3^4 = 81$
63.
$\ln(e^5) = 5$
$e^5 = e^5$
65.
$\log_3\left(\dfrac{1}{9}\right) = -2$
$3^{-2} = \dfrac{1}{9}$
67.
$\log_5\left(\dfrac{1}{25}\right) = -2$
$5^{-2} = \dfrac{1}{25}$
69.
$2^{10} = 1024$
$\log_2 1024 = 10$
71.
$5^4 = 625$
$\log_5 625 = 4$
73.
$25^{0.5} = 5$
$\log_{25} 5 = 0.5$
75.
$10^5 = 100{,}000$
$\log 100{,}000 = 5$

77.

$$\left(\frac{1}{5}\right)^4 = \frac{1}{625}$$

$$\log_{\frac{1}{5}}\left(\frac{1}{625}\right) = 4$$

79.

$$\left(\frac{1}{2}\right)^3 = \frac{1}{8}$$

$$\log_{\frac{1}{2}}\left(\frac{1}{8}\right) = 3$$

81.

$$3^{2x} = 729$$

$$\log_3 729 = 2x$$

83.

$$f(x) = 7^x$$

$$y = 7^x$$

$$\log_7 y = x$$

$$\log_7 x = y$$

$$\frac{\log x}{\log 7} = y$$

$$f^{-1}(x) = \log_7 x$$

$$f^{-1}(x) = \frac{\log x}{\log 7}$$

85.

$$h(c) = 10^c$$

$$h = 10^c$$

$$\log h = c$$

$$\log c = h$$

$$h^{-1}(c) = \log c$$

87.

$$m(r) = 5^r$$

$$m = 5^r$$

$$\log_5 m = r$$

$$\log_5 r = m$$

$$\frac{\log r}{\log 5} = m$$

$$m^{-1}(r) = \log_5 r$$

$$m^{-1}(r) = \frac{\log r}{\log 5}$$

89.

$$f(x) = \log_3 x$$

$$y = \log_3 x$$

$$3^y = x$$

$$3^x = y$$

$$f^{-1}(x) = 3^x$$

91.

$$g(x) = \ln x$$

$$y = \ln x$$

$$e^y = x$$

$$e^x = y$$

$$g^{-1}(x) = e^x$$

93.

$$f(x) = 3(4)^x$$

$$y = 3(4)^x$$

$$\frac{y}{3} = (4)^x$$

$$\log_4\left(\frac{y}{3}\right) = x$$

$$\log_4\left(\frac{x}{3}\right) = y$$

$$\frac{\log\left(\frac{x}{3}\right)}{\log 4} = y$$

$$f^{-1}(x) = \log_4\left(\frac{x}{3}\right)$$

$$f^{-1}(x) = \frac{\log\left(\frac{x}{3}\right)}{\log 4}$$

95.

$$h(x) = \frac{1}{2}(9)^x$$

$$y = \frac{1}{2}(9)^x$$

$$2y = (9)^x$$

$$\log_9(2y) = x$$

$$\log_9(2x) = y$$

$$\frac{\log(2x)}{\log 9} = y$$

$$h^{-1}(x) = \log_9(2x)$$

$$h^{-1}(x) = \frac{\log(2x)}{\log 9}$$

97.

$$\log x = 3$$

$$10^3 = x$$

$$x = 1000$$

99.

$\ln t = 2$

$e^2 = t$

$t \approx 7.389$

101.

$\log_3 w = 2.4$

$3^{2.4} = w$

$w \approx 13.967$

103.

$\log_{16} m = \dfrac{1}{2}$

$16^{\frac{1}{2}} = m$

$m = 4$

105.

$\log_{12} x = \dfrac{1}{5}$

$12^{\frac{1}{5}} = x$

$x \approx 1.644$

107.

$\log(8t) = 2$

$10^2 = 8t$

$\dfrac{10^2}{8} = t$

$\dfrac{100}{8} = t$

$t = \dfrac{25}{2}$

$t = 12.5$

109.

$\ln(3x) = 4$

$e^4 = 3x$

$\dfrac{e^4}{3} = x$

$x \approx 18.199$

Section 6.3

1. a) $f(50) = 2$ b) $x = 125$

3. a) $h(250) = 4$ b) $x = 25$

5. a) $g(40) = -3$ b) $x = 13$

7. a) $h(3) = -11$ b) $x = 5$

9. The base is greater than 1, because the graph is increasing.

11. The base is less than 1, because the graph is decreasing.

13.

x	$f(x) = \log x$
1	0
2	0.301
5	0.699
10	1

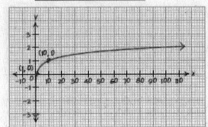

15.

x	$f(x) = \log_4 x$
1	0
2	.5
3	0.792
4	1

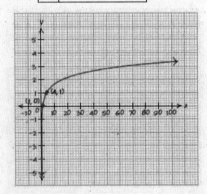

17.

x	$f(x) = \log_9 x$
1	0
2	0.315
3	0.5
9	1

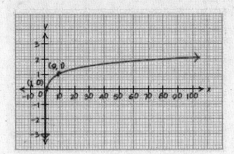

19.

x	$f(x) = \log_{20} x$
1	0
5	0.537
10	0.769
20	1

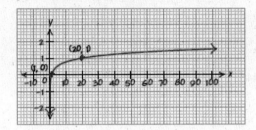

21.

x	$f(x) = \log_{0.2} x$
0.2	1
0.4	0.569
1	0
2	-0.431

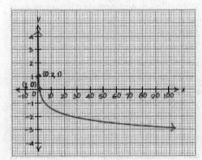

23.

x	$f(x) = \log_{0.9} x$
0.3	11.427
0.9	1
1	0
3	-10.427

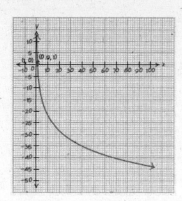

25.

x	$f(x) = \log_{0.6} x$
0.3	2.357
0.6	1
1	0
3	-2.151

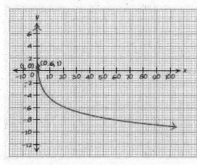

27. Domain: $(0, \infty)$ Range: $(-\infty, \infty)$

29. Domain: $(0, \infty)$ Range: $(-\infty, \infty)$

31. Domain: $(0, \infty)$ Range: $(-\infty, \infty)$

33. Domain: $(0, \infty)$ Range: $(-\infty, \infty)$

35.

$f(x) = \log_{20} x$

Domain: $(0, \infty)$

Range: $(-\infty, \infty)$

37.

$g(x) = \log_{0.2} x$

Domain: $(0, \infty)$

Range: $(-\infty, \infty)$

39.

$f(x) = 2x + 12$

$y = 2x + 12$

$y - 12 = 2x$

$\frac{1}{2}(y - 12) = \frac{1}{2}(2x)$

$\frac{1}{2}y - 6 = x$

$\frac{1}{2}x - 6 = y$

$f^{-1}(x) = \frac{1}{2}x - 6$

41.

$f(x) = 4^x$

$y = 4^x$

$\log_4 y = x$

$\log_4 x = y$

$\frac{\log x}{\log 4} = y$

$f^{-1}(x) = \log_4 x$

or

$f^{-1}(x) = \frac{\log x}{\log 4}$

43.

$g(x) = \log_9 x$

$y = \log_9 x$

$9^y = x$

$9^x = y$

$g^{-1}(x) = 9^x$

45. $f(x) = \frac{2}{3}x - 9$

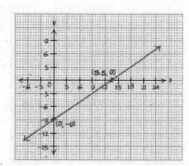

47. $f(x) = 5(1.2)^x$

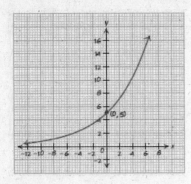

49. $g(x) = 2(x+3)^2 - 8$

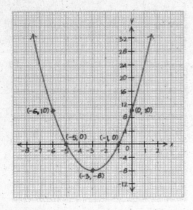

51. $f(x) = \log_5(x)$

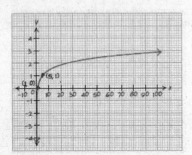

53. $\log 300 \approx 2.477$

55. $\ln 22 \approx 3.091$

57. $\log_5 630 = \dfrac{\log 630}{\log 5} \approx 4.005$

59. $\log_{4.5} 1256 = \dfrac{\log 1256}{\log 4.5} \approx 4.744$

61. $\log_6 0.45 = \dfrac{\log 0.45}{\log 6} \approx -0.446$

63.
$\log_8 x = 3$
$8^3 = x$
$x = 512$

65.
$\log x = 7$
$10^7 = x$
$x = 10,000,000$

67.
$\log_4 x = -2$
$4^{-2} = x$
$x = \dfrac{1}{4^2}$
$x = \dfrac{1}{16}$

Section 6.4

1. $\log(5x) = \log 5 + \log x$

3. $\ln(xy) = \ln x + \ln y$

5.
$$\ln(4ab^3) = \ln 4 + \ln a + \ln b^3$$
$$= \ln 4 + \ln a + 3\ln b$$

7.
$$\ln(2h^2k^3) = \ln 2 + \ln h^2 + \ln k^3$$
$$= \ln 2 + 2\ln h + 3\ln k$$

9.
$$\log_3\left(\frac{1}{2}ab^2c^3\right) = \log_3 1 - \log_3 2 + \log_3 a + \log_3 b^2 + \log_3 c^3$$
$$= 0 - \log_3 2 + \log_3 a + 2\log_3 b + 3\log_3 c$$
$$= -\log_3 2 + \log_3 a + 2\log_3 b + 3\log_3 c$$

11. $\log\left(\dfrac{x}{y}\right) = \log x - \log y$

13. $\log\left(\dfrac{12}{m}\right) = \log 12 - \log m$

15.
$$\log\left(\frac{2x^2}{y}\right) = \log 2 + \log x^2 - \log y$$
$$= \log 2 + 2\log x - \log y$$

17.
$$\ln\left(\frac{3x^4 y^3}{z}\right) = \ln 3 + \ln x^4 + \ln y^3 - \ln z$$
$$= \ln 3 + 4\ln x + 3\ln y - \ln z$$

19.
$$\log\left(\sqrt{5x}\right) = \log(5x)^{\frac{1}{2}}$$
$$= \frac{1}{2}(\log 5x)$$
$$= \frac{1}{2}(\log 5 + \log x)$$
$$= \frac{1}{2}\log 5 + \frac{1}{2}\log x$$

21.
$$\ln\left(\sqrt{7ab}\right) = \ln(7ab)^{\frac{1}{2}}$$
$$= \frac{1}{2}\ln(7ab)$$
$$= \frac{1}{2}(\ln 7 + \ln a + \ln b)$$
$$= \frac{1}{2}\ln 7 + \frac{1}{2}\ln a + \frac{1}{2}\ln b$$

23.
$$\log_9\left(\sqrt{2w^2z^3}\right) = \log_9\left(2w^2z^3\right)^{\frac{1}{2}}$$
$$= \frac{1}{2}\log_9\left(2w^2z^3\right)$$
$$= \frac{1}{2}\left(\log_9 2 + \log_9 w^2 + \log_9 z^3\right)$$
$$= \frac{1}{2}\left(\log_9 2 + 2\log_9 w + 3\log_9 z\right)$$
$$= \frac{1}{2}\log_9 2 + \log_9 w + \frac{3}{2}\log_9 z$$

25.
$$\ln\left(\sqrt[5]{m^2p^3}\right) = \ln\left(m^2p^3\right)^{\frac{1}{5}}$$
$$= \frac{1}{5}\ln\left(m^2p^3\right)$$
$$= \frac{1}{5}\left(\ln m^2 + \ln p^3\right)$$
$$= \frac{1}{5}(2\ln m + 3\ln p)$$
$$= \frac{2}{5}\ln m + \frac{3}{5}\ln p$$

27.
$$\log_{15}\left(\frac{\sqrt{3x^4y^3}}{z^5}\right) = \log_{15}\left(\frac{(3x^4y^3)^{\frac{1}{2}}}{z^5}\right)$$
$$= \log_{15}\left(\frac{3^{\frac{1}{2}}x^2 y^{\frac{3}{2}}}{z^5}\right)$$
$$= \log_{15} 3^{\frac{1}{2}} + \log_{15} x^2 + \log_{15} y^{\frac{3}{2}} - \log_{15} z^5$$
$$= \frac{1}{2}\log_{15} 3 + 2\log_{15} x + \frac{3}{2}\log_{15} y - 5\log_{15} z$$

29. $\ln x + \ln y = \ln(xy)$

31. $\log a^2 + \log b^3 = \log(a^2 b^3)$

33. $\ln x - \ln y = \ln\left(\dfrac{x}{y}\right)$

35.
$$\log 5 + 2\log x + \log y + 3\log z = \log 5 + \log x^2 + \log y + \log z^3$$
$$= \log(5x^2 yz^3)$$

37.
$$\log_3 a + 2\log_3 b - 5\log_3 c = \log_3 a + \log_3 b^2 - \log_3 c^5$$
$$= \log_3\left(\frac{ab^2}{c^5}\right)$$

Chapter 6 Logarithmic Functions

39.
$$\log_5 7 + \frac{1}{2}\log_5 x + \frac{1}{2}\log_5 y = \log_5 7 + \log_5 x^{1/2} + \log_5 y^{1/2}$$
$$= \log_5 \left(7\sqrt{xy}\right)$$

41.
$$\frac{1}{2}\ln 7 + \frac{1}{2}\ln a + \frac{3}{2}\ln b - 4\ln c = \ln 7^{1/2} + \ln a^{1/2} + \ln b^{3/2} - \ln c^4$$
$$= \ln\left(\frac{\sqrt{7ab^3}}{c^4}\right)$$

43.
$$5\ln 7 + 2\ln a + 4\ln b - 3\ln c - 2\ln d = \ln 7^5 + \ln a^2 + \ln b^4 - \ln c^3 - \ln d^2$$
$$= \ln\left(\frac{7^5 a^2 b^4}{c^3 d^2}\right)$$
$$= \ln\left(\frac{16,807 a^2 b^4}{c^3 d^2}\right)$$

45.
$$\log_3 a^2 + 2\log_3 bc - 5\log_3 3c = \log_3 a^2 + \log_3 (bc)^2 - \log_3 (3c)^5$$
$$= \log_3 \left(\frac{a^2 b^2 c^2}{3^5 c^5}\right)$$
$$= \log_3 \left(\frac{a^2 b^2 c^{2-5}}{243}\right)$$
$$= \log_3 \left(\frac{a^2 b^2 c^{-3}}{243}\right)$$
$$= \log_3 \left(\frac{a^2 b^2}{243 c^3}\right)$$

47.
$$\log_5 7 + 2\log_5 xy + \log_5 xy = \log_5 7 + 3\log_5 xy$$
$$= \log_5 7 + \log_5 (xy)^3$$
$$= \log_5 \left(7x^3 y^3\right)$$

49.
$$\frac{1}{2}\log 5 + \frac{1}{2}\log 3x + \frac{5}{2}\log y - 4\log z = \log 5^{1/2} + \log (3x)^{1/2} + \log y^{5/2} - \log z^4$$
$$= \log\left(\frac{\sqrt{5 \cdot 3x y^5}}{z^4}\right)$$
$$= \log\left(\frac{\sqrt{15 x y^5}}{z^4}\right)$$

51.
$$\log(x+6) + \log(x-2) = \log((x+6)(x-2))$$
$$= \log(x^2 + 4x - 12)$$

53.
$$\ln(x-3) + \ln(x-7) = \ln((x-3)(x-7))$$
$$= \ln(x^2 - 10x + 21)$$

55. There is no expansion for the expression $\log(x+5)$. We can expand when there is multiplication or division inside the logarithm, not for addition or subtraction inside the logarithm.

57. The student incorrectly multiplied two separate log functions instead of multiplying just inside one log function. The expression $\log x + \log y$ comes together as $\log(xy)$.

59.
Given $\log_b 2 = 3$ and $\log_b 16 = 4$
$$\log_b 32 = \log_b (2 \cdot 16)$$
$$= \log_b 2 + \log_b 16$$
$$= 3 + 4$$
$$= 7$$

61.
Given $\log_b 2 = 3$ and $\log_b 16 = 4$
$$\log_b 8 = \log_b \left(\frac{16}{2}\right)$$
$$= \log_b 16 - \log_b 2$$
$$= 4 - 3$$
$$= 1$$

63.
Given $\log_b 30 = 8$ and $\log_b 2 = 14$
$$\log_b 15 = \log_b \left(\frac{30}{2}\right)$$
$$= \log_b 30 - \log_b 2$$
$$= 8 - 14$$
$$= -6$$

65.
Given $\log_b 100 = 4$ and $\log_b 500 = 6$
$$\log_b 50,000 = \log_b (100 \cdot 500)$$
$$= \log_b 100 + \log_b 500$$
$$= 4 + 6$$
$$= 10$$

67.
Given $\log_b 5 = 10$ and $\log_b 20 = 4$
$$\log_b 0.25 = \log_b \left(\frac{5}{20}\right)$$
$$= \log_b 5 - \log_b 20$$
$$= 10 - 4$$
$$= 6$$

Section 6.5

1.
$3^w = 125$
$\log_3 125 = w$
$\dfrac{\log 125}{\log 3} = w$
$w \approx 4.395$

3.
$5^x = 373$
$\log_5 373 = x$
$\dfrac{\log 373}{\log 5} = x$
$x \approx 3.679$

5.
$6^x = 0.48$
$\log_6 0.48 = x$
$\dfrac{\log 0.48}{\log 6} = x$
$x \approx -0.410$

7.
$0.4^x = 0.13$
$\log_{0.4} 0.13 = x$
$\dfrac{\log 0.13}{\log 0.4} = x$
$x \approx 2.227$

9.
$0.6^x = 24$
$\log_{0.6} 24 = x$
$\dfrac{\log 24}{\log 0.6} = x$
$x \approx -6.221$

11.

a. Let $C(t)$ be the number of hours the Cray C90 has been used per academic year t years since 1980.

t	$C(t)$
7	32
8	100
9	329
10	831
11	1685
12	2233
13	3084
14	8517
15	15,584
16	27,399

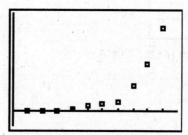

An exponential model is appropriate so choose a point from the flatter end and one from the steeper end.

Note: Your model may differ, but it must follow the trend of the data.

Chapter 6 Logarithmic Functions

$C(t) = ab^t$

To find a and b choose two points (8,100) and (15,15584).

$100 = ab^8$ and $15584 = ab^{15}$

$\dfrac{15584}{100} = \dfrac{ab^{15}}{ab^8}$

$155.84 = b^7$

$(155.84)^{\frac{1}{7}} = (b^7)^{\frac{1}{7}}$

$b \approx 2.057$

Use b and one of the above equations to find a.

$100 = a(2.057)^8$

$\dfrac{100}{(2.057)^8} = \dfrac{a(2.057)^8}{(2.057)^8}$

$a \approx 0.311$

Then $C(t) = 0.311(2.057)^t$

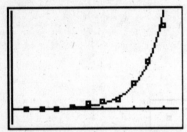

b. Domain: $[7, 20]$ Range: $[48.5, 572066]$

c.

$C(17) = 0.311(2.057)^{17}$

$C(17) = 65,726.9$

The Cray C90 was in use approximately 65,727 hours in the 1997 academic school year.

d.

$500,000 = 0.311(2.057)^t$

$\dfrac{500,000}{0.311} = \dfrac{0.311(2.057)^t}{0.311}$

$\left(\dfrac{500,000}{0.311}\right) = (2.057)^t$

$\log_{2.057}\left(\dfrac{500,000}{0.311}\right) = t$

$\dfrac{\log\left(\dfrac{500,000}{0.311}\right)}{\log 2.057} = t$

$t \approx 19.813$

The Cray C90 was in use approximately 500,000 hours in the academic school year of 2000.

13.

$7^{x-3} = 16807$

$7^{x-3} = 7^5$

$x - 3 = 5$

$x = 8$

15.

$4^{x+7} = 3$

$\ln(4^{x+7}) = \ln 3$

$(x+7)\ln 4 = \ln 3$

$\dfrac{(x+7)\ln 4}{\ln 4} = \dfrac{\ln 3}{\ln 4}$

$x + 7 = \dfrac{\ln 3}{\ln 4}$

$x = \dfrac{\ln 3}{\ln 4} - 7$

$x \approx -6.208$

17.

$604(0.4)^x = 0.158$

$\dfrac{604(0.4)^x}{604} = \dfrac{0.158}{604}$

$(0.4)^x = \dfrac{0.158}{604}$

$\ln(0.4)^x = \ln\left(\dfrac{0.158}{604}\right)$

$x\ln(0.4) = \ln\left(\dfrac{0.158}{604}\right)$

$\dfrac{x\ln(0.4)}{\ln(0.4)} = \dfrac{\ln\left(\dfrac{0.158}{604}\right)}{\ln(0.4)}$

$x = \dfrac{\ln\left(\dfrac{0.158}{604}\right)}{\ln(0.4)}$

$x \approx 9.002$

19.

$5(3.2)^x = 74.2$

$\dfrac{5(3.2)^x}{5} = \dfrac{74.2}{5}$

$(3.2)^x = 14.84$

$\ln(3.2)^x = \ln 14.84$

$x\ln 3.2 = \ln 14.84$

$\dfrac{x\ln 3.2}{\ln 3.2} = \dfrac{\ln 14.84}{\ln 3.2}$

$x = \dfrac{\ln 14.84}{\ln 3.2}$

$x \approx 2.319$

6.5 Exercises

21.

$$2500\left(\frac{1}{5}\right)^x = 0.032$$

$$\frac{2500\left(\frac{1}{5}\right)^x}{2500} = \frac{0.032}{2500}$$

$$\left(\frac{1}{5}\right)^x = 0.0000128$$

$$\ln\left(\frac{1}{5}\right)^x = \ln 0.0000128$$

$$x \ln\left(\frac{1}{5}\right) = \ln 0.0000128$$

$$\frac{x \ln\left(\frac{1}{5}\right)}{\ln\left(\frac{1}{5}\right)} = \frac{\ln 0.0000128}{\ln\left(\frac{1}{5}\right)}$$

$$x = \frac{\ln 0.0000128}{\ln\left(\frac{1}{5}\right)}$$

$$x = 7$$

23.

$$2^t - 58 = 6$$
$$2^t = 64$$
$$2^t = 2^6$$
$$t = 6$$

25.

$$1.5^h + 20 = 106.5$$
$$1.5^h = 86.5$$
$$\ln 1.5^h = \ln 86.5$$
$$h \ln 1.5 = \ln 86.5$$
$$\frac{h \ln 1.5}{\ln 1.5} = \frac{\ln 86.5}{\ln 1.5}$$
$$h = \frac{\ln 86.5}{\ln 1.5}$$
$$h \approx 11.000$$

27.

$$-3(2.5)^m - 89 = -3262$$
$$-3(2.5)^m = -3173$$
$$\frac{-3(2.5)^m}{-3} = \frac{-3173}{-3}$$
$$(2.5)^m = \frac{3173}{3}$$
$$\ln(2.5)^m = \ln\left(\frac{3173}{3}\right)$$
$$m \ln(2.5) = \ln\left(\frac{3173}{3}\right)$$
$$\frac{m \ln(2.5)}{\ln(2.5)} = \frac{\ln\left(\frac{3173}{3}\right)}{\ln(2.5)}$$
$$m = \frac{\ln\left(\frac{3173}{3}\right)}{\ln(2.5)}$$
$$m \approx 7.600$$

29.

$$3(2)^{n+2} = 96$$
$$\frac{3(2)^{n+2}}{3} = \frac{96}{3}$$
$$2^{n+2} = 32$$
$$2^{n+2} = 2^5$$
$$n + 2 = 5$$
$$n = 3$$

31.

$$10 = 4(1.019)^t$$
$$\frac{10}{4} = \frac{4(1.019)^t}{4}$$
$$2.5 = (1.019)^t$$
$$\log 2.5 = \log(1.019)^t$$
$$\log 2.5 = t \log(1.019)$$
$$\frac{\log 2.5}{\log(1.019)} = t$$
$$t \approx 48.68$$

The world population will reach 10 billion, midway through the year 2023 (48.7 years after 1975).

Chapter 6 Logarithmic Functions

33.

$$5 = 1.19(1.08)^t$$

$$\frac{5}{1.19} = \frac{1.19(1.08)^t}{1.19}$$

$$\frac{5}{1.19} = (1.08)^t$$

$$\log\left(\frac{5}{1.19}\right) = \log(1.08)^t$$

$$\log\left(\frac{5}{1.19}\right) = t\log(1.08)$$

$$\frac{\log\left(\frac{5}{1.19}\right)}{\log(1.08)} = t$$

$$t \approx 18.7$$

The white-tailed deer population will reach 5 million late in the year 1998 (18.7 years past 1980).

35.

a.

$$40 = 43.8(0.995)^t$$

$$\frac{40}{43.8} = \frac{43.8(0.995)^t}{43.8}$$

$$\frac{40}{43.8} = (0.995)^t$$

$$\log\left(\frac{40}{43.8}\right) = \log(0.995)^t$$

$$\log\left(\frac{40}{43.8}\right) = t\log(0.995)$$

$$\frac{\log\left(\frac{40}{43.8}\right)}{\log(0.995)} = t$$

$$t \approx 18.1$$

South Africa's population will be down to 40 million by the year 2026 (18.1 years past 2008).

b. South Africa's population in 2008 was 43.8 million. Half of that is 21.9 million.

$$21.9 = 43.8(0.995)^t$$

$$\frac{21.9}{43.8} = \frac{43.8(0.995)^t}{43.8}$$

$$0.5 = (0.995)^t$$

$$\log(0.5) = \log(0.995)^t$$

$$\log(0.5) = t\log(0.995)$$

$$\frac{\log(0.5)}{\log(0.995)} = t$$

$$t \approx 138.3$$

South Africa's population will be down to 21.9 million by the year 2146 (138.3 years past 2008) if the trend continues.

37.

$$3^{5x-4} = 729$$

$$3^{5x-4} = 3^6$$

$$5x - 4 = 6$$

$$5x - 4 + 4 = 6 + 4$$

$$5x = 10$$

$$x = 2$$

39.

$$3(4.6)^{2m+7} = 17.3$$

$$\frac{3(4.6)^{2m+7}}{3} = \frac{17.3}{3}$$

$$(4.6)^{2m+7} = \frac{17.3}{3}$$

$$2m + 7 = \log_{4.6}\left(\frac{17.3}{3}\right)$$

$$2m + 7 = \frac{\log\left(\frac{17.3}{3}\right)}{\log(4.6)}$$

$$2m = \frac{\log\left(\frac{17.3}{3}\right)}{\log(4.6)} - 7$$

$$2m \approx -5.85188$$

$$m \approx \frac{-5.85188}{2}$$

$$m \approx -2.926$$

41.

$$2^{x^2+6} = 1024$$

$$2^{x^2+6} = 2^{10}$$

$$x^2 + 6 = 10$$

$$x^2 = 10 - 6$$

$$x^2 = 4$$

$$x^2 = \pm\sqrt{4}$$

$$x = \pm 2$$

43.

$\left(\dfrac{1}{2}\right)^{-x^2+4} = 4$

$\left(2^{-1}\right)^{-x^2+4} = 2^2$

$2^{(-1)(-x^2+4)} = 2^2$

$2^{x^2-4} = 2^2$

$x^2 - 4 = 2$

$x^2 - 4 + 4 = 2 + 4$

$x^2 = 6$

$x = \pm\sqrt{6}$

$x = \pm 2.449$

45.

$10(2)^t - 450 = 6(2)^t + 15{,}934$

$10(2)^t - 6(2)^t = 15{,}934 + 450$

$4(2)^t = 16{,}384$

$\dfrac{4(2)^t}{4} = \dfrac{16{,}384}{4}$

$(2)^t = 4096$

$(2)^t = 2^{12}$

$t = 12$

47.

$5(4.3)^x + 7 = 2(4.3)^x + 89$

$5(4.3)^x - 2(4.3)^x = 89 - 7$

$3(4.3)^x = 82$

$\dfrac{3(4.3)^x}{3} = \dfrac{82}{3}$

$(4.3)^x = \dfrac{82}{3}$

$x = \log_{4.3}\left(\dfrac{82}{3}\right)$

$x = \dfrac{\log\left(\dfrac{82}{3}\right)}{\log 4.3}$

$x \approx 2.268$

49.

Use $A = P\left(1 + \dfrac{r}{n}\right)^{nt}$ with $n = 365$

$80{,}000 = 40{,}000\left(1 + \dfrac{r}{365}\right)^{365 \times 8}$

$\dfrac{80{,}000}{40{,}000} = \dfrac{40{,}000\left(1 + \dfrac{r}{365}\right)^{2920}}{40{,}000}$

$2 = \left(1 + \dfrac{r}{365}\right)^{2920}$

$2^{\frac{1}{2920}} = \left(\left(1 + \dfrac{r}{365}\right)^{2920}\right)^{\frac{1}{2920}}$

$2^{\frac{1}{2920}} = 1 + \dfrac{r}{365}$

$2^{\frac{1}{2920}} - 1 = 1 + \dfrac{r}{365} - 1$

$365\left(2^{\frac{1}{2920}} - 1\right) = 365\left(\dfrac{r}{365}\right)$

$365\left(2^{\frac{1}{2920}} - 1\right) = r$

$r \approx 0.08665$

$r \approx 8.67\%$

To double their money in 8 years, they would need to invest at 8.67% compounded daily.

51.

Use $A = Pe^{rt}$

$10{,}000 = 5000e^{12r}$

$\dfrac{10{,}000}{5000} = \dfrac{5000e^{12r}}{5000}$

$2 = e^{12r}$

$\ln 2 = 12r$

$\dfrac{\ln 2}{12} = \dfrac{12r}{12}$

$0.0578 \approx r$

$r = 5.78\%$

To double your money in 12 years, you would need to invest at 5.78% compounded continuously.

Chapter 6 Logarithmic Functions

53.

Use $A = P\left(1 + \dfrac{r}{n}\right)^{nt}$ with $n = 12$

$100{,}000 = 50{,}000\left(1 + \dfrac{r}{12}\right)^{12 \times 7}$

$\dfrac{100{,}000}{50{,}000} = \dfrac{50{,}000\left(1 + \dfrac{r}{12}\right)^{84}}{50{,}000}$

$2 = \left(1 + \dfrac{r}{12}\right)^{84}$

$(2)^{\frac{1}{84}} = \left(\left(1 + \dfrac{r}{12}\right)^{84}\right)^{\frac{1}{84}}$

$1.008286 \approx 1 + \dfrac{r}{12}$

$1.008286 - 1 \approx \dfrac{r}{12}$

$0.008286 \approx \dfrac{r}{12}$

$12(0.008286) \approx 12\left(\dfrac{r}{12}\right)$

$0.0994 \approx r$

$r = 9.94\%$

To double your money in 7 years, you would need to invest at 9.94% compounded monthly.

55.

Use $A = P\left(1 + \dfrac{r}{n}\right)^{nt}$ with $n = 12$

$10{,}000 = 5000\left(1 + \dfrac{0.07}{12}\right)^{12t}$

$\dfrac{10{,}000}{5000} = \dfrac{5000(1.005833)^{12t}}{5000}$

$2 = (1.005833)^{12t}$

$\log 2 = \log(1.005833)^{12t}$

$\log 2 = 12t \log(1.005833)$

$\dfrac{\log 2}{12 \log(1.005833)} = \dfrac{12t \log(1.005833)}{12 \log(1.005833)}$

$t \approx 9.9$

It takes approximately 9.9 years to double your money at 7% compounded monthly.

57.

Use $A = Pe^{rt}$

$20{,}000 = 10{,}000e^{0.09t}$

$\dfrac{20{,}000}{10{,}000} = \dfrac{10{,}000e^{0.09t}}{10{,}000}$

$2 = e^{0.09t}$

$\ln 2 = 0.09t$

$\dfrac{\ln 2}{0.09} = \dfrac{0.09t}{0.09}$

$t \approx 7.7$

It takes approximately 7.7 years to double your money at 9% compounded continuously.

59.

Use $A = P\left(1 + \dfrac{r}{n}\right)^{nt}$ with $n = 365$

$24{,}000 = 8000\left(1 + \dfrac{0.02}{365}\right)^{365t}$

$\dfrac{24{,}000}{8000} = \dfrac{8000(1.0000548)^{365t}}{8000}$

$3 = (1.0000548)^{365t}$

$\log 3 = \log(1.0000548)^{365t}$

$\log 3 = 365t \log(1.0000548)$

$\dfrac{\log 3}{365 \log(1.0000548)} = \dfrac{365t \log(1.0000548)}{365 \log(1.0000548)}$

$t \approx 54.9$

It takes approximately 54.9 years to triple your money at 2% compounded daily.

61.

Use $A = Pe^{rt}$

$21{,}000 = 7000e^{0.025t}$

$\dfrac{21{,}000}{7000} = \dfrac{7000e^{0.025t}}{7000}$

$3 = e^{0.025t}$

$\ln 3 = 0.025t$

$\dfrac{\ln 3}{0.025} = \dfrac{0.025t}{0.025}$

$t \approx 43.9$

It takes approximately 43.9 years to triple your money at 2.5% compounded continuously.

63.

$f(x) = 5(3)^x$ Replace $f(x)$ with y.

$y = 5(3)^x$ Solve for x.

$\dfrac{y}{5} = \dfrac{5(3)^x}{5}$

$\dfrac{y}{5} = 3^x$

$\log_3\left(\dfrac{y}{5}\right) = \log_3 3^x$

$\log_3\left(\dfrac{y}{5}\right) = x \log_3 3$

$\log_3\left(\dfrac{y}{5}\right) = x$

$x = \log_3\left(\dfrac{y}{5}\right)$ Interchange the variables x and y.

$y = \log_3\left(\dfrac{x}{5}\right)$

$y = \dfrac{\log\left(\dfrac{x}{5}\right)}{\log(3)}$ Replace y with $f^{-1}(x)$.

$f^{-1}(x) = \dfrac{\log\left(\dfrac{x}{5}\right)}{\log(3)}$

65.

$h(x) = -2.4(4.7)^x$ Replace $h(x)$ with y.

$y = -2.4(4.7)^x$ Solve for x.

$\dfrac{y}{-2.4} = \dfrac{-2.4(4.7)^x}{-2.4}$

$\dfrac{y}{-2.4} = 4.7^x$

$\ln\left(\dfrac{y}{-2.4}\right) = \ln 4.7^x$

$\ln\left(\dfrac{y}{-2.4}\right) = x \ln 4.7$

$\dfrac{\ln\left(\dfrac{y}{-2.4}\right)}{\ln 4.7} = \dfrac{x \ln 4.7}{\ln 4.7}$

$x = \dfrac{\ln\left(\dfrac{y}{-2.4}\right)}{\ln 4.7}$ Interchange the variables x and y.

$y = \dfrac{\ln\left(\dfrac{x}{-2.4}\right)}{\ln 4.7}$ Replace y with $h^{-1}(x)$.

$h^{-1}(x) = \dfrac{\ln\left(\dfrac{x}{-2.4}\right)}{\ln 4.7}$

67.

$g(x) = -3.4 e^x$ Replace $g(x)$ with y.

$y = -3.4 e^x$ Solve for x.

$\dfrac{y}{-3.4} = \dfrac{-3.4 e^x}{-3.4}$

$\dfrac{y}{-3.4} = e^x$

$\ln\left(\dfrac{y}{-3.4}\right) = \ln e^x$

$\ln\left(\dfrac{y}{-3.4}\right) = x \ln e$

$\ln\left(\dfrac{y}{-3.4}\right) = x$

$x = \ln\left(\dfrac{y}{-3.4}\right)$ Interchange the variables x and y.

$y = \ln\left(\dfrac{x}{-3.4}\right)$ Replace y with $g^{-1}(x)$.

$g^{-1}(x) = \ln\left(\dfrac{x}{-3.4}\right)$

69.

$4n^2 + 3n - 8 = 0$

$a = 4, b = 3, c = -8$

$n = \dfrac{-3 \pm \sqrt{(-3)^2 - 4(4)(-8)}}{2(4)}$

$n = \dfrac{-3 \pm \sqrt{137}}{8}$

$n = \dfrac{-3 + \sqrt{137}}{8}, n = \dfrac{-3 - \sqrt{137}}{8}$

$n \approx 1.088, n \approx -1.838$

71.

$7d^5 + 20 = 300$

$7d^5 = 280$

$d^5 = 40$

$(d^5)^{\frac{1}{5}} = 40^{\frac{1}{5}}$

$d = 40^{\frac{1}{5}}$

$d \approx 2.091$

73.

$3(1.6)^x - 20 = 400$

$3(1.6)^x = 420$

$\dfrac{3(1.6)^x}{3} = \dfrac{420}{3}$

$(1.6)^x = 140$

$\ln(1.6)^x = \ln(140)$

$x \ln(1.6) = \ln(140)$

$\dfrac{x \ln(1.6)}{\ln(1.6)} = \dfrac{\ln(140)}{\ln(1.6)}$

$x = \dfrac{\ln(140)}{\ln(1.6)}$

$x \approx 10.514$

75.

$4(n-3)^2 + 10 = 30$

$4(n-3)(n-3) + 10 = 30$

$4(n^2 - 6n + 9) + 10 = 30$

$4n^2 - 24n + 36 + 10 = 30$

$4n^2 - 24n + 46 = 30$

$4n^2 - 24n + 16 = 0$

$\dfrac{4n^2}{4} - \dfrac{24n}{4} + \dfrac{16}{4} = \dfrac{0}{4}$

$n^2 - 6n + 4 = 0$

$a = 1, b = -6, c = 4$

$n = \dfrac{6 \pm \sqrt{(-6)^2 - 4(1)(4)}}{2(1)}$

$n = \dfrac{6 \pm \sqrt{20}}{2}$

$n = \dfrac{6 \pm 2\sqrt{5}}{2}$

$n = 3 \pm \sqrt{5}$

$n = 3 + \sqrt{5}, n = 3 - \sqrt{5}$

$n \approx 5.236, n \approx 0.764$

77. $(3.098, 75.145)$

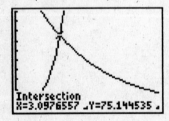

79. $(-4.130, -12.566)$

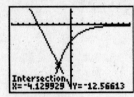

81. $(7.214, 8.013)$ and $(1.859, 54.111)$

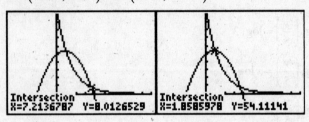

Section 6.6

1.
$\log x = 4$
$x = 10^4$
$x = 10,000$
Check your answer:
$\log(10,000) = \log(10^4)$
$ = 4$

3.
$\ln x = 5$
$x = e^5$
$x \approx 148.413$
Check your answer:
$\ln(e^5) = 5$
or
$\ln(148.413) \approx 4.99999$

5.
$\log_2 4x = 6$
$4x = 2^6$
$4x = 64$
$x = \dfrac{64}{4}$
$x = 16$
Check your answer:
$\log_2 4(16) = \log_2 64$
$ = \log_2 2^6$
$ = 6$

7.
$\log(5x+2) = 2$
$5x + 2 = 10^2$
$5x + 2 = 100$
$5x = 100 - 2$
$x = \dfrac{98}{5}$
$x = 19.6$
Check your answer:
$\log(5(19.6)+2) = \log(100)$
$ = 2$

9.
$\ln(3x-5) = 2$
$3x - 5 = e^2$
$3x = e^2 + 5$
$x = \dfrac{e^2 + 5}{3}$
$x \approx 4.1297$
Check your answer:
$\ln\left(3\left(\dfrac{e^2+5}{3}\right) - 5\right) = \ln\left((e^2+5) - 5\right)$
$\phantom{\ln\left(3\left(\dfrac{e^2+5}{3}\right) - 5\right)} = \ln(e^2)$
$\phantom{\ln\left(3\left(\dfrac{e^2+5}{3}\right) - 5\right)} = 2$
or
$\ln(3(4.1297) - 5) \approx 2.0000059$

11.
$M = \log\left(\dfrac{2000}{10^{-4}}\right)$
$M \approx 7.3$
An earthquake that has an intensity of 2000 cm has a magnitude of about 7.3.

13.
$M = \log\left(\dfrac{500,000}{10^{-4}}\right)$
$M \approx 9.7$
An earthquake that has an intensity of 500,000 cm has a magnitude of about 9.7.

15.
$8.0 = \log\left(\dfrac{I}{10^{-4}}\right)$
$\dfrac{I}{10^{-4}} = 10^8$
$10^{-4}\left(\dfrac{I}{10^{-4}}\right) = 10^{-4}(10^8)$
$I = 10^{-4+8} = 10^4$
$I = 10,000 \, cm$

The intensity of this 8.0 magnitude earthquake was 10,000 cm.

Chapter 6 Logarithmic Functions

17.

$6.6 = \log\left(\dfrac{I}{10^{-4}}\right)$

$\dfrac{I}{10^{-4}} = 10^{6.6}$

$10^{-4}\left(\dfrac{I}{10^{-4}}\right) = 10^{-4}\left(10^{6.6}\right)$

$I = 10^{-4+6.6} = 10^{2.6}$

$I = 398.11\,cm$

The intensity of this 6.6 magnitude earthquake was 398.11 cm.

19.

$pH = -\log\left(1.0 \times 10^{-7}\right)$

$pH = -\log\left(10^{-7}\right) = 7$

Scope mouthwash, with a hydrogen ion concentration of 1.0×10^{-7} M, has a pH of 7.

21.

$1 = -\log\left(H^+\right)$

$-1 = \log\left(H^+\right)$

$10^{-1} = H^+$

$H^+ = 1.0 \times 10^{-1}$

Car battery acid, with a pH of 1, has a hydrogen ion concentration of 1.0×10^{-1} M.

23.

$14 = -\log\left(H^+\right)$

$-14 = \log\left(H^+\right)$

$10^{-14} = H^+$

$H^+ = 1.0 \times 10^{-14}$

Oven cleaner, with a pH of 14, has a hydrogen ion concentration of 1.0×10^{-14} M.

25.

$7.35 = -\log\left(H^+\right)$ $7.45 = -\log\left(H^+\right)$

$-7.35 = \log\left(H^+\right)$ $-7.45 = \log\left(H^+\right)$

$10^{-7.35} = H^+$ $10^{-7.45} = H^+$

$H^+ = 4.47 \times 10^{-8}$ $H^+ = 3.55 \times 10^{-8}$

Blood plasma, with a pH level between 7.35 and 7.45, has a hydrogen ion concentration between 3.55×10^{-8} M and 4.47×10^{-8} M.

27.

$L\left(10^{-1}\right) = 10\log\left(\dfrac{10^{-1}}{10^{-12}}\right)$

$= 10\log\left(10^{-1} \cdot 10^{12}\right)$

$= 10\log\left(10^{11}\right)$

$= 10(11)$

$= 110$ dB

The chain saw, with an intensity of 10^{-1} watts per m^2, has a decibel level of 110 dB.

29.

$L\left(10^{-8}\right) = 10\log\left(\dfrac{10^{-8}}{10^{-12}}\right)$

$= 10\log\left(10^{-8} \cdot 10^{12}\right)$

$= 10\log\left(10^{4}\right)$

$= 10(4)$

$= 40$ dB

Raindrops, with an intensity of 10^{-8} watts per m^2, has a decibel level of 40 dB.

31.

$120 = 10\log\left(\dfrac{I}{10^{-12}}\right)$

$\dfrac{120}{10} = \log\left(\dfrac{I}{10^{-12}}\right)$

$12 = \log\left(\dfrac{I}{10^{-12}}\right)$

$10^{12} = \dfrac{I}{10^{-12}}$

$\left(10^{12}\right)10^{-12} = \left(\dfrac{I}{10^{-12}}\right)10^{-12}$

$I = 1$

A jack hammer, with a decibel level of 120 dB, has an intensity of 1 watt per m^2.

33.

$\log(2x) + 5 = 7$

$\log(2x) = 2$

$10^2 = 2x$

$100 = 2x$

$x = 50$

35.
$\log_4(-2h) + 40 = 47$
$\log_4(-2h) = 7$
$4^7 = -2h$
$16384 = -2h$
$h = -8192$

37.
$\ln(3t) + 5 = 3$
$\ln(3t) = -2$
$e^{-2} = 3t$
$\dfrac{e^{-2}}{3} = t$
$t \approx 0.045$

39.
$\ln(r+4) = 2$
$e^2 = r + 4$
$e^2 - 4 = r$
$r \approx 3.389$

41.
$\log(2p+1) = -0.5$
$10^{-0.5} = 2p + 1$
$10^{-0.5} - 1 = 2p$
$\dfrac{10^{-0.5} - 1}{2} = \dfrac{2p}{2}$
$p = \dfrac{10^{-0.5} - 1}{2}$
$p \approx -0.342$

43.
$\log_5(3x) + \log_5 x = 4$
$\log_5(3x^2) = 4$
$5^4 = 3x^2$
$625 = 3x^2$
$x^2 = \dfrac{625}{3}$
$x = \pm\sqrt{\dfrac{625}{3}}$
$x = \sqrt{\dfrac{625}{3}}, x = -\sqrt{\dfrac{625}{3}}$
$x \approx 14.434, x \approx -14.434$
The negative answer is not part of the domain.
The final answer is:
$x \approx 14.434$

45.
$\log_3(x+2) + \log_3(x-3) = 2$
$\log_3((x+2)(x-3)) = 2$
$3^2 = (x+2)(x-3)$
$9 = x^2 - x - 6$
$0 = x^2 - x - 15$
$x = \dfrac{1 \pm \sqrt{(-1)^2 - 4(1)(-15)}}{2(1)}$
$x = \dfrac{1 \pm \sqrt{61}}{2}$
$x = \dfrac{1 + \sqrt{61}}{2}, x = \dfrac{1 - \sqrt{61}}{2}$
$x \approx 4.405, x \approx -3.405$
The negative answer is not part of the domain.
The final answer is:
$x \approx 4.405$

47. Matt made the base 10 on the second step when it should have been e.

49. Frank added 3 and 5 on the second step when the student should have subtracted 5 from both sides of the equation.

51.
$\log(5t^3) - \log(2t) = 2$
$\log\left(\dfrac{5t^3}{2t}\right) = 2$
$\log(2.5t^2) = 2$
$10^2 = 2.5t^2$
$100 = 2.5t^2$
$40 = t^2$
$t = \pm\sqrt{40}$
$t = \pm 2\sqrt{10}$
$t = 2\sqrt{10}, t = -2\sqrt{10}$
$t \approx 6.325, t \approx -6.325$
The negative answer is not part of the domain.
The final answer is:
$x \approx 6.325$

Chapter 6 Logarithmic Functions

53.
$\log_5(x) + \log_5(2x+7) - 3 = 0$
$\log_5(x) + \log_5(2x+7) = 3$
$\log_5(x(2x+7)) = 3$
$5^3 = x(2x+7)$
$125 = 2x^2 + 7x$
$0 = 2x^2 + 7x - 125$
$x = \dfrac{-7 \pm \sqrt{7^2 - 4(2)(-125)}}{2(2)}$
$x = \dfrac{-7 \pm \sqrt{1049}}{4}$
$x = \dfrac{-7 + \sqrt{1049}}{4}, x = \dfrac{-7 - \sqrt{1049}}{4}$
$x \approx 6.347, x \approx -9.847$

The negative answer is not part of the domain.

The final answer is:

$x \approx 6.347$

55.
$\log_4(-2x+5) + \log_4(x+21.5) = 4$
$\log_4((-2x+5)(x+21.5)) = 4$
$4^4 = (-2x+5)(x+21.5)$
$256 = -2x^2 - 38x + 107.5$
$0 = -2x^2 - 38x - 148.5$
$x = \dfrac{38 \pm \sqrt{(-38)^2 - 4(-2)(-148.5)}}{2(-2)}$
$x = \dfrac{38 \pm \sqrt{256}}{-4}$
$x = \dfrac{38 + 16}{-4}, x = \dfrac{38 - 16}{-4}$
$x = -13.5, x = -5.5$

57.
$3\log_2(2x) + \log_2(16) = 7$
$3\log_2(2x) + \log_2(2^4) = 7$
$3\log_2(2x) + 4 = 7$
$3\log_2(2x) = 3$
$\dfrac{3\log_2(2x)}{3} = \dfrac{3}{3}$
$\log_2(2x) = 1$
$2^1 = 2x$
$2 = 2x$
$x = 1$

59.
$\ln(3x^2 + 5x) - 3 = 2$
$\ln(3x^2 + 5x) = 5$
$e^5 = 3x^2 + 5x$
$0 = 3x^2 + 5x - e^5$
$x = \dfrac{-5 \pm \sqrt{5^2 - 4(3)(-e^5)}}{2(3)}$
$x = \dfrac{-5 \pm \sqrt{25 + 12e^5}}{6}$
$x = \dfrac{-5 + \sqrt{25 + 12e^5}}{6}, x = \dfrac{-5 - \sqrt{25 + 12e^5}}{6}$
$x \approx 6.249, x \approx -7.916$

61.
$2\log(3x) + 8 = 2$
$2\log(3x) = -6$
$\dfrac{2\log(3x)}{2} = \dfrac{-6}{2}$
$\log(3x) = -3$
$10^{-3} = 3x$
$\dfrac{1}{1000} = 3x$
$\dfrac{1}{1000}\left(\dfrac{1}{3}\right) = 3x\left(\dfrac{1}{3}\right)$
$x = \dfrac{1}{3000}$
$x \approx 0.0003$

63.
$3^x = 27$
$3^x = 3^3$
$x = 3$

65.
$4x^2 + 6x - 10 = 0$
$\dfrac{4x^2}{2} + \dfrac{6x}{2} - \dfrac{10}{2} = \dfrac{0}{2}$
$2x^2 + 3x - 5 = 0$
$(2x+5)(x-1) = 0$
$2x+5 = 0, x-1 = 0$
$x = -2.5, x = 1$

67.
$t^3 + 6t^2 = -5t$
$t^3 + 6t^2 + 5t = 0$
$t(t^2 + 6t + 5) = 0$
$t(t+5)(t+1) = 0$
$t = 0, t+5 = 0, t+1 = 0$
$t = 0, t = -5, t = -1$

69.

$3(d+4)-14=0$

$3d+12-14=0$

$3d-2=0$

$d=\dfrac{2}{3}$

$d\approx 0.667$

Chapter 6 Logarithmic Functions
Chapter 6 Review

1.
$f(x) = 1.4x - 7$
$y = 1.4x - 7$
$y + 7 = 1.4x$
$y + 7 = \frac{14}{10}x$
$\left(\frac{10}{14}\right)(y+7) = \left(\frac{10}{14}\right)\left(\frac{14}{10}x\right)$
$\frac{5}{7}y + 5 = x$
$\frac{5}{7}x + 5 = y$
$f^{-1}(x) = \frac{5}{7}x + 5$

2.
$g(t) = -2t + 6$
$g = -2t + 6$
$g - 6 = -2t$
$\left(-\frac{1}{2}\right)(g - 6) = \left(-\frac{1}{2}\right)(-2t)$
$-\frac{1}{2}g + 3 = t$
$-\frac{1}{2}t + 3 = g$
$g^{-1}(x) = -\frac{1}{2}t + 3$

3.
$h(x) = 5^x$
$y = 5^x$
$\log_5 y = x$
$\log_5 x = y$
$\frac{\log x}{\log 5} = y$
$h^{-1}(x) = \log_5 x$
$h^{-1}(x) = \frac{\log x}{\log 5}$

4.
$f(x) = 3.5(6)^x$
$y = 3.5(6)^x$
$\frac{y}{3.5} = \frac{3.5(6)^x}{3.5}$
$\frac{y}{3.5} = (6)^x$
$\log_6\left(\frac{y}{3.5}\right) = x$
$\log_6\left(\frac{x}{3.5}\right) = y$
$\frac{\log\left(\frac{x}{3.5}\right)}{\log 6} = y$
$f^{-1}(x) = \log_6\left(\frac{x}{3.5}\right)$
$f^{-1}(x) = \frac{\log\left(\frac{x}{3.5}\right)}{\log 6}$

5.
$g(x) = 2e^x$
$y = 2e^x$
$\frac{y}{2} = \frac{2e^x}{2}$
$\frac{y}{2} = e^x$
$\ln\left(\frac{y}{2}\right) = x$
$\ln\left(\frac{x}{2}\right) = y$
$g^{-1}(x) = \ln\left(\frac{x}{2}\right)$

6.
$f(x) = \log_5 x$
$y = \log_5 x$
$5^y = x$
$5^x = y$
$f^{-1}(x) = 5^x$

7.
$h(t) = \log_{0.2} t$
$h = \log_{0.2} t$
$0.2^h = t$
$0.2^t = h$
$h^{-1}(t) = 0.2^t$

8.
a.
$D(e) = 400e + 50$
$D = 400e + 50$
$D - 50 = 400e$
$\dfrac{D-50}{400} = \dfrac{400e}{400}$
$\dfrac{D}{400} - \dfrac{50}{400} = e$
$0.0025D - 0.125 = e$
$e(D) = 0.0025D - 0.125$

b. $D(10) = 400(10) + 50 = 4050$

The AEM Toy company can manufacture 4050 dolls when they have 10 employees working.

c. $e(1000) = 0.0025(1000) - 0.125 = 2.375$

The AEM Toy company must have 3 employees working on a particular day to manufacture 1000 dolls.

d. Let the range of the inverse function be $5 \le e \le 30$, then
$D(5) = 400(5) + 50 = 2050$
$D(30) = 400(30) + 50 = 12,050$

Therefore, the domain of the inverse function is $2050 \le D \le 12,050$.

9.
a.
$P(d) = 4.5d - 300$
$P = 4.5d - 300$
$P + 300 = 4.5d$
$\dfrac{P+300}{4.5} = \dfrac{4.5d}{4.5}$
$\dfrac{P}{4.5} + \dfrac{300}{4.5} = d$
$0.222P + 66.667 = d$
$d(P) = 0.222P + 66.667$

b.
$d(12,000) = 0.222(12,000) + 66.667$
$d(12,000) = 2730.667$

The AEM Toy Company must sell 2731 dolls to make a $12,000 profit.

c. Let the range of the inverse function be $500 \le d \le 4000$, then
$P(500) = 4.5(500) - 300 = 1950$
$P(4000) = 4.5(4000) - 300 = 17,700$

Therefore, the domain of the inverse function is $1950 \le P \le 17,700$.

10. This function passes the horizontal line test because each horizontal line that you could draw across the graph would hit the function only once. Therefore, it is a one-to-one function.

11. This function fails the horizontal line test because many of the horizontal lines that you could draw through this graph hit the graph more than once. Therefore, this is not a one-to-one function.

12. $\log_4 64 = \log_4 4^3 = 3$

13. $\log_3 1 = 0$

14. $\log_8 8^5 = 5$

15.
$\log_2 \left(\dfrac{1}{8}\right) = \log_2 \left(\dfrac{1}{2^3}\right)$
$= \log_2 2^{-3}$
$= -3$

16. $f(x) = \log_{2.5}(x)$. Find points using $2.5^y = x$.

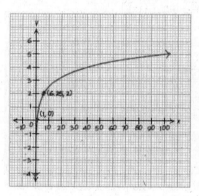

17. $f(x) = \log_{0.4}(x)$. Find points using $0.4^y = x$.

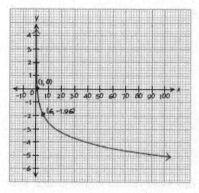

18.
$f(x) = \log_{2.5}(x)$
Domain: $(0, \infty)$
Range: $(-\infty, \infty)$

Chapter 6 Logarithmic Functions

19.

$f(x) = \log_{0.4}(x)$

Domain: $(0, \infty)$

Range: $(-\infty, \infty)$

20.

$\log(7xy) = \log 7 + \log x + \log y$

21.

$\log_5(3a^3b^4) = \log_5 3 + \log_5(a^3) + \log_5(b^4)$
$= \log_5 3 + 3\log_5 a + 4\log_5 b$

22.

$\ln(2x^3y^4) = \ln 2 + \ln(x^3) + \ln(y^4)$
$= \ln 2 + 3\ln x + 4\ln y$

23.

$\log_7\left(\sqrt[5]{3x^3y}\right) = \log_7(3x^3y)^{\frac{1}{5}}$
$= \frac{1}{5}\log_7(3x^3y)$
$= \frac{1}{5}\left(\log_7 3 + \log_7(x^3) + \log_7 y\right)$
$= \frac{1}{5}\left(\log_7 3 + 3\log_7 x + \log_7 y\right)$
$= \frac{1}{5}\log_7 3 + \frac{3}{5}\log_7 x + \frac{1}{5}\log_7 y$

24.

$\log_3\left(\sqrt{5ab^2c^3}\right)$
$= \log_3(5ab^2c^3)^{\frac{1}{2}}$
$= \frac{1}{2}\log_3(5ab^2c^3)$
$= \frac{1}{2}\left(\log_3 5 + \log_3 a + \log_3(b^2) + \log_3(c^3)\right)$
$= \frac{1}{2}\left(\log_3 5 + \log_3 a + 2\log_3 b + 3\log_3 c\right)$
$= \frac{1}{2}\log_3 5 + \frac{1}{2}\log_3 a + \log_3 b + \frac{3}{2}\log_3 c$

25.

$\log\left(\frac{3x^3y^5}{z^4}\right) = \log 3 + \log(x^3) + \log(y^5) - \log(z^4)$
$= \log 3 + 3\log x + 5\log y - 4\log z$

26.

$\log\left(\frac{4a^5b}{c^3}\right) = \log 4 + \log(a^5) + \log b - \log(c^3)$
$= \log 4 + 5\log a + \log b - 3\log c$

27.

$\log 4 + 3\log x + 2\log y + \log z$
$= \log 4 + \log x^3 + \log y^2 + \log z$
$= \log(4x^3y^2z)$

28.

$\log 2 + 4\log x + 5\log y - 2\log z$
$= \log 2 + \log x^4 + \log y^5 - \log z^2$
$= \log\left(\frac{2x^4y^5}{z^2}\right)$

29.

$\log_3 a + 3\log_3 b - 2\log_3 c$
$= \log_3 a + \log_3 b^3 - \log_3 c^2$
$= \log_3\left(\frac{ab^3}{c^2}\right)$

30.

$\ln 7 + 2\ln x + \ln z - 5\ln y - 3\ln z$
$= \ln 7 + 2\ln x - 5\ln y - 2\ln z$
$= \ln 7 + \ln x^2 - \ln y^5 - \ln z^2$
$= \ln\left(\frac{7x^2}{y^5z^2}\right)$

31.

$\log(x+2) + \log(x-7) = \log((x+2)(x-7))$
$= \log(x^2 - 5x - 14)$

32.

a.

$2005 \rightarrow t = 5$

$F(5) = 750(0.9405)^5$

$F(5) \approx 551.89$

In 2005 the population of Franklin's gulls was approximately 551.89 thousand.

b.

$$500 = 750(0.9405)^t$$

$$\frac{500}{750} = \frac{750(0.9405)^t}{750}$$

$$\frac{2}{3} = (0.9405)^t$$

$$\log\left(\frac{2}{3}\right) = \log(0.9405)^t$$

$$\log\left(\frac{2}{3}\right) = t\log(0.9405)$$

$$\frac{\log\left(\frac{2}{3}\right)}{\log(0.9405)} = \frac{t\log(0.9405)}{\log(0.9405)}$$

$$t \approx 6.6$$

According to the model, the population of Franklin's gulls reached 500,000 midway through the year 2006.

c. This population might stop declining at this rate due to environmental intervention.

33.

Use $A = P\left(1 + \frac{r}{n}\right)^{nt}$ with $n = 12$

$$14,000 = 7000\left(1 + \frac{0.06}{12}\right)^{12t}$$

$$\frac{14,000}{7000} = \frac{7000(1.005)^{12t}}{7000}$$

$$2 = (1.005)^{12t}$$

$$\log 2 = \log(1.005)^{12t}$$

$$\log 2 = 12t \log(1.005)$$

$$\frac{\log 2}{12\log(1.005)} = \frac{12t \log(1.005)}{12\log(1.005)}$$

$$t \approx 11.6$$

It takes approximately 11.6 years to double your money at 6% compounded monthly.

34.

Use $A = Pe^{rt}$

$$200,000 = 100,000 e^{0.085t}$$

$$\frac{200,000}{100,000} = \frac{100,000 e^{0.085t}}{100,000}$$

$$2 = e^{0.085t}$$

$$\ln 2 = 0.085t$$

$$\frac{\ln 2}{0.085} = \frac{0.085t}{0.085}$$

$$t \approx 8.2$$

It takes approximately 8.2 years to double your money at 8.5% compounded continuously.

35.

$$4^w = 12.5$$

$$\log 4^w = \log 12.5$$

$$w \log 4 = \log 12.5$$

$$\frac{w \log 4}{\log 4} = \frac{\log 12.5}{\log 4}$$

$$w = \frac{\log 12.5}{\log 4}$$

$$w \approx 1.822$$

36.

$$3^t = \frac{1}{243}$$

$$3^t = \frac{1}{3^5}$$

$$3^t = 3^{-5}$$

$$t = -5$$

37.

$$6^{3x-5} = 204$$

$$\ln 6^{3x-5} = \ln 204$$

$$(3x-5)\ln 6 = \ln 204$$

$$\frac{(3x-5)\ln 6}{\ln 6} = \frac{\ln 204}{\ln 6}$$

$$3x - 5 = \frac{\ln 204}{\ln 6}$$

$$3x = \frac{\ln 204}{\ln 6} + 5$$

$$\frac{3x}{3} = \frac{\frac{\ln 204}{\ln 6} + 5}{3}$$

$$x = \frac{\frac{\ln 204}{\ln 6} + 5}{3}$$

$$x \approx 2.656$$

38.

$$7^{x+2} = 47$$

$$\log 7^{x+2} = \log 47$$

$$(x+2)\log 7 = \log 47$$

$$\frac{(x+2)\log 7}{\log 7} = \frac{\log 47}{\log 7}$$

$$x + 2 = \frac{\log 47}{\log 7}$$

$$x = \frac{\log 47}{\log 7} - 2$$

$$x \approx -0.021$$

39.

$17(1.9)^x = 55.4$

$\dfrac{17(1.9)^x}{17} = \dfrac{55.4}{17}$

$(1.9)^x = \dfrac{55.4}{17}$

$\ln(1.9)^x = \ln\left(\dfrac{55.4}{17}\right)$

$x\ln(1.9) = \ln\left(\dfrac{55.4}{17}\right)$

$\dfrac{x\ln(1.9)}{\ln(1.9)} = \dfrac{\ln\left(\dfrac{55.4}{17}\right)}{\ln(1.9)}$

$x = \dfrac{\ln\left(\dfrac{55.4}{17}\right)}{\ln(1.9)}$

$x \approx 1.841$

40.

$4\left(\dfrac{1}{3}\right)^x - 8 = 28$

$4\left(\dfrac{1}{3}\right)^x = 36$

$\dfrac{4\left(\dfrac{1}{3}\right)^x}{4} = \dfrac{36}{4}$

$\left(\dfrac{1}{3}\right)^x = 9$

$\left(3^{-1}\right)^x = 3^2$

$3^{-x} = 3^2$

$-x = 2$

$x = -2$

41.

$3(5)^{3x+2} - 700 = 11$

$3(5)^{3x+2} = 711$

$\dfrac{3(5)^{3x+2}}{3} = \dfrac{711}{3}$

$5^{3x+2} = 237$

$\log 5^{3x+2} = \log 237$

$(3x+2)\log 5 = \log 237$

$\dfrac{(3x+2)\log 5}{\log 5} = \dfrac{\log 237}{\log 5}$

$3x + 2 = \dfrac{\log 237}{\log 5}$

$3x = \dfrac{\log 237}{\log 5} - 2$

$\dfrac{3x}{3} = \dfrac{\dfrac{\log 237}{\log 5} - 2}{3}$

$x = \dfrac{\dfrac{\log 237}{\log 5} - 2}{3}$

$x \approx 0.466$

42.

$5^{x^2+4} = 15625$

$5^{x^2+4} = 5^6$

$x^2 + 4 = 6$

$x^2 = 2$

$x = \pm\sqrt{2}$

$x \approx \pm 1.414$

43.

$5.7 = \log\left(\dfrac{I}{10^{-4}}\right)$

$\dfrac{I}{10^{-4}} = 10^{5.7}$

$10^{-4}\left(\dfrac{I}{10^{-4}}\right) = 10^{-4}\left(10^{5.7}\right)$

$I = 10^{-4+5.7} = 10^{1.7}$

$I \approx 50.12\, cm$

The intensity of this 5.7 magnitude earthquake was 50.12 cm.

44.

$pH = -\log\left(3.7 \times 10^{-3}\right)$

$pH \approx 2.4$

Cranberry juice, with a hydrogen ion concentration of 3.7×10^{-3} M, has a pH of 2.4.

45.
$5.6 = -\log(H^+)$
$-5.6 = \log(H^+)$
$10^{-5.6} = H^+$
$H^+ = 2.5 \times 10^{-6}$
Shampoo, with a pH of 5.6, has a hydrogen ion concentration of 2.5×10^{-6} M.

46.
$\log x = -4$
$10^{-4} = x$
$x = 0.0001$

47.
$\log_8 t = 5$
$8^5 = t$
$t = 32768$

48.
$\log(10x + 20) = 5$
$10^5 = 10x + 20$
$100000 = 10x + 20$
$99980 = 10x$
$\dfrac{99980}{10} = \dfrac{10x}{10}$
$x = 9998$

49.
$\ln(4.3t) + 7.5 = 3$
$\ln(4.3t) = -4.5$
$e^{-4.5} = 4.3t$
$\dfrac{e^{-4.5}}{4.3} = \dfrac{4.3t}{4.3}$
$t = \dfrac{e^{-4.5}}{4.3}$
$t \approx 0.003$

50.
$\log_5(2x) + \log_5 x = 4$
$\log_5(2x^2) = 4$
$5^4 = 2x^2$
$625 = 2x^2$
$312.5 = x^2$
$x = \pm\sqrt{312.5}$
$x \approx \pm 17.678$
The negative answer is not part of the domain.
The final answer is:
$x \approx 17.678$

51.
$\log_3(x-4) + \log_3(x+1) = 3$
$\log_3((x-4)(x+1)) = 3$
$3^3 = (x-4)(x+1)$
$27 = x^2 - 3x - 4$
$0 = x^2 - 3x - 31$
$x = \dfrac{3 \pm \sqrt{(-3)^2 - 4(1)(-31)}}{2(1)}$
$x = \dfrac{3 \pm \sqrt{133}}{2}$
$x = \dfrac{3 + \sqrt{133}}{2}, x = \dfrac{3 - \sqrt{133}}{2}$
$x \approx 7.266, x \approx -4.266$
The negative answer is not part of the domain.
The final answer is:
$x \approx 7.266$

52.
$\log_4(x+4) + \log_4(x+7) = 1$
$\log_4((x+4)(x+7)) = 1$
$4^1 = (x+4)(x+7)$
$4 = x^2 + 11x + 28$
$0 = x^2 + 11x + 24$
$x = \dfrac{-11 \pm \sqrt{11^2 - 4(1)(24)}}{2(1)}$
$x = \dfrac{-11 \pm \sqrt{25}}{2}$
$x = \dfrac{-11 \pm 5}{2}$
$x = \dfrac{-11 + 5}{2}, x = \dfrac{-11 - 5}{2}$
$x = -3, x = -8$
The answer -8 is not part of the domain.
The final answer is:
$x = -3$

Chapter 6 Logarithmic Functions

53.

$\log(8x^5) - \log(2x) = 3$

$\log\left(\dfrac{8x^5}{2x}\right) = 3$

$\log(4x^4) = 3$

$10^3 = 4x^4$

$1000 = 4x^4$

$250 = x^4$

$(x^4)^{\frac{1}{4}} = \pm 250^{\frac{1}{4}}$

$x \approx \pm 3.976$

The negative answer is not part of the domain.

The final answer is:

$x \approx 3.976$

Chapter 6 Test

1.

Use $A = P\left(1 + \dfrac{r}{n}\right)^{nt}$ with $n = 12$

$2000 = 1000\left(1 + \dfrac{0.035}{12}\right)^{12t}$

$\dfrac{2000}{1000} = \dfrac{1000(1.002917)^{12t}}{1000}$

$2 = (1.002917)^{12t}$

$\log 2 = \log(1.002917)^{12t}$

$\log 2 = 12t \log(1.002917)$

$\dfrac{\log 2}{12\log(1.002917)} = \dfrac{12t\log(1.002917)}{12\log(1.002917)}$

$19.8 \approx t$

It takes approximately 19.8 years to double your money at 3.5% compounded monthly.

2.

$7.0 = \log\left(\dfrac{I}{10^{-4}}\right)$

$\dfrac{I}{10^{-4}} = 10^{7.0}$

$10^{-4}\left(\dfrac{I}{10^{-4}}\right) = 10^{-4}(10^{7.0})$

$I = 10^{-4+7} = 10^3$

$I = 1000\, cm$

The intensity of this 7.0 magnitude earthquake was 1000 cm.

3.

$\log_{16}(2x+1) = -0.5$

$16^{-0.5} = 2x+1$

$0.25 = 2x + 1$

$-0.75 = 2x$

$x = -0.375$

4.

$\log_7 x + \log_7(2x+7) - 3 = 0$

$\log_7 x + \log_7(2x+7) = 3$

$\log_7(x(2x+7)) = 3$

$7^3 = x(2x+7)$

$343 = 2x^2 + 7x$

$0 = 2x^2 + 7x - 343$

$x = \dfrac{-7 \pm \sqrt{7^2 - 4(2)(-343)}}{2(2)}$

$x = \dfrac{-7 + \sqrt{2793}}{4},\, x = \dfrac{-7 - \sqrt{2793}}{4}$

$x \approx 11.462,\, x \approx -14.962$

The negative answer is not part of the domain.

The final answer is: $x \approx 11.462$

5.

$20 = -3 + 4(2^x)$

$23 = 4(2^x)$

$\dfrac{23}{4} = \dfrac{4(2^x)}{4}$

$2^x = \dfrac{23}{4}$

$\ln 2^x = \ln\left(\dfrac{23}{4}\right)$

$x \ln 2 = \ln\left(\dfrac{23}{4}\right)$

$\dfrac{x \ln 2}{\ln 2} = \dfrac{\ln\left(\dfrac{23}{4}\right)}{\ln 2}$

$x = \dfrac{\ln\left(\dfrac{23}{4}\right)}{\ln 2}$

$x \approx 2.524$

6.

$8 = 3^{7x-1}$

$\ln 8 = \ln 3^{7x-1}$

$\ln 8 = (7x-1)\ln 3$

$\dfrac{\ln 8}{\ln 3} = \dfrac{(7x-1)\ln 3}{\ln 3}$

$\dfrac{\ln 8}{\ln 3} = 7x - 1$

$\dfrac{\ln 8}{\ln 3} + 1 = 7x$

$\dfrac{\dfrac{\ln 8}{\ln 3} + 1}{7} = \dfrac{7x}{7}$

$x = \dfrac{\dfrac{\ln 8}{\ln 3} + 1}{7}$

$x \approx 0.413$

7. No, the graph is not one-to-one because it does not pass the horizontal line test.

8.

$4\log(2a^3) + 5\log(ab^3) = \log(2a^3)^4 + \log(ab^3)^5$

$= \log(2^4 a^{12}) + \log(a^5 b^{15})$

$= \log(2^4 a^{12} \cdot a^5 b^{15})$

$= \log(16 a^{17} b^{15})$

9.

$\log 5 + \log x + 3\log y - 4\log z$

$= \log 5 + \log x + \log y^3 - \log z^4$

$= \log\left(\dfrac{5xy^3}{z^4}\right)$

10.

$\ln(4a^3 bc^4) = \ln 4 + \ln(a^3) + \ln b + \ln(c^4)$

$= \ln 4 + 3\ln a + \ln b + 4\ln c$

11.

$\log\left(\dfrac{3xy^5}{\sqrt{z}}\right) = \log 3 + \log x + \log y^5 - \log \sqrt{z}$

$= \log 3 + \log x + \log y^5 - \log z^{\frac{1}{2}}$

$= \log 3 + \log x + 5\log y - \dfrac{1}{2}\log z$

12.

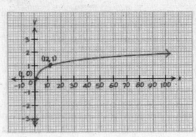

13.

$f(x) = \log_{12}(x)$

Domain: $(0, \infty)$ Range: $(-\infty, \infty)$

14.

a. Let $C(t)$ be the number of CD singles shipped (in millions) in the year t years since 1990.

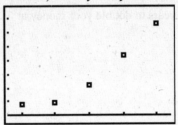

An exponential model is appropriate so choose a point from the flatter end and one from the steeper end.

Note: Your model may differ, but it must follow the trend of the data.

$C(t) = ab^t$

To find a and b choose two points $(3, 7.8)$ and $(7, 66.7)$.

$7.8 = ab^3$ and $66.7 = ab^7$

$\dfrac{66.7}{7.8} = \dfrac{ab^7}{ab^3}$

$\dfrac{66.7}{7.8} = b^4$

$\left(\dfrac{66.7}{7.8}\right)^{\frac{1}{4}} = (b^4)^{\frac{1}{4}}$

NOTE: We do not need the $\pm$ on the answer because $b > 0$.

$b \approx 1.71$

Use b and one of the above equations to find a.

$7.8 = ab^3$

$\dfrac{7.8}{(1.71)^3} = \dfrac{a(1.71)^3}{(1.71)^3}$

$a \approx 1.56$

Then $C(t) = 1.56(1.71)^t$

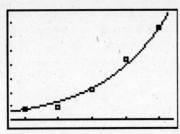

b. Domain: $[0,10]$ Range: $[1.56, 333.49]$

c. $C(10) = 1.56(1.71)^{10} \approx 333.49$ In 2000 there were approximately 333.49 million CD singles shipped.

d.
$$5 = 1.56(1.71)^t$$
$$\frac{5}{1.56} = \frac{1.56(1.71)^t}{1.56}$$
$$\frac{5}{1.56} = (1.71)^t$$
$$\log\left(\frac{5}{1.56}\right) = \log(1.71)^t$$
$$\log\left(\frac{5}{1.56}\right) = t \log(1.71)$$
$$\frac{\log\left(\frac{5}{1.56}\right)}{\log(1.71)} = \frac{t \log(1.71)}{\log(1.71)}$$
$$t \approx 2.17$$

In 1992 there were approximately 5 million CD singles shipped.

15.
$$f(x) = 5x + 2$$
$$y = 5x + 2$$
$$y - 2 = 5x$$
$$\frac{1}{5}(y - 2) = \frac{1}{5}(5x)$$
$$\frac{1}{5}y - \frac{2}{5} = x$$
$$\frac{1}{5}x - \frac{2}{5} = y$$
$$f^{-1}(x) = \frac{1}{5}x - \frac{2}{5}$$

16.
$$g(x) = \log_{15} x$$
$$y = \log_{15} x$$
$$15^y = x$$
$$15^x = y$$
$$g^{-1}(x) = 15^x$$

17.
$$f(x) = 5^x$$
$$y = 5^x$$
$$\log_5 y = x$$
$$\log_5 x = y$$
$$\frac{\log x}{\log 5} = y$$
$$f^{-1}(x) = \log_5 x$$
$$f^{-1}(x) = \frac{\log x}{\log 5}$$

18.
$$h(x) = 2(7)^x$$
$$y = 2(7)^x$$
$$\frac{y}{2} = (7)^x$$
$$\log_7\left(\frac{y}{2}\right) = x$$
$$\log_7\left(\frac{x}{2}\right) = y$$
$$\frac{\log\left(\frac{x}{2}\right)}{\log 7} = y$$
$$h^{-1}(x) = \log_7\left(\frac{x}{2}\right)$$
$$h^{-1}(x) = \frac{\log\left(\frac{x}{2}\right)}{\log 7}$$

19.
$$2.58 = -\log(H^+)$$
$$-2.58 = \log(H^+)$$
$$10^{-2.58} = H^+$$
$$H^+ = 2.63 \times 10^{-3}$$

A solution with a pH of 2.58 has a hydrogen ion concentration of 2.63×10^{-3} M.

20.
$$\log_3 200 = \frac{\log 200}{\log 3}$$
$$\approx 4.8227$$

Chapter 6 Logarithmic Functions
Chapter 6 Cumulative Review

1.
$3(4.2)^x + 7 = 23$
$3(4.2)^x = 16$
$(4.2)^x = \dfrac{16}{3}$
$\ln(4.2)^x = \ln\left(\dfrac{16}{3}\right)$
$x\ln(4.2) = \ln\left(\dfrac{16}{3}\right)$
$\dfrac{x\ln(4.2)}{\ln(4.2)} = \dfrac{\ln\left(\dfrac{16}{3}\right)}{\ln(4.2)}$
$x = \dfrac{\ln\left(\dfrac{16}{3}\right)}{\ln(4.2)}$
$x \approx 1.166$

2.
$3x + 18 = 7x - 6$
$3x = 7x - 24$
$-4x = -24$
$x = 6$

3.
$3(t-5)^2 - 18 = 30$
$3(t-5)^2 = 48$
$(t-5)^2 = 16$
$\sqrt{(t-5)^2} = \pm\sqrt{16}$
$t - 5 = \pm 4$
$t - 5 = -4, t - 5 = 4$
$t = 1, t = 9$

4.
$\ln(2c - 7) = 5$
$e^5 = 2c - 7$
$e^5 + 7 = 2c$
$c = \dfrac{e^5 + 7}{2}$

5.
$\dfrac{2}{5}b + \dfrac{4}{5} = \dfrac{1}{5}(b + 3)$
$\dfrac{2}{5}b + \dfrac{4}{5} = \dfrac{1}{5}b + \dfrac{3}{5}$
$5\left(\dfrac{2}{5}b + \dfrac{4}{5}\right) = 5\left(\dfrac{1}{5}b + \dfrac{3}{5}\right)$
$2b + 4 = b + 3$
$b + 4 = 3$
$b = -1$

6.
$2.5n^2 + 1.4n - 3.8 = 0$
$n = \dfrac{-1.4 \pm \sqrt{(1.4)^2 - 4(2.5)(-3.8)}}{2(2.5)}$
$n = \dfrac{-1.4 \pm \sqrt{39.96}}{5}$
$n \approx 0.984, n \approx -1.544$

7.
$4h^3 + 16h^2 - 32h = 0$
$4h(h^2 + 4h - 8) = 0$
$4h = 0, h^2 + 4h - 8 = 0$
$h = 0, h = \dfrac{-4 \pm \sqrt{4^2 - 4(1)(-8)}}{2(1)}$
$h = 0, h = \dfrac{-4 \pm \sqrt{48}}{2}$
$h = 0, h = \dfrac{-4 \pm 4\sqrt{3}}{2}$
$h = 0, h = -2 \pm 2\sqrt{3}$

8.
$2^{3x-7} = 45$
$\ln(2^{3x-7}) = \ln(45)$
$(3x - 7)\ln(2) = \ln(45)$
$\dfrac{(3x - 7)\ln(2)}{\ln(2)} = \dfrac{\ln(45)}{\ln(2)}$
$3x - 7 = \dfrac{\ln(45)}{\ln(2)}$
$3x = \dfrac{\ln(45)}{\ln(2)} + 7$
$x = \dfrac{\dfrac{\ln 45}{\ln 2} + 7}{3}$
$x \approx 4.164$

9.
$8.4g + 7.6 = 3.2(g - 4.2)$
$8.4g + 7.6 = 3.2g - 13.44$
$8.4g = 3.2g - 21.04$
$5.2g = -21.04$
$g \approx -4.046$

10.
$(x+5)(x-4) = -6$
$x^2 + x - 20 = -6$
$x^2 + x - 14 = 0$
$x = \dfrac{-1 \pm \sqrt{1^2 - 4(1)(-14)}}{2(1)}$
$x = \dfrac{-1 \pm \sqrt{57}}{2}$
$x \approx 3.275, \; x \approx -4.275$

11.
$\log(x+2) + \log(x-7) = 2$
$\log[(x+2)(x-7)] = 2$
$10^2 = (x+2)(x-7)$
$100 = x^2 - 5x - 14$
$0 = x^2 - 5x - 114$
$x = \dfrac{5 \pm \sqrt{(-5)^2 - 4(1)(-114)}}{2(1)}$
$x = \dfrac{5 \pm \sqrt{481}}{2}$

12.
$e^{x+5} = 89$
$\ln(e^{x+5}) = \ln(89)$
$(x+5)\ln(e) = \ln(89)$
$(x+5) \cdot 1 = \ln(89)$
$x + 5 = \ln 89$
$x = -5 + \ln 89$

13.
$|b+7| = 20$
$b + 7 = 20, \; b + 7 = -20$
$b = 13, \; b = -27$

14.
$|2t - 12| = 6$
$2t - 12 = 6, \; 2t - 12 = -6$
$2t = 18, \; 2t = 6$
$t = 9, \; t = 3$

15.
$4(3^x) + 25 = 8(3^x) - 299$
$4(3^x) + 324 = 8(3^x)$
$324 = 4(3^x)$
$\dfrac{324}{4} = \dfrac{4(3^x)}{4}$
$81 = 3^x$
$3^4 = 3^x$
$x = 4$

16.
$2.5x^8 - 3000 = -2360$
$2.5x^8 = 640$
$x^8 = 256$
$(x^8)^{\frac{1}{8}} = \pm(256)^{\frac{1}{8}}$
$x = \pm 2$

17.
$\begin{cases} y = 5x - 20 \\ 2x - 3y = -31 \end{cases}$
$2x - 3(5x - 20) = -31$
$2x - 15x + 60 = -31$
$-13x + 60 = -31$
$-13x = -91$
$x = 7$

Substituting x back into either equation solve for y:
$y = 5(7) - 20$
$y = 35 - 20$
$y = 15$

The solution to the system is $(7, 15)$.

18.

$$\begin{cases} 5x+12y=66.4 \\ -4x+10y=29.2 \end{cases}$$

$$\begin{cases} 4(5x+12y)=4(66.4) \\ 5(-4x+10y)=5(29.2) \end{cases}$$

$$\begin{cases} 20x+48y=265.6 \\ -20x+50y=146 \end{cases}$$

$$98y=411.6$$

$y=4.2$

Substituting y back into either equation solve for x:

$5x+12(4.2)=66.4$

$5x+50.4=66.4$

$5x=16$

$x=3.2$

The solution to the system is $(3.2, 4.2)$.

19.

$$\begin{cases} y=0.6x+3 \\ 3x-5y=-15 \end{cases}$$

$3x-5(0.6x+3)=-15$

$3x-3x-15=-15$

$-15=-15$

Infinite number of solutions.

20. The solutions to the system are $(-6, 18)$ and $(4, 48)$.

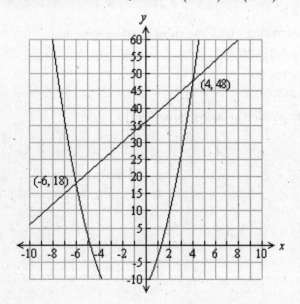

21. The solutions to the system are $(3, 6)$ and $\left(\dfrac{1}{3}, -\dfrac{98}{9}\right)$.

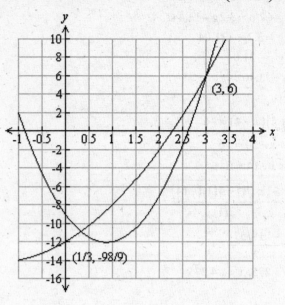

22. No solution

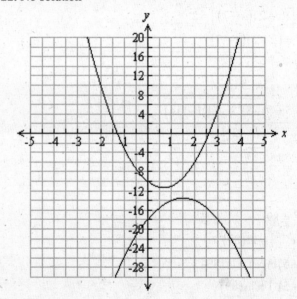

23. Let $P(t)$ be the gross profit for UTstarcom, Inc. (in millions of dollars) t years since 1990.

a.

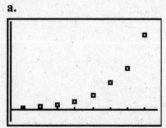

An exponential model is appropriate so choose a point from the flatter end and one from the steeper end.

Note: Your model may differ, but it must follow the trend of the data.

$P(t) = ab^t$

To find a and b choose two points $(6, 13.2)$ and $(13, 636.2)$.

$13.2 = ab^6$ and $636.2 = ab^{13}$

$\dfrac{636.2}{13.2} = \dfrac{ab^{13}}{ab^6}$

$\dfrac{636.2}{13.2} = b^7$

$\left(\dfrac{636.2}{13.2}\right)^{\frac{1}{7}} = (b^7)^{\frac{1}{7}}$

$b \approx 1.74$

Use b and one of the above equations to find a.

$13.2 = a(1.74)^6$

$\dfrac{1.3}{(1.74)^6} = \dfrac{a(1.74)^6}{(1.74)^6}$

$a \approx 0.48$

Then $P(t) = 0.48(1.74)^t$

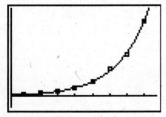

b. Domain: $[4, 15]$

$P(4) = 0.48(1.74)^4 = 4.4$

$P(15) = 0.48(1.74)^{15} = 1947.5$

Range: $[4.4, 1947.5]$

c.

$2 billion → $P(t) = 2000$

$2000 = 0.48(1.74)^t$

$\dfrac{2000}{0.48} = \dfrac{0.48(1.74)^t}{0.48}$

$\left(\dfrac{2000}{0.48}\right) = 1.74^t$

$t = \log_{1.74}\left(\dfrac{2000}{0.48}\right)$

$t = \dfrac{\log\left(\dfrac{2000}{0.48}\right)}{\log 1.74}$

$t \approx 15$

In 2005, the gross profit for UTstarcom, Inc reached $2 billion.

24. Let $C(t)$ be the number of cocaine-related emergency department episodes (for people over the age of 35) t years since 1990.

a.

t	$C(t)$
0	23,054
1	30,582
5	57,341
6	68,717
7	74,600
8	83,730
9	85,869
10	93,357
11	106,810

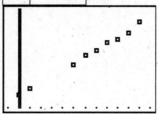

$C(t) = mt + b$

Using two points: $(1, 30582)$ and $(7, 74600)$

Find the slope: $m = \dfrac{74,600 - 30,582}{7 - 1} \approx 7336.3$

$C(t) = mt + b$

$b = 30,582 - 7336.3(1)$

$b = 23,245.7$

$C(t) = 7336.3t + 23,245.7$

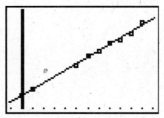

b.

$2003 → t = 13$

$C(13) = 7336.3(13) + 23,245.7$

$C(13) = 118,617.6$

In the year 2003, the number of cocaine-related emergency department episodes (for people over the age of 35) was 118,618.

Chapter 6 Logarithmic Functions

c.
$$140000 = 7336.3t + 23245.7$$
$$116754.3 = 7336.3t$$
$$\frac{116754.3}{7336.3} = \frac{7336.3t}{7336.3}$$
$$t \approx 16$$

In 2006, the number of cocaine-related emergency department episodes (for people over the age of 35) reached 140,000.

d. The slope is 7336.3. The number of cocaine-related emergency department episodes (for people over the age of 35) increases by 7336.3 cases each year.

e. Domain: $[-1, 15]$

$$C(-1) = 7336.3(-1) + 23,245.7 = 15,909.4$$
$$C(15) = 7336.3(15) + 23,245.7 = 133,290.2$$

Range: $[15909, 133290]$

25. $f(x) = 4x - 7$

a.
$$f(9) = 4(9) - 7$$
$$f(9) = 29$$

b.
$$-2 = 4x - 7$$
$$5 = 4x$$
$$x = \frac{5}{4}$$

Therefore, $f\left(\frac{5}{4}\right) = -2$.

c. Domain: $(-\infty, \infty)$ Range: $(-\infty, \infty)$

26.

a.
$$f(x) + g(x) = (3.5x + 4) + (-2x + 5.5)$$
$$= 3.5x - 2x + 4 + 5.5$$
$$= 1.5x + 9.5$$

b.
$$f(g(x)) = 3.5(-2x + 5.5) + 4$$
$$= -7x + 19.25 + 4$$
$$= -7x + 23.25$$

c.
$$f(x)g(x) = (3.5x + 4)(-2x - 5.5)$$
$$= -7x^2 + 19.25x - 8x + 22$$
$$= -7x^2 + 11.25x + 22$$

27.

a.
$$f(3) - g(3) = [4(3) - 15] - [-6(3) + 3]$$
$$= (12 - 15) + (18 - 3)$$
$$= (-3) + (15)$$
$$= 12$$

b.
$$g(5) = -6(5) + 3 = -27$$

$$f(g(5)) = f(-27)$$
$$= 4(-27) - 15$$
$$= -123$$

c.
$$f(4) = 4(4) - 15 \text{ and } g(4) = -6(4) + 3$$
$$= 1 \qquad\qquad\qquad = -21$$

$$f(4)g(4) = (1)(-21) = -21$$

28.

a.
$$f(x) + g(x) = (3x^2 + 5x - 2) + (4x - 7)$$
$$= 3x^2 + 9x - 9$$

b.
$$f(g(x)) = 3(4x - 7)^2 + 5(4x - 7) - 2$$
$$= 3(16x^2 - 56x + 49) + 20x - 35 - 2$$
$$= 48x^2 - 168x + 147 + 20x - 37$$
$$= 48x^2 - 148x + 110$$

c.
$$f(x)g(x) = (3x^2 + 5x - 2)(4x - 7)$$
$$= 12x^3 - 21x^2 + 20x^2 - 35x - 8x + 14$$
$$= 12x^3 - x^2 - 43x + 14$$

29.

a.
$$f(5) = -4(5)^2 + 7(5) - 3 \text{ and } g(5) = 5(5) + 8$$
$$= -4(25) + 35 - 3 \qquad\qquad = 25 + 8$$
$$= -100 + 32 \qquad\qquad\qquad = 33$$
$$= -68$$

$$f(5) - g(5) = (-68) - (33)$$
$$= -101$$

b.

$g(2) = 5(2) + 8 = 18$

$f(g(2)) = f(18)$
$= -4(18)^2 + 7(18) - 3$
$= -4(324) + 126 - 3$
$= -1173$

c.

$f(3) = -4(3)^2 + 7(3) - 3$ and $g(3) = 5(3) + 8$
$= -4(9) + 21 - 3$ $\quad\quad = 15 + 8$
$= -36 + 18$ $\quad\quad\quad\quad = 23$
$= -18$

$f(3)g(3) = (-18)(23)$
$= -414$

30.
V is the number of visitors to Yellowstone Natural Park
t is the year
L is the amount of litter left in the park (in tons)
p is the number of visitors to Yellowstone Natural Park
Therefore $p = V(t)$ and $L(p) = L(V(t))$

$L(V(t))$ is the amount of litter (in tons) left in Yellowstone National Park in year t.

31.
a.
$W(5) = 1.31(5) + 13.94$
$\quad\quad = 6.55 + 13.94$
$\quad\quad = 20.49$

In 1995, about 20.5% of the USA triathlon members were women.

b.
$W(t) > M(t)$
$1.31y + 13.94 > -1.31t + 86.06$
$\underline{+1.31t - 13.94 \quad +1.31t - 13.94}$
$\quad\quad 2.62t > 72.12$
$\quad\quad \dfrac{2.62t}{2.62} > \dfrac{72.12}{2.62}$
$\quad\quad\quad t > 27.5$

In the year 2017 the percent of USA Triathlon members that are women will be greater that that of men.

32.
$2x - 7 < 5x - 15$
$\underline{-5x + 7 \quad -5x + 7}$
$\quad -3x < -8$
$\quad \dfrac{-3x}{-3} > \dfrac{-8}{-3}$
$\quad\quad x > \dfrac{8}{3}$

33.
$3(4x - 8) \geq 2x + 8$
$12x - 24 \geq 2x + 8$
$\underline{-2x + 24 \quad -2x + 24}$
$\quad 10x \geq 32$
$\quad\quad x \geq 3.2$

34.
$|x - 7| > 5$
$x - 7 < -5$ or $x - 7 > 5$
$\underline{+7 \quad +7} \quad\quad \underline{+7 + 7}$
$x < 2$ or $\quad x > 12$

35.
$|3x + 5| \leq 8$
$-8 \leq 3x + 5 \leq 8$
$\underline{-5 \quad\quad -5 \quad -5}$
$-13 \leq 3x \leq 3$
$\dfrac{-13}{3} \leq \dfrac{3x}{3} \leq \dfrac{3}{3}$
$-\dfrac{13}{3} \leq x \leq 1$

36.
a.

x	y	Slope
-2	10	$\dfrac{5-10}{0-(-2)} = -\dfrac{5}{2}$
0	5	
2	0	$\dfrac{0-5}{2-0} = -\dfrac{5}{2}$
4	-5	$\dfrac{-5-0}{4-2} = -\dfrac{5}{2}$
6	-10	$\dfrac{-10-(-5)}{6-4} = -\dfrac{5}{2}$

b. x-intercept: $(2, 0)$

c. y-intercept: $(0, 5)$

d.

$y = mx + b$

$y = -\dfrac{5}{2}x + 5$

37.

Use the slope of the line

$2x - 3y = 30$

and the point $(24, 4)$:

$-3y = -2x + 30$

$\dfrac{-3y}{-3} = \dfrac{-2x}{-3} + \dfrac{30}{-3}$

$y = \dfrac{2}{3}x - 10 \to m = \dfrac{2}{3}$

$y - 4 = \dfrac{2}{3}(x - 24)$

$y - 4 = \dfrac{2}{3}x - 16$

$+4 +4$

$y = \dfrac{2}{3}x - 12$

38.

$(0, -9)$ and $(9, -3)$

Dashed line, shaded below, means to use the $<$ symbol.

$m = \dfrac{-3 - (-9)}{9 - 0} = \dfrac{6}{9} = \dfrac{2}{3}$

$y < \dfrac{2}{3}x - 9$

39. Let $P(t)$ be the population of Detroit Michigan in thousands t years since 2000.

a.

$(6, 4487)$ and $(8, 4425)$

$P(t) = mt + b$

$m = \dfrac{4425 - 4487}{8 - 6} = \dfrac{-62}{2} = -31$

$4487 = (-31)(6) + b$

$4487 = -186 + b$

$b = 4673$

$P(t) = -31t + 4673$

b.

$P(15) = -31(15) + 4673$

$ = 4208$

In 2015, the population of Detroit Michigan will be 4208 thousand.

c. The slope is -31. The population of Detroit Michigan is decreasing by 31 thousand each year.

40. Let x be the number of ml of 15% HCl solution and y be the number of ml of 60% HCl solution used to get the 30 ml of a 40% HCl solution. Then:

$\begin{cases} x + y = 30 \\ 0.15x + 0.60y = (0.40)(30) = 12 \end{cases}$

$\begin{cases} x + y = 30 \\ \left(\dfrac{-1}{0.15}\right)(0.15x + 0.60y) = \left(\dfrac{-1}{0.15}\right)(12) \end{cases}$

$\begin{cases} x + y = 30 \\ -x - 4y = -80 \end{cases}$

$-3y = -50$

$\dfrac{-3y}{-3} = \dfrac{-50}{-3}$

$y = 16\tfrac{2}{3}$

Now solve for x:

$x = 30 - 16\tfrac{2}{3}$

$x = 13\tfrac{1}{3}$

Sarah needs $13\tfrac{1}{3}$ ml of 15% HCl solution and $16\tfrac{2}{3}$ ml of 60% HCl solution to make 30 ml of 40% HCl solution.

41. Let $W(t)$ be the number of women triathletes in the U.S. t years since 1990.

a.

t	$W(t)$
0	1,100
5	4,600
10	13,000
13	19,100

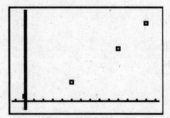

Choose $(0, 1100)$ as the vertex and $(13, 19100)$ as the additional point used to solve for "a". Then:

$W(t) = a(t-0)^2 + 1{,}100$

$19{,}100 = a(13)^2 + 1{,}100$

$\underline{-1{,}100 -1{,}100}$

$18{,}000 = 169a$

$\dfrac{18{,}000}{169} = \dfrac{169a}{169} \Big\} \ a \approx 106.5$

$W(t) = 106.5t^2 + 1{,}100$

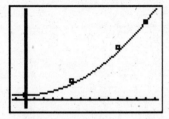

b. Domain: $[0, 20]$

$W(0) = 106.5(0)^2 + 1{,}100 = 1{,}100$

$W(20) = 106.5(20)^2 + 1{,}100 = 43{,}700$

Range: $[1100, 43700]$

c.

$W(15) = 106.5(15)^2 + 1{,}100$

$ \approx 25{,}062.5$

In 2005 there were approximately 25,063 women triathletes in the United States.

d. The vertex is $(0, 1100)$. In 1990, there were 1100 women triathletes in the United States. This was the lowest number during this time.

e.

$40{,}000 = 106.5t^2 + 1{,}100$

$\underline{-1{,}100 -1{,}100}$

$ 38{,}900 = 106.5t^2$

$ \dfrac{38{,}900}{106.5} = \dfrac{106.5t^2}{106.5}$

$ \dfrac{38{,}900}{106.5} = t^2$

$ \sqrt{\dfrac{38{,}900}{106.5}} = t^2$

$ 19.1 \approx t$

In 2009, there were about 40,000 women triathletes in the United States.

42.

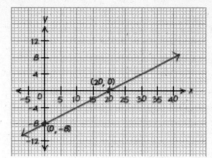

43.

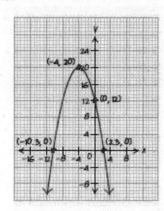

44.

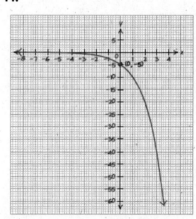

45.

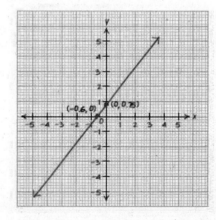

46.

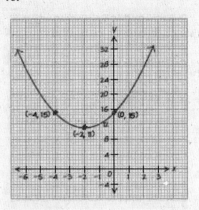

47.

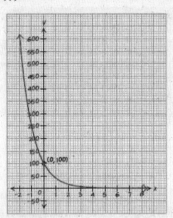

48.

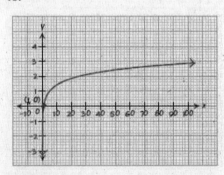

49.

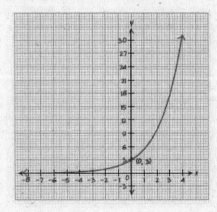

50.

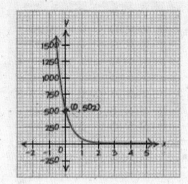

51.

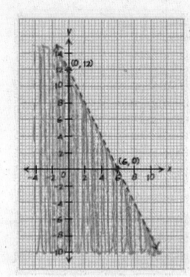

52.

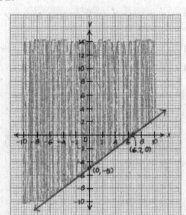

53.

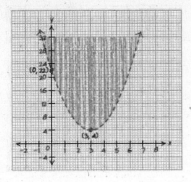

54.
$$(4d-8)+(7d+2) = 4d+7d-8+2$$
$$= 11d-6$$

55.
$$(2x+5)(7x-8) = 14x^2 -16x+35x-40$$
$$= 14x^2 +19x-40$$

56.
$$(3x^2+4x-12)-(x^2+9x-20) = 3x^2-x^2+4x-9x-12+20$$
$$= 2x^2-5x+8$$

57.
$$(3t-4)^2 = (3t-4)(3t-4)$$
$$= 9t^2-12t-12t+16$$
$$= 9t^2-24t+16$$

58.
$$4p^2+(2p+3)(5p-8) = 4p^2+10p^2-16p+15p-24$$
$$= 14p^2-p-24$$

59.

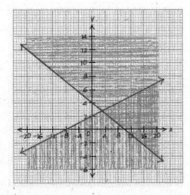

60.
$$2x^2-11x-21 = (2x+3)(x-7)$$

61.
$$25r^2-64t^2 = (5r+8t)(5r-8t)$$

62.
$$42g^3-7g^2-280g = 7g(6g^2-g-40)$$
$$= 7g(3g-8)(2g+5)$$

63.
$$t^3+343 = t^3+7^3$$
$$= (t+7)(t^2-7t+7^2)$$
$$= (t+7)(t^2-7t+49)$$

64.
$$4x^2-12x+9 = (2x-3)(2x-3)$$
$$= (2x-3)^2$$

65. 2nd degree polynomial

66.
$$f(x) = 3x^2+24x-10$$
$$f(x) = 3(x^2+8x)-10$$
$$f(x) = 3(x^2+8x+\Box-\Box)-10$$

Completing the square: $\left(\dfrac{8}{2}\right)^2 = 16$

$$f(x) = 3(x^2+8x+\boxed{16}-\boxed{16})-10$$
$$f(x) = 3(x^2+8x+\boxed{16})-(3)\boxed{16}-10$$
$$f(x) = 3(x^2+8x+16)-48-10$$
$$f(x) = 3(x+4)^2-58$$

67. $P(t) = 0.0237t^2+0.507t+8.384$

a.
$$P(20) = 0.0237(20)^2+0.507(20)+8.384$$
$$P(20) = 28.004$$

The average price for baseball tickets in 2010 is $28.00.

b. Vertex (h,k) with $h = \dfrac{-b}{2a}$ and $k = P(h)$.

$$h = \dfrac{-0.507}{2(0.0237)} \approx -10.69$$

$$k = P(-10.69)$$
$$k = 0.0237(-10.69)^2+0.507(10.69)+8.384$$
$$k \approx 5.67$$

The vertex is $(-10.7, 5.67)$ In 1979, the average price for baseball tickets was $5.67. This was a minimum for this model.

Chapter 6 Logarithmic Functions

c.

$$40.77 = 0.0237t^2 + 0.507t + 8.384$$
$$\underline{-40.77 \qquad\qquad\qquad -40.77}$$
$$0 = 0.0237t^2 + 0.507t - 32.386$$

$$t = \frac{-0.507 \pm \sqrt{(0.507)^2 - 4(0.0237)(-32.386)}}{2(0.0237)}$$

$$t = \frac{-0.507 - \sqrt{3.3272418}}{0.0474} \approx \cancel{-49.18}$$

$$t = \frac{-0.507 + \sqrt{3.3272418}}{0.0474} \approx 27.79$$

In 2017, the average price for baseball tickets will be $40.77.

68. Let $T(m)$ be the number of thorium-233 atoms after m minutes. Let n be the number of 22-minute time intervals.

a.

m	$T(m)$
0	5000
22	$5000\left(\frac{1}{2}\right)$
2(22)	$5000\left(\frac{1}{2}\right)^2$
3(22)	$5000\left(\frac{1}{2}\right)^3$
$n(22)$	$5000\left(\frac{1}{2}\right)^n$

$$n(22) = m$$
$$\frac{n(22)}{22} = \frac{m}{22}$$
$$n = \frac{m}{22}$$
$$T(m) = 5000\left(\frac{1}{2}\right)^{m/22}$$

b.

$$1000 = 5000\left(\frac{1}{2}\right)^{m/22}$$

$$\frac{1000}{5000} = \frac{5000\left(\frac{1}{2}\right)^{m/22}}{5000}$$

$$0.2 = (0.5)^{m/22}$$

$$\frac{m}{22} = \log_{0.5}(0.2)$$

$$m = 22\log_{0.5}(0.2)$$

$$m = \frac{22\log(0.2)}{\log(0.5)} \approx 51.08$$

After about 51 minutes there will be only 1000 of the thorium-233 atoms left.

c.

3 hours → 180 minutes

$$T(180) = 5000\left(\frac{1}{2}\right)^{180/22}$$

$$T(180) \approx 17.22$$

By the end of the experiment, there will be only 17 of the thorium-233 atoms left.

69.

$$f(x) = 3x + 12$$
$$y = 3x + 12$$
$$y - 12 = 3x$$
$$\frac{1}{3}(y - 12) = \frac{1}{3}(3x)$$
$$\frac{1}{3}y - 4 = x$$
$$\frac{1}{3}x - 4 = y$$
$$f^{-1}(x) = \frac{1}{3}x - 4$$

70.
$$g(x) = 4(3)^x$$
$$y = 4(3)^x$$
$$\frac{y}{4} = \frac{4(3)^x}{4}$$
$$\frac{y}{4} = (3)^x$$
$$\log_3\left(\frac{y}{4}\right) = x$$
$$\log_3\left(\frac{x}{4}\right) = y$$
$$g^{-1}(x) = \log_3\left(\frac{x}{4}\right)$$
$$g^{-1}(x) = \frac{\log\left(\frac{x}{4}\right)}{\log 3}$$

71.
$$6 = -\log(H^+)$$
$$-6 = \log(H^+)$$
$$10^{-6} = H^+$$
$$H^+ = 1 \times 10^{-6} \text{ M}$$

72. Let $P(t)$ be the Muslim population in Israel (in millions) t years since 2008.

a.
$$P(t) = ab^t$$
$a = 1.24$ and $r = 2.8\% = 0.028$
$b = 1 + r$ so $b = 1.028$
Then $P(t) = 1.24(1.028)^t$

b.
$2012 \to t = 4$
$$P(4) = 1.24(1.028)^4$$
$$P(4) \approx 1.38$$

The Muslim population in Israel in 2012 will be 1.38 million.

c.
$$2 = 1.24(1.028)^t$$
$$\frac{2}{1.24} = \frac{1.24(1.028)^t}{1.24}$$
$$\frac{2}{1.24} = (1.028)^t$$
$$t = \log_{1.028}\left(\frac{2}{1.24}\right)$$
$$t = \frac{\log\left(\frac{2}{1.24}\right)}{\log(1.028)} \approx 17.3$$

Seventeen years past 2008 is the year 2025.

The Muslim population in Israel will reach 2 million by the year 2025.

73.
$$(4m^5 n^3)(15mn^2) = 4 \cdot 15 \cdot m^{5+1} \cdot n^{3+2}$$
$$= 60m^6 n^5$$

74.
$$(-3g^3 h^{-4})^4 = (-3)^4 (g^3)^4 (h^{-4})^4$$
$$= 81 g^{12} h^{-16}$$
$$= \frac{81 g^{12}}{h^{16}}$$

75.
$$3(17x^3 y^7)^0 + 5(12x^4 y^6)^0 = 3(1) + 5(1) = 8$$

76.
$$\frac{300 x^4 y^{-3} z^2}{125 x^2 y^4 z^{-5}} = \frac{12 \cdot \cancel{25} x^{4-2} y^{-3-4} z^{2-(-5)}}{5 \cdot \cancel{25}}$$
$$= \frac{12 x^2 y^{-7} z^7}{5}$$
$$= \frac{12 x^2 z^7}{5 y^7}$$

77.
$$(125 a^3 b^9)^{-\frac{1}{3}} (4ab^{-5})^2 = (125)^{-\frac{1}{3}} (a^3)^{-\frac{1}{3}} (b^9)^{-\frac{1}{3}} (4)^2 (a)^2 (b^{-5})^2$$
$$= \frac{1}{5} a^{-1} b^{-3} \cdot 16 a^2 b^{-10}$$
$$= \frac{16 ab^{-13}}{5}$$
$$= \frac{16a}{5b^{13}}$$

Chapter 6 Logarithmic Functions

78.

$$\left(\frac{16m^{-5}n^3}{81m^{-1}n^{-5}}\right)^{-\frac{1}{4}} = \left(\frac{16m^{-5-(-1)}n^{3-(-5)}}{81}\right)^{-\frac{1}{4}}$$

$$= \left(\frac{16m^{-4}n^8}{81}\right)^{-\frac{1}{4}}$$

$$= \frac{16^{-\frac{1}{4}}\left(m^{-4}\right)^{-\frac{1}{4}}\left(n^8\right)^{-\frac{1}{4}}}{81^{-\frac{1}{4}}}$$

$$= \frac{3mn^{-2}}{2} = \frac{3m}{2n^2}$$

79. $a > 0$. The graph is above the x-axis.

80. $b > 1$. The graph is increasing and $a > 0$.

81. Domain: $(-\infty, \infty)$ Range: $(0, \infty)$

82. $a > 0$. The graph is facing upward.

83. $(h, k) = (-1, -2)$, where (h, k) is the vertex of the parabola.

84. $f(2.5) = 18$

85.
$f(2) = 12$
$f(-4) = 12$

86. Domain: $(-\infty, \infty)$ Range: $[-2, \infty)$

87. $m > 0$. The slope is increasing.

88. $b = -1$. The y-intercept is $(0, -1)$.

89. $m = \dfrac{2}{3}$

90. $f(3) = 1$

91. Domain: $(-\infty, \infty)$ Range: $(-\infty, \infty)$

Section 7.1

1. Let $C(p)$ be the per player cost in dollars for a charity poker tournament if p players participate.

a. $C(p) = \dfrac{600}{p}$

b. $C(75) = \dfrac{600}{75} = 8$ The per person cost for 75 players to participate is $8 per person.

c. Domain: $2 \le p \le 100$ Range: $6 \le C \le 300$

3.

a.
$C(p) = 4.55p + 365.00$
$C(100) = 4.55(100) + 365.00$
$C(100) = 820$

The total cost for 100 people to attend the event is $820.

b. $\dfrac{C(100)}{100} = \dfrac{820}{100} = 8.20$ The per person cost for 100 people to attend the event is $8.20 per person.

c. Let $F(p)$ be the per person cost in dollars if p people attend the charity event. Then $F(p) = \dfrac{4.55p + 365.00}{p}$.**d.**

$F(150) = \dfrac{4.55(150) + 365.00}{150}$ The per person cost for 150
$F(150) \approx 6.98$

people to attend the event is $6.98 per person.

e. Domain: $1 \le p \le 250$ Range: $6.01 \le F \le 369.55$

5. Let $C(n)$ be the cost in dollars for a room at this particular hotel if you stay for n nights.

a.
$C(n) = kn$
Use $C(3) = 450$
to solve for k.
$450 = k(3)$
$\dfrac{450}{3} = \dfrac{k(3)}{3}$
$150 = k$
so
$C(n) = 150n$

b.
$C(7) = 150(7)$
$C(7) = 1050$

The total cost for a 7 night stay at this hotel is $1050.

c.
$800 = 150n$
$\dfrac{800}{150} = \dfrac{150n}{150}$
$5.33 \approx n$

With a budget of $800, you can stay 5 nights at this hotel.

7. Let $V(t)$ be the volume in liters it takes to store the helium if the temperature is t Kelvin.

a.
$V(t) = kt$
Use $V(400) = 0.821$
to solve for k.
$0.821 = k(400)$
$\dfrac{0.821}{400} = \dfrac{k(400)}{400}$
$0.0020525 = k$
so
$V(t) = 0.0020525t$

b.
$V(250) = 0.0020525(250)$
$V(250) \approx 0.513$

It takes 0.513 L to store 0.50 moles of helium at 250 K.

c.
$0.75 = 0.0020525t$
$\dfrac{0.75}{0.0020525} = \dfrac{0.0020525t}{0.0020525}$ If you have 0.75 L to store the
$t \approx 365.4$

helium, the temperature needs to be at 365.4 K.

9. Let $I(d)$ be the illumination (in foot-candles) of a light source d feet from the source.

a.
$I(d) = \dfrac{k}{d^2}$
Use $I(10) = 25.5$ to solve for k.
$25.5 = \dfrac{k}{(10)^2}$
$(100)(25.5) = \left(\dfrac{k}{100}\right)(100)$
$2550 = k$
so $I(d) = \dfrac{2550}{d^2}$

b.

$I(5) = \dfrac{2550}{(5)^2}$

$I(5) = \dfrac{2550}{25} = 102$

The illumination of this light at a distance of 5 feet from the source is 102 foot-candles.

c.

$I(30) = \dfrac{2550}{(30)^2}$

$I(30) = \dfrac{2550}{900}$

$I(30) \approx 2.83$

The illumination of this light at a distance of 30 feet from the source is 2.83 foot-candles.

d.

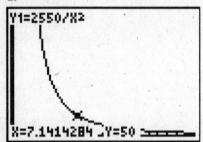

An illumination of 50 foot-candles occurs at a distance of approximately 7.1 feet from the source.

11. Let $W(d)$ be the weight of a body (in pounds) that is a distance d miles from the center of the earth.

$W(d) = \dfrac{k}{d^2}$

Use $W(4000) = 220$ to solve for k.

$220 = \dfrac{k}{(4000)^2}$

$((4000)^2)(220) = \left(\dfrac{k}{(4000)^2}\right)((4000)^2)$

$3.52 \times 10^9 = k$

so

$W(d) = \dfrac{3.52 \times 10^9}{d^2}$

$2000 + 4000 = 6000$ from the center of the earth

$W(6000) = \dfrac{3.52 \times 10^9}{(6000)^2} \approx 97.78$

A 220 pound man weighs only 97.78 pounds when he is 2000 miles above the earth.

13. Let $I(r)$ be the current in amps that flows through an electrical circuit with resistance r ohms.

$I(r) = \dfrac{k}{r}$

Use $I(200) = 1.2$ to solve for k.

$1.2 = \dfrac{k}{200}$

$(200)(1.2) = \left(\dfrac{k}{200}\right)(200)$

$240 = k$

so

$I(r) = \dfrac{240}{r}$

If $r = 130$, then $I(130) = \dfrac{240}{130} \approx 1.8$.

The current is 1.8 amps when the resistance is 130 ohms.

15.

a. $P(2) = \dfrac{30}{2} = 15$ The pressure of the gas is 15 pounds per square inch when the volume of the balloon is 2 cubic inches.

b. $P(10) = \dfrac{30}{10} = 3$ The pressure of the gas is 3 pounds per square inch when the volume of the balloon is 10 cubic inches.

17.

a.

$y = kx^3$

Use $y = 150.4$ when $x = 4$ to solve for k.

$150.4 = k4^3$

$\dfrac{150.4}{64} = \dfrac{k(64)}{64}$

$2.35 = k$

so

$y = 2.35x^3$

b.

$y = 2.35(8)^3$

$y = 1203.2$

19.
a.
$$y = \frac{k}{x^3}$$
Use $y = 405$ when $x = 3$ to solve for k.
$$405 = \frac{k}{3^3} = \frac{k}{27}$$
$$(27)(405) = \frac{k}{27}(27)$$
$$10935 = k$$
so
$$y = \frac{10{,}935}{x^3}$$
b.
$$y = \frac{10935}{5^3}$$
$$y = 87.48$$

21.
a.
$$M = \frac{k}{5\sqrt{t}}$$
Use $M = 115$ when $t = 9$ to solve for k.
$$115 = \frac{k}{5\sqrt{9}} = \frac{k}{5(3)}$$
$$(15)(115) = \frac{k}{15}(15)$$
$$1725 = k$$
so
$$M = \frac{1725}{5\sqrt{t}}$$
b.
$$M = \frac{1725}{5\sqrt{25}}$$
$$M = 69$$

23. $B(t) = \dfrac{-470{,}001t^2 + 4{,}110{,}992t + 14{,}032{,}612}{-469.4t^2 + 3745t + 19{,}774}$

a.
$$B(5) = \frac{-470{,}001(5)^2 + 4{,}110{,}992(5) + 14{,}032{,}612}{-469.4(5)^2 + 3745(5) + 19{,}774}$$
$$B(5) \approx 853.293$$

The average benefit for a person participating in the food stamp program in 1995 was $853.29.

b.

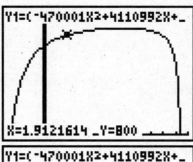

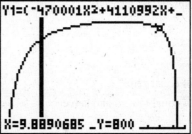

The average benefit for a person participating in the food stamp program was approximately $800 in 1992 and again in 2000.

25.

a. Let $N(t)$ be the average amount of national debt (per person in dollars) in the United States t years since 1970.
$$N(t) = \frac{D(t)}{P(t)}$$
$$N(t) = \frac{8215.1t^2 - 23{,}035.4t + 413{,}525.6}{0.226t^2 + 1.885t + 204.72}$$

b.
$$N(30) = \frac{8215.1(30)^2 - 23{,}035.4(30) + 413{,}525.6}{0.226(30)^2 + 1.885(30) + 204.72}$$
$$N(30) \approx 15{,}314.209$$

The average amount of national debt per person in 2000 was $15,314.21.

Chapter 7 Rational Functions

c.

```
  X      Y1
20.446  9998.2
20.447  9998.8
20.448  9999.4
20.449  10000
20.45   10001
20.451  10001
20.452  10002
X=20.449
```

The average amount of national debt per person was $10,000 midway through 1990.

27.

$g(x) = \dfrac{25}{x}$

The denominator cannot equal zero.

Set the linear factors not equal to zero and solve:

$x \neq 0$

Domain: x is all real numbers except $x \neq 0$.

29.

$f(x) = \dfrac{x+5}{x-3}$

The denominator cannot equal zero.

Set the linear factors not equal to zero and solve:

$x - 3 \neq 0$

$x \neq 3$

Domain: x is all real numbers except $x \neq 3$.

31.

$m(b) = \dfrac{b+9}{b+11}$

The denominator cannot equal zero.

Set the linear factors not equal to zero and solve:

$b + 11 \neq 0$

$b \neq -11$

Domain: b is all real numbers except $b \neq -11$.

33.

$p(t) = \dfrac{3t-7}{4t+5}$

The denominator cannot equal zero.

Set the linear factors not equal to zero and solve:

$4t + 5 \neq 0$

$t \neq -\dfrac{5}{4}$

Domain: t is all real numbers except $t \neq -\dfrac{5}{4}$.

35.

$h(a) = \dfrac{3a-1}{(2a+7)(a-3)}$

The denominator cannot equal zero.

Set the linear factors not equal to zero and solve:

$2a + 7 \neq 0,\ a - 3 \neq 0$

$a \neq -\dfrac{7}{2},\ a \neq 3$

Domain: a is all real numbers except $a \neq -\dfrac{7}{2},\ a \neq 3$.

37.

$C(n) = \dfrac{n+10}{(n+4)(n+10)}$

The denominator cannot equal zero.

Set the linear factors not equal to zero and solve:

$n + 4 \neq 0,\ n + 10 \neq 0$

$n \neq -4,\ n \neq -10$

Domain: n is all real numbers except $n \neq -4,\ n \neq -10$.

39.

$h(m) = \dfrac{4m^2 + 2m - 9}{m^2 + 7m + 12} = \dfrac{4m^2 + 2m - 9}{(m+3)(m+4)}$

The denominator cannot equal zero.

Set the linear factors not equal to zero and solve:

$m + 3 \neq 0$ and $m + 4 \neq 0$

$m \neq -3$ and $m \neq -4$

Domain: m is all real numbers except $m \neq -3$ and $m \neq -4$.

41.

$f(x) = \dfrac{2x+1}{x^2 + 3x + 19}$

The denominator cannot equal zero.

Set the linear factors not equal to zero and solve:

$x^2 + 3x + 19 \neq 0$

$x \neq \dfrac{-3 \pm \sqrt{3^2 - 4(1)(19)}}{2(1)}$

$x \neq \dfrac{-3 \pm \sqrt{-67}}{2}$

This is an imaginary number, meaning there is no real number such that the denominator will equal zero.

Domain: x is all real numbers.

43.

$h(t) = \dfrac{t+7}{t^2-49} = \dfrac{t+7}{(t+7)(t-7)}$

The denominator cannot equal zero.
Set the linear factors not equal to zero and solve:
$t+7 \neq 0$ and $t-7 \neq 0$
$t \neq -7$ and $t \neq 7$
Domain: t is all real numbers except $t \neq -7$ and $t \neq 7$.

45.

$b(r) = \dfrac{r+3}{r^2+5r+6} = \dfrac{r+3}{(r+3)(r+2)}$

The denominator cannot equal zero.
Set the linear factors not equal to zero and solve:
$r+3 \neq 0$ and $r+2 \neq 0$
$r \neq -3$ and $r \neq -2$
Domain: r is all real numbers except $r \neq -3$ and $r \neq -2$.

47.

$h(t) = \dfrac{t^2-3t+5}{t^2-7t+10} = \dfrac{t^2-3t+5}{(t-5)(t-2)}$

The denominator cannot equal zero.
Set the linear factors not equal to zero and solve:
$t-5 \neq 0$ and $t-2 \neq 0$
$t \neq 5$ and $t \neq 2$
Domain: t is all real numbers except $t \neq 5$ and $t \neq 2$.

49.

$P(a) = \dfrac{a^2+2a+1}{a^2+9}$

The denominator cannot equal zero.
Set the linear factors not equal to zero and solve:
$a^2 + 9 \neq 0$

$a \neq \dfrac{0 \pm \sqrt{0^2 - 4(1)(9)}}{2(1)}$

$a \neq \dfrac{\pm\sqrt{-36}}{2}$

This is an imaginary number, meaning there is no real number such that the denominator will equal zero.
Domain: a is all real numbers.

51.
 a. Domain: x is all real numbers but $x \neq -5$ and $x \neq 7$.
 b. $f(0) = 0$
 c. $f(x) = 5$ when $x \approx -4.5$ and $x \approx 7.5$.

53.
 a. Domain: x is all real numbers but $x \neq -3$ and $x \neq 4$.
 b. $h(1) \approx -1$
 c. $h(-4) \approx 1.9$
 d. $h(x) = -2$ when $x \approx -1.9$ and $x \approx 2.8$.

55. Domain: x is all real numbers but $x \neq -3$ and $x \neq -5$.

57. Domain: x is all real numbers but $x \neq -4$ and $x \neq 6$.

59. Domain: x is all real numbers but $x \neq -3$ and $x \neq -5$.

61. Domain: x is all real numbers but $x \neq -3$.
The value represents a vertical asymptote because the factor $(x+3)$ does not simplify with the numerator.

63. Domain: x is all real numbers but $x \neq -3$ and $x \neq -1$.
The values represent vertical asymptotes because the factors $(x+3)$ and $(x+1)$ do not simplify with the numerator.

65. Domain: x is all real numbers but $x \neq -5$ and $x \neq -2$.
The value $x = -5$ represents a vertical asymptote because the factor $(x+5)$ does not simplify with the numerator.
The value $x = -2$ represents a hole because the factor $(x+2)$ simplifies with the numerator.

67.

$g(x) = \dfrac{3x+9}{6x^2+7x-20} = \dfrac{3(x+3)}{(3x-4)(2x+5)}$

Domain: x is all real numbers but $x \neq \dfrac{4}{3}$ and $x \neq -\dfrac{5}{2}$.

The values represent vertical asymptotes because the factors $(3x-4)$ and $(2x+5)$ do not simplify with the numerator.

69.

$f(x) = \dfrac{x+2}{x^2+3x+2} = \dfrac{x+2}{(x+1)(x+2)}$

Domain: x is all real numbers but $x \neq -1$ and $x \neq -2$.

The value $x = -1$ represents a vertical asymptote because the factor $(x+1)$ does not simplify with the numerator.
The value $x = -2$ represents a hole because the factor $(x+2)$ simplifies with the numerator.

Chapter 7 Rational Functions

71.

$$h(x) = \frac{5x}{x^3 - 2x^2 - 8x} = \frac{5x}{x(x+2)(x-4)}$$

Domain: x is all real numbers but $x \neq 0, x \neq -2$ and $x \neq 4$.

The values $x = -2$ and $x = 4$ represent vertical asymptotes because the factors $(x + 2)$ and $(x - 4)$ do not simplify with the numerator. The value $x = 0$ represents a hole because the factor (x) simplifies with the numerator.

73.

$$g(x) = \frac{3x + 12}{x^2 - 16} = \frac{3(x+4)}{(x-4)(x+4)}$$

Domain: x is all real numbers but $x \neq 4$ and $x \neq -4$.

The value $x = 4$ represents a vertical asymptote because the factor $(x - 4)$ does not simplify with the numerator.
The value $x = -4$ represents a hole because the factor $(x + 4)$ simplifies with the numerator.

75.

$$g(x) = \frac{x+5}{x^2 + 25}$$

The denominator cannot equal zero.
Set the linear factors not equal to zero and solve:

$$x^2 + 25 \neq 0$$

$$x \neq \frac{0 \pm \sqrt{0^2 - 4(1)(25)}}{2(1)}$$

$$x \neq \frac{\pm\sqrt{-100}}{2}$$

This is an imaginary number, meaning there is no real number such that the denominator will equal zero.
Domain: x is all real numbers.

Section 7.2

1.
$$\frac{20x^3}{14x} = \frac{2 \cdot 10 \cdot x \cdot x \cdot x}{2 \cdot 7 \cdot x}$$
$$= \frac{\cancel{2} \cdot 10 \cdot \cancel{x} \cdot x \cdot x}{\cancel{2} \cdot 7 \cdot \cancel{x}}$$
$$= \frac{10x^2}{7}$$

3.
$$\frac{12(x+9)}{(x+5)(x+9)} = \frac{12\cancel{(x+9)}}{(x+5)\cancel{(x+9)}}$$
$$= \frac{12}{x+5}$$

5.
$$\frac{x+3}{2x+6} = \frac{x+3}{2(x+3)}$$
$$= \frac{\cancel{(x+3)}}{2\cancel{(x+3)}}$$
$$= \frac{1}{2}$$

7.
$$\frac{8-x}{2x-16} = \frac{-1(x-8)}{2(x-8)}$$
$$= \frac{-1\cancel{(x-8)}}{2\cancel{(x-8)}}$$
$$= -\frac{1}{2}$$

9.
$$\frac{4x-24}{18-3x} = \frac{4(x-6)}{-3(x-6)}$$
$$= \frac{4\cancel{(x-6)}}{-3\cancel{(x-6)}}$$
$$= -\frac{4}{3}$$

11.
$$\frac{(x+2)(x-5)}{(5-x)(x-7)} = \frac{(x+2)(x-5)}{-1(x-5)(x-7)}$$
$$= \frac{(x+2)\cancel{(x-5)}}{-1\cancel{(x-5)}(x-7)}$$
$$= \frac{(x+2)}{-1(x-7)}$$
$$= \frac{x+2}{7-x}$$

13.
$$\frac{(x+5)(x-3)}{(x-3)(x-7)} = \frac{(x+5)\cancel{(x-3)}}{\cancel{(x-3)}(x-7)}$$
$$= \frac{x+5}{x-7}$$

15.
$$\frac{x^2+6x+9}{(x+2)(x+3)} = \frac{(x+3)(x+3)}{(x+2)(x+3)}$$
$$= \frac{(x+3)\cancel{(x+3)}}{(x+2)\cancel{(x+3)}}$$
$$= \frac{x+3}{x+2}$$

17.
$$\frac{t^2+2t-15}{t^2-9t+18} = \frac{(t+5)(t-3)}{(t-6)(t-3)}$$
$$= \frac{(t+5)\cancel{(t-3)}}{(t-6)\cancel{(t-3)}}$$
$$= \frac{t+5}{t-6}$$

19.
$$\frac{w^2-16}{w^2+w-12} = \frac{(w+4)(w-4)}{(w+4)(w-3)}$$
$$= \frac{\cancel{(w+4)}(w-4)}{\cancel{(w+4)}(w-3)}$$
$$= \frac{w-4}{w-3}$$

21.
$$\frac{5(x+7)}{(x+7)+10} = \frac{5(x+7)}{x+17}$$

Chapter 7 Rational Functions

23.
$$\frac{m+3}{2(m+3)-5} = \frac{m+3}{2m+6-5}$$
$$= \frac{m+3}{2m+1}$$

25.
$$\frac{2x^2-23x-70}{x^2-21x+98} = \frac{(2x+5)(x-14)}{(x-7)(x-14)}$$
$$= \frac{(2x+5)\cancel{(x-14)}}{(x-7)\cancel{(x-14)}}$$
$$= \frac{2x+5}{x-7}$$

27.
$$\frac{10x^3+5x^2-4x}{5x} = \frac{10x^3}{5x}+\frac{5x^2}{5x}-\frac{4x}{5x}$$
$$= 2x^2+x-\frac{4}{5}$$

29.
$$\frac{3x^4-8x^3+x^2}{4x^3} = \frac{3x^4}{4x^3}-\frac{8x^3}{4x^3}+\frac{x^2}{4x^3}$$
$$= \frac{3x}{4}-2+\frac{1}{4x}$$

31.
$$\frac{10a^3b^2+12a^2b^3-14ab^2}{2ab} = \frac{10a^3b^2}{2ab}+\frac{12a^2b^3}{2ab}-\frac{14ab^2}{2ab}$$
$$= 5a^2b+6ab^2-7b$$

33.
$$\frac{10g^5h^3-30g^4h^2+20g^2h}{5g^2h^2} = \frac{10g^5h^3}{5g^2h^2}-\frac{30g^4h^2}{5g^2h^2}+\frac{20g^2h}{5g^2h^2}$$
$$= 2g^3h-6g^2+\frac{4}{h}$$

35. Tom should have factored the numerator.
$$\frac{x^2+6x+8}{x+4} = \frac{(x+4)(x+2)}{x+4}$$
$$= \frac{\cancel{(x+4)}(x+2)}{\cancel{(x+4)}}$$
$$= \frac{x+2}{1}$$
$$= x+2$$

37. Amy did not distribute the negative through the binomial on the second step.
The binomial $-6x^2-15x$ should have been $-6x^2-15x$.

$$\begin{array}{r} 3x+7 \\ 2x-5\overline{)6x^2-x-35} \\ \underline{-6x^2+15x} \\ 14x-35 \\ \underline{-14x+35} \\ 0 \end{array}$$

39. The length is $7x-24$.
$$\begin{array}{r} 7x-24 \\ 3x+5\overline{)21x^2-37x-120} \\ \underline{-21x^2-35x} \\ -72x-120 \\ \underline{72x+120} \\ 0 \end{array}$$

41. The length is $5x^2+7x+3$.

$$\begin{array}{r|rrrr} -4 & 5 & 27 & 31 & 12 \\ & & -20 & -28 & -12 \\ \hline & 5 & 7 & 3 & 0 \end{array}$$

$$(5x^3+27x^2+31x+12)\div(x+4) = 5x^2+7x+3$$

43.
$$\begin{array}{r} 6x+5 \\ x+4\overline{)6x^2+29x+20} \\ \underline{-6x^2-24x} \\ 5x+20 \\ \underline{-5x-20} \\ 0 \end{array}$$

$$(6x^2+29x+20)\div(x+4) = 6x+5$$

45.
$$\begin{array}{r} x^2+5x-4 \\ x+2\overline{)x^3+7x^2+6x-8} \\ \underline{-x^3-2x^2} \\ 5x^2+6x \\ \underline{-5x^2-10x} \\ -4x-8 \\ \underline{4x+8} \\ 0 \end{array}$$

$$(x^3+7x^2+6x-8)\div(x+2) = x^2+5x-4$$

47.

$$x^2+2\overline{)3x^3+5x^2+6x+10}3x+5$$

$$\underline{-(3x^3+0x^2+6x)}$$
$$5x^2+0x+10$$
$$\underline{-(5x^2+0x+10)}$$
$$0$$

$(3x^3+5x^2+6x+10)\div(x^2+2)=3x+5$

49.

$$b^2-7\overline{)4b^4-3b^3-23b^2+21b-35}4b^2-3b+5$$
$$\underline{-(4b^4+0b^3-28b^2)}$$
$$-3b^3+5b^2+21b$$
$$\underline{-(-3b^3+0b^2+21b)}$$
$$5b^2+0b-35$$
$$\underline{-(5b^2+0b-35)}$$
$$0$$

$(4b^4-3b^3-23b^2+21b-35)\div(b^2-7)$
$=4b^2-3b+5$

51.

$$n^2+3n+2\overline{)5n^4+19n^3+14n^2-16n-16}5n^2+4n-8$$
$$\underline{-(5n^4+15n^3+10n^2)}$$
$$4n^3+4n^2-16n$$
$$\underline{-(4n^3+12n^2+8n)}$$
$$-8n^2-24n-16$$
$$\underline{-(-8n^2-24n-16)}$$
$$0$$

$(5n^4+19n^3+14n^2-16n-16)\div(n^2+3n+2)$
$=5n^2+4n-8$

53.

$$\begin{array}{r|rrr} 7 & 1 & -3 & -28 \\ & & 7 & 28 \\ \hline & 1 & 4 & 0 \end{array}$$

$(x^2-3x-28)\div(x-7)=x+4$

55.

$$\begin{array}{r|rrr} 6 & 2 & -5 & -42 \\ & & 12 & 42 \\ \hline & 2 & 7 & 0 \end{array}$$

$(2m^2-5m-42)\div(m-6)=2m+7$

57.

$$\begin{array}{r|rrr} -3 & 4 & 17 & 15 \\ & & -12 & -15 \\ \hline & 4 & 5 & 0 \end{array}$$

$(4x^2+17x+15)\div(x+3)=4x+5$

59.

$$\begin{array}{r|rrr} -4 & 3 & 17 & 25 \\ & & -12 & -20 \\ \hline & 3 & 5 & 5 \end{array}$$

$(3x^2+17x+25)\div(x+4)=3x+5+\dfrac{5}{x+4}$

61.

$$2x+3\overline{)12x^2+32x+21}6x+7$$
$$\underline{-(12x^2+18x)}$$
$$14x+21$$
$$\underline{-(14x+21)}$$
$$0$$

$(12x^2+32x+21)\div(2x+3)=6x+7$

63.

$$\begin{array}{r|rrrr} -4 & 2 & 11 & 17 & 20 \\ & & -8 & -12 & -20 \\ \hline & 2 & 3 & 5 & 0 \end{array}$$

$(2x^3+11x^2+17x+20)\div(x+4)$
$=2x^2+3x+5$

65.

$$x^2-5 \overline{\smash{\big)}\, x^4+2x^3+0x^2-10x-25}$$
$$\underline{-(x^4+0x^3-5x^2)}$$
$$2x^3+5x^2-10x$$
$$\underline{-(2x^3+0x^2-10x)}$$
$$5x^2+0x-25$$
$$\underline{-(5x^2+0x-25)}$$
$$0$$

Quotient: x^2+2x+5

$(x^4+2x^3-10x-25) \div (x^2-5)$
$= x^2+2x+5$

67.

$$x^2+4 \overline{\smash{\big)}\, x^4-3x^3+8x^2-12x+16}$$
$$\underline{-(x^4+0x^3+4x^2)}$$
$$-3x^3+4x^2-12x$$
$$\underline{-(-3x^3+0x^2-12x)}$$
$$4x^2+0x+16$$
$$\underline{-(4x^2+0x+16)}$$
$$0$$

$(x^4-3x^3+8x^2-12x+16) \div (x^2+4)$
$= x^2-3x+4$

69.

$$\begin{array}{r|rrrr} -3 & 1 & 3 & -12 & 16 \\ & & -3 & 0 & 36 \\ \hline & 1 & 0 & -12 & 52 \end{array}$$

$(t^3+3t^2-12t+16) \div (t+3)$
$= t^2-12+\dfrac{52}{t+3}$

71.

$$a^2+5 \overline{\smash{\big)}\, a^4+2a^3+8a^2+10a+15}$$
$$\underline{-(a^4+0a^3+5a^2)}$$
$$2a^3+3a^2+10a$$
$$\underline{-(2a^3+0a^2+10a)}$$
$$3a^2+0a+15$$
$$\underline{-(3a^2+0a+15)}$$
$$0$$

$(a^4+2a^3+8a^2+10a+15) \div (a^2+5)$
$= a^2+2a+3$

73.

$$\begin{array}{r|rrrr} 2 & 1 & 4 & 0 & -24 \\ & & 2 & 12 & 24 \\ \hline & 1 & 6 & 12 & 0 \end{array}$$

$(t^3+4t^2-24) \div (t-2)$
$= t^2+6t+12$

75.

$$\begin{array}{r|rrrr} -2 & 4 & 0 & 2 & 40 \\ & & -8 & 16 & -36 \\ \hline & 4 & -8 & 18 & 4 \end{array}$$

$(4a^3+2a+40) \div (a+2)$
$= 4a^2-8a+18+\dfrac{4}{a+2}$

77.

a. $65 \div 5 = 13$

b. $65 = 5 \cdot 13$

79.

a.

$$\begin{array}{r|rrr} 7 & 4 & -25 & -21 \\ & & 28 & 21 \\ \hline & 4 & 3 & 0 \end{array}$$

$(4x^2-25x-21) \div (x-7) = 4x+3$

b. $(x-7)(4x+3)$

7.2 Exercises

81.

a.

$$\begin{array}{r|rrr} -3 & 4 & 5 & -21 \\ & & -12 & 21 \\ \hline & 4 & -7 & 0 \end{array}$$

$(4x^2 + 5x - 21) \div (x+3) = 4x - 7$

b. $(x+3)(4x-7)$

83.

a.

$$\begin{array}{r|rrrr} -5 & 2 & 1 & -63 & -90 \\ & & -10 & 45 & 90 \\ \hline & 2 & -9 & -18 & 0 \end{array}$$

$(2x^3 + x^2 - 63x - 90) \div (x+5)$
$= 2x^2 - 9x - 18$

b. $(x+5)(2x^2 - 9x - 18) = (x+5)(x-6)(2x+3)$

85.

a.

$$\require{enclose}
\begin{array}{r}
3t^2 - 17t + 20 \\
t^2 + 9 \enclose{longdiv}{3t^4 - 17t^3 + 47t^2 - 153t + 180}
\end{array}$$

$$\begin{array}{r}
-(3t^4 + 0t^3 + 27t^2) \\ \hline
-17t^3 + 20t^2 - 153t \\
-(-17t^3 + 0t^2 - 153t) \\ \hline
20t^2 + 0t + 180 \\
-(20t^2 + 0t + 180) \\ \hline
0
\end{array}$$

$(3t^4 - 17t^3 + 47t^2 - 153t + 180) \div (t^2 + 9)$
$= 3t^2 - 17t + 20$

b. $(t^2 + 9)(3t^2 - 17t + 20)$

Section 7.3

1.

$$\frac{15x^2}{9y^3} \cdot \frac{21y^5}{35x} = \frac{3 \cdot 5 \cdot x^{2-1}}{3 \cdot 3} \cdot \frac{3 \cdot 7 \cdot y^{5-3}}{5 \cdot 7}$$
$$= xy^2$$

3.

$$\frac{35a^2}{12b^3c} \cdot \frac{40b}{a^5c} = \frac{5 \cdot 7 \cdot a^{2-5}}{2 \cdot 2 \cdot 3 \cdot c} \cdot \frac{2 \cdot 2 \cdot 2 \cdot 5 \cdot b^{1-3}}{c}$$
$$= \frac{350a^{2-5}b^{1-3}}{3c^{1+1}}$$
$$= \frac{350a^{-3}b^{-2}}{3c^2}$$
$$= \frac{350}{3a^3b^2c^2}$$

5.

$$\frac{x+3}{x+7} \cdot \frac{x+7}{x-5} = \frac{(x+3)}{(x+7)} \cdot \frac{(x+7)}{(x-5)}$$
$$= \frac{x+3}{x-5}$$

7.

$$\frac{7-x}{x+3} \cdot \frac{2x+3}{x-7} = \frac{-1(7-x)}{(x+3)} \cdot \frac{(2x+3)}{(x-7)}$$
$$= -\frac{2x+3}{x+3}$$

9.

$$\frac{2(x-4)}{3x-12} \cdot \frac{3(x+5)}{x-7} = \frac{2(x-4)}{3(x-4)} \cdot \frac{3(x+5)}{x-7}$$
$$= \frac{2(x+5)}{x-7}$$

11.

$$\frac{2(x-4)}{4x+20} \cdot \frac{3(x+5)}{12-3x} = \frac{2(x-4)}{4(x+5)} \cdot \frac{3(x+5)}{-3(-4+x)}$$
$$= \frac{2(x-4)}{2 \cdot 2(x+5)} \cdot \frac{1 \cdot 3(x+5)}{-1 \cdot 3(x-4)}$$
$$= -\frac{1}{2}$$

13.

$$\frac{(x+3)(x+7)}{(x-2)(x+3)} \cdot \frac{(x+7)(x-2)}{(x-3)(x-9)}$$
$$= \frac{(x+3)(x+7)}{(x-2)(x+3)} \cdot \frac{(x+7)(x-2)}{(x-3)(x-9)}$$
$$= \frac{(x+7)^2}{(x-3)(x-9)}$$

15.

$$\frac{(k+5)(7-k)}{(k-7)(k-3)} \cdot \frac{(k-3)(k+6)}{(k+9)(k+5)}$$
$$= \frac{-1(k+5)(7-k)}{(k-7)(k-3)} \cdot \frac{(k-3)(k+6)}{(k+9)(k+5)}$$
$$= -\frac{k+6}{k+9}$$

17.

$$\frac{m^2+8m+7}{m^2-2m-3} \cdot \frac{m^2-9}{m^2+9m+14}$$
$$= \frac{(m+7)(m+1)}{(m-3)(m+1)} \cdot \frac{(m+3)(m-3)}{(m+7)(m+2)}$$
$$= \frac{(m+7)(m+1)}{(m-3)(m+1)} \cdot \frac{(m+3)(m-3)}{(m+7)(m+2)}$$
$$= \frac{m+3}{m+2}$$

19.

$$\frac{x^2-16x+55}{x^2-x-12} \cdot \frac{x^2+12x+27}{x^2-9x+20}$$
$$= \frac{(x-11)(x-5)}{(x-4)(x+3)} \cdot \frac{(x+9)(x+3)}{(x-5)(x-4)}$$
$$= \frac{(x-11)(x-5)}{(x-4)(x+3)} \cdot \frac{(x+9)(x+3)}{(x-5)(x-4)}$$
$$= \frac{(x-11)(x+9)}{(x-4)^2}$$

21. Frank missed the factor of (-1) when dividing out $(7-x)$ and $(x-7)$.

$$\frac{3x+5}{x-7} \cdot \frac{7-x}{x+2} = \frac{(3x+5)}{(x-7)} \cdot \frac{(7-x)(-1)}{(x+2)}$$
$$= -\frac{3x+5}{x+2}$$

23.

$$\frac{x^2 y}{z^3} \div \frac{x}{z} = \frac{x^2 y}{z^3} \cdot \frac{z}{x}$$

$$= \frac{x^{2-1} y}{1} \cdot \frac{z^{1-3}}{1}$$

$$= \frac{xyz^{-2}}{1}$$

$$= \frac{xy}{z^2}$$

25.

$$\frac{x+5}{x-3} \div \frac{x-5}{x-7} = \frac{(x+5)}{(x-3)} \cdot \frac{(x-7)}{(x-5)}$$

$$= \frac{(x+5)(x-7)}{(x-3)(x-5)}$$

27.

$$\frac{5x+20}{x-4} \div \frac{3x+12}{x+7} = \frac{5(x+4)}{(x-4)} \cdot \frac{(x+7)}{3(x+4)}$$

$$= \frac{5\cancel{(x+4)}}{(x-4)} \cdot \frac{(x+7)}{3\cancel{(x+4)}}$$

$$= \frac{5(x+7)}{3(x-4)}$$

29.

$$\frac{5-x}{x+7} \div \frac{x-5}{x+9} = \frac{5-x}{x+7} \cdot \frac{x+9}{x-5}$$

$$= \frac{-1(x-5)}{(x+7)} \cdot \frac{(x+9)}{(x-5)}$$

$$= \frac{-1\cancel{(x-5)}}{(x+7)} \cdot \frac{(x+9)}{\cancel{(x-5)}}$$

$$= -\frac{x+9}{x+7}$$

31.

$$\frac{8b-10}{3b+12} \div \frac{5-4b}{5b-20} = \frac{8b-10}{3b+12} \cdot \frac{5b-20}{5-4b}$$

$$= \frac{2(4b-5)}{3(b+4)} \cdot \frac{5(b-4)}{(-1)(4b-5)}$$

$$= \frac{2\cancel{(4b-5)}}{3(b+4)} \cdot \frac{5(b-4)}{(-1)\cancel{(4b-5)}}$$

$$= -\frac{10(b-4)}{3(b+4)}$$

33.

$$\frac{(x+3)(x+2)}{(x-8)(x-7)} \div \frac{(x+2)(x-5)}{(x-7)(x-5)}$$

$$= \frac{(x+3)(x+2)}{(x-8)(x-7)} \cdot \frac{(x-7)(x-5)}{(x+2)(x-5)}$$

$$= \frac{(x+3)\cancel{(x+2)}}{(x-8)\cancel{(x-7)}} \cdot \frac{\cancel{(x-7)}\cancel{(x-5)}}{\cancel{(x+2)}\cancel{(x-5)}}$$

$$= \frac{x+3}{x-8}$$

35.

$$\frac{w^2+8w+15}{w^2+12w+35} \div \frac{w^2-5w-24}{w^2+3w-28}$$

$$= \frac{w^2+8w+15}{w^2+12w+35} \cdot \frac{w^2+3w-28}{w^2-5w-24}$$

$$= \frac{(w+5)(w+3)}{(w+7)(w+5)} \cdot \frac{(w+7)(w-4)}{(w+3)(w-8)}$$

$$= \frac{\cancel{(w+5)}\cancel{(w+3)}}{\cancel{(w+7)}\cancel{(w+5)}} \cdot \frac{\cancel{(w+7)}(w-4)}{\cancel{(w+3)}(w-8)}$$

$$= \frac{w-4}{w-8}$$

37.

$$\frac{c^2-c-20}{c^2+6c+8} \div \frac{2c^2+11c+12}{3c^2+c-10}$$

$$= \frac{c^2-c-20}{c^2+6c+8} \cdot \frac{3c^2+c-10}{2c^2+11c+12}$$

$$= \frac{(c+4)(c-5)}{(c+4)(c+2)} \cdot \frac{(3c-5)(c+2)}{(2c+3)(c+4)}$$

$$= \frac{\cancel{(c+4)}(c-5)}{\cancel{(c+4)}\cancel{(c+2)}} \cdot \frac{(3c-5)\cancel{(c+2)}}{(2c+3)(c+4)}$$

$$= \frac{(c-5)(3c-5)}{(2c+3)(c+4)}$$

39.

$$\frac{6t^2+t-35}{10t^2+17t-20} \div \frac{12t^2-t-63}{20t^2+29t-36}$$

$$= \frac{6t^2+t-35}{10t^2+17t-20} \cdot \frac{20t^2+29t-36}{12t^2-t-63}$$

$$= \frac{(3t-7)(2t+5)}{(2t+5)(5t-4)} \cdot \frac{(4t+9)(5t-4)}{(4t+9)(3t-7)}$$

$$= \frac{\cancel{(3t-7)}\cancel{(2t+5)}}{\cancel{(2t+5)}\cancel{(5t-4)}} \cdot \frac{\cancel{(4t+9)}\cancel{(5t-4)}}{\cancel{(4t+9)}\cancel{(3t-7)}}$$

$$= 1$$

41.

$$\frac{x^2-25}{x^2+7x+10} \div \frac{x^2-4}{x^2+9x+8}$$

$$= \frac{x^2-25}{x^2+7x+10} \cdot \frac{x^2+9x+8}{x^2-4}$$

$$= \frac{(x+5)(x-5)}{(x+5)(x+2)} \cdot \frac{(x+8)(x+1)}{(x+2)(x-2)}$$

$$= \frac{\cancel{(x+5)}(x-5)}{\cancel{(x+5)}(x+2)} \cdot \frac{(x+8)(x+1)}{(x+2)(x-2)}$$

$$= \frac{(x-5)(x+8)(x+1)}{(x+2)^2(x-2)}$$

43.

$$\frac{a^2-25}{a^2+6a+5} \cdot \frac{a^2-1}{a^2+3a-4}$$

$$= \frac{(a+5)(a-5)}{(a+5)(a+1)} \cdot \frac{(a+1)(a-1)}{(a+4)(a-1)}$$

$$= \frac{\cancel{(a+5)}(a-5)}{\cancel{(a+5)}\cancel{(a+1)}} \cdot \frac{\cancel{(a+1)}\cancel{(a-1)}}{(a+4)\cancel{(a-1)}}$$

$$= \frac{a-5}{a+4}$$

45.

$$\frac{h^2-7h+12}{h^2+2h-8} \div \frac{h^2+4h-32}{6h^2-12h}$$

$$= \frac{h^2-7h+12}{h^2+2h-8} \cdot \frac{6h^2-12h}{h^2+4h-32}$$

$$= \frac{(h-3)(h-4)}{(h+4)(h-2)} \cdot \frac{6h(h-2)}{(h+8)(h-4)}$$

$$= \frac{(h-3)\cancel{(h-4)}}{(h+4)\cancel{(h-2)}} \cdot \frac{6h\cancel{(h-2)}}{(h+8)\cancel{(h-4)}}$$

$$= \frac{6h(h-3)}{(h+4)(h+8)}$$

47.

$$\frac{6x^2+x-35}{10x^2+39x+35} \div \frac{12x^2-37x+21}{15x^2+11x-14}$$

$$= \frac{6x^2+x-35}{10x^2+39x+35} \cdot \frac{15x^2+11x-14}{12x^2-37x+21}$$

$$= \frac{(3x-7)(2x+5)}{(2x+5)(5x+7)} \cdot \frac{(3x-2)(5x+7)}{(4x-3)(3x-7)}$$

$$= \frac{\cancel{(3x-7)}\cancel{(2x+5)}}{\cancel{(2x+5)}\cancel{(5x+7)}} \cdot \frac{(3x-2)\cancel{(5x+7)}}{(4x-3)\cancel{(3x-7)}}$$

$$= \frac{3x-2}{4x-3}$$

49.

$$\frac{10t^2-29t-21}{6t^2-5t-56} \cdot \frac{3t^2-4t-32}{8t^2-27t-20}$$

$$= \frac{(5t+3)(2t-7)}{(3t+8)(2t-7)} \cdot \frac{(t-4)(3t+8)}{(8t+5)(t-4)}$$

$$= \frac{(5t+3)\cancel{(2t-7)}}{\cancel{(3t+8)}\cancel{(2t-7)}} \cdot \frac{\cancel{(t-4)}\cancel{(3t+8)}}{(8t+5)\cancel{(t-4)}}$$

$$= \frac{5t+3}{8t+5}$$

51.

$$\frac{x^3-8}{x^2+2x-8} \div \frac{x^2+2x+4}{x^2+7x+12}$$

$$= \frac{x^3-8}{x^2+2x-8} \cdot \frac{x^2+7x+12}{x^2+2x+4}$$

$$= \frac{\cancel{(x-2)}\cancel{(x^2+2x+4)}}{\cancel{(x+4)}\cancel{(x-2)}} \cdot \frac{\cancel{(x+4)}(x+3)}{\cancel{(x^2+2x+4)}}$$

$$= x+3$$

53.

$$\frac{x^4-81}{5x^3+45x} \cdot \frac{4x^2-15x+9}{4x^2+9x-9}$$

$$= \frac{\cancel{(x^2+9)}(x^2-9)}{(5x)\cancel{(x^2+9)}} \cdot \frac{\cancel{(4x-3)}(x-3)}{\cancel{(4x-3)}(x+3)}$$

$$= \frac{(x^2-9)(x-3)}{5x(x+3)}$$

$$= \frac{\cancel{(x+3)}(x-3)(x-3)}{5x\cancel{(x+3)}}$$

$$= \frac{(x-3)^2}{5x}$$

Section 7.4

1.

$\dfrac{7}{12x^2y}$ and $\dfrac{3}{28xy}$

$\dfrac{7}{2^2 \cdot 3 \cdot x^2 \cdot y}$ and $\dfrac{3}{2^2 \cdot 7 \cdot x \cdot y}$

LCD $= 2^2 \cdot 3 \cdot 7 \cdot x^2 \cdot y = 84x^2y$

$\dfrac{7}{12x^2y} \cdot \dfrac{7}{7}$ and $\dfrac{3}{28xy} \cdot \dfrac{3x}{3x}$

$\dfrac{49}{84x^2y}$ and $\dfrac{9x}{84x^2y}$

3.

$\dfrac{7m}{180n^2p}$ and $\dfrac{4n}{150p}$

$\dfrac{7m}{2^2 \cdot 3^2 \cdot 5 \cdot n^2 \cdot p}$ and $\dfrac{4n}{2 \cdot 3 \cdot 5^2 \cdot p}$

LCD $= 2^2 \cdot 3^2 \cdot 5^2 \cdot n^2 \cdot p = 900n^2p$

$\dfrac{7m}{180n^2p} \cdot \dfrac{5}{5}$ and $\dfrac{4n}{150p} \cdot \dfrac{6n^2}{6n^2}$

$\dfrac{35m}{900n^2p}$ and $\dfrac{24n^3}{900n^2p}$

5.

$\dfrac{x+2}{x+8}$ and $\dfrac{x-4}{x+7}$

LCD $= (x+8)(x+7)$

$\dfrac{(x+2)}{(x+8)} \cdot \dfrac{(x+7)}{(x+7)}$ and $\dfrac{(x-4)}{(x+7)} \cdot \dfrac{(x+8)}{(x+8)}$

$\dfrac{(x+2)(x+7)}{(x+8)(x+7)}$ and $\dfrac{(x+8)(x-4)}{(x+8)(x+7)}$

7.

$\dfrac{x-5}{x+6}$ and $\dfrac{x-5}{x+5}$

LCD $= (x+6)(x+5)$

$\dfrac{(x-5)}{(x+6)} \cdot \dfrac{(x+5)}{(x+5)}$ and $\dfrac{(x-5)}{(x+5)} \cdot \dfrac{(x+6)}{(x+6)}$

$\dfrac{(x-5)(x+5)}{(x+6)(x+5)}$ and $\dfrac{(x+6)(x-5)}{(x+6)(x+5)}$

9.

$\dfrac{2x}{5-x}$ and $\dfrac{3x}{x-5}$

LCD $= (x-5)$

$\dfrac{2x}{(5-x)} \cdot \dfrac{(-1)}{(-1)}$ and $\dfrac{3x}{(x-5)}$

$\dfrac{-2x}{(x-5)}$ and $\dfrac{3x}{(x-5)}$

11.

$2x$ and $\dfrac{5}{x+1}$

LCD $= (x+1)$

$\dfrac{2x}{1} \cdot \dfrac{(x+1)}{(x+1)}$ and $\dfrac{5}{(x+1)}$

$\dfrac{2x(x+1)}{(x+1)}$ and $\dfrac{5}{(x+1)}$

13.

$\dfrac{h+2}{(h+3)(h+7)}$ and $\dfrac{h-4}{(h+3)(h+5)}$

LCD $= (h+3)(h+7)(h+5)$

$\dfrac{h+2}{(h+3)(h+7)} \cdot \dfrac{(h+5)}{(h+5)}$ and $\dfrac{h-4}{(h+3)(h+5)} \cdot \dfrac{(h+7)}{(h+7)}$

$\dfrac{(h+2)(h+5)}{(h+3)(h+7)(h+5)}$ and $\dfrac{(h-4)(h+7)}{(h+3)(h+7)(h+5)}$

15.

$\dfrac{n+1}{(n-7)(n-2)}$ and $\dfrac{n+2}{(n-8)(n-2)}$

LCD $= (n-7)(n-2)(n-8)$

$\dfrac{n+1}{(n-7)(n-2)} \cdot \dfrac{(n-8)}{(n-8)}$ and $\dfrac{n+2}{(n-8)(n-2)} \cdot \dfrac{(n-7)}{(n-7)}$

$\dfrac{(n+1)(n-8)}{(n-7)(n-2)(n-8)}$ and $\dfrac{(n+2)(n-7)}{(n-7)(n-2)(n-8)}$

17.

$\dfrac{x+1}{x^2+5x+6}$ and $\dfrac{x-3}{x^2-2x-8}$

$\dfrac{x+1}{(x+3)(x+2)}$ and $\dfrac{x-3}{(x-4)(x+2)}$

LCD $= (x+3)(x+2)(x-4)$

$\dfrac{x+1}{(x+3)(x+2)} \cdot \dfrac{(x-4)}{(x-4)}$ and $\dfrac{x-3}{(x-4)(x+2)} \cdot \dfrac{(x+3)}{(x+3)}$

$\dfrac{(x+1)(x-4)}{(x+3)(x+2)(x-4)}$ and $\dfrac{(x-3)(x+3)}{(x+3)(x+2)(x-4)}$

19.

$$\frac{t+3}{t^2-2t-35} \text{ and } \frac{t-7}{t^2+t-20}$$

$$\frac{t+3}{(t-7)(t+5)} \text{ and } \frac{t-7}{(t+5)(t-4)}$$

$$\text{LCD} = (t-7)(t+5)(t-4)$$

$$\frac{t+3}{(t-7)(t+5)} \cdot \frac{(t-4)}{(t-4)} \text{ and } \frac{t-7}{(t+5)(t-4)} \cdot \frac{(t-7)}{(t-7)}$$

$$\frac{(t+3)(t-4)}{(t-7)(t+5)(t-4)} \text{ and } \frac{(t-7)(t-7)}{(t-7)(t+5)(t-4)}$$

21.

$$\frac{x+2}{x+7} + \frac{x+8}{x+7} = \frac{x+2+x+8}{x+7}$$

$$= \frac{2x+10}{x+7}$$

$$= \frac{2(x+5)}{x+7}$$

23.

$$\frac{m+7}{m+5} - \frac{4m+22}{m+5} = \frac{m+7-4m-22}{m+5}$$

$$= \frac{-3m-15}{m+5}$$

$$= \frac{-3(m+5)}{m+5}$$

$$= \frac{-3\cancel{(m+5)}}{\cancel{(m+5)}}$$

$$= -3$$

25.

$$\frac{d+5}{d+3} + \frac{d-2}{d+7} = \frac{(d+5)}{(d+3)} \cdot \frac{(d+7)}{(d+7)} + \frac{(d-2)}{(d+7)} \cdot \frac{(d+3)}{(d+3)}$$

$$= \frac{d^2+12d+35}{(d+3)(d+7)} + \frac{d^2+d-6}{(d+7)(d+3)}$$

$$= \frac{d^2+12d+35+d^2+d-6}{(d+3)(d+7)}$$

$$= \frac{2d^2+13d+29}{(d+3)(d+7)}$$

27.

$$\frac{c+2}{c+3} - \frac{c+4}{c+7} = \frac{(c+2)}{(c+3)} \cdot \frac{(c+7)}{(c+7)} - \frac{(c+4)}{(c+7)} \cdot \frac{(c+3)}{(c+3)}$$

$$= \frac{c^2+9c+14}{(c+3)(c+7)} - \frac{c^2+7c+12}{(c+7)(c+3)}$$

$$= \frac{c^2+9c+14-(c^2+7c+12)}{(c+3)(c+7)}$$

$$= \frac{c^2+9c+14-c^2-7c-12}{(c+3)(c+7)}$$

$$= \frac{2c+2}{(c+3)(c+7)}$$

$$= \frac{2(c+1)}{(c+3)(c+7)}$$

29.

$$\frac{n+6}{n-2} + \frac{n-5}{2-n} = \frac{n+6}{n-2} + \frac{(n-5)}{(2-n)} \cdot \frac{(-1)}{(-1)}$$

$$= \frac{n+6}{n-2} + \frac{5-n}{n-2}$$

$$= \frac{n+6+5-n}{n-2}$$

$$= \frac{11}{n-2}$$

31.

$$\frac{x+7}{x-9} + \frac{x-8}{9-x} = \frac{x+7}{x-9} + \frac{(x-8)}{(9-x)} \cdot \frac{(-1)}{(-1)}$$

$$= \frac{x+7}{x-9} + \frac{8-x}{x-9}$$

$$= \frac{x+7+8-x}{x-9}$$

$$= \frac{15}{x-9}$$

33.

$$5 + \frac{2}{x+7} = \frac{5}{1} \cdot \frac{(x+7)}{(x+7)} + \frac{2}{x+7}$$

$$= \frac{5x+35}{x+7} + \frac{2}{x+7}$$

$$= \frac{5x+35+2}{x+7}$$

$$= \frac{5x+37}{x+7}$$

35.

$$2x + \frac{7}{xy^2} = \frac{2x}{1} \cdot \frac{(xy^2)}{(xy^2)} + \frac{7}{xy^2}$$

$$= \frac{2x^2y^2}{xy^2} + \frac{7}{xy^2}$$

$$= \frac{2x^2y^2 + 7}{xy^2}$$

37. Marie added the numerator together and then added the denominators together, but she needed to get a common denominator and only add the numerators.

$$\frac{x+5}{x-7} + \frac{x+3}{x+2} = \frac{(x+5)}{(x-7)} \cdot \frac{(x+2)}{(x+2)} + \frac{(x+3)}{(x+2)} \cdot \frac{(x-7)}{(x-7)}$$

$$= \frac{x^2+7x+10}{(x-7)(x+2)} + \frac{x^2-4x-21}{(x+2)(x-7)}$$

$$= \frac{x^2+7x+10+x^2-4x-21}{(x-7)(x+2)}$$

$$= \frac{2x^2+3x-11}{(x-7)(x+2)}$$

39. Matt cross-canceled the denominator in the first expression with the numerator in the second expression. He should only cross-cancel when multiplying two fractions.

$$\frac{x+1}{x-7} + \frac{x-7}{x+4} = \frac{(x+1)}{(x-7)} \cdot \frac{(x+4)}{(x+4)} + \frac{(x-7)}{(x+4)} \cdot \frac{(x-7)}{(x-7)}$$

$$= \frac{x^2+5x+4}{(x-7)(x+4)} + \frac{x^2-14x+49}{(x+4)(x-7)}$$

$$= \frac{x^2+5x+4+x^2-14x+49}{(x-7)(x+4)}$$

$$= \frac{2x^2-9x+53}{(x-7)(x+4)}$$

41.

a.

$$\frac{50}{d^2} + \frac{75}{d^2} = \frac{50+75}{d^2}$$

$$= \frac{125}{d^2}$$

b. $\frac{125}{(5)^2} = \frac{125}{25} = 5$

If both lights are placed a distance of 5 feet away, the illumination is 5 foot candles.

43.

a.

$$\frac{3.5p+250}{p} + \frac{2p+600}{p} = \frac{3.5p+250+2p+600}{p}$$

$$= \frac{5.5p+850}{p}$$

b.

$$\frac{5.5(100)+850}{100} = \frac{550+850}{100}$$

$$= \frac{1400}{100}$$

$$= 14$$

If 100 people attend this event, the average cost will be $14 per person.

45.

$$\frac{r+2}{(r+5)(r+3)} + \frac{r+4}{(r+5)(r+3)} = \frac{r+2+r+4}{(r+5)(r+3)}$$

$$= \frac{2r+6}{(r+5)(r+3)}$$

$$= \frac{2(r+3)}{(r+5)(r+3)}$$

$$= \frac{2\cancel{(r+3)}}{(r+5)\cancel{(r+3)}}$$

$$= \frac{2}{r+5}$$

47.

$$\frac{x+1}{(x-7)(x-2)} + \frac{x+2}{(x-8)(x-2)}$$

$$= \left(\frac{x-8}{x-8}\right)\frac{(x+1)}{(x-7)(x-2)} + \frac{(x+2)}{(x-8)(x-2)}\left(\frac{x-7}{x-7}\right)$$

$$= \frac{x^2-7x-8}{(x-8)(x-7)(x-2)} + \frac{x^2-5x-14}{(x-8)(x-7)(x-2)}$$

$$= \frac{x^2-7x-8+x^2-5x-14}{(x-8)(x-7)(x-2)}$$

$$= \frac{2x^2-12x-22}{(x-8)(x-7)(x-2)}$$

$$= \frac{2(x^2-6x-11)}{(x-8)(x-7)(x-2)}$$

Chapter 7 Rational Functions

49.

$$\frac{w-2}{(w+4)(w-6)} - \frac{3w+1}{(w+3)(w-6)}$$

$$= \left(\frac{w+3}{w+3}\right)\frac{(w-2)}{(w+4)(w-6)} - \frac{(3w+1)}{(w+3)(w-6)}\left(\frac{w+4}{w+4}\right)$$

$$= \frac{w^2+w-6}{(w+3)(w+4)(w-6)} - \frac{3w^2+13w+4}{(w+3)(w+4)(w-6)}$$

$$= \frac{(w^2+w-6)-(3w^2+13w+4)}{(w+3)(w+4)(w-6)}$$

$$= \frac{w^2+w-6-3w^2-13w-4}{(w+3)(w+4)(w-6)}$$

$$= \frac{-2w^2-12w-10}{(w+3)(w+4)(w-6)}$$

$$= \frac{-2(w^2+6w+5)}{(w+3)(w+4)(w-6)}$$

$$= \frac{-2(w+5)(w+1)}{(w+3)(w+4)(w-6)}$$

51.

$$\frac{h+3}{h^2-9h+14} - \frac{h-4}{h^2-10h+16}$$

$$= \frac{h+3}{(h-7)(h-2)} - \frac{h-4}{(h-8)(h-2)}$$

$$= \left(\frac{h-8}{h-8}\right)\frac{(h+3)}{(h-7)(h-2)} - \frac{(h-4)}{(h-8)(h-2)}\left(\frac{h-7}{h-7}\right)$$

$$= \frac{h^2-5h-24}{(h-8)(h-7)(h-2)} - \frac{h^2-11h+28}{(h-8)(h-7)(h-2)}$$

$$= \frac{(h^2-5h-24)-(h^2-11h+28)}{(h-8)(h-7)(h-2)}$$

$$= \frac{h^2-5h-24-h^2+11h-28}{(h-8)(h-7)(h-2)}$$

$$= \frac{6h-52}{(h-8)(h-7)(h-2)}$$

$$= \frac{2(3h-26)}{(h-8)(h-7)(h-2)}$$

53.

$$\frac{t-2}{t^2+6t-27} + \frac{t+7}{t^2+t-12}$$

$$= \left(\frac{t+4}{t+4}\right)\frac{(t-2)}{(t+9)(t-3)} + \frac{(t+7)}{(t-3)(t+4)}\left(\frac{t+9}{t+9}\right)$$

$$= \frac{t^2+2t-8}{(t+4)(t+9)(t-3)} + \frac{t^2+16t+63}{(t+4)(t+9)(t-3)}$$

$$= \frac{t^2+2t-8+t^2+16t+63}{(t+4)(t+9)(t-3)}$$

$$= \frac{2t^2+18t+55}{(t+4)(t+9)(t-3)}$$

55.

$$\frac{k+2}{2k^2+7k+3} + \frac{k-4}{4k^2+7k-15}$$

$$= \frac{k+2}{(2k+1)(k+3)} + \frac{k-4}{(4k-5)(k+3)}$$

$$= \left(\frac{4k-5}{4k-5}\right)\frac{k+2}{(2k+1)(k+3)} + \frac{k-4}{(4k-5)(k+3)}\left(\frac{2k+1}{2k+1}\right)$$

$$= \frac{4k^2+3k-10}{(4k-5)(2k+1)(k+3)} + \frac{2k^2-7k-4}{(4k-5)(2k+1)(k+3)}$$

$$= \frac{4k^2+3k-10+2k^2-7k-4}{(4k-5)(2k+1)(k+3)}$$

$$= \frac{6k^2-4k-14}{(4k-5)(2k+1)(k+3)}$$

$$= \frac{2(3k^2-2k-7)}{(4k-5)(2k+1)(k+3)}$$

57.

$$\frac{r^2+10r}{r^2+16r+63} - \frac{4r+27}{r^2+16r+63}$$

$$= \frac{(r^2+10r)-(4r+27)}{r^2+16r+63}$$

$$= \frac{r^2+10r-4r-27}{r^2+16r+63}$$

$$= \frac{r^2+6r-27}{r^2+16r+63}$$

$$= \frac{\cancel{(r+9)}(r-3)}{\cancel{(r+9)}(r+7)}$$

$$= \frac{r-3}{r+7}$$

59.

$$\frac{d+4}{d+2} - \frac{2}{d^2+5d+6}$$

$$= \frac{d+4}{d+2} - \frac{2}{(d+2)(d+3)}$$

$$= \left(\frac{d+3}{d+3}\right)\left(\frac{d+4}{d+2}\right) - \frac{2}{(d+2)(d+3)}$$

$$= \frac{d^2+7d+12}{(d+2)(d+3)} - \frac{2}{(d+2)(d+3)}$$

$$= \frac{d^2+7d+10}{(d+2)(d+3)}$$

$$= \frac{\cancel{(d+2)}(d+5)}{\cancel{(d+2)}(d+3)}$$

$$= \frac{d+5}{d+3}$$

61.

$$\left(\frac{3+\frac{6}{x}}{2-\frac{5}{x}}\right)\left(\frac{x}{x}\right) = \frac{3x+6}{2x-5}$$

63.

$$\left(\frac{7+\frac{4}{a^2}}{5+\frac{2}{a}}\right)\left(\frac{a^2}{a^2}\right) = \frac{7a^2+4}{5a^2+2a}$$

65.

$$\left(\frac{3+\frac{75}{r^2}}{2-\frac{10}{r}}\right)\left(\frac{r^2}{r^2}\right) = \frac{3r^2+75}{2r^2-10r}$$

$$= \frac{3(r^2+25)}{2r(r-5)}$$

67. Use the LCD: x^4y^2 to multiply by "1".

$$\left(\frac{\frac{3}{x}+\frac{5}{y^2}}{\frac{7}{y}-\frac{6}{x^4}}\right)\left(\frac{x^4y^2}{x^4y^2}\right) = \frac{\frac{3x^4y^2}{x}+\frac{5x^4y^2}{y^2}}{\frac{7x^4y^2}{y}-\frac{6x^4y^2}{x^4}}$$

$$= \frac{3x^3y^2+5x^4}{7x^4y-6y^2}$$

69. Use the LCD: $15xy^2$ to multiply by "1".

$$\left(\frac{\frac{2x}{5y^2}+\frac{6}{x}}{\frac{8}{x}+\frac{2}{3y}}\right)\left(\frac{15xy^2}{15xy^2}\right) = \frac{\frac{2x \cdot 15xy^2}{5y^2}+\frac{6 \cdot 15xy^2}{x}}{\frac{8 \cdot 15xy^2}{x}+\frac{2 \cdot 15xy^2}{3y}}$$

$$= \frac{\frac{2x \cdot 3 \cdot \cancel{5}x\cancel{y^2}}{\cancel{5y^2}}+\frac{2 \cdot 3 \cdot 3 \cdot 5\cancel{x}y^2}{\cancel{x}}}{\frac{2 \cdot 2 \cdot 2 \cdot 3 \cdot 5\cancel{x}y^2}{\cancel{x}}+\frac{2 \cdot \cancel{3} \cdot 5xy^{2-1}}{\cancel{3}}}$$

$$= \frac{6x^2+90y^2}{120y^2+10xy}$$

$$= \frac{\cancel{2}(3x^2+45y^2)}{\cancel{2}(60y^2+5xy)}$$

$$= \frac{3x^2+45y^2}{60y^2+5xy}$$

71. Use the LCD: $(x+5)(x+1)$ to multiply by "1".

$$\left(\frac{\frac{3}{x+5}}{\frac{2}{x+1}+4x}\right)\frac{(x+5)(x+1)}{(x+5)(x+1)}$$

$$= \frac{\frac{3\cancel{(x+5)}(x+1)}{\cancel{(x+5)}}}{\frac{2(x+5)\cancel{(x+1)}}{\cancel{(x+1)}}+4x(x+5)(x+1)}$$

$$= \frac{3(x+1)}{2(x+5)+4x(x+5)(x+1)}$$

$$= \frac{3(x+1)}{2(x+5)(1+2x(x+1))}$$

$$= \frac{3(x+1)}{2(x+5)(2x^2+2x+1)}$$

Chapter 7 Rational Functions

73. Use the LCD: $(x+8)(x-3)$ to multiply by "1".

$$\left(\frac{7+\frac{2}{x+8}}{\frac{1}{x-3}}\right)\left(\frac{(x+8)(x-3)}{(x+8)(x-3)}\right)$$

$$= \frac{7(x+8)(x-3) + \frac{2(x+8)(x-3)}{(x+8)}}{\frac{1(x+8)(x-3)}{(x-3)}}$$

$$= \frac{7(x+8)(x-3) + 2(x-3)}{x+8}$$

$$= \frac{(x-3)(7(x+8)+2)}{x+8}$$

$$= \frac{(x-3)(7x+58)}{x+8}$$

75. Use the LCD: $(x+3)(x+2)(x-3)$ to multiply by "1".

$$\left(\frac{\frac{2}{x+3} - \frac{5x}{x+2}}{\frac{4}{x+2} + \frac{7x}{x-3}}\right)\left(\frac{(x+3)(x+2)(x-3)}{(x+3)(x+2)(x-3)}\right)$$

$$= \frac{\frac{2(x+3)(x+2)(x-3)}{(x+3)} - \frac{5x(x+3)(x+2)(x-3)}{(x+2)}}{\frac{4(x+3)(x+2)(x-3)}{(x+2)} + \frac{7x(x+3)(x+2)(x-3)}{(x-3)}}$$

$$= \frac{2(x+2)(x-3) - 5x(x+3)(x-3)}{4(x+3)(x-3) + 7x(x+3)(x+2)}$$

$$= \frac{(x-3)(2(x+2) - 5x(x+3))}{(x+3)(4(x-3) + 7x(x+2))}$$

$$= \frac{(x-3)(-5x^2 - 13x + 4)}{(x+3)(7x^2 + 18x - 12)}$$

77.

a. Use the LCD: $R_1 R_2 R_3$ to multiply by "1".

$$R = \left(\frac{1}{\frac{1}{R_1} + \frac{1}{R_2} + \frac{1}{R_3}}\right)\left(\frac{R_1 R_2 R_3}{R_1 R_2 R_3}\right)$$

$$R = \frac{R_1 R_2 R_3}{\frac{R_1 R_2 R_3}{R_1} + \frac{R_1 R_2 R_3}{R_2} + \frac{R_1 R_2 R_3}{R_3}}$$

$$R = \frac{R_1 R_2 R_3}{R_2 R_3 + R_1 R_3 + R_1 R_2}$$

b. $R = \frac{(3)(6)(9)}{(6)(9) + (3)(9) + (3)(6)} \approx 1.636$ ohms

79.

a.

$$f = \left(\frac{1}{\frac{1}{d} + \frac{1}{c}}\right)\left(\frac{cd}{cd}\right)$$

$$f = \frac{cd}{\frac{cd}{d} + \frac{cd}{c}}$$

$$f = \frac{cd}{c+d}$$

b. $f = \frac{(0.75)(5)}{0.75 + 5} \approx 0.65$ feet

81.

$$\frac{a+2}{a^2 + 6a + 9} \cdot \frac{a+3}{a^2 - 4a + 3} = \frac{(a+2)}{(a+3)(a+3)} \cdot \frac{(a+3)}{(a-3)(a-1)}$$

$$= \frac{(a+2)}{(a+3)} \cdot \frac{1}{(a-3)(a-1)}$$

$$= \frac{a+2}{(a+3)(a-3)(a-1)}$$

83.

$$\frac{m^2 - 25}{m^2 + 8m + 7} - \frac{m+7}{m+1} = \frac{m^2 - 25}{(m+1)(m+7)} - \frac{m+7}{m+1}$$

$$= \frac{m^2 - 25}{(m+1)(m+7)} - \frac{(m+7)}{(m+1)} \cdot \frac{(m+7)}{(m+7)}$$

$$= \frac{m^2 - 25}{(m+1)(m+7)} - \frac{(m+7)(m+7)}{(m+1)(m+7)}$$

$$= \frac{m^2 - 25}{(m+1)(m+7)} - \frac{m^2 + 14m + 49}{(m+1)(m+7)}$$

$$= \frac{m^2 - 25 - (m^2 + 14m + 49)}{(m+1)(m+7)}$$

$$= \frac{m^2 - 25 - m^2 - 14m - 49}{(m+1)(m+7)}$$

$$= \frac{-14m - 74}{(m+1)(m+7)}$$

$$= \frac{-2(7m + 37)}{(m+1)(m+7)}$$

7.4 Exercises

85.

$$\frac{6-x}{x+2} \cdot \frac{x+7}{x-6} = \frac{-1\cancel{(x-6)}}{(x+2)} \cdot \frac{(x+7)}{\cancel{(x-6)}}$$

$$= -\frac{(x+7)}{(x+2)}$$

87.

$$\frac{7x+8}{x^2+5x+6} - \frac{x-4}{x^2+5x+6} = \frac{(7x+8)-(x-4)}{x^2+5x+6}$$

$$= \frac{7x+8-x+4}{x^2+5x+6}$$

$$= \frac{6x+12}{x^2+5x+6}$$

$$= \frac{6\cancel{(x+2)}}{\cancel{(x+2)}(x+3)}$$

$$= \frac{6}{x+3}$$

89.

$$\frac{5x}{x^2-16} - \frac{7}{x^2+7x+12}$$

$$= \frac{(x+3)5x}{(x+3)(x+4)(x-4)} - \frac{7(x-4)}{(x+4)(x+3)(x-4)}$$

$$= \frac{(x+3)5x - 7(x-4)}{(x+3)(x+4)(x-4)}$$

$$= \frac{5x^2+15x-7x+28}{(x+3)(x+4)(x-4)}$$

$$= \frac{5x^2+8x+28}{(x+3)(x+4)(x-4)}$$

Chapter 7 Rational Functions
Section 7.5

1.
$$6 = \frac{30}{x+2}$$
$$6(x+2) = \left(\frac{30}{x+2}\right)(x+2)$$
$$6(x+2) = \left(\frac{30}{x+2}\right)(x+2)$$
$$6x + 12 = 30$$
$$6x = 18$$
$$x = 3$$

3.
$$5 = \frac{4}{2w+3}$$
$$5(2w+3) = \left(\frac{4}{2w+3}\right)(2w+3)$$
$$5(2w+3) = \left(\frac{4}{2w+3}\right)(2w+3)$$
$$10w + 15 = 4$$
$$10x = -11$$
$$x = -\frac{11}{10}$$

5.
$$2t = \frac{8}{t-3}$$
$$2t(t-3) = \left(\frac{8}{t-3}\right)(t-3)$$
$$2t(t-3) = \left(\frac{8}{t-3}\right)(t-3)$$
$$2t^2 - 6t = 8$$
$$2t^2 - 6t - 8 = 0$$
$$\frac{2t^2}{2} - \frac{6t}{2} - \frac{8}{2} = \frac{0}{2}$$
$$t^2 - 3t - 4 = 0$$
$$(t+1)(t-4) = 0$$
$$t+1 = 0, t-4 = 0$$
$$t = -1, t = 4$$

7.
$$3m = \frac{-9}{m+4}$$
$$3m(m+4) = \left(\frac{-9}{m+4}\right)(m+4)$$
$$3m(m+4) = \left(\frac{-9}{m+4}\right)(m+4)$$
$$3m^2 + 12m = -9$$
$$3m^2 + 12m + 9 = 0$$
$$\frac{3m^2}{3} + \frac{12m}{3} + \frac{9}{3} = \frac{0}{3}$$
$$m^2 + 4m + 3 = 0$$
$$(m+3)(m+1) = 0$$
$$m+3 = 0, m+1 = 0$$
$$x = -3, x = -1$$

9.
a.
$$20 = \frac{600}{p}$$
$$(p)(20) = \left(\frac{600}{p}\right)(p)$$
$$20p = 600$$
$$\frac{20p}{20} = \frac{600}{20}$$
$$p = 30$$

There must be 30 players participating for the cost to be $20 per player.

b.
$$7 = \frac{600}{p}$$
$$(p)(7) = \left(\frac{600}{p}\right)(p)$$
$$7p = 600$$
$$\frac{7p}{7} = \frac{600}{7}$$
$$p \approx 85.7$$

There must be 86 players participating for the cost to be about $7 per player.

11.
a. Let $C(p)$ be the per person cost if p people attend the charity event. Then $C(p) = \frac{3400}{p}$.

7.5 Exercises

b.

$$40 = \frac{3400}{p}$$

$$(p)(40) = \left(\frac{3400}{p}\right)(p)$$

$$40p = 3400$$

$$\frac{40p}{40} = \frac{3400}{40}$$

$$p = 85$$

To keep the cost down to $40 per person, 85 people need to attend the charity dinner.

13.

$$2 - \frac{6}{x} = \frac{4}{x}$$

$$(x)\left(2 - \frac{6}{x}\right) = \left(\frac{4}{x}\right)(x)$$

$$2x - 6 = 4$$

$$2x = 10$$

$$\frac{2x}{2} = \frac{10}{2}$$

$$x = 5$$

15.

$$6 + \frac{2}{5m} = \frac{3}{m}$$

$$(5m)\left(6 + \frac{2}{5m}\right) = \left(\frac{3}{m}\right)(5m)$$

$$30m + 2 = 15$$

$$30m = 13$$

$$\frac{30m}{30} = \frac{13}{30}$$

$$m = \frac{13}{30}$$

17.

$$\frac{4}{x+2} = -1 + \frac{8}{x+2}$$

$$\frac{4}{x+2} - \frac{8}{x+2} = -1 + \frac{8}{x+2} - \frac{8}{x+2}$$

$$\frac{-4}{x+2} = -1$$

$$(x+2)\left(\frac{-4}{x+2}\right) = (-1)(x+2)$$

$$-4 = -x - 2$$

$$x = 4 - 2$$

$$x = 2$$

19.

$$4 + \frac{6}{h-3} = 7 + \frac{12}{h-3}$$

$$(h-3)(4) + (h-3)\left(\frac{6}{h-3}\right) = (h-3)(7) + \left(\frac{12}{h-3}\right)(h-3)$$

$$4h - 12 + 6 = 7h - 21 + 12$$

$$4h - 6 = 7h - 9$$

$$-3h - 6 = -9$$

$$-3h = -3$$

$$\frac{-3h}{-3} = \frac{-3}{-3}$$

$$h = 1$$

21.

a.

$$P(5) = \frac{216.89(5)^2 - 2129.2(5) + 19{,}114.9}{33.67(5) + 1444.29}$$

$$P(5) \approx 8.61$$

Approximately 8.61% of the housing units in Colorado were vacant in 1995.

b.

$$11.5 = \frac{216.89t^2 - 2129.2t + 19{,}114.9}{33.67t + 1444.29}$$

$$11.5(33.67t + 1444.29) = 216.89t^2 - 2129.2t + 19{,}114.9$$

$$387.205t + 16{,}609.335 = 216.89t^2 - 2129.2t + 19{,}114.9$$

$$-387.205t - 16{,}609.335 \qquad -387.205t - 16{,}609.335$$

$$0 = 216.89t^2 - 2516.405t + 2505.565$$

Use the quadratic formula to solve:

$$t = \frac{-(-2516.405) \pm \sqrt{(-2516.405)^2 - 4(216.89)(2505.565)}}{2(216.89)}$$

$$t = \frac{2516.405 \pm \sqrt{4{,}158{,}566.153}}{2(216.89)}$$

$$t \approx 10.5 \quad \text{or} \quad t \approx 1.1$$

Chapter 7 Rational Functions

X	Y₁
1	11.639
1.1	11.5
1.2	11.364
1.3	11.232
1.4	11.103
1.5	10.977
1.6	10.856

X=1.1

X	Y₁
10	10.956
10.1	11.06
10.2	11.166
10.3	11.275
10.4	11.385
10.5	11.497
10.6	11.612

X=10.5

There was an 11.5% vacancy rate in Colorado housing units in 1991 and again in mid 2000.

23.

a. Let $N(t)$ be the average amount California spent in dollars per person t years since 1900.

$$N(t) = \frac{55.125t^2 - 6435.607t + 186,914.286}{0.464t - 12.47}$$

b.

$$N(95) = \frac{55.125(95)^2 - 6435.607(95) + 186,914.286}{0.464(95) - 12.47}$$

$$N(95) \approx 2310.49$$

The per capita spending in California in 1995 was $2310.49.

c.

$$3000 = \frac{55.125t^2 - 6435.607t + 186,914.286}{0.464t - 12.47}$$

$$3000(0.464t - 12.47) = 55.125t^2 - 6435.607t + 186,914.286$$

$$1392t - 37,410 = 55.125t^2 - 6435.607t + 186,914.286$$

$$\underline{-1392t + 37,410 \qquad\qquad -1392t \quad + 37,410}$$

$$0 = 55.125t^2 - 7827.607t + 224,324.286$$

Use the quadratic formula to solve:

$$t = \frac{-(-7827.607) \pm \sqrt{(-7827.607)^2 - 4(55.125)(224,324.286)}}{2(55.125)}$$

$$t = \frac{7827.607 \pm \sqrt{11,807,926.28}}{2(55.125)}$$

$t \approx 39.8$ $\qquad t \approx 102$

X	Y₁
101.8	2964
101.9	2973.8
102	2983.6
102.1	2993.4
102.2	3003.3
102.3	3013.1
102.4	3023

X=102.2

The per capita spending in California reached $3000 in 2002. We exclude the answer $t \approx 39.8$ because it would not be in a reasonable domain for this function.

25.

a.

$$S(8) = \frac{1100(8) + 10,600}{0.007(8) + 3.453}$$

$$S(8) \approx 5528.64$$

The per capita spending in Connecticut in 2008 was $5528.64.

b.

$$4000 = \frac{1100t + 10,600}{0.007t + 3.453}$$

$$4000(0.007t + 3.453) = 1100t + 10,600$$

$$28t + 13,812 = 1100t + 10,600$$

$$\underline{-28t - 10,600 \qquad -28t - 10,600}$$

$$3212 = 1072t$$

$$\frac{3212}{1072} = \frac{1072t}{1072}$$

$$3.00 \approx t$$

X	Y₁
2.99	3998.1
3	4001.2
3.01	4004.2
3.02	4007.3
3.03	4010.4
3.04	4013.5
3.05	4016.6

X=3

The per capita spending in Connecticut reached $4000 in 2003.

27.

$I = \dfrac{k}{R}$

Use $I = 1.2$ when 200 to solve for k.

$1.2 = \dfrac{k}{200} \rightarrow (200)(1.2) = k$

$240 = k$ so $I = \dfrac{240}{R}$

$0.9 = \dfrac{240}{R}$

$0.9R = 240$

$\dfrac{0.9R}{0.9} = \dfrac{240}{0.9}$

$R \approx 266.7$

The resistance is 266.7 ohms when the current is 0.9 amps.

29.

a.

$y = \dfrac{k}{x^3}$

Use $y = 11$ when $x = 5$ to solve for k.

$11 = \dfrac{k}{5^3} = \dfrac{k}{125}$

$(125)(11) = \dfrac{k}{125}(125)$

$1375 = k$

so

$y = \dfrac{1375}{x^3}$

b.

$4 = \dfrac{1375}{x^3}$

$(x^3)(4) = \left(\dfrac{1375}{x^3}\right)(x^3)$

$4x^3 = 1375$

$\dfrac{4x^3}{4} = \dfrac{1375}{4}$

$x^3 = 343.75$

$(x^3)^{1/3} = (343.75)^{1/3}$

$x \approx 7.005$

31.

a.

$W = \dfrac{k}{p^4}$

Use $W = 135$ when $p = 3$ to solve for k.

$135 = \dfrac{k}{3^4} = \dfrac{k}{81}$

$(81)(135) = 10{,}935 = k$

so

$W = \dfrac{10{,}935}{p^4}$

b. $W = \dfrac{10{,}935}{10^4} = 1.0935$

c.

$400 = \dfrac{10{,}935}{p^4}$

$(p^4)(400) = \left(\dfrac{10{,}935}{p^4}\right)(p^4)$

$400p^4 = 10{,}935$

$\dfrac{400p^4}{400} = \dfrac{10{,}935}{400}$

$p^4 = 27.3375$

$(p^4)^{1/4} = (27.3375)^{1/4}$

$p \approx 2.287$

33.

$\dfrac{x+5}{3x-6} = \dfrac{2}{3}$

$\left(\dfrac{x+5}{3x-6}\right)(3(3x-6)) = \left(\dfrac{2}{3}\right)(3(3x-6))$

$(x+5)3 = 2(3x-6)$

$3x + 15 = 6x - 12$

$-3x = -27$

$x = 9$

35.

$\dfrac{x^2}{x+5} = \dfrac{25}{x+5}$

$\left(\dfrac{x^2}{x+5}\right)(x+5) = \left(\dfrac{25}{x+5}\right)(x+5)$

$x^2 = 25$

$\sqrt{x^2} = \pm\sqrt{25}$

$x = \pm 5$

When x equals -5 the denominators are zero so the original fractions are undefined.

The final solution is $x = 5$.

Chapter 7 Rational Functions

37.
$$\frac{w^2}{w-3} = \frac{9}{w-3}$$
$$\left(\frac{w^2}{w-3}\right)(w-3) = \left(\frac{9}{w-3}\right)(w-3)$$
$$w^2 = 9$$
$$\sqrt{w^2} = \pm\sqrt{9}$$
$$w = \pm 3$$

When w equals 3 the denominators are zero so the original fractions are undefined.
The final solution is $w = -3$.

39.
$$\frac{4}{x+5} = \frac{6}{x+8}$$
$$\left(\frac{4}{x+5}\right)(x+5)(x+8) = \left(\frac{6}{x+8}\right)(x+5)(x+8)$$
$$4(x+8) = 6(x+5)$$
$$4x + 32 = 6x + 30$$
$$-2x = -2$$
$$x = 1$$

41.
$$\frac{2r}{r-3} = \frac{9}{r-7}$$
$$\left(\frac{2r}{r-3}\right)(r-3)(r-7) = \left(\frac{9}{r-7}\right)(r-3)(r-7)$$
$$2r(r-7) = 9(r-3)$$
$$2r^2 - 14r = 9r - 27$$
$$2r^2 - 23r + 27 = 0$$
$$r = \frac{23 \pm \sqrt{(-23)^2 - 4(2)(27)}}{2(2)}$$
$$r = \frac{23 \pm \sqrt{313}}{4}$$

43.
$$\frac{12}{x+5} - 2 = \frac{-10}{x+3}$$
$$\left(\frac{12}{x+5}\right)(x+5)(x+3) - 2(x+5)(x+3) = \left(\frac{-10}{x+3}\right)(x+5)(x+3)$$
$$12(x+3) - 2(x+5)(x+3) = -10(x+5)$$
$$12x + 36 - 2(x^2 + 8x + 15) = -10x - 50$$
$$12x + 36 - 2x^2 - 16x - 30 = -10x - 50$$
$$-2x^2 - 4x + 6 = -10x - 50$$
$$0 = 2x^2 - 6x - 56$$
$$0 = \frac{2x^2}{2} - \frac{6x}{2} - \frac{56}{2}$$
$$0 = x^2 - 3x - 28$$
$$0 = (x-7)(x+4)$$
$$x - 7 = 0, x + 4 = 0$$
$$x = 7, x = -4$$

45.
$$\frac{2}{a+6} + \frac{19}{5} = \frac{8}{a-2}$$
$$\left(\frac{2}{a+6}\right)(5)(a+6)(a-2) + \frac{19}{5}(5)(a+6)(a-2) = \left(\frac{8}{a-2}\right)(5)(a+6)(a-2)$$
$$(2)(5)(a-2) + (19)(a+6)(a-2) = (8)(5)(a+6)$$
$$10(a-2) + 19(a^2 + 4a - 12) = 40(a+6)$$
$$10a - 20 + 19a^2 + 76a - 228 = 40a + 240$$
$$19a^2 + 86a - 248 = 40a + 240$$
$$19a^2 + 46a - 488 = 0$$
$$(19a + 122)(a - 4) = 0$$
$$19a + 122 = 0, a - 4 = 0$$
$$a = -\frac{122}{19}, a = 4$$
$$t - 33 = 0, t - 5 = 0$$
$$t = 33, t = 5$$

47.
$$\frac{v+6}{v-2} = \frac{v-1}{v-8}$$
$$\left(\frac{v+6}{v-2}\right)(v-2)(v-8) = \left(\frac{v-1}{v-8}\right)(v-2)(v-8)$$
$$(v+6)(v-8) = (v-1)(v-2)$$
$$v^2 - 2v - 48 = v^2 - 3v + 2$$
$$v^2 - 2v - 48 - v^2 = v^2 - 3v + 2 - v^2$$
$$-2v - 48 = -3v + 2$$
$$v = 50$$

7.5 Exercises

49.
$$\frac{d+3}{d+5} = \frac{d+8}{d+2}$$
$$\left(\frac{d+3}{d+5}\right)(d+5)(d+2) = \left(\frac{d+8}{d+2}\right)(d+5)(d+2)$$
$$(d+3)(d+2) = (d+8)(d+5)$$
$$d^2 + 5d + 6 = d^2 + 13d + 40$$
$$d^2 + 5d + 6 - d^2 = d^2 + 13d + 40 - d^2$$
$$5d + 6 = 13d + 40$$
$$-8d = 34$$
$$d = -4.25$$

51.
$$\frac{2x+5}{x-7} = \frac{x-2}{x+3}$$
$$\left(\frac{2x+5}{x-7}\right)(x-7)(x+3) = \left(\frac{x-2}{x+3}\right)(x-7)(x+3)$$
$$(2x+5)(x+3) = (x-2)(x-7)$$
$$2x^2 + 11x + 15 = x^2 - 9x + 14$$
$$x^2 + 20x + 1 = 0$$
$$x = \frac{-20 \pm \sqrt{20^2 - 4(1)(1)}}{2(1)}$$
$$x = \frac{-20 \pm \sqrt{396}}{2}$$
$$x = \frac{-20 \pm 6\sqrt{11}}{2}$$
$$x = -10 \pm 3\sqrt{11}$$

53.
$$\frac{x}{x-14} + \frac{9}{x-7} = \frac{x^2}{x^2 - 21x + 98}$$
$$\frac{x}{x-14} + \frac{9}{x-7} = \frac{x^2}{(x-14)(x-7)}$$
$$\left(\frac{x}{x-14}\right)(x-14)(x-7) + \left(\frac{9}{x-7}\right)(x-14)(x-7)$$
$$= \frac{x^2}{(x-14)(x-7)}(x-14)(x-7)$$
$$x(x-7) + 9(x-14) = x^2$$
$$x^2 - 7x + 9x - 126 = x^2$$
$$x^2 + 2x - 126 = x^2$$
$$x^2 + 2x - 126 - x^2 = x^2 - x^2$$
$$2x - 126 = 0$$
$$2x = 126$$
$$x = 63$$

55.
$$\frac{k-2}{k+1} = \frac{7k+22}{k^2 + 6k + 5}$$
$$\frac{k-2}{k+1} = \frac{7k+22}{(k+5)(k+1)}$$
$$\left(\frac{k-2}{k+1}\right)(k+5)(k+1) = \frac{7k+22}{(k+5)(k+1)}(k+5)(k+1)$$
$$(k-2)(k+5) = 7k + 22$$
$$k^2 + 3k - 10 = 7k + 22$$
$$k^2 - 4k - 32 = 0$$
$$(k+4)(k-8) = 0$$
$$k + 4 = 0, k - 8 = 0$$
$$k = -4, k = 8$$

57.
$$\frac{t}{t-6} + \frac{5}{t-3} = \frac{t^2}{t^2 - 9t + 18}$$
$$\frac{t}{t-6} + \frac{5}{t-3} = \frac{t^2}{(t-6)(t-3)}$$
$$\left(\frac{t}{t-6}\right)(t-6)(t-3) + \left(\frac{5}{t-3}\right)(t-6)(t-3) = \frac{t^2}{(t-6)(t-3)}(t-6)(t-3)$$
$$t(t-3) + 5(t-6) = t^2$$
$$t^2 - 3t + 5t - 30 = t^2$$
$$t^2 + 2t - 30 = t^2$$
$$t^2 + 2t - 30 - t^2 = t^2 - t^2$$
$$2t - 30 = 0$$
$$2t = 30$$
$$t = 15$$

59.
$$\frac{25}{w^2 + w - 12} = \frac{w+4}{w-3}$$
$$\frac{25}{(w-3)(w+4)} = \frac{w+4}{w-3}$$
$$\left(\frac{25}{(w-3)(w+4)}\right)(w-3)(w+4) = \frac{w+4}{w-3}(w-3)(w+4)$$
$$25 = (w+4)(w+4)$$
$$25 = w^2 + 8w + 16$$
$$0 = w^2 + 8w - 9$$
$$0 = (w+9)(w-1)$$
$$w + 9 = 0, w - 1 = 0$$
$$w = -9, w = 1$$

Chapter 7 Rational Functions

61. If Gina takes 12 hours to solve all the problems in a chapter, her rate of work would be $\frac{1}{12}$ of the job done in one hour. If Karen takes 16 hours to solve all the problems in a chapter, her rate of work would be $\frac{1}{16}$ of the job done in one hour. If it takes t hours working together to solve all of the problems in a chapter then:

$$\frac{t}{12}+\frac{t}{16}=1$$

$$(48)\left(\frac{t}{12}+\frac{t}{16}\right)=(48)(1)$$

$$4t+3t=48$$

$$7t=48$$

$$t=\frac{48}{7}$$

$$t\approx 6.9$$

It would take about 6.9 hours for Gina and Karen working together to solve all the problems in the chapter.

63. If Rosemary takes 4 hours to mow a large lawn, her rate of work would be $\frac{1}{4}$ of the job done in one hour. If Will takes 12 hours to mow a large lawn, his rate of work would be $\frac{1}{12}$ of the job done in one hour. If it takes t hours working together to mow a large lawn then:

$$\frac{t}{4}+\frac{t}{12}=1$$

$$(12)\left(\frac{t}{4}+\frac{t}{12}\right)=(12)(1)$$

$$3t+t=12$$

$$4t=12$$

$$t=\frac{12}{4}$$

$$t=3$$

It would take 3 hours for Rosemary and Will working together to mow the lawn.

65.
$$\frac{x+3}{x+2}+\frac{x+7}{(x+2)(x+5)}=\frac{(x+3)}{(x+2)}\cdot\frac{(x+5)}{(x+5)}+\frac{x+7}{(x+2)(x+5)}$$

$$=\frac{x^2+8x+15}{(x+2)(x+5)}+\frac{x+7}{(x+2)(x+5)}$$

$$=\frac{x^2+8x+15+x+7}{(x+2)(x+5)}$$

$$=\frac{x^2+9x+22}{(x+2)(x+5)}$$

67.
$$\frac{4m+2}{2m^2-5m-3}\cdot\frac{5m-15}{m+6}=\frac{2(2m+1)}{(2m+1)(m-3)}\cdot\frac{5(m-3)}{m+6}$$

$$=\frac{10}{m+6}$$

69.
$$\frac{x+8}{x-7}=\frac{x+2}{x-6}$$

$$\frac{(x+8)}{(x-7)}(x-7)(x-6)=\frac{(x+2)}{(x-6)}(x-7)(x-6)$$

$$(x+8)(x-6)=(x+2)(x-7)$$

$$x^2+2x-48=x^2-5x-14$$

$$7x-34=0$$

$$7x=34$$

$$x=\frac{34}{7}$$

71.
$$\frac{c+5}{c+4}=\frac{12}{c^2+6c+8}$$

$$\frac{c+5}{c+4}=\frac{12}{(c+4)(c+2)}$$

$$\frac{(c+5)}{(c+4)}(c+4)(c+2)=\frac{12}{(c+4)(c+2)}(c+4)(c+2)$$

$$(c+5)(c+2)=12$$

$$c^2+7c+10=12$$

$$c^2+7c-2=0$$

$$c=\frac{-7\pm\sqrt{7^2-4(1)(-2)}}{2(1)}$$

$$c=\frac{-7\pm\sqrt{57}}{2}$$

73.

$$\frac{4x}{x+3} = 5 + \frac{x}{x+3}$$

$$\frac{4x}{\cancel{x+3}}\cancel{(x+3)} = 5(x+3) + \frac{x}{\cancel{x+3}}\cancel{(x+3)}$$

$$4x = 5x + 15 + x$$
$$0 = 2x + 15$$
$$-2x = 15$$
$$x = -7.5$$

75.

This is a quadratic equation.

$$3x^2 + 5x - 10 = 5$$
$$ -5 -5$$
$$3x^2 + 5x - 15 = 0$$
$$x = \frac{-5 \pm \sqrt{5^2 - 4(3)(-15)}}{2(3)}$$

The two solutions are:

$$x = \frac{-5 + \sqrt{205}}{6} \approx 1.553$$
$$x = \frac{-5 - \sqrt{205}}{6} \approx -3.220$$

77.

This is a linear equation.

$$\frac{2}{3}(x+5) = \frac{5}{6} + \frac{2}{5}x$$

Multiply both sides by LCD: 30.

$$(30)\left(\frac{2}{3}\right)(x+5) = (30)\left(\frac{5}{6} + \frac{2}{5}x\right)$$

$$20(x+5) = 25 + 12x$$
$$20x + \cancel{100} = \cancel{12x} + 25$$
$$\underline{-12x } \underline{-12x - 100}$$
$$8x = -75$$
$$x = \frac{-75}{8} = -9.375$$

79.

This is a logarithmic equation.

$$\ln(x+3) = 4$$
$$e^4 = x + 3$$
$$x = e^4 - 3$$
$$x \approx 51.598$$

This answer checks.

81.

This is a rational equation.

$$\frac{3x}{x+2} - \frac{5}{x-7} = \frac{x^2 + 2}{x^2 - 5x - 14}$$

Multiply both sides by LCD: $(x+2)(x-7)$

$$(x+2)(x-7)\left(\frac{3x}{x+2} - \frac{5}{x-7}\right) = \left(\frac{x^2+2}{x^2-5x-14}\right)(x+2)(x-7)$$

$$3x(x-7) - 5(x+2) = x^2 + 2$$

$$3x^2 - 21x - 5x - 10 = \cancel{x^2} \cancel{+2}$$
$$\underline{-x^2 -2 \cancel{-x^2}\cancel{-2}}$$
$$2x^2 - 26x - 12 = 0$$
$$\frac{2x^2 - 26x - 12}{2} = \frac{0}{2}$$
$$x^2 - 13x - 6 = 0$$
$$x = \frac{13 \pm \sqrt{(-13)^2 - 4(1)(-6)}}{2(1)}$$

The two solutions are:

$$x = \frac{13 + \sqrt{193}}{2} \approx 13.446$$
$$x = \frac{13 - \sqrt{193}}{2} \approx -0.446$$

Both answers check.

83.

This is a logarithmic equation.

$$\log(x+2) + \log(x-5) = 1$$
$$\log((x+2)(x-5)) = 1$$
$$\log(x^2 - 3x - 10) = 1$$
$$x^2 - 3x - 10 = 10^1$$
$$ -10 -10$$
$$x^2 - 3x - 20 = 0$$
$$x = \frac{3 \pm \sqrt{(-3)^2 - 4(1)(-20)}}{2(1)}$$

This quadratic equation has two solutions:

$$x = \frac{3 + \sqrt{89}}{2} \approx 6.217$$
$$x = \frac{3 - \sqrt{89}}{2} \approx \cancel{-3.217}$$

Only $x = 6.217$ checks in the original logarithmic equation.

Chapter 7 Rational Functions
Chapter 7 Review

1. Let $I(d)$ be the illumination (in foot-candles) of a light source d feet from the source.

a.

$$I(d) = \frac{k}{d^2}$$

Use $I(30) = 40$ to solve for k.

$$40 = \frac{k}{(30)^2}$$

$$(900)(40) = \left(\frac{k}{900}\right)(900)$$

$$36{,}000 = k$$

so

$$I(d) = \frac{36{,}000}{d^2}$$

b.

$$I(40) = \frac{36{,}000}{40^2}$$

$$I(40) = 22.5$$

The illumination of this light at a distance of 40 feet from the source is 22.5 foot-candles.

c.

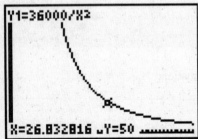

An illumination of 50 foot-candles occurs at a distance of approximately 26.8 feet from the source.

2. Let $C(m)$ be the cost in dollars for an international call lasting m minutes.

a.

$$C(m) = km$$

Use $C(7) = 1.89$ to solve for k.

$$1.89 = k(7)$$

$$\frac{1.89}{7} = \frac{k(7)}{7}$$

$$0.27 = k$$

so

$$C(m) = 0.27m$$

b.

$$C(30) = 0.27(30)$$

$$C(30) = 8.1$$

The total cost for a 30 minute international phone call is $8.10.

c.

$$20 = 0.27m$$

$$\frac{20}{0.27} = \frac{0.27m}{0.27}$$

$$m \approx 74.07$$

You could make a 74 minute call for $20 with this calling card.

3.

a.

$$y = kx^2$$

Use $y = 144$ when $x = 6$ to solve for k.

$$144 = k6^2$$

$$\frac{144}{36} = \frac{k(36)}{36}$$

$$4 = k$$

so

$$y = 4x^2$$

b. $y = 4(2)^2 = 16$

4.

a.

$$y = \frac{k}{x^5}$$

Use $y = 7$ when $x = 2$ to solve for k.

$$7 = \frac{k}{2^5} = \frac{k}{32}$$

$$(32)(7) = \frac{k}{32}(32)$$

$$224 = k$$

so

$$y = \frac{224}{x^5}$$

b.

$$y = \frac{224}{4^5} = 0.21875$$

5. Let $E(p)$ be the entrance fee in dollars for the math competition when p people compete.

a. There will be p people coming, 10 of which will not be paying.

$$E(p) = \frac{7.50p + 500}{p - 10}$$

b.

$$E(100) = \frac{7.50(100)+500}{100-10}$$

$$E(100) = \frac{750+500}{90} = \frac{1250}{90}$$

$$E(100) \approx 13.89$$

If 100 people attend, then each non-scholarship person will pay $13.89.

c.

$$10 = \frac{7.50p+500}{p-10}$$

$$(p-10)(10) = \left(\frac{7.50p+500}{p-10}\right)(p-10)$$

$$10p - 100 = 7.50p + 500$$

$$2.50p - 100 = 500$$

$$2.50p = 600$$

$$\frac{2.50p}{2.50} = \frac{600}{2.50}$$

$$p = 240$$

For each non-scholarship person pay $10.00, there needs to be 240 people participating.

6.

a.

$$C(5) = \frac{637.325(5)+7428.649}{0.226(5)^2 + 10.925(5) + 332.82}$$

$$C(5) \approx 27.0$$

Americans ate about 27 pounds of cheese per capita in the year 1995.

b.

$$30 = \frac{637.325t + 7428.649}{0.226t^2 + 10.925t + 332.82}$$

$$(0.226t^2 + 10.925t + 332.82)30 = 637.325t + 7428.649$$

$$6.78t^2 + 327.75t + 9984.6 = 637.325t + 7428.649$$

$$6.78t^2 - 309.575t + 2555.951 = 0$$

Use the quadratic formula to solve:

$$t = \frac{-b \pm \sqrt{b^2 - 4ac}}{2a}$$

$t \approx 34.84 \quad t \approx 10.82$

According to the model, Americans ate about 30 pounds of cheese per capita in the year 2000 and will do so again in 2024.

7. $f(x) = \dfrac{3x+2}{x-9}$ Domain: All real numbers such that $x \neq 9$.

8. $g(x) = \dfrac{x+5}{(x+5)(x-3)}$ Domain: All real numbers such that $x \neq -5$ and $x \neq 3$

9.

$$h(x) = \frac{5x+2}{2x^2 - 7x - 15}$$

$$h(x) = \frac{5x+2}{(2x+3)(x-5)}$$

$2x + 3 = 0 \qquad x - 5 = 0$

$2x = -3 \qquad x = 5$

$\dfrac{2x}{2} = \dfrac{-3}{2}$

$x = \dfrac{-3}{2}$

Domain: All real numbers such that $x \neq -\dfrac{3}{2}$ and $x \neq 5$

10.

$$f(x) = \frac{x+3}{7x^2 + 5x + 25}$$

$7x^2 + 5x + 25 \neq 0$

Domain: All real numbers

11. Domain: All real numbers such that $x \neq -6$ and $x \neq 4$

12. Domain: All real numbers such that $x \neq -3$ and $x \neq 6$

13.

$$\frac{24x^8 + 16x^5 - 30x^2}{4x^3} = \frac{24x^8}{4x^3} + \frac{16x^5}{4x^3} - \frac{30x^2}{4x^3}$$

$$= 6x^5 + 4x^2 - \frac{15x^{-1}}{2}$$

$$= 6x^5 + 4x^2 - \frac{15}{2x}$$

14.

$$\frac{4-m}{(m+5)(m-4)} = \frac{-1\cancel{(4-m)}}{(m+5)\cancel{(m-4)}}$$

$$= \frac{-1}{m+5}$$

15.

$$\frac{4a+8}{a^2+6a+8} = \frac{4\cancel{(a+2)}}{(a+4)\cancel{(a+2)}}$$

$$= \frac{4}{a+4}$$

16.

$$\frac{6p^2+25p+14}{2p^2+25p+63} = \frac{(3p+2)(2p+7)}{(2p+7)(p+9)}$$

$$= \frac{3p+2}{p+9}$$

17.

$$\begin{array}{r} 4h-3 \\ h+5 \overline{\smash{\big)}\,4h^2+17h-15} \\ \underline{-(4h^2+20h)} \\ -3h-15 \\ \underline{-(-3h-15)} \\ 0 \end{array}$$

$$(4h^2+17h-15) \div (h+5) = 4h-3$$

18.

$$\begin{array}{r} b+7 \\ b-4 \overline{\smash{\big)}\,b^2+3b+5} \\ \underline{-(b^2-4b)} \\ 7b+5 \\ \underline{-(7b-28)} \\ 33 \end{array}$$

$$(b^2+3b+5) \div (b-4) = b+7+\frac{33}{b-4}$$

19.

$$\begin{array}{r} x^2+6x-5 \\ 3x+2 \overline{\smash{\big)}\,3x^3+20x^2-3x-10} \\ \underline{-(3x^3+2x^2)} \\ 18x^2-3x \\ \underline{-(18x^2+12x)} \\ -15x-10 \\ \underline{-(-15x-10)} \\ 0 \end{array}$$

$$(3x^3+20x^2-3x-10) \div (3x+2)$$
$$= x^2+6x-5$$

20.

$$\begin{array}{r} c^2+3c-4 \\ c^2+5 \overline{\smash{\big)}\,c^4+3c^3+c^2+15c-20} \\ \underline{-(c^4\quad\quad+5c^2)} \\ 3c^3-4c^2+15c \\ \underline{-(3c^3\quad\quad+15c)} \\ -4c^2\quad\quad-20 \\ \underline{-(-4c^2\quad\quad-20)} \\ 0 \end{array}$$

$$(c^4+3c^3+c^2+15c-20) \div (c^2+5)$$
$$= c^2+3c-4$$

21.

$$\begin{array}{r|rrr} 3 & 2 & -1 & -15 \\ & & 6 & 15 \\ \hline & 2 & 5 & 0 \end{array}$$

$$(2t^2-t-15) \div (t-3) = 2t+5$$

22.

$$\begin{array}{r|rrr} -4 & 5 & 17 & -10 \\ & & -20 & 12 \\ \hline & 5 & -3 & 2 \end{array}$$

$$(5r^2+17r-10) \div (r+4) = 5r-3+\frac{2}{r+4}$$

23.

$$\begin{array}{r|rrrr} -2 & 4 & 15 & 12 & -4 \\ & & -8 & -14 & 4 \\ \hline & 4 & 7 & -2 & 0 \end{array}$$

$$(4x^3+15x^2+12x-4) \div (x+2)$$
$$= 4x^2+7x-2$$

24.

$$\begin{array}{r|rrrr} 3 & 5 & 0 & -20 & -75 \\ & & 15 & 45 & 75 \\ \hline & 5 & 15 & 25 & 0 \end{array}$$

$$(5y^3-20y-75) \div (y-3)$$
$$= 5y^2+15y+25$$

25.

$$\frac{x+3}{x-15} \cdot \frac{x-15}{x-2} = \frac{(x+3)}{\cancel{(x-15)}} \cdot \frac{\cancel{(x-15)}}{(x-2)}$$

$$= \frac{x+3}{x-2}$$

26.

$$\frac{4}{m+3} \cdot \frac{m+9}{m-7} = \frac{4(m+9)}{(m+3)(m-7)}$$

27.

$$\frac{x+2}{x-3} \div \frac{x+7}{x-3} = \frac{(x+2)}{\cancel{(x-3)}} \cdot \frac{\cancel{(x-3)}}{(x+7)}$$

$$= \frac{x+2}{x+7}$$

28.

$$\frac{h-5}{h+2} \div \frac{5-h}{h-4} = \frac{(h-5)}{(h+2)} \cdot \frac{(h-4)}{(5-h)}$$

$$= \frac{\cancel{(h-5)}}{(h+2)} \cdot \frac{(h-4)}{(-1)\cancel{(h-5)}}$$

$$= -\frac{h-4}{h+2}$$

29.

$$\frac{(d+3)(d-4)}{(d-4)(d+2)} \cdot \frac{(d+2)(d-1)}{(d+2)(d-3)} = \frac{(d+3)\cancel{(d-4)}}{\cancel{(d-4)}(d+2)} \cdot \frac{\cancel{(d+2)}(d-1)}{\cancel{(d+2)}(d-3)}$$

$$= \frac{(d+3)(d-1)}{(d+2)(d-3)}$$

30.

$$\frac{(x+4)(x+5)}{(x-5)(x+2)} \div \frac{(x+5)(x-7)}{(x+2)(x-7)} = \frac{(x+4)\cancel{(x+5)}}{(x-5)\cancel{(x+2)}} \cdot \frac{\cancel{(x+2)}\cancel{(x-7)}}{\cancel{(x+5)}\cancel{(x-7)}}$$

$$= \frac{x+4}{x-5}$$

31.

$$\frac{2n+6}{n^2+8n+15} \cdot \frac{3n+15}{n^2+11n+28} = \frac{2(n+3)}{(n+5)(n+3)} \cdot \frac{3(n+5)}{(n+4)(n+7)}$$

$$= \frac{2\cancel{(n+3)}}{\cancel{(n+5)}\cancel{(n+3)}} \cdot \frac{3\cancel{(n+5)}}{(n+4)(n+7)}$$

$$= \frac{6}{(n+4)(n+7)}$$

32.

$$\frac{v+5}{v^2-6v-55} \div \frac{v-8}{v^2-4v-77} = \frac{(v+5)}{(v+5)(v-11)} \cdot \frac{(v+7)(v-11)}{(v-8)}$$

$$= \frac{\cancel{(v+5)}}{\cancel{(v+5)}\cancel{(v-11)}} \cdot \frac{(v+7)\cancel{(v-11)}}{(v-8)}$$

$$= \frac{v+7}{v-8}$$

33.

$$\frac{12}{x+6} + \frac{4x+3}{x+6} = \frac{12+4x+3}{x+6}$$

$$= \frac{4x+15}{x+6}$$

34.

$$\frac{3}{h-5} - \frac{7}{h+2} = \frac{3}{(h-5)} \cdot \frac{(h+2)}{(h+2)} - \frac{7}{(h+2)} \cdot \frac{(h-5)}{(h-5)}$$

$$= \frac{3(h+2)}{(h-5)(h+2)} - \frac{7(h-5)}{(h+2)(h-5)}$$

$$= \frac{3h+6}{(h-5)(h+2)} - \frac{7h-35}{(h+2)(h-5)}$$

$$= \frac{3h+6-(7h-35)}{(h-5)(h+2)}$$

$$= \frac{3h+6-7h+35}{(h-5)(h+2)}$$

$$= \frac{-4h+41}{(h-5)(h+2)}$$

35.

$$\frac{x+2}{(x+3)(x+5)} + \frac{6}{(x+3)(x-7)}$$

$$= \frac{(x+2)}{(x+3)(x+5)} \cdot \frac{(x-7)}{(x-7)} + \frac{6}{(x+3)(x-7)} \cdot \frac{(x+5)}{(x+5)}$$

$$= \frac{(x+2)(x-7)}{(x+3)(x+5)(x-7)} + \frac{6(x+5)}{(x+3)(x-7)(x+5)}$$

$$= \frac{x^2-5x-14}{(x+3)(x+5)(x-7)} + \frac{6x+30}{(x+3)(x-7)(x+5)}$$

$$= \frac{x^2-5x-14+6x+30}{(x+3)(x+5)(x-7)}$$

$$= \frac{x^2+x+16}{(x+3)(x+5)(x-7)}$$

Chapter 7 Review

36.

$$\frac{2a+5}{(a+1)(a-4)} - \frac{3}{(a+3)(a+1)} = \frac{(2a+5)}{(a+1)(a-4)} \cdot \frac{(a+3)}{(a+3)} - \frac{3}{(a+3)(a+1)} \cdot \frac{(a-4)}{(a-4)}$$

$$= \frac{(2a+5)(a+3)}{(a+1)(a-4)(a+3)} - \frac{3(a-4)}{(a+3)(a+1)(a-4)}$$

$$= \frac{2a^2+11a+15}{(a+1)(a-4)(a+3)} - \frac{3a-12}{(a+3)(a+1)(a-4)}$$

$$= \frac{2a^2+11a+15-(3a-12)}{(a+1)(a-4)(a+3)}$$

$$= \frac{2a^2+11a+15-3a+12}{(a+1)(a-4)(a+3)}$$

$$= \frac{2a^2+8a+27}{(a+1)(a-4)(a+3)}$$

37.

$$\frac{2t}{3t^2-13t-10} - \frac{t+1}{3t^2+14t+8} = \frac{2t}{(3t+2)(t-5)} - \frac{t+1}{(3t+2)(t+4)}$$

$$= \frac{2t}{(3t+2)(t-5)} \cdot \frac{(t+4)}{(t+4)} - \frac{(t+1)}{(3t+2)(t+4)} \cdot \frac{(t-5)}{(t-5)}$$

$$= \frac{2t(t+4)}{(3t+2)(t-5)(t+4)} - \frac{(t+1)(t-5)}{(3t+2)(t+4)(t-5)}$$

$$= \frac{2t^2+8t}{(3t+2)(t-5)(t+4)} - \frac{t^2-4t-5}{(3t+2)(t+4)(t-5)}$$

$$= \frac{2t^2+8t-(t^2-4t-5)}{(3t+2)(t-5)(t+4)}$$

$$= \frac{2t^2+8t-t^2+4t+5}{(3t+2)(t-5)(t+4)}$$

$$= \frac{t^2+12t+5}{(3t+2)(t-5)(t+4)}$$

38.

$$\frac{2x-7}{x^2-5x-14} + \frac{x+5}{x^2-4x-21} = \frac{2x-7}{(x-7)(x+2)} + \frac{x+5}{(x-7)(x+3)}$$

$$= \frac{(2x-7)}{(x-7)(x+2)} \cdot \frac{(x+3)}{(x+3)} + \frac{(x+5)}{(x-7)(x+3)} \cdot \frac{(x+2)}{(x+2)}$$

$$= \frac{(2x-7)(x+3)}{(x-7)(x+2)(x+3)} + \frac{(x+5)(x+2)}{(x-7)(x+3)(x+2)}$$

$$= \frac{2x^2-x-21}{(x-7)(x+2)(x+3)} + \frac{x^2+7x+10}{(x-7)(x+3)(x+2)}$$

$$= \frac{2x^2-x-21+x^2+7x+10}{(x-7)(x+2)(x+3)}$$

$$= \frac{3x^2+6x-11}{(x-7)(x+2)(x+3)}$$

39.

$$\frac{3+\dfrac{4}{a}}{4-\dfrac{7}{a}} = \frac{\left(3+\dfrac{4}{a}\right)(a)}{\left(4-\dfrac{7}{a}\right)(a)}$$

$$= \frac{3a+\dfrac{4}{a}(a)}{4a-\dfrac{7}{a}(a)}$$

$$= \frac{3a+4}{4a-7}$$

40.

$$\frac{5-\dfrac{7}{4n}}{2+\dfrac{6}{n}} = \frac{\left(5-\dfrac{7}{4n}\right)(4n)}{\left(2+\dfrac{6}{n}\right)(4n)}$$

$$= \frac{5(4n)-\dfrac{7}{4n}(4n)}{2(4n)+\dfrac{6}{n}(4n)}$$

$$= \frac{20n-7}{8n+24}$$

41.

$$\frac{2+\dfrac{4}{x+3}}{\dfrac{5}{x-2}} = \frac{\left(2+\dfrac{4}{x+3}\right)(x+3)(x-2)}{\left(\dfrac{5}{x-2}\right)(x+3)(x-2)}$$

$$= \frac{2(x+3)(x-2)+\left(\dfrac{4}{x+3}\right)(x+3)(x-2)}{\left(\dfrac{5}{x-2}\right)(x+3)(x-2)}$$

$$= \frac{2(x+3)(x-2)+4(x-2)}{5(x+3)}$$

$$= \frac{2(x^2+x-6)+4x-8}{5(x+3)}$$

$$= \frac{2x^2+2x-12+4x-8}{5(x+3)}$$

$$= \frac{2x^2+6x-20}{5(x+3)}$$

$$= \frac{2(x^2+3x-10)}{5(x+3)}$$

$$= \frac{2(x+5)(x-2)}{5(x+3)}$$

42.

$$\frac{\frac{5}{t+2}-\frac{3}{t+5}}{\frac{1}{t+5}+\frac{3}{t-2}} = \frac{\left(\frac{5}{t+2}-\frac{3}{t+5}\right)(t+2)(t+5)(t-2)}{\left(\frac{1}{t+5}+\frac{3}{t-2}\right)(t+2)(t+5)(t-2)}$$

$$= \frac{\left(\frac{5}{t+2}\right)(t+2)(t+5)(t-2)-\left(\frac{3}{t+5}\right)(t+2)(t+5)(t-2)}{\left(\frac{1}{t+5}\right)(t+2)(t+5)(t-2)+\left(\frac{3}{t-2}\right)(t+2)(t+5)(t-2)}$$

$$= \frac{5(t+5)(t-2)-3(t+2)(t-2)}{(t+2)(t-2)+3(t+2)(t+5)}$$

$$= \frac{5(t^2+3t-10)-3(t^2-4)}{t^2-4+3(t^2+7t+10)}$$

$$= \frac{5t^2+15t-50-3t^2+12}{t^2-4+3t^2+21t+30}$$

$$= \frac{2t^2+15t-38}{4t^2+21t+26}$$

$$= \frac{(2t+19)(t-2)}{(4t+13)(t+2)}$$

43.

$$\frac{6}{m+2} = \frac{2m}{m+2}$$

$$\left(\frac{6}{m+2}\right)(m+2) = \left(\frac{2m}{m+2}\right)(m+2)$$

$$6 = 2m$$

$$m = 3$$

44.

$$\frac{5}{c-3} = \frac{2}{c+7}$$

$$\left(\frac{5}{c-3}\right)(c-3)(c+7) = \left(\frac{2}{c+7}\right)(c-3)(c+7)$$

$$5(c+7) = 2(c-3)$$

$$5c+35 = 2c-6$$

$$3c = -41$$

$$c = -\frac{41}{3}$$

45.

$$\frac{4x}{x+3} = \frac{2x}{x-9}$$

$$\left(\frac{4x}{x+3}\right)(x+3)(x-9) = \left(\frac{2x}{x-9}\right)(x+3)(x-9)$$

$$4x(x-9) = 2x(x+3)$$

$$4x^2-36x = 2x^2+6x$$

$$2x^2-42x = 0$$

$$2x(x-21) = 0$$

$$2x = 0, x-21 = 0$$

$$x = 0, x = 21$$

46.

$$\frac{2}{b} = 3+\frac{7}{4b}$$

$$\left(\frac{2}{b}\right)(4b) = (3)(4b)+\left(\frac{7}{4b}\right)(4b)$$

$$8 = 12b+7$$

$$1 = 12b$$

$$b = \frac{1}{12}$$

47.

$$\frac{2}{x+5}+6 = \frac{-5}{x-3}$$

$$\left(\frac{2}{x+5}\right)(x+5)(x-3)+6(x+5)(x-3) = \left(\frac{-5}{x-3}\right)(x+5)(x-3)$$

$$2(x-3)+6(x+5)(x-3) = -5(x+5)$$

$$2x-6+6(x^2+2x-15) = -5x-25$$

$$2x-6+6x^2+12x-90 = -5x-25$$

$$6x^2+14x-96 = -5x-25$$

$$6x^2+19x-71 = 0$$

$$x = \frac{-19\pm\sqrt{19^2-4(6)(-71)}}{2(6)}$$

$$x = \frac{-19\pm\sqrt{2065}}{12}$$

48.

$$\frac{3}{h+2}+4 = \frac{-20}{h-5}$$

$$\left(\frac{3}{h+2}\right)(h+2)(h-5)+4(h+2)(h-5) = \left(\frac{-20}{h-5}\right)(h+2)(h-5)$$

$$3(h-5)+4(h+2)(h-5) = -20(h+2)$$

$$3h-15+4(h^2-3h-10) = -20h-40$$

$$3h-15+4h^2-12h-40 = -20h-40$$

$$4h^2-9h-55 = -20h-40$$

$$4h^2+11h-15 = 0$$

$$(4h+15)(h-1) = 0$$

$$4h+15 = 0, h-1 = 0$$

$$h = -\frac{15}{4}, h = 1$$

Chapter 7 Rational Functions

49.

$\frac{x}{x+5} + \frac{7}{x-3} = \frac{56}{(x+5)(x-3)}$

$\left(\frac{x}{x+5}\right)(x+5)(x-3) + \left(\frac{7}{x-3}\right)(x+5)(x-3) = \frac{56}{(x+5)(x-3)}(x+5)(x-3)$

$x(x-3) + 7(x+5) = 56$

$x^2 - 3x + 7x + 35 = 56$

$x^2 + 4x + 35 = 56$

$x^2 + 4x - 21 = 0$

$(x+7)(x-3) = 0$

$x+7 = 0, x-3 = 0$

$x = -7, x = 3$

When x equals 3 the denominators are zero so the original fractions are undefined.

The final solution is $x = -7$.

50.

$\frac{w}{w-7} - \frac{3}{w-4} = \frac{9}{(w-7)(w-4)}$

$\left(\frac{w}{w-7}\right)(w-7)(w-4) - \left(\frac{3}{w-4}\right)(w-7)(w-4) = \frac{9}{(w-7)(w-4)}(w-7)(w-4)$

$w(w-4) - 3(w-7) = 9$

$w^2 - 4w - 3w + 21 = 9$

$w^2 - 7w + 21 = 9$

$w^2 - 7w + 12 = 0$

$(w-4)(w-3) = 0$

$w-4 = 0, w-3 = 0$

$w = 4, w = 3$

When w equals 4 the denominators are zero so the original fractions are undefined.

The final solution is $w = 3$.

51.

$\frac{5x+3}{x^2+7x-9} = \frac{10x-12}{x^2+7x-9}$

$\left(\frac{5x+3}{x^2+7x-9}\right)(x^2+7x-9) = \left(\frac{10x-12}{x^2+7x-9}\right)(x^2+7x-9)$

$5x+3 = 10x-12$

$-5x = -15$

$x = 3$

52.

$\frac{3k}{k+2} + \frac{2}{k+5} = \frac{50}{k^2+7k+10}$

$\left(\frac{3k}{k+2}\right)(k+2)(k+5) + \left(\frac{2}{k+5}\right)(k+2)(k+5)$

$= \frac{50}{(k+5)(k+2)}(k+2)(k+5)$

$3k(k+5) + 2(k+2) = 50$

$3k^2 + 15k + 2k + 4 = 50$

$3k^2 + 17k + 4 = 50$

$3k^2 + 17k - 46 = 0$

$(3k+23)(k-2) = 0$

$3k+23 = 0, k-2 = 0$

$k = -\frac{23}{3}, k = 2$

53.

$\frac{5}{x^2+3x-28} = \frac{3x}{x^2-8x+16}$

$\frac{5}{(x+7)(x-4)} = \frac{3x}{(x-4)(x-4)}$

$\frac{5}{(x+7)(x-4)}(x-4)(x-4)(x+7) = \frac{3x}{(x-4)(x-4)}(x-4)(x-4)(x+7)$

$5(x-4) = 3x(x+7)$

$5x - 20 = 3x^2 + 21x$

$0 = 3x^2 + 16x + 20$

$0 = (3x+10)(x+2)$

$3x+10 = 0, x+2 = 0$

$x = -\frac{10}{3}, x = -2$

54.

$\frac{b}{2b^2+9b+10} = \frac{1}{b^2+b-2}$

$\frac{b}{(2b+5)(b+2)} = \frac{1}{(b+2)(b-1)}$

$\frac{b}{(2b+5)(b+2)}(b+2)(b-1)(2b+5)$

$= \frac{1}{(b+2)(b-1)}(b+2)(b-1)(2b+5)$

$b(b-1) = 2b+5$

$b^2 - b = 2b + 5$

$b^2 - 3b - 5 = 0$

$b = \frac{3 \pm \sqrt{(-3)^2 - 4(1)(-5)}}{2(1)}$

$b = \frac{3 \pm \sqrt{29}}{2}$

55.

Amount of job done in one hour:

Sam → 20 hours → $\left(\frac{1}{20}\right)$th of job done

Craig → 16 hours → $\left(\frac{1}{16}\right)$th of job done

Let t be the number of hours needed to do the job if they are working together.

$$\frac{1}{20} + \frac{1}{16} = \frac{1}{t}$$

$$(80t)\left(\frac{1}{20} + \frac{1}{16}\right) = \left(\frac{1}{t}\right)(80t)$$

$$4t + 5t = 80$$

$$9t = 80$$

$$\frac{9t}{9} = \frac{80}{9}$$

$$t \approx 8.9$$

It would take about 8.9 hours for Sam and Craig working together to put up the fence.

56.

Amount of job done in one hour:

Mark → 6 years → $\left(\frac{1}{6}\right)$th of job done

Cindy → 4 years → $\left(\frac{1}{4}\right)$th of job done

Let t be the number of hours needed to do the job if they are working together.

$$\frac{1}{6} + \frac{1}{4} = \frac{1}{t}$$

$$(12t)\left(\frac{1}{6} + \frac{1}{4}\right) = \left(\frac{1}{t}\right)(12t)$$

$$2t + 3t = 12$$

$$5t = 12$$

$$\frac{5t}{5} = \frac{12}{5}$$

$$t = 2.4$$

It would take 2.4 years for Mark and Cindy working together to write the book.

Chapter 7 Rational Functions
Chapter 7 Test

1. Let $I(d)$ be the illumination (in foot-candles) of a light source d feet from the source.

a.

$I(d) = \dfrac{k}{d^2}$

Use $I(12) = 30$ to solve for k.

$30 = \dfrac{k}{(12)^2}$

$(144)(30) = \left(\dfrac{k}{144}\right)(144)$

$4320 = k$

so

$I(d) = \dfrac{4320}{d^2}$

b.

$I(7) = \dfrac{4320}{7^2}$

$I(7) = 88.2$

The illumination of this light at a distance of 7 feet from the source is 88.2 foot-candles.

c.

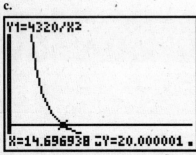

An illumination of 20 foot-candles occurs at a distance of approximately 14.7 feet from the source.

2. Let $L(d)$ be the length of a wall (in feet) if you know the length of the drawing d (in inches) on the blueprint.

a.

$L(d) = kd$

Use $L(3) = 7.5$ to solve for k.

$7.5 = k(3)$

$\dfrac{7.5}{3} = \dfrac{k(3)}{3}$

$2.5 = k$

so

$L(d) = 2.5d$

b. $L(5) = 2.5(5) = 12.5$ A 2.5 inch drawing represents a wall 12.5 feet long.

c.

$20 = 2.5d$

$\dfrac{20}{2.5} = \dfrac{2.5d}{2.5}$

$d = 8$

A 20 foot wall is represented by an 8 inch drawing.

3.

$\dfrac{2x}{x+7} \cdot \dfrac{x+3}{x-4} = \dfrac{2x(x+3)}{(x+7)(x-4)}$

4.

$\dfrac{(x+3)(x+5)}{(x-7)(x+3)} \div \dfrac{(x+5)(x+2)}{(x+4)(x-7)} = \dfrac{(x+3)(x+5)}{(x-7)(x+3)} \cdot \dfrac{(x+4)(x-7)}{(x+5)(x+2)}$

$= \dfrac{\cancel{(x+3)}\cancel{(x+5)}}{\cancel{(x-7)}\cancel{(x+3)}} \cdot \dfrac{(x+4)\cancel{(x-7)}}{\cancel{(x+5)}(x+2)}$

$= \dfrac{x+4}{x+2}$

5.

$\dfrac{5w+3}{w^2-4w-21} \cdot \dfrac{2w-14}{5w^2-17w-12} = \dfrac{5w+3}{(w+3)(w-7)} \cdot \dfrac{2(w-7)}{(5w+3)(w-4)}$

$= \dfrac{\cancel{(5w+3)}}{(w+3)\cancel{(w-7)}} \cdot \dfrac{2\cancel{(w-7)}}{\cancel{(5w+3)}(w-4)}$

$= \dfrac{2}{(w+3)(w-4)}$

6.

$\dfrac{2m+4}{m^2-m-20} \div \dfrac{m+2}{m^2-4m-5} = \dfrac{2m+4}{m^2-m-20} \cdot \dfrac{m^2-4m-5}{m+2}$

$= \dfrac{2(m+2)}{(m+4)(m-5)} \cdot \dfrac{(m-5)(m+1)}{(m+2)}$

$= \dfrac{2\cancel{(m+2)}}{(m+4)\cancel{(m-5)}} \cdot \dfrac{\cancel{(m-5)}(m+1)}{\cancel{(m+2)}}$

$= \dfrac{2(m+1)}{(m+4)}$

7.

$\dfrac{5x}{2x-7} + \dfrac{3x-8}{2x-7} = \dfrac{5x+3x-8}{2x-7}$

$= \dfrac{8x-8}{2x-7}$

$= \dfrac{8(x-1)}{2x-7}$

8.

$$\frac{x+5}{x-3} - \frac{x-2}{x+4} = \frac{(x+5)}{(x-3)} \cdot \frac{(x+4)}{(x+4)} - \frac{(x-2)}{(x+4)} \cdot \frac{(x-3)}{x-3}$$

$$= \frac{(x+5)(x+4)}{(x-3)(x+4)} - \frac{(x-2)(x-3)}{(x+4)(x-3)}$$

$$= \frac{x^2+9x+20}{(x-3)(x+4)} - \frac{x^2-5x+6}{(x+4)(x-3)}$$

$$= \frac{x^2+9x+20-(x^2-5x+6)}{(x-3)(x+4)}$$

$$= \frac{x^2+9x+20-x^2+5x-6}{(x-3)(x+4)}$$

$$= \frac{14x+14}{(x-3)(x+4)}$$

$$= \frac{14(x+1)}{(x-3)(x+4)}$$

9.

$$\frac{5x+2}{x^2+5x+6} + \frac{x-4}{x^2+9x+14} = \frac{5x+2}{(x+3)(x+2)} + \frac{x-4}{(x+7)(x+2)}$$

$$= \frac{(5x+2)}{(x+3)(x+2)} \cdot \frac{(x+7)}{(x+7)} + \frac{(x-4)}{(x+7)(x+2)} \cdot \frac{(x+3)}{(x+3)}$$

$$= \frac{5x^2+37x+14}{(x+3)(x+2)(x+7)} + \frac{x^2-x-12}{(x+7)(x+2)(x+3)}$$

$$= \frac{6x^2+36x+2}{(x+3)(x+2)(x+7)}$$

$$= \frac{2(3x^2+18x+1)}{(x+3)(x+2)(x+7)}$$

10.

$$\frac{2+\dfrac{5}{6d}}{4-\dfrac{8}{d}} = \frac{\left(2+\dfrac{5}{6d}\right) \cdot 6d}{\left(4-\dfrac{8}{d}\right) \cdot 6d}$$

$$= \frac{2 \cdot 6d + \dfrac{5}{6d} \cdot 6d}{4 \cdot 6d - \dfrac{8}{d} \cdot 6d}$$

$$= \frac{12d+5}{24d-48}$$

$$= \frac{12d+5}{24(d-2)}$$

11.

$$\frac{\dfrac{2}{x+1}+3}{\dfrac{5}{x+1}-\dfrac{4}{x-1}} = \frac{\left(\dfrac{2}{x+1}+3\right) \cdot (x+1)(x-1)}{\left(\dfrac{5}{x+1}-\dfrac{4}{x-1}\right) \cdot (x+1)(x-1)}$$

$$= \frac{\dfrac{2}{(x+1)} \cdot (x+1)(x-1) + 3 \cdot (x+1)(x-1)}{\dfrac{5}{(x+1)} \cdot (x+1)(x-1) - \dfrac{4}{(x-1)} \cdot (x+1)(x-1)}$$

$$= \frac{2(x-1)+3(x+1)(x-1)}{5(x-1)-4(x+1)}$$

$$= \frac{2x-2+3(x^2-1)}{5x-5-4x-4}$$

$$= \frac{2x-2+3x^2-3}{x-9}$$

$$= \frac{3x^2+2x-5}{x-9}$$

$$= \frac{(3x+5)(x-1)}{x-9}$$

12.

a.

$$h(x) = \frac{3x-7}{x^2+6x-27} = \frac{3x-7}{(x+9)(x-3)}$$

The denominator cannot equal zero.

Set the linear factors not equal to zero and solve:

$x+9 \neq 0$ and $x-3 \neq 0$

$x \neq -9$ and $x \neq 3$

Domain: x is all real numbers except $x \neq -9$ and $x \neq 3$.

b. Domain: x is all real numbers except $x \neq -2$ and $x \neq 4$.

13.

$$\begin{array}{r} x^2+4x-5 \\ 5x+2 \overline{\smash{\big)}\, 5x^3+22x^2-17x-10} \\ \underline{-(5x^3+2x^2)} \\ 20x^2-17x \\ \underline{-(20x^2+8x)} \\ -25x-10 \\ \underline{-(-25x-10)} \\ 0 \end{array}$$

$(5x^3+22x^2-17x-10) \div (5x+2)$

$= x^2+4x-5$

Chapter 7 Rational Functions

14.

$$\begin{array}{r|rrrrr} -5 & 1 & 5 & 3 & 19 & 20 \\ & & -5 & 0 & -15 & -20 \\ \hline & 1 & 0 & 3 & 4 & 0 \end{array}$$

$(x^4+5x^3+3x^2+19x+20)\div(x+5)$
$= x^3+3x+4$

15.

$$\frac{2x+5}{x-7}=\frac{3x-12}{x-7}$$

$\left(\dfrac{2x+5}{\cancel{x-7}}\right)(\cancel{x-7}) = \left(\dfrac{3x-12}{\cancel{x-7}}\right)(\cancel{x-7})$

$2x+5 = 3x-12$
$-x = -17$
$x = 17$

16.

$$\frac{5}{x+6} = \frac{3}{x-4}$$

$\left(\dfrac{5}{\cancel{x+6}}\right)(\cancel{x+6})(x-4) = \left(\dfrac{3}{\cancel{x-4}}\right)(x+6)(\cancel{x-4})$

$5(x-4) = 3(x+6)$
$5x-20 = 3x+18$
$2x = 38$
$x = 19$

17.

$$\frac{5}{x+2}+7 = \frac{-16}{x-5}$$

$\left(\dfrac{5}{\cancel{x+2}}\right)(\cancel{x+2})(x-5)+7(x+2)(x-5) = \left(\dfrac{-16}{\cancel{x-5}}\right)(x+2)(\cancel{x-5})$

$5(x-5)+7(x^2-3x-10) = -16(x+2)$
$5x-25+7x^2-21x-70 = -16x-32$
$7x^2-16x-95 = -16x-32$
$7x^2 = 63$
$\dfrac{7x^2}{7} = \dfrac{63}{7}$
$x^2 = 9$
$x = \pm\sqrt{9}$
$x = \pm 3$

18.

$$\frac{x^2}{(x+2)(x+3)}+\frac{5}{x+3} = \frac{4}{x+2}$$

$\left(\dfrac{x^2}{\cancel{(x+2)(x+3)}}\right)\cancel{(x+2)(x+3)} + \left(\dfrac{5}{\cancel{x+3}}\right)(x+2)\cancel{(x+3)}$
$= \left(\dfrac{4}{\cancel{x+2}}\right)\cancel{(x+2)}(x+3)$

$x^2+5(x+2) = 4(x+3)$
$x^2+5x+10 = 4x+12$
$x^2+x-2 = 0$
$(x-1)(x+2) = 0$
$x-1 = 0,\; x+2 = 0$
$x = 1,\; x = -2$

When x equals -2 the denominators are zero so some of the original fractions are undefined.
The final solution is $x = 1$.

19.

$$\frac{5.6}{x^2-3x-10} = \frac{2x}{x^2+9x+14}$$

$$\frac{5.6}{(x+2)(x-5)} = \frac{2x}{(x+2)(x+7)}$$

$\left(\dfrac{5.6}{\cancel{(x+2)(x-5)}}\right)\cancel{(x+2)(x-5)}(x+7)$
$= \left(\dfrac{2x}{\cancel{(x+2)(x+7)}}\right)\cancel{(x+2)}(x-5)\cancel{(x+7)}$

$5.6(x+7) = 2x(x-5)$
$5.6x+39.2 = 2x^2-10x$
$0 = 2x^2-15.6x-39.2$

$x = \dfrac{15.6\pm\sqrt{(-15.6)^2-4(2)(-39.2)}}{2(2)}$

$x = \dfrac{15.6\pm\sqrt{556.96}}{4}$

$x = \dfrac{15.6+\sqrt{556.96}}{4},\quad x = \dfrac{15.6-\sqrt{556.96}}{4}$

$x = 9.8,\quad x = -2$

When x equals -2 the denominators are zero so the original fractions are undefined.
The final solution is $x = 9.8$.

20.

a. Let $U(t)$ be the unemployment rate for the state of Florida t years since 2000. Then $U(t) = \dfrac{80.5t+287.5}{141.5t+7826.8}$

b.

$$U(3) = \frac{80.5(3) + 287.5}{141.5(3) + 7826.8}$$

$U(3) \approx 0.064$

$U(3) = 6.4\%$

In 2003 the unemployment rate for the state of Florida was about 6.4%.

c.

$$0.08 = \frac{80.5t + 287.5}{141.5t + 7826.8}$$

$$0.08(141.5t + 7826.8) = \left(\frac{80.5t + 287.5}{141.5t + 7826.8}\right)(141.5t + 7826.8)$$

$$11.32t + 626.144 = 80.5t + 287.5$$

$$-69.18t = -338.644$$

$$\frac{-69.18t}{-69.18} = \frac{-338.644}{-69.18}$$

$$t \approx 4.895$$

The unemployment rate for the state of Florida reached 8% in 2004.

Another method for estimating the solution is by looking at the below graph:

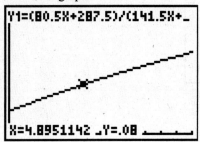

The unemployment rate for the state of Florida reached 8% in 2005.

Chapter 8 Radical Functions
Section 8.1

1.

a.

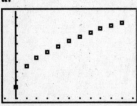

A radical function seems appropriate here because the scatterplot increases rapidly and curves slightly.

b.

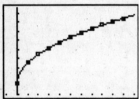

The graph fits the data well and seems to follow the pattern.

c. The cost to produce 550,000 heaters is approximately $8.57 million.

d. The cost to produce 1,500,000 heaters is approximately $12.07 million.

e. Domain: $[0,15]$ Range: $[3.2, 12.07]$

3.

a. The graph fits the data well and seems to follow the pattern.

b. A basilisk lizard with body mass of 50 grams will have a leg length of 0.061 meters.

c. A basilisk lizard with body mass of 100 grams will have a leg length of 0.077 meters.

d.

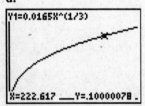

A basilisk lizard with leg length of 0.1 meter will have a body mass of 222.6 grams.

5.

a. The airspeed of a butterfly with a thoracic mass of 0.05 gram is 2.92 meters per second.

b. The airspeed of a butterfly with a thoracic mass of 0.12 gram is 4.93 meters per second.

c. Range: $[0.28, 12.95]$

7.

a.

$f(x) = 2.5\sqrt{x}$

$f(81) = 2.5\sqrt{81}$

$f(81) = 2.5(9)$

$f(81) = 22.5$

b. $f(25) = 12.5$

X	Y1
21	11.456
22	11.726
23	11.99
24	12.247
25	12.5
26	12.748
27	12.99

X=25

9.

a.

$f(x) = 6\sqrt[5]{x}$

$f(45) = 6\sqrt[5]{45}$

$f(45) \approx 12.847$

b. $f(32) = 12$

X	Y1
30	11.846
31	11.924
32	12
33	12.074
34	12.146
35	12.217
36	12.286

X=32

11.

a.

$f(x) = \sqrt{x+8}$

$f(28) = \sqrt{28+8}$

$f(28) = \sqrt{36}$

$f(28) = 6$

b. $f(113) = 11$

X	Y1
111	10.909
112	10.954
113	11
114	11.045
115	11.091
116	11.136
117	11.18

X=113

13.

a.

$f(x) = \sqrt{x-10}$

$f(50) = \sqrt{50-10}$

$f(50) = \sqrt{40}$

$f(50) \approx 6.325$

b. $f(59) = 7$

[calculator table: X=59, Y1=7; surrounding values 56→6.7823, 57→6.8557, 58→6.9282, 60→7.0711, 61→7.1414, 62→7.2111]

15.

a.

$f(x) = \sqrt[7]{x+20}$

$f(-60) = \sqrt[7]{-60+20}$

$f(-60) = \sqrt[7]{-40}$

$f(-60) \approx -1.694$

b. $f(-2207) = -3$

[calculator table: X=-2207, Y1=-3; surrounding values -2237→-3.006, -2227→-3.004, -2217→-3.002, -2197→-2.998, -2187→-2.996, -2177→-2.994]

17.

a.

$f(x) = \sqrt[4]{x^3}$

$f(5) = \sqrt[4]{5^3}$

$f(5) = \sqrt[4]{125}$

$f(5) \approx 3.344$

b. $f(27.473) \approx 12$

[calculator table: X=27.473, Y1=12; surrounding values 27.453→11.993, 27.463→11.997, 27.483→12.003, 27.493→12.007, 27.503→12.01, 27.513→12.013]

19.

a. $f(5) = 3$

b. $f(x) = 1$ when $x = -3$

21.

a. $f(4) = 2$

b. $f(x) = -1.5$ when $x = -7.5$

23.

a. This is an odd root.

b. Domain: $(-\infty, \infty)$

c. Range: $(-\infty, \infty)$

25.

a. This is an even root.

b. Domain: $[-4, \infty)$

c. Range: $[0, \infty)$

27.

a. This is an even root.

b. Domain: $(-\infty, -5]$

c. Range: $[0, \infty)$

29.

a. This is an odd root.

b. Domain: $(-\infty, \infty)$

c. Range: $(-\infty, \infty)$

31.

a. This is an even root.

b. Domain: $(-\infty, -2]$

c. Range: $(-\infty, 0]$

33.

Since it is an odd root:

Domain: All real numbers

Range: All real numbers

35.

Since it is a even root there is a restricted domain and range:

Domain: $x \geq 0$

Range: $[0, \infty)$

37.

Since it is an odd root:

Domain: All real numbers

Range: All real numbers

39.

Since it is a even root there is a restricted domain and range:

Domain: $x \geq 0$

Range: $[0, \infty)$

41.
Since it is a even root there is a restricted domain and range:
$x + 12 \geq 0$
$x + 12 - 12 \geq 0 - 12$
$x \geq -12$
Domain: $x \geq -12$
Range: $[0, \infty)$

43.
Since it is a even root there is a restricted domain and range:
$x + 7 \geq 0$
$x + 7 - 7 \geq 0 - 7$
$x \geq -7$
Domain: $x \geq -7$
Range: $[0, \infty)$

45.
Since it is an odd root:
Domain: All real numbers
Range: All real numbers

47.
Since it is a even root there is a restricted domain and range:
$x + 2 \geq 0$
$x + 2 - 2 \geq 0 - 2$
$x \geq -2$
Domain: $x \geq -2$
Range: $[0, \infty)$

49.
Since it is a even root there is a restricted domain and range:
$x - 23 \geq 0$
$x - 23 + 23 \geq 0 + 23$
$x \geq 23$
Domain: $x \geq 23$
Range: $[0, \infty)$

51.
Since it is a even root there is a restricted domain and range:
$-x + 3 \geq 0$
$-x + 3 - 3 \geq 0 - 3$
$-x \geq -3$
$\dfrac{-x}{-1} \leq \dfrac{-3}{-1}$
$x \leq 3$
Domain: $x \leq 3$
Range: $[0, \infty)$

53.
Since it is a even root there is a restricted domain and range:
$-x - 10 \geq 0$
$-x - 10 + 10 \geq 0 + 10$
$-x \geq 10$
$\dfrac{-x}{-1} \leq \dfrac{10}{-1}$
$x \leq -10$
Domain: $x \leq -10$
Range: $[0, \infty)$

55.
Since it is an odd root:
Domain: All real numbers
Range: All real numbers

57.
Since it is a even root there is a restricted domain and range:
$3x + 5 \geq 0$
$3x + 5 - 5 \geq 0 - 5$
$3x \geq -5$
$\dfrac{3x}{3} \geq \dfrac{-5}{3}$
$x \geq -\dfrac{5}{3}$
Domain: $x \geq -\dfrac{5}{3}$
Range: $[0, \infty)$

59.
Since it is a even root there is a restricted domain and range:
$-4x+9 \geq 0$
$-4x+9-9 \geq 0-9$
$-4x \geq -9$
$\dfrac{-4x}{-4} \leq \dfrac{-9}{-4}$
$x \leq \dfrac{9}{4}$

Domain: $x \leq \dfrac{9}{4}$
Range: $[0,\infty)$

61. $f(x) = \sqrt{x+7}$

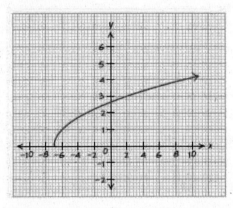

63. $g(x) = \sqrt{x-8}$

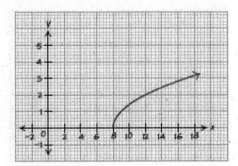

65. $f(x) = -\sqrt{x+3}$

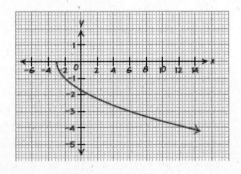

67. $f(x) = -\sqrt{x-6}$

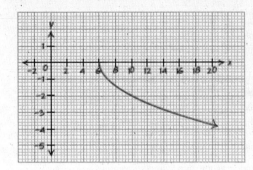

69. $f(x) = \sqrt[3]{x+2}$

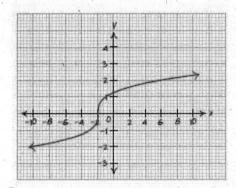

71. $h(x) = \sqrt[3]{x-11}$

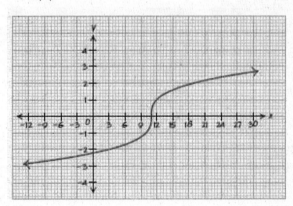

73. $f(x) = \sqrt{9-x}$

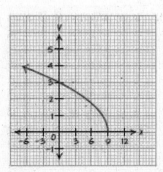

Chapter 8 Radical Functions

75. $g(x) = -\sqrt{x-3}$

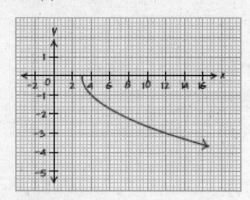

77. $h(x) = \sqrt[5]{2x}$

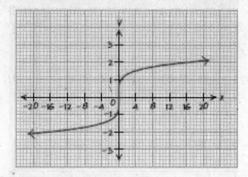

79. $g(x) = 2\sqrt{x+3}$

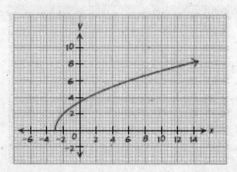

Section 8.2

1.
$$\sqrt{100} = \sqrt{10^2}$$
$$= 10$$

3.
$$\sqrt{121} = \sqrt{11^2}$$
$$= 11$$

5.
$$\sqrt[3]{64} = \sqrt[3]{4^3}$$
$$= 4$$

7.
$$\sqrt[3]{-8} = \sqrt[3]{(-2)^3}$$
$$= -2$$

9. $\sqrt{-9}$ = Not a real number

11.
$$\sqrt{50} = \sqrt{25} \cdot \sqrt{2}$$
$$= \sqrt{5^2} \cdot \sqrt{2}$$
$$= 5\sqrt{2}$$

13.
$$\sqrt{180} = \sqrt{36} \cdot \sqrt{5}$$
$$= \sqrt{6^2} \cdot \sqrt{5}$$
$$= 6\sqrt{5}$$

15.
$$\sqrt[3]{40} = \sqrt[3]{8} \cdot \sqrt[3]{5}$$
$$= \sqrt[3]{2^3} \cdot \sqrt[3]{5}$$
$$= 2\sqrt[3]{5}$$

17.
$$\sqrt[3]{-16} = \sqrt[3]{-8} \cdot \sqrt[3]{2}$$
$$= \sqrt[3]{(-2)^3} \cdot \sqrt[3]{2}$$
$$= -2\sqrt[3]{2}$$

19.
$$\sqrt{49x^2} = \sqrt{7^2} \sqrt{x^2}$$
$$= 7x$$

21.
$$\sqrt[3]{125y^6} = \left(5^3 y^6\right)^{\frac{1}{3}}$$
$$= 5y^2$$

23. The student simplified the square root of 10 to 5.
$$\sqrt{10x^2y^8} = \sqrt{x^2y^8} \cdot \sqrt{10}$$
$$= xy^4\sqrt{10}$$

25.
$$\sqrt{196m^4n^2} = \left(14^2 m^4 n^2\right)^{\frac{1}{2}}$$
$$= 14m^2n$$

27.
$$\sqrt{36a^3} = \sqrt{36a^2} \cdot \sqrt{a}$$
$$= \sqrt{6^2 a^2} \cdot \sqrt{a}$$
$$= 6a\sqrt{a}$$

29.
$$\sqrt{1296a^5b^8c^{15}} = \sqrt{1296a^4b^8c^{14}} \cdot \sqrt{ac}$$
$$= \sqrt{36^2 a^4 b^8 c^{14}} \cdot \sqrt{ac}$$
$$= 36a^2b^4c^7\sqrt{ac}$$

31.
$$\sqrt[3]{8x^3y^6} = \left(2^3 x^3 y^6\right)^{\frac{1}{3}}$$
$$= 2xy^2$$

33.
$$\sqrt[3]{3888x^3y^5z^8} = \sqrt[3]{216x^3y^3z^6} \cdot \sqrt[3]{18y^2z^2}$$
$$= \sqrt[3]{6^3 x^3 y^3 z^6} \cdot \sqrt[3]{2 \cdot 3^2 y^2 z^2}$$
$$= 6xyz^2 \sqrt[3]{18y^2z^2}$$

35.
$$\sqrt[4]{960m^3n^4p^5} = \sqrt[4]{16n^4p^4} \cdot \sqrt[4]{60m^3p}$$
$$= \sqrt[4]{2^4 n^4 p^4} \cdot \sqrt[4]{2^2 \cdot 3 \cdot 5 \cdot m^3 p}$$
$$= 2np\sqrt[4]{60m^3p}$$

37.
$$\sqrt[5]{-32c^5d^{10}} = \left[(-2)^5 c^5 d^{10}\right]^{\frac{1}{5}}$$
$$= -2cd^2$$

39.
$$\sqrt[5]{32m^5n^{10}} = \left(2^5 m^5 n^{10}\right)^{\frac{1}{5}}$$
$$= 2mn^2$$

41. $\sqrt{5x} + 3\sqrt{5x} = 4\sqrt{5x}$

43. $5\sqrt{t} - 3\sqrt{t} = 2\sqrt{t}$

45. $10\sqrt[3]{5b} + 4\sqrt[3]{5b} = 14\sqrt[3]{5b}$

47.
$$3n\sqrt{6} + 2n\sqrt{6} + 8n\sqrt{6} = (3+2+8)n\sqrt{6}$$
$$= 13n\sqrt{6}$$

49. $5t\sqrt{11r} - 8t\sqrt{11r} = -3t\sqrt{11r}$

Chapter 8 Radical Functions

51.
$$2\sqrt{4x^2y} - 3x\sqrt{y} = 2\cdot 2x\sqrt{y} - 3x\sqrt{y}$$
$$= 4x\sqrt{y} - 3x\sqrt{y}$$
$$= x\sqrt{y}$$

53.
$$\sqrt[3]{8a} + 4\sqrt[3]{a} = 2\sqrt[3]{a} + 4\sqrt[3]{a}$$
$$= 6\sqrt[3]{a}$$

55. The student added the radicands.
$$8\sqrt{5} + 3\sqrt{5} = 11\sqrt{5}$$

57. The student added terms that are not like terms. These can not be added. The correct answer is just $4\sqrt{7} + 10\sqrt[3]{7}$.

59.
$$7\sqrt{50a^3b^4c} - 2ab\sqrt{162ab^2c}$$
$$= 7\sqrt{25\cdot 2a^2ab^4c} - 2ab\sqrt{81\cdot 2ab^2c}$$
$$= 7\sqrt{25a^2b^4\cdot 2ac} - 2ab\sqrt{81b^2\cdot 2ac}$$
$$= 7\cdot 5ab^2\sqrt{2ac} - 2ab\cdot 9b\sqrt{2ac}$$
$$= 35ab^2\sqrt{2ac} - 18ab^2\sqrt{2ac}$$
$$= 17ab^2\sqrt{2ac}$$

61.
$$7\sqrt[3]{125x^6y^9z^2} - 9x^2y\sqrt[3]{216y^3z^2}$$
$$= 7\sqrt[3]{5^3x^6y^9\cdot z^2} - 9x^2y\sqrt[3]{6^3y^3\cdot z^2}$$
$$= 7\cdot 5x^2y^3\sqrt[3]{z^2} - 9x^2y\cdot 6y\sqrt[3]{z^2}$$
$$= 35x^2y^3\sqrt[3]{z^2} - 54x^2y^2\sqrt[3]{z^2}$$

63.
$$5\sqrt{13x} + 7x\sqrt{2} - 12\sqrt{13x} + 3x\sqrt{2}$$
$$= 5\sqrt{13x} - 12\sqrt{13x} + 7x\sqrt{2} + 3x\sqrt{2}$$
$$= -7\sqrt{13x} + 10x\sqrt{2}$$

65.
$$4\sqrt{18} + 5\sqrt{2} - 8\sqrt{75} + 3\sqrt{48}$$
$$= 4\sqrt{9\cdot 2} + 5\sqrt{2} - 8\sqrt{25\cdot 3} + 3\sqrt{16\cdot 3}$$
$$= 4\cdot 3\sqrt{2} + 5\sqrt{2} - 8\cdot 5\sqrt{3} + 3\cdot 4\sqrt{3}$$
$$= 12\sqrt{2} + 5\sqrt{2} - 40\sqrt{3} + 12\sqrt{3}$$
$$= 17\sqrt{2} - 28\sqrt{3}$$

67.
$$3xy\sqrt{9z^5} + 7xz^2\sqrt{yz} - 2xyz^2\sqrt{z}$$
$$= 3xy\sqrt{9z^4\cdot z} + 7xz^2\sqrt{yz} - 2xyz^2\sqrt{z}$$
$$= 3xy\cdot 3z^2\sqrt{z} + 7xz^2\sqrt{yz} - 2xyz^2\sqrt{z}$$
$$= 9xyz^2\sqrt{z} - 2xyz^2\sqrt{z} + 7xz^2\sqrt{yz}$$
$$= 7xyz^2\sqrt{z} + 7xz^2\sqrt{yz}$$

69.
$$5\sqrt{2x} + 4\sqrt[3]{2x} + 7\sqrt{2x} = 5\sqrt{2x} + 7\sqrt{2x} + 4\sqrt[3]{2x}$$
$$= 12\sqrt{2x} + 4\sqrt[3]{2x}$$

71.
$$3\sqrt[5]{64a^5b^{10}c^3} - 10ab\sqrt[5]{2b^5c^3}$$
$$= 3\sqrt[5]{32\cdot 2a^5b^{10}c^3} - 10ab\sqrt[5]{2b^5c^3}$$
$$= 3\sqrt[5]{2^5a^5b^{10}\cdot 2c^3} - 10ab\sqrt[5]{b^5\cdot 2c^3}$$
$$= 3\cdot 2ab^2\sqrt[5]{2c^3} - 10ab\cdot b\sqrt[5]{2c^3}$$
$$= 6ab^2\sqrt[5]{2c^3} - 10ab^2\sqrt[5]{2c^3}$$
$$= -4ab^2\sqrt[5]{2c^3}$$

73.
$$\sqrt[4]{16x^5y^8} + 7xy^2\sqrt[4]{81x} = \sqrt[4]{16x^4xy^8} + 7xy^2\sqrt[4]{81x}$$
$$= \sqrt[4]{2^4x^4y^8\cdot x} + 7xy^2\sqrt[4]{3^4x}$$
$$= 2xy^2\sqrt[4]{x} + 7xy^2\cdot 3\sqrt[4]{x}$$
$$= 2xy^2\sqrt[4]{x} + 21xy^2\sqrt[4]{x}$$
$$= 23xy^2\sqrt[4]{x}$$

75.
$$\sqrt[3]{7xy} + \sqrt[5]{7xy} - \sqrt[3]{448xy} + \sqrt[5]{224xy}$$
$$= \sqrt[3]{7xy} + \sqrt[5]{7xy} - \sqrt[3]{64\cdot 7xy} + \sqrt[5]{32\cdot 7xy}$$
$$= \sqrt[3]{7xy} + \sqrt[5]{7xy} - \sqrt[3]{4^3\cdot 7xy} + \sqrt[5]{2^5\cdot 7xy}$$
$$= \sqrt[3]{7xy} + \sqrt[5]{7xy} - 4\sqrt[3]{7xy} + 2\sqrt[5]{7xy}$$
$$= \sqrt[3]{7xy} - 4\sqrt[3]{7xy} + \sqrt[5]{7xy} + 2\sqrt[5]{7xy}$$
$$= -3\sqrt[3]{7xy} + 3\sqrt[5]{7xy}$$

77. Add the sides of the triangle to get the perimeter.
$$\text{Perimeter} = \sqrt{32}\text{ in} + \sqrt{18}\text{ in} + \sqrt{50}\text{ in}$$
$$= \sqrt{16\cdot 2}\text{ in} + \sqrt{9\cdot 2}\text{ in} + \sqrt{25\cdot 2}\text{ in}$$
$$= 4\sqrt{2}\text{ in} + 3\sqrt{2}\text{ in} + 5\sqrt{2}\text{ in}$$
$$= 12\sqrt{2}\text{ inches}$$

79. Yes.
$$8^2 = 64 \rightarrow \sqrt{64} = 8$$
$$4^3 = 64 \rightarrow \sqrt[3]{64} = 4$$
$$2^6 = 64 \rightarrow \boxed{\sqrt[6]{64} = 2}$$

81.
$$\sqrt{64} = 8$$
$$\sqrt[3]{64} = 4$$

82.
$$\sqrt[3]{4096} = 16$$
$$\sqrt[4]{4096} = 8$$

Section 8.3

1.
$$\sqrt{7} \cdot \sqrt{3} = \sqrt{7 \cdot 3}$$
$$= \sqrt{21}$$

3.
$$\sqrt{6} \cdot \sqrt{10} = \sqrt{6 \cdot 10}$$
$$= \sqrt{60}$$
$$= \sqrt{4 \cdot 15}$$
$$= 2\sqrt{15}$$

5.
$$\sqrt{5x} \cdot \sqrt{10x} = \sqrt{5x \cdot 10x}$$
$$= \sqrt{5 \cdot 5 \cdot 2 \cdot x^2}$$
$$= \sqrt{5^2 x^2 \cdot 2}$$
$$= 5x\sqrt{2}$$

7.
$$\sqrt{3m} \cdot \sqrt{12n} = \sqrt{3m \cdot 12n}$$
$$= \sqrt{36mn}$$
$$= 6\sqrt{mn}$$

9.
$$\sqrt[3]{4x} \cdot \sqrt[3]{2x^2} = \sqrt[3]{4x \cdot 2x^2}$$
$$= \sqrt[3]{8x^3} = 2x$$

11.
$$\sqrt{7m} \cdot \sqrt{14mn} = \sqrt{7m \cdot 14mn}$$
$$= \sqrt{7^2 \cdot 2 \cdot m^2 n}$$
$$= 7m\sqrt{2n}$$

13.
$$\sqrt{12a^3 b} \cdot \sqrt{15a^2 b^5} = \sqrt{12a^3 b \cdot 15a^2 b^5}$$
$$= \sqrt{4 \cdot 3 \cdot 3 \cdot 5 a^5 b^6}$$
$$= \sqrt{36 a^4 b^6 \cdot 5a}$$
$$= 6a^2 b^3 \sqrt{5a}$$

15.
$$5\sqrt{3xy} \cdot 7x\sqrt{2y} = 35x\sqrt{3xy \cdot 2y}$$
$$= 35x\sqrt{6xy^2}$$
$$= 35xy\sqrt{6x}$$

17.
$$5n^3 \sqrt[4]{m^2 n^2} \cdot 7m^2 n \sqrt[4]{m^3 n^7} = 35 m^2 n^4 \sqrt[4]{m^2 n^2 \cdot m^3 n^7}$$
$$= 35 m^2 n^4 \sqrt[4]{m^5 n^9}$$
$$= 35 m^2 n^4 \sqrt[4]{m^4 n^8 \cdot mn}$$
$$= 35 m^2 n^4 \cdot mn^2 \sqrt[4]{mn}$$
$$= 35 m^3 n^6 \sqrt[4]{mn}$$

19.
$$2x^2 \sqrt[3]{5x^2 y} \cdot 7y \sqrt[3]{4xy} = 14x^2 y \sqrt[3]{5x^2 y \cdot 4xy}$$
$$= 14x^2 y \sqrt[3]{20 x^3 y^2}$$
$$= 14x^2 y \cdot x \sqrt[3]{20 y^2}$$
$$= 14x^3 y \sqrt[3]{20 y^2}$$

21.
$$12x^2 y \sqrt[5]{7x^3 y^4} \cdot 3xy^3 \sqrt[5]{98 x^4 y^6} = 36 x^3 y^4 \sqrt[5]{7x^3 y^4 \cdot 98 x^4 y^6}$$
$$= 36 x^3 y^4 \sqrt[5]{686 x^7 y^{10}}$$
$$= 36 x^3 y^4 \sqrt[5]{x^5 y^{10} \cdot 686 x^2}$$
$$= 36 x^3 y^4 \cdot xy^2 \sqrt[5]{686 x^2}$$
$$= 36 x^4 y^6 \sqrt[5]{686 x^2}$$

23.
$$(4+\sqrt{5})(3+\sqrt{2}) = 12 + 4\sqrt{2} + 3\sqrt{5} + \sqrt{10}$$

25.
$$(5+\sqrt{6})(3-\sqrt{15}) = 15 - 5\sqrt{15} + 3\sqrt{6} - \sqrt{90}$$
$$= 15 - 5\sqrt{15} + 3\sqrt{6} - \sqrt{9 \cdot 10}$$
$$= 15 - 5\sqrt{15} + 3\sqrt{6} - 3\sqrt{10}$$

27.
$$(3+\sqrt{2x})(6+\sqrt{6x}) = 18 + 3\sqrt{6x} + 6\sqrt{2x} + \sqrt{12x^2}$$
$$= 18 + 3\sqrt{6x} + 6\sqrt{2x} + \sqrt{4x^2 \cdot 3}$$
$$= 18 + 3\sqrt{6x} + 6\sqrt{2x} + 2x\sqrt{3}$$

29.
$$(\sqrt{3} - \sqrt{c})(\sqrt{6} - \sqrt{3c}) = \sqrt{18} - \sqrt{9c} - \sqrt{6c} + \sqrt{3c^2}$$
$$= \sqrt{9 \cdot 2} - \sqrt{9c} - \sqrt{6c} + \sqrt{3c^2}$$
$$= 3\sqrt{2} - 3\sqrt{c} - \sqrt{6c} + c\sqrt{3}$$

31.
$$(2+\sqrt{5})^2 = (2+\sqrt{5})(2+\sqrt{5})$$
$$= 4 + 2\sqrt{5} + 2\sqrt{5} + 5$$
$$= 9 + 4\sqrt{5}$$

33.
$$(4+\sqrt{3m})^2 = (4+\sqrt{3m})(4+\sqrt{3m})$$
$$= 16 + 4\sqrt{3m} + 4\sqrt{3m} + 3m$$
$$= 16 + 8\sqrt{3m} + 3m$$

35.
$$(3+\sqrt{7})(3-\sqrt{7}) = 9 - 3\sqrt{7} + 3\sqrt{7} - 7$$
$$= 2$$

Chapter 8 Radical Functions

37.

$(8+5\sqrt{7ab})(8-5\sqrt{7ab}) = 64 - 40\sqrt{7ab} + 40\sqrt{7ab} - 25\cdot 7ab$
$\phantom{(8+5\sqrt{7ab})(8-5\sqrt{7ab})} = 64 - 175ab$

39. The student distributed the exponent to each term. The correct method would be to use the FOIL method. The correct answer is:

$(5+\sqrt{3})^2 = (5+\sqrt{3})(5+\sqrt{3})$
$\phantom{(5+\sqrt{3})^2} = 25 + 5\sqrt{3} + 5\sqrt{3} + 3$
$\phantom{(5+\sqrt{3})^2} = 28 + 10\sqrt{3}$

41. $\sqrt{\dfrac{20}{5}} = \sqrt{4} = 2$

43. $\sqrt{\dfrac{72}{8}} = \sqrt{9} = 3$

45.

$\sqrt{\dfrac{7x^5}{25x}} = \sqrt{\dfrac{7x^4}{25}}$
$\phantom{\sqrt{\dfrac{7x^5}{25x}}} = \dfrac{\sqrt{7x^4}}{\sqrt{25}}$
$\phantom{\sqrt{\dfrac{7x^5}{25x}}} = \dfrac{x^2\sqrt{7}}{5}$

47.

$\sqrt{\dfrac{5x^3}{45xy^2}} = \sqrt{\dfrac{x^2}{9y^2}}$
$\phantom{\sqrt{\dfrac{5x^3}{45xy^2}}} = \dfrac{x}{3y}$

49.

$\dfrac{7}{\sqrt{5n}} = \dfrac{7\cdot\sqrt{5n}}{\sqrt{5n}\cdot\sqrt{5n}}$
$\phantom{\dfrac{7}{\sqrt{5n}}} = \dfrac{7\sqrt{5n}}{5n}$

51.

$\sqrt{\dfrac{7cd^2}{3c}} = \sqrt{\dfrac{7d^2}{3}}$
$\phantom{\sqrt{\dfrac{7cd^2}{3c}}} = \dfrac{\sqrt{7d^2}}{\sqrt{3}}$
$\phantom{\sqrt{\dfrac{7cd^2}{3c}}} = \dfrac{d\sqrt{7}\cdot\sqrt{3}}{\sqrt{3}\cdot\sqrt{3}}$
$\phantom{\sqrt{\dfrac{7cd^2}{3c}}} = \dfrac{d\sqrt{21}}{3}$

53.

$\dfrac{5\sqrt{2x}}{3\sqrt{7y}} = \dfrac{5\sqrt{2x}\cdot\sqrt{7y}}{3\sqrt{7y}\cdot\sqrt{7y}}$
$\phantom{\dfrac{5\sqrt{2x}}{3\sqrt{7y}}} = \dfrac{5\sqrt{14xy}}{3\cdot 7y}$
$\phantom{\dfrac{5\sqrt{2x}}{3\sqrt{7y}}} = \dfrac{5\sqrt{14xy}}{21y}$

55.

$\sqrt[3]{\dfrac{5x}{7x^2y^2}} = \sqrt[3]{\dfrac{5}{7xy^2}}$
$\phantom{\sqrt[3]{\dfrac{5x}{7x^2y^2}}} = \dfrac{\sqrt[3]{5}\cdot\sqrt[3]{7^2x^2y}}{\sqrt[3]{7xy^2}\cdot\sqrt[3]{7^2x^2y}}$
$\phantom{\sqrt[3]{\dfrac{5x}{7x^2y^2}}} = \dfrac{\sqrt[3]{5\cdot 49\cdot x^2\cdot y}}{\sqrt[3]{7^3x^3y^3}}$
$\phantom{\sqrt[3]{\dfrac{5x}{7x^2y^2}}} = \dfrac{\sqrt[3]{245x^2y}}{7xy}$

57.

$\dfrac{\sqrt[3]{9x^2y^2}}{7x\sqrt[3]{2xy^3}} = \dfrac{\sqrt[3]{9xy^2}}{7x\sqrt[3]{2y^3}}$
$\phantom{\dfrac{\sqrt[3]{9x^2y^2}}{7x\sqrt[3]{2xy^3}}} = \dfrac{\sqrt[3]{9xy^2}\cdot\sqrt[3]{2^2}}{7x\sqrt[3]{2y^3}\cdot\sqrt[3]{2^2}}$
$\phantom{\dfrac{\sqrt[3]{9x^2y^2}}{7x\sqrt[3]{2xy^3}}} = \dfrac{\sqrt[3]{4\cdot 9\cdot xy^2}}{7x\sqrt[3]{2^3y^3}}$
$\phantom{\dfrac{\sqrt[3]{9x^2y^2}}{7x\sqrt[3]{2xy^3}}} = \dfrac{\sqrt[3]{36xy^2}}{7x\cdot 2y}$
$\phantom{\dfrac{\sqrt[3]{9x^2y^2}}{7x\sqrt[3]{2xy^3}}} = \dfrac{\sqrt[3]{36xy^2}}{14xy}$

59.

$$\sqrt[4]{\frac{5ab}{8a^3b^6}} = \sqrt[4]{\frac{5}{8a^2b^5}}$$

$$= \frac{\sqrt[4]{5}}{\sqrt[4]{2^3 a^2 b^5}}$$

$$= \frac{\sqrt[4]{5}}{\sqrt[4]{b^4}\sqrt[4]{2^3 a^2 b}}$$

$$= \frac{\sqrt[4]{5}}{b\sqrt[4]{2^3 a^2 b}}$$

$$= \frac{\sqrt[4]{5} \cdot \sqrt[4]{2a^2 b^3}}{b\sqrt[4]{2^3 a^2 b} \cdot \sqrt[4]{2a^2 b^3}}$$

$$= \frac{\sqrt[4]{2 \cdot 5 \cdot a^2 b^3}}{b\sqrt[4]{2^4 a^4 b^4}}$$

$$= \frac{\sqrt[4]{10a^2 b^3}}{b \cdot 2ab}$$

$$= \frac{\sqrt[4]{10a^2 b^3}}{2ab^2}$$

61.

$$\frac{8\sqrt[4]{4xy}}{\sqrt[4]{144xy^3 z^2}} = \frac{8}{\sqrt[4]{36y^2 z^2}}$$

$$= \frac{8 \cdot \sqrt[4]{2^2 \cdot 3^2 \cdot y^2 \cdot z^2}}{\sqrt[4]{2^2 \cdot 3^2 \cdot y^2 \cdot z^2} \cdot \sqrt[4]{2^2 \cdot 3^2 \cdot y^2 \cdot z^2}}$$

$$= \frac{8\sqrt[4]{4 \cdot 9 \cdot y^2 z^2}}{\sqrt[4]{2^4 3^4 y^4 z^4}}$$

$$= \frac{8\sqrt[4]{36y^2 z^2}}{2 \cdot 3 \cdot y \cdot z}$$

$$= \frac{8\sqrt[4]{36y^2 z^2}}{6yz}$$

$$= \frac{4\sqrt[4]{36y^2 z^2}}{3yz}$$

63. The student multiplied by the reciprocal but the student was supposed to simplify first and then multiply the numerator and the denominator by $\sqrt{2}$.

$$\frac{8\sqrt{7xy}}{\sqrt{2x}} = \frac{8\sqrt{7y}}{\sqrt{2}}$$

$$= \frac{8\sqrt{7y} \cdot \sqrt{2}}{\sqrt{2} \cdot \sqrt{2}}$$

$$= \frac{8\sqrt{14y}}{2}$$

$$= 4\sqrt{14y}$$

65.

$$\frac{2+\sqrt{6}}{\sqrt{3}} = \frac{(2+\sqrt{6}) \cdot \sqrt{3}}{\sqrt{3} \cdot \sqrt{3}}$$

$$= \frac{2\sqrt{3} + \sqrt{18}}{3}$$

$$= \frac{2\sqrt{3} + 3\sqrt{2}}{3}$$

67.

$$\frac{2+\sqrt{3}}{8\sqrt{15}} = \frac{(2+\sqrt{3}) \cdot \sqrt{15}}{8\sqrt{15} \cdot \sqrt{15}}$$

$$= \frac{2\sqrt{15} + \sqrt{45}}{8 \cdot 15}$$

$$= \frac{2\sqrt{15} + 3\sqrt{5}}{120}$$

69.

$$\frac{5}{2+\sqrt{3}} = \frac{5(2-\sqrt{3})}{(2+\sqrt{3})(2-\sqrt{3})}$$

$$= \frac{10 - 5\sqrt{3}}{4 - 2\sqrt{3} + 2\sqrt{3} - \sqrt{9}}$$

$$= \frac{10 - 5\sqrt{3}}{4 - 3}$$

$$= \frac{10 - 5\sqrt{3}}{1}$$

$$= 10 - 5\sqrt{3}$$

71.

$$\frac{2+\sqrt{3}}{5-\sqrt{7}} = \frac{(2+\sqrt{3})(5+\sqrt{7})}{(5-\sqrt{7})(5+\sqrt{7})}$$

$$= \frac{10 + 2\sqrt{7} + 5\sqrt{3} + \sqrt{21}}{25 + 5\sqrt{7} - 5\sqrt{7} - \sqrt{49}}$$

$$= \frac{10 + 2\sqrt{7} + 5\sqrt{3} + \sqrt{21}}{25 - 7}$$

$$= \frac{10 + 2\sqrt{7} + 5\sqrt{3} + \sqrt{21}}{18}$$

73.

$$\frac{9}{3+\sqrt{2x}} = \frac{9(3-\sqrt{2x})}{(3+\sqrt{2x})(3-\sqrt{2x})}$$

$$= \frac{27 - 9\sqrt{2x}}{9 - 3\sqrt{2x} + 3\sqrt{2x} - \sqrt{4x^2}}$$

$$= \frac{27 - 9\sqrt{2x}}{9 - 2x}$$

75.

$$\frac{4+2\sqrt{7x}}{2-3\sqrt{12x}} = \frac{(4+2\sqrt{7x})(2+3\sqrt{12x})}{(2-3\sqrt{12x})(2+3\sqrt{12x})}$$

$$= \frac{8+12\sqrt{12x}+4\sqrt{7x}+6\sqrt{84x^2}}{4+6\sqrt{12x}-6\sqrt{12x}-9\sqrt{144x^2}}$$

$$= \frac{8+12\sqrt{12x}+4\sqrt{7x}+6\cdot 2x\sqrt{21}}{4-9\cdot 12x}$$

$$= \frac{8+12\sqrt{12x}+4\sqrt{7x}+12x\sqrt{21}}{4-108x}$$

$$= \frac{4(2+3\sqrt{12x}+\sqrt{7x}+3x\sqrt{21})}{4(1-27x)}$$

$$= \frac{2+3\sqrt{12x}+\sqrt{7x}+3x\sqrt{21}}{1-27x}$$

77.

$$\frac{2+5\sqrt{m}}{3-\sqrt{mn}} = \frac{(2+5\sqrt{m})(3+\sqrt{mn})}{(3-\sqrt{mn})(3+\sqrt{mn})}$$

$$= \frac{6+2\sqrt{mn}+15\sqrt{m}+5\sqrt{m^2n}}{9+3\sqrt{mn}-3\sqrt{mn}-\sqrt{m^2n^2}}$$

$$= \frac{6+2\sqrt{mn}+15\sqrt{m}+5m\sqrt{n}}{9-mn}$$

79.

$$\frac{4+8\sqrt{5x}}{2+\sqrt{30x}} = \frac{(4+8\sqrt{5x})(2-\sqrt{30x})}{(2+\sqrt{30x})(2-\sqrt{30x})}$$

$$= \frac{8-4\sqrt{30x}+16\sqrt{5x}-8\sqrt{150x^2}}{4-2\sqrt{30x}+2\sqrt{30x}-\sqrt{900x^2}}$$

$$= \frac{8-4\sqrt{30x}+16\sqrt{5x}-8\cdot 5x\sqrt{6}}{4-30x}$$

$$= \frac{8-4\sqrt{30x}+16\sqrt{5x}-40x\sqrt{6}}{4-30x}$$

81.

$$\sqrt{5x}+7\sqrt{2}-8\sqrt{5x} = 7\sqrt{2}-7\sqrt{5x}$$

83.

$$(\sqrt{3}+\sqrt{5})^2 = (\sqrt{3}+\sqrt{5})(\sqrt{3}+\sqrt{5})$$

$$= \sqrt{9}+\sqrt{15}+\sqrt{15}+\sqrt{25}$$

$$= 3+2\sqrt{15}+5$$

$$= 8+2\sqrt{15}$$

85.

$$\frac{8}{3+\sqrt{5x}} = \frac{8(3-\sqrt{5x})}{(3+\sqrt{5x})(3-\sqrt{5x})}$$

$$= \frac{24-8\sqrt{5x}}{9-3\sqrt{5x}+3\sqrt{5x}-\sqrt{25x^2}}$$

$$= \frac{24-8\sqrt{5x}}{9-5x}$$

87.

$$\sqrt{3x}+(4+\sqrt{2x})(7+\sqrt{6})$$

$$= \sqrt{3x}+28+4\sqrt{6}+7\sqrt{2x}+\sqrt{12x}$$

$$= \sqrt{3x}+28+4\sqrt{6}+7\sqrt{2x}+2\sqrt{3x}$$

$$= 28+4\sqrt{6}+7\sqrt{2x}+3\sqrt{3x}$$

89.

$$\frac{4\sqrt[3]{6m^2n}}{55\sqrt[3]{4m^4n^2}} = \frac{4\sqrt[3]{3}}{55\sqrt[3]{2m^2n}}$$

$$= \frac{4\sqrt[3]{3}\cdot\sqrt[3]{2^2m n^2}}{55\sqrt[3]{2m^2n}\cdot\sqrt[3]{2^2mn^2}}$$

$$= \frac{4\sqrt[3]{4\cdot 3mn^2}}{55\sqrt[3]{2^3m^3n^3}}$$

$$= \frac{4\sqrt[3]{12mn^2}}{55\cdot 2mn}$$

$$= \frac{2\sqrt[3]{12mn^2}}{55mn}$$

Section 8.4

1.
$$\sqrt{x+2} = 5$$
$$\left(\sqrt{x+2}\right)^2 = 5^2$$
$$x+2 = 25$$
$$x+2-2 = 25-2$$
$$x = 23$$

3.
$$\sqrt{3x+4} = 15$$
$$\left(\sqrt{3x+4}\right)^2 = 15^2$$
$$3x+4 = 225$$
$$3x+4-4 = 225-4$$
$$3x = 221$$
$$\frac{3x}{3} = \frac{221}{3}$$
$$x = \frac{221}{3}$$

5.
$$\sqrt{-2t+4} = 7$$
$$\left(\sqrt{-2t+4}\right)^2 = 7^2$$
$$-2t+4 = 49$$
$$-2t+4-4 = 49-4$$
$$-2t = 45$$
$$\frac{-2t}{-2} = \frac{45}{-2}$$
$$t = -\frac{45}{2}$$

7.
$$3\sqrt{2x+5} + 12 = 24$$
$$3\sqrt{2x+5} = 12$$
$$\frac{3\sqrt{2x+5}}{3} = \frac{12}{3}$$
$$\sqrt{2x+5} = 4$$
$$\left(\sqrt{2x+5}\right)^2 = 4^2$$
$$2x+5 = 16$$
$$2x+5-5 = 16-5$$
$$2x = 11$$
$$\frac{2x}{2} = \frac{11}{2}$$
$$x = \frac{11}{2}$$

9.
Use $s = \sqrt{30fd}$ with $d = 150$ and $f = 1$
$$s = \sqrt{30(1)(150)}$$
$$s = \sqrt{4500}$$
$$s \approx 67.08$$
The speed of the car is approximately 67 mph.

11.
Use $s = \sqrt{30fd}$ with $s = 55$ and $d = 115$
$$55 = \sqrt{30f(115)}$$
$$55 = \sqrt{3450f}$$
$$(55)^2 = \left(\sqrt{3450f}\right)^2$$
$$3025 = 3450f$$
$$\frac{3025}{3450} = \frac{3450f}{3450}$$
$$f \approx 0.877$$
The coefficient of friction for this road is 0.877.

13.
Use $T(L) = 2\pi\sqrt{\dfrac{L}{32}}$ with $L = 3$
$$T(3) = 2\pi\sqrt{\dfrac{3}{32}} = 1.92$$
The period of a pendulum of length 3 feet is 1.92 seconds.

15.
Use $T(L) = 2\pi\sqrt{\dfrac{L}{32}}$ with $T(L) = 1$
$$1 = 2\pi\sqrt{\dfrac{L}{32}}$$
$$\frac{1}{2\pi} = \frac{2\pi\sqrt{\dfrac{L}{32}}}{2\pi}$$
$$\frac{1}{2\pi} = \sqrt{\dfrac{L}{32}}$$
$$\left(\frac{1}{2\pi}\right)^2 = \left(\sqrt{\dfrac{L}{32}}\right)^2$$
$$\left(\frac{1}{2\pi}\right)^2 = \frac{L}{32}$$
$$32\left(\frac{1}{2\pi}\right)^2 = L$$
$$L \approx 0.81$$
For a period of 1 second, the pendulum needs to be about 0.81 feet (9.7 inches) in length.

Chapter 8 Radical Functions

17.

Use $t = \sqrt{\dfrac{2h}{32}}$ with $h = 100$

$t = \sqrt{\dfrac{2(100)}{32}}$

$t = \sqrt{\dfrac{200}{16}}$

$t = 2.5$

It takes an object 2.5 seconds to fall 100 feet.

19.

Use $t = \sqrt{\dfrac{2h}{32}}$ with $t = 10$

$10 = \sqrt{\dfrac{2h}{32}}$

$10 = \sqrt{\dfrac{h}{16}}$

$(10)^2 = \left(\sqrt{\dfrac{h}{16}}\right)^2$

$100 = \dfrac{h}{16}$

$h = 1600$

An object should be dropped from a height of 1600 feet if you want it to fall for 10 seconds.

21.

Use $d = \sqrt{1.5h}$ with $h = 6$

$d = \sqrt{1.5(6)}$

$d = \sqrt{9}$

$d = 3$

A person can see a distance of 3 miles if their eye is 6 feet off the ground.

23.

Use $d = \sqrt{1.5h}$ with $d = 175$

$175 = \sqrt{1.5h}$

$175^2 = \left(\sqrt{1.5h}\right)^2$

$30,625 = 1.5h$

$\dfrac{30,625}{1.5} = \dfrac{1.5h}{1.5}$

$h \approx 20,416.7$

A person is at an altitude of 20,417 feet to see a distance of 175 miles.

25.

$5\sqrt{2x} = 40$

$\dfrac{5\sqrt{2x}}{5} = \dfrac{40}{5}$

$\sqrt{2x} = 8$

$\left(\sqrt{2x}\right)^2 = 8^2$

$2x = 64$

$\dfrac{2x}{2} = \dfrac{64}{2}$

$x = 32$

27.

$2.4\sqrt{2.5g + 4} - 7.5 = 4.6$

$2.4\sqrt{2.5g + 4} - 7.5 + 7.5 = 4.6 + 7.5$

$2.4\sqrt{2.5g + 4} = 12.1$

$\dfrac{2.4\sqrt{2.5g + 4}}{2.4} = \dfrac{12.1}{2.4}$

$\sqrt{2.5g + 4} \approx 5.04167$

$\left(\sqrt{2.5g + 4}\right)^2 \approx 5.04167^2$

$2.5g + 4 \approx 25.4184$

$2.5g + 4 - 4 \approx 25.4184 - 4$

$2.5g \approx 21.4184$

$\dfrac{2.5g}{2.5} \approx \dfrac{21.4184}{2.5}$

$g \approx 8.567$

29.

$\sqrt{2x + 5} = \sqrt{3x - 4}$

$\left(\sqrt{2x + 5}\right)^2 = \left(\sqrt{3x - 4}\right)^2$

$2x + 5 = 3x - 4$

$-x + 5 = -4$

$-x = -9$

$x = 9$

31.

$\sqrt{4 + 2z} = \sqrt{3 - 7z}$

$\left(\sqrt{4 + 2z}\right)^2 = \left(\sqrt{3 - 7z}\right)^2$

$4 + 2z = 3 - 7z$

$4 + 9z = 3$

$9z = -1$

$z = -\dfrac{1}{9}$

33.
$$c = 340.3\sqrt{\frac{28 + 273.15}{288.15}}$$
$$c = 340.3\sqrt{\frac{301.15}{288.15}}$$
$$c \approx 347.89$$

When the temperature is $28°C$, the speed of sound is 347.89 meters per second.

35. $C(h) = 2.29\sqrt{h} + 3.2$

a.
$$15 = 2.29\sqrt{h} + 3.2$$
$$15 - 3.2 = 2.29\sqrt{h} + 3.2 - 3.2$$
$$11.8 = 2.29\sqrt{h}$$
$$\frac{11.8}{2.29} = \frac{2.29\sqrt{h}}{2.29}$$
$$\frac{11.8}{2.29} = \sqrt{h}$$
$$\left(\frac{11.8}{2.29}\right)^2 = \left(\sqrt{h}\right)^2$$
$$h \approx 26.552$$

Jim Bob's can make 26.552 hundred thousand space heaters with a budget of $15 million.

b.
$$12 = 2.29\sqrt{h} + 3.2$$
$$12 - 3.2 = 2.29\sqrt{h} + 3.2 - 3.2$$
$$8.8 = 2.29\sqrt{h}$$
$$\frac{8.8}{2.29} = \frac{2.29\sqrt{h}}{2.29}$$
$$\frac{8.8}{2.29} = \sqrt{h}$$
$$\left(\frac{8.8}{2.29}\right)^2 = \left(\sqrt{h}\right)^2$$
$$h \approx 14.767$$

Jim Bob's can make 14.767 hundred thousand space heaters with a budget of $12 million..

37. $L(M) = 0.330\sqrt[3]{M}$

a.
$$1 = 0.330\sqrt[3]{M}$$
$$\frac{1}{0.330} = \frac{0.330\sqrt[3]{M}}{0.330}$$
$$\left(\frac{1}{0.330}\right)^3 = \left(\sqrt[3]{M}\right)^3$$
$$M \approx 27.8$$

The body mass of a mammal with a body length of 1 meter is 27.8 kg.

b.
$$2.4 = 0.330\sqrt[3]{M}$$
$$\frac{2.4}{0.330} = \frac{0.330\sqrt[3]{M}}{0.330}$$
$$\left(\frac{2.4}{0.330}\right)^3 = \left(\sqrt[3]{M}\right)^3$$
$$M \approx 384.7$$

The body mass of a mammal with a body length of 2.4 meters is 384.7 kg.

39.

$a = \sqrt[3]{P^2}$ with $P = 84$

$a = \sqrt[3]{84^2} \approx 19.18$

The average distance from the Sun to Uranus is approximately 19.2 AU.

41.

$a = \sqrt[3]{P^2}$ with $a = 39.5$

$$39.5 = \sqrt[3]{P^2}$$
$$39.5 = P^{2/3}$$
$$(39.5)^{3/2} = \left(P^{2/3}\right)^{3/2}$$
$$P \approx 248.3$$

It takes Pluto approximately 248.3 years to orbit the sun.

43.
$$\sqrt{x-7} - \sqrt{x} = -1$$
$$\sqrt{x-7} - \sqrt{x} = -1$$
$$\sqrt{x-7} = \sqrt{x} - 1$$
$$\left(\sqrt{x-7}\right)^2 = \left(\sqrt{x} - 1\right)^2$$
$$x - 7 = \left(\sqrt{x} - 1\right)\left(\sqrt{x} - 1\right)$$
$$x - 7 = x - 2\sqrt{x} + 1$$
$$-7 = -2\sqrt{x} + 1$$
$$-8 = -2\sqrt{x}$$
$$\frac{-8}{-2} = \frac{-2\sqrt{x}}{-2}$$
$$4 = \sqrt{x}$$
$$4^2 = \left(\sqrt{x}\right)^2$$
$$x = 16$$

Chapter 8 Radical Functions

45.

$\sqrt{2x+3} = 1 - \sqrt{x+5}$

$\left(\sqrt{2x+3}\right)^2 = \left(1 - \sqrt{x+5}\right)^2$

$2x+3 = \left(1 - \sqrt{x+5}\right)\left(1 - \sqrt{x+5}\right)$

$2x+3 = 1 - 2\sqrt{x+5} + x + 5$

$2x+3 = -2\sqrt{x+5} + x + 6$

$x+3 = -2\sqrt{x+5} + 6$

$x-3 = -2\sqrt{x+5}$

$(x-3)^2 = \left(-2\sqrt{x+5}\right)^2$

$x^2 - 6x + 9 = 4(x+5)$

$x^2 - 6x + 9 = 4x + 20$

$x^2 - 10x - 11 = 0$

$(x-11)(x+1) = 0$

$x = 11, \, x = -1$

Both solutions do not check out, therefore there is no solution.

47.

$\sqrt{4x^2 + 9x + 5} = 5 + x$

$\left(\sqrt{4x^2 + 9x + 5}\right)^2 = (5+x)^2$

$4x^2 + 9x + 5 = (5+x)(5+x)$

$4x^2 + 9x + 5 = 25 + 10x + x^2$

$3x^2 - x - 20 = 0$

$x = \dfrac{1 \pm \sqrt{(-1)^2 - 4(3)(-20)}}{2(3)}$

$x = \dfrac{1 \pm \sqrt{241}}{6}$

49.

$\sqrt{w+2} = 4 - w$

$\left(\sqrt{w+2}\right)^2 = (4-w)^2$

$w+2 = (4-w)(4-w)$

$w+2 = 16 - 8w + w^2$

$0 = w^2 - 9w + 14$

$0 = (w-7)(w-2)$

$w = 7, w = 2$

The answer $w = 7$ does not check out, therefore the solution is $w = 2$.

51.

$\sqrt{6x+1} - \sqrt{9x} = -1$

$\sqrt{6x+1} = \sqrt{9x} - 1$

$\left(\sqrt{6x+1}\right)^2 = \left(\sqrt{9x} - 1\right)^2$

$6x+1 = \left(\sqrt{9x} - 1\right)\left(\sqrt{9x} - 1\right)$

$6x+1 = 9x - 2\sqrt{9x} + 1$

$6x = 9x - 2\sqrt{9x}$

$(-3x)^2 = \left(-2\sqrt{9x}\right)^2$

$9x^2 = 4(9x)$

$9x^2 = 36x$

$9x^2 - 36x = 0$

$9x(x-4) = 0$

$9x = 0, x - 4 = 0$

$x = 0, x = 4$

The answer $x = 0$ does not check out, therefore the solution is $x = 4$.

53.

$\sqrt{6x+5} + \sqrt{3x+2} = 5$

$\sqrt{6x+5} = 5 - \sqrt{3x+2}$

$\left(\sqrt{6x+5}\right)^2 = \left(5 - \sqrt{3x+2}\right)^2$

$6x+5 = \left(5 - \sqrt{3x+2}\right)\left(5 - \sqrt{3x+2}\right)$

$6x+5 = 25 - 10\sqrt{3x+2} + 3x + 2$

$6x+5 = 27 + 3x - 10\sqrt{3x+2}$

$3x+5 = 27 - 10\sqrt{3x+2}$

$3x - 22 = -10\sqrt{3x+2}$

$(3x-22)^2 = \left(-10\sqrt{3x+2}\right)^2$

$(3x-22)(3x-22) = 100(3x+2)$

$9x^2 - 132x + 484 = 300x + 200$

$9x^2 - 432x + 284 = 0$

$(3x - 142)(3x - 2) = 0$

$x = \dfrac{142}{3}, x = \dfrac{2}{3}$

The answer $x = \dfrac{142}{3}$ does not check out, therefore the solution is $x = \dfrac{2}{3}$.

55.

$\sqrt{8x+8} - \sqrt{8x-4} = -2$
$\sqrt{8x+8} = -2 + \sqrt{8x-4}$
$\left(\sqrt{8x+8}\right)^2 = \left(-2 + \sqrt{8x-4}\right)^2$
$8x+8 = \left(-2+\sqrt{8x-4}\right)\left(-2+\sqrt{8x-4}\right)$
$8x+8 = 4 - 4\sqrt{8x-4} + 8x - 4$
$8x+8 = 8x - 4\sqrt{8x-4}$
$8 = -4\sqrt{8x-4}$
$\dfrac{8}{-4} = \dfrac{-4\sqrt{8x-4}}{-4}$
$-2 = \sqrt{8x-4}$
$(-2)^2 = \left(\sqrt{8x-4}\right)^2$
$4 = 8x - 4$
$8 = 8x$
$x = 1$

The answer $x=1$ does not check out, therefore there is no solution.

57.

$\sqrt{2x+3} = 1 + \sqrt{x+1}$
$\left(\sqrt{2x+3}\right)^2 = \left(1+\sqrt{x+1}\right)^2$
$2x+3 = \left(1+\sqrt{x+1}\right)\left(1+\sqrt{x+1}\right)$
$2x+3 = 1 + 2\sqrt{x+1} + x + 1$
$2x+3 = 2\sqrt{x+1} + x + 2$
$x+1 = 2\sqrt{x+1}$
$(x+1)^2 = \left(2\sqrt{x+1}\right)^2$
$(x+1)(x+1) = 4(x+1)$
$x^2 + 2x + 1 = 4x + 4$
$x^2 - 2x - 3 = 0$
$(x+1)(x-3) = 0$
$x+1 = 0, x-3 = 0$
$x = -1, x = 3$

59. When the student squared both sides the operation was done incorrectly on the left hand side. $x = 12$

$\sqrt{x+4} - \sqrt{3x} = -2$
$\sqrt{x+4} = \sqrt{3x} - 2$
$\left(\sqrt{x+4}\right)^2 = \left(\sqrt{3x}-2\right)^2$
$x+4 = \left(\sqrt{3x}-2\right)\left(\sqrt{3x}-2\right)$
$x+4 = 3x - 4\sqrt{3x} + 4$
$-2x = -4\sqrt{3x}$
$(-2x)^2 = \left(-4\sqrt{3x}\right)^2$
$4x^2 = 16(3x)$
$4x^2 = 48x$
$4x^2 - 48x = 0$
$4x(x-12) = 0$
$4x = 0, x-12 = 0$
$x = 0, x = 12$

The answer $x=0$ does not check out, therefore the solution is $x=12$.

61.

$\sqrt[3]{2x+5} = 6$
$\left(\sqrt[3]{2x+5}\right)^3 = (6)^3$
$2x+5 = 216$
$2x = 211$
$x = 105.5$

63.

$\sqrt[4]{5x-9} = 2$
$\left(\sqrt[4]{5x-9}\right)^4 = (2)^4$
$5x-9 = 16$
$5x = 25$
$x = 5$

65.

$\sqrt[3]{5x+2} + 7 = 12$
$\sqrt[3]{5x+2} = 5$
$\left(\sqrt[3]{5x+2}\right)^3 = (5)^3$
$5x+2 = 125$
$5x = 123$
$x = 24.6$

67.

$\sqrt[3]{-4x+3} = -5$
$\left(\sqrt[3]{-4x+3}\right)^3 = (-5)^3$
$-4x+3 = -125$
$-4x = -128$
$x = 32$

69.
$\sqrt{7x+2} = 3$
$\left(\sqrt{7x+2}\right)^2 = (3)^2$
$7x+2 = 9$
$7x = 7$
$x = 1$

71.
$\sqrt[3]{2x} \cdot \sqrt[3]{20x} = \sqrt[3]{40x^2}$
$= \sqrt[3]{2^3 \cdot 5 \cdot x^2}$
$= 2\sqrt[3]{5x^2}$

73.
$(6+\sqrt{2x})(6-\sqrt{2x}) = 36 - 6\sqrt{2x} + 6\sqrt{2x} - 2x$
$= 36 - 2x$

75.
$5\sqrt{x-8} = \sqrt{x+4} + 6$
$\left(5\sqrt{x-8}\right)^2 = \left(\sqrt{x+4}+6\right)^2$
$25(x-8) = \left(\sqrt{x+4}+6\right)\left(\sqrt{x+4}+6\right)$
$25x - 200 = x + 4 + 12\sqrt{x+4} + 36$
$25x - 200 = x + 40 + 12\sqrt{x+4}$
$24x - 240 = 12\sqrt{x+4}$
$\dfrac{24x}{12} - \dfrac{240}{12} = \dfrac{12\sqrt{x+4}}{12}$
$2x - 20 = \sqrt{x+4}$
$(2x-20)^2 = \left(\sqrt{x+4}\right)^2$
$(2x-20)(2x-20) = x+4$
$4x^2 - 80x + 400 = x + 4$
$4x^2 - 81x + 396 = 0$
$(4x-33)(x-12) = 0$
$4x - 33 = 0, x - 12 = 0$
$x = 8.25, x = 12$

The answer $x = 8.25$ does not check out, therefore the solution is $x = 12$.

77.
$\sqrt[6]{4ab^2} + 5b\sqrt{32a} + \sqrt[6]{4ab^2} + 7\sqrt{2ab^2}$
$= 2\sqrt[6]{4ab^2} + 5b\sqrt{32a} + 7\sqrt{2ab^2}$
$= 2\sqrt[6]{4ab^2} + 5b\sqrt{2^5 a} + 7\sqrt{2ab^2}$
$= 2\sqrt[6]{4ab^2} + 5b\sqrt{2^4}\sqrt{2a} + 7\sqrt{b^2}\sqrt{2a}$
$= 2\sqrt[6]{4ab^2} + 5b \cdot 2^2 \sqrt{2a} + 7 \cdot b\sqrt{2a}$
$= 2\sqrt[6]{4ab^2} + 20b\sqrt{2a} + 7b\sqrt{2a}$
$= 2\sqrt[6]{4ab^2} + 27b\sqrt{2a}$

79.
$\sqrt{6a+1} = 1 + \sqrt{5a-4}$
$\left(\sqrt{6a+1}\right)^2 = \left(1+\sqrt{5a-4}\right)^2$
$6a+1 = \left(1+\sqrt{5a-4}\right)\left(1+\sqrt{5a-4}\right)$
$6a+1 = 1 + 2\sqrt{5a-4} + 5a - 4$
$6a+1 = 2\sqrt{5a-4} + 5a - 3$
$a + 4 = 2\sqrt{5a-4}$
$(a+4)^2 = \left(2\sqrt{5a-4}\right)^2$
$(a+4)(a+4) = 4(5a-4)$
$a^2 + 8a + 16 = 20a - 16$
$a^2 - 12a + 32 = 0$
$(a-4)(a-8) = 0$
$a = 4, a = 8$

81. Quadratic
$4t^2 + 3t = 232$
$4t^2 + 3t - 232 = 0$
$(4t-29)(t+8) = 0$
$4t - 29 = 0, t + 8 = 0$
$4t = 29, t = -8$
$t = 7.25, t = -8$

83. Linear
$6g + 8 = 3(g-4)$
$6g + 8 = 3g - 12$
$3g + 8 = -12$
$3g = -20$
$g = -\dfrac{20}{3}$

85. Other

$2x^3 + 8x^2 = 24x$

$2x^3 + 8x^2 - 24x = 0$

$2x(x^2 + 4x - 12) = 0$

$2x(x+6)(x-2) = 0$

$2x = 0, x + 6 = 0, x - 2 = 0$

$x = 0, x = -6, x = 2$

87. Radical

$\sqrt{2n+6} + \sqrt{n-4} = 5$

$\sqrt{2n+6} = 5 - \sqrt{n-4}$

$(\sqrt{2n+6})^2 = (5 - \sqrt{n-4})^2$

$2n + 6 = (5 - \sqrt{n-4})(5 - \sqrt{n-4})$

$2n + 6 = 25 - 10\sqrt{n-4} + n - 4$

$2n + 6 = 21 - 10\sqrt{n-4} + n$

$n - 15 = -10\sqrt{n-4}$

$(n-15)^2 = (-10\sqrt{n-4})^2$

$(n-15)(n-15) = 100(n-4)$

$n^2 - 30n + 225 = 100n - 400$

$n^2 - 130n + 625 = 0$

$(n-125)(n-5) = 0$

$n = 125, n = 5$

The answer $n = 125$ does not check out, therefore the solution is $n = 5$.

89. Other

$7d^4 = 45927$

$d^4 = 6561$

$d^4 - 6561 = 0$

$(d^2 - 81)(d^2 + 81) = 0$

$d^2 = 81, d^2 = -81$

$d = \pm\sqrt{81}, \cancel{d = \pm\sqrt{-81}}$

$x = \pm 9$

Chapter 8 Radical Functions
Section 8.5

1.
$$\sqrt{-64} = \sqrt{-1} \cdot \sqrt{64}$$
$$= i \cdot 8$$
$$= 8i$$

3.
$$\sqrt{-81} = \sqrt{-1} \cdot \sqrt{81}$$
$$= i \cdot 9$$
$$= 9i$$

5. $-\sqrt{36} = -6$

7.
$$\sqrt{-20} = \sqrt{-1} \cdot \sqrt{20}$$
$$= i \cdot \sqrt{4} \cdot \sqrt{5}$$
$$= 2i\sqrt{5}$$
$$\approx 4.472i$$

9.
$$\sqrt{-200} = \sqrt{-1} \cdot \sqrt{200}$$
$$= i \cdot \sqrt{100} \cdot \sqrt{2}$$
$$= 10i\sqrt{2}$$
$$\approx 14.142i$$

11.
$$12 - \sqrt{-100} = 12 - \sqrt{-1} \cdot \sqrt{100}$$
$$= 12 - i \cdot 10$$
$$= 12 - 10i$$

13.
$$\sqrt{-4} + \sqrt{-36} = \sqrt{-1} \cdot \sqrt{4} + \sqrt{-1} \cdot \sqrt{36}$$
$$= i \cdot 2 + i \cdot 6$$
$$= 2i + 6i$$
$$= 8i$$

15.
$$\sqrt{9} + \sqrt{-16} = \sqrt{9} + \sqrt{-1} \cdot \sqrt{16}$$
$$= 3 + i \cdot 4$$
$$= 3 + 4i$$

17.
$$\sqrt{-3} + \sqrt{-5.6} = \sqrt{-1} \cdot \sqrt{3} + \sqrt{-1} \cdot \sqrt{5.6}$$
$$= i\sqrt{3} + i\sqrt{5.6}$$
$$\approx 1.732i + 2.366i$$
$$\approx 4.098i$$

19.
$$-4(3 + \sqrt{-63}) = -12 - 4\sqrt{-63}$$
$$= -12 - 4\sqrt{-1} \cdot \sqrt{63}$$
$$= -12 - 4i \cdot 3\sqrt{7}$$
$$= -12 - 4 \cdot 3i\sqrt{7}$$
$$= -12 - 12i\sqrt{7}$$
$$\approx -12 - 31.749i$$

21. Real part = 2   Imaginary part = 5

23. Real part = 2.3   Imaginary part = −4.9

25. Real part = 3   Imaginary part = 0

27. Real part = 0   Imaginary part = 7

29.
$$(2 + 5i) + (6 + 4i) = 2 + 6 + 5i + 4i$$
$$= 8 + 9i$$

31.
$$(5.6 + 3.2i) + (2.3 - 4.9i) = 5.6 + 2.3 + 3.2i - 4.9i$$
$$= 7.9 - 1.7i$$

33.
$$(5i) + (3 - 9i) = 3 + 5i - 9i$$
$$= 3 - 4i$$

35.
$$(5 + 7i) - 10 = 5 - 10 + 7i$$
$$= -5 + 7i$$

37.
$$(3 + 5i) - (7 + 8i) = 3 - 7 + 5i - 8i$$
$$= -4 - 3i$$

39.
$$(4.7 - 3.5i) + (1.8 - 5.7i) = 4.7 + 1.8 - 3.5i - 5.7i$$
$$= 6.5 - 9.2i$$

41.
$$(2 + 8i) + (3 - 8i) = 2 + 3 + 8i - 8i$$
$$= 5$$

43.
$$(2 + 3i)(4 + 7i) = 8 + 14i + 12i + 21i^2$$
$$= 8 + 26i + 21(-1)$$
$$= 8 + 26i - 21$$
$$= -13 + 26i$$

45.
$$(2 - 4i)(3 - 5i) = 6 - 10i - 12i + 20i^2$$
$$= 6 - 22i + 20(-1)$$
$$= 6 - 22i - 20$$
$$= -14 - 22i$$

47.
$$(7.5+3i)(4.5-9i) = 33.75 - 67.5i + 13.5i - 27i^2$$
$$= 33.75 - 67.5i + 13.5i - 27(-1)$$
$$= 33.75 - 54i + 27$$
$$= 60.75 - 54i$$

49.
$$(2+8i)(2+8i) = 4 + 16i + 16i + 64i^2$$
$$= 4 + 32i + 64(-1)$$
$$= 4 + 32i - 64$$
$$= -60 + 32i$$

51.
$$(3-7i)(3+7i) = 9 + 21i - 21i - 49i^2$$
$$= 9 - 49(-1)$$
$$= 9 + 49$$
$$= 58$$

53.
$$(7+6i)^2 = (7+6i)(7+6i)$$
$$= 49 + 42i + 42i + 36i^2$$
$$= 49 + 84i + 36(-1)$$
$$= 49 + 84i - 36$$
$$= 13 + 84i$$

55.
$$\frac{12+9i}{3} = \frac{12}{3} + \frac{9i}{3}$$
$$= 4 + 3i$$

57.
$$\frac{4+7i}{2i} = \left(\frac{4+7i}{2i}\right)\left(\frac{i}{i}\right)$$
$$= \frac{4i + 7i^2}{2i^2}$$
$$= \frac{4i + 7(-1)}{2(-1)}$$
$$= \frac{4i - 7}{-2}$$
$$= \frac{-7}{-2} + \frac{4i}{-2}$$
$$= \frac{7}{2} - 2i$$

59.
$$\frac{2.4+3.7i}{7.2i} = \left(\frac{2.4+3.7i}{7.2i}\right)\left(\frac{i}{i}\right)$$
$$= \frac{2.4i + 3.7i^2}{7.2i^2}$$
$$= \frac{2.4i + 3.7(-1)}{7.2(-1)}$$
$$= \frac{2.4i - 3.7}{-7.2}$$
$$= \frac{-3.7}{-7.2} + \frac{2.4i}{-7.2}$$
$$= \frac{37}{72} - \frac{1}{3}i$$

61.
$$\frac{5+2i}{2+3i} = \left(\frac{5+2i}{2+3i}\right)\left(\frac{2-3i}{2-3i}\right)$$
$$= \frac{10 - 11i - 6i^2}{4 - 9i^2}$$
$$= \frac{10 - 11i - 6(-1)}{4 - 9(-1)}$$
$$= \frac{10 - 11i + 6}{4 + 9}$$
$$= \frac{16 - 11i}{13}$$
$$= \frac{16}{13} - \frac{11}{13}i$$

63.
$$\frac{2-4i}{3-7i} = \left(\frac{2-4i}{3-7i}\right)\left(\frac{3+7i}{3+7i}\right)$$
$$= \frac{6 + 2i - 28i^2}{9 - 49i^2}$$
$$= \frac{6 + 2i - 28(-1)}{9 - 49(-1)}$$
$$= \frac{6 + 2i + 28}{9 + 49}$$
$$= \frac{34 + 2i}{58}$$
$$= \frac{17}{29} + \frac{1}{29}i$$

Chapter 8 Radical Functions

65.

$$\frac{-8-5i}{-2+9i} = \left(\frac{-8-5i}{-2+9i}\right)\left(\frac{-2-9i}{-2-9i}\right)$$
$$= \frac{16+82i+45i^2}{4-81i^2}$$
$$= \frac{16+82i+45(-1)}{4-81(-1)}$$
$$= \frac{16+82i-45}{4+81}$$
$$= \frac{16+82i-45}{4+81}$$
$$= -\frac{29}{85} + \frac{82}{85}i$$

67.

$$\frac{-12-73i}{-5-14i} = \left(\frac{-12-73i}{-5-14i}\right)\left(\frac{-5+14i}{-5+14i}\right)$$
$$= \frac{60+197i-1022i^2}{25-196i^2}$$
$$= \frac{60+197i-1022(-1)}{25-196(-1)}$$
$$= \frac{60+197i+1022}{25+196}$$
$$= \frac{1082+197i}{221}$$
$$= \frac{1082}{221} + \frac{197}{221}i$$

69.

$x^2 + 9 = 0$
$x^2 = -9$
$x = \pm\sqrt{-9}$
$x = \pm 3i$

71.

$3t^2 + 45 = -3$
$3t^2 = -48$
$\frac{3t^2}{3} = \frac{-48}{3}$
$t^2 = -16$
$t = \pm\sqrt{-16}$
$t = \pm 4i$

73.

$3(x-8)^2 + 12 = 4$
$3(x-8)^2 = -8$
$\frac{3(x-8)^2}{3} = -\frac{8}{3}$
$(x-8)^2 = -\frac{8}{3}$
$x - 8 = \pm\sqrt{-\frac{8}{3}}$
$x = 8 \pm i\sqrt{\frac{8}{3}} \cdot \sqrt{\frac{3}{3}}$
$x = 8 \pm i\sqrt{\frac{24}{9}}$
$x = 8 \pm \frac{2\sqrt{6}}{3}i$

75.

$2x^2 + 5x + 4 = -4$
$2x^2 + 5x + 8 = 0$
$x = \frac{-5 \pm \sqrt{5^2 - 4(2)(8)}}{2(2)}$
$x = \frac{-5 \pm \sqrt{-39}}{4}$
$x = -\frac{5}{4} \pm \frac{\sqrt{39}}{4}i$

77.

$5x^3 + 2x^2 - 7x = 0$
$x(5x^2 + 2x - 7) = 0$
$x(5x+7)(x-1) = 0$
$x = 0, \ 5x + 7 = 0, \ x - 1 = 0$
$x = 0, \ x = -\frac{7}{5}, \ x = 1$

79.

$0.25x^3 - 3.5x^2 + 16.25x = 0$
$x(0.25x^2 - 3.5x + 16.25) = 0$
$x = 0, \ 0.25x^2 - 3.5x + 16.25 = 0$
$x = 0, \ x = \frac{3.5 \pm \sqrt{(-3.5)^2 - 4(0.25)(16.25)}}{2(0.25)}$
$x = 0, \ x = \frac{3.5 \pm \sqrt{-4}}{0.5}$
$x = 0, \ x = \frac{3.5 \pm 2i}{0.5}$
$x = 0, \ x = 7 \pm 4i$

81.

$2x^2 - 28x + 45 = -54$

$2x^2 - 28x + 99 = 0$

$x = \dfrac{28 \pm \sqrt{(-28)^2 - 4(2)(99)}}{2(2)}$

$x = \dfrac{28 \pm \sqrt{-8}}{4}$

$x = 7 \pm \dfrac{\sqrt{2}}{2}i$

Chapter 8 Radical Functions
Chapter 8 Review

1.
a.
$P(3) = 10.3 + 2\sqrt{3}$
$P(3) \approx 13.76$
In the third month of the year Big Jim's Mart had a profit of about $13.76 thousand dollars.

b.
$P(8) = 10.3 + 2\sqrt{8}$
$P(8) \approx 15.96$
In August Big Jim's Mart had a profit of about $15.96 thousand dollars.

2.
a.
$W(4) = 22.37\sqrt[5]{4^3}$
$W(4) \approx 51.39$
A four-year-old alpaca weighs about 51.39 kilograms.

3. $f(x) = -5\sqrt{x}$
Domain: $[0, \infty)$
Range: $(-\infty, 0]$

4. $h(x) = 5\sqrt{x}$
Domain: $[0, \infty)$
Range: $[0, \infty)$

5. $f(x) = \sqrt{3x+5}$
Domain: $\left[-\dfrac{5}{3}, \infty\right)$
Range: $[0, \infty)$

6. $g(x) = \sqrt{2x-7}$
Domain: $\left[\dfrac{7}{2}, \infty\right)$
Range: $[0, \infty)$

7. $f(x) = \sqrt{4-x}$
Domain: $(-\infty, 4]$
Range: $[0, \infty)$

8. $h(x) = \sqrt{7-x}$
Domain: $(-\infty, 7]$
Range: $[0, \infty)$

9. $f(x) = -2\sqrt[3]{x}$
Domain: All real numbers
Range: All real numbers

10. $f(x) = 4\sqrt[3]{x} + 3$
Domain: All real numbers
Range: All real numbers

11. $k(a) = 3\sqrt[4]{a}$
Domain: $[0, \infty)$
Range: $[0, \infty)$

12. $M(p) = -2.3\sqrt[6]{p}$
Domain: $[0, \infty)$
Range: $(-\infty, 0]$

13. $g(x) = 3.4\sqrt[5]{x}$
Domain: All real numbers
Range: All real numbers

14. $f(x) = 8\sqrt[9]{x-7}$
Domain: All real numbers
Range: All real numbers

15.
Domain: $[5, \infty)$
Range: $(-\infty, 0]$

16.
Domain: $(-\infty, 18]$
Range: $(-\infty, 0]$

17.
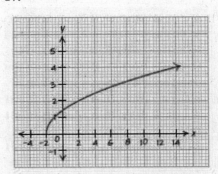

18.
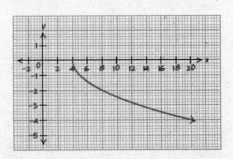

19.

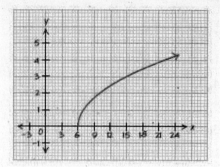

20.

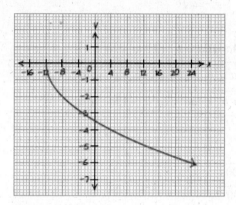

21.

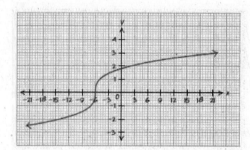

22.

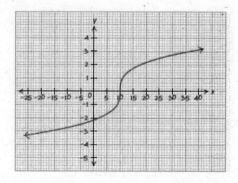

23.

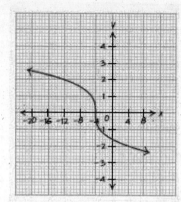

24.

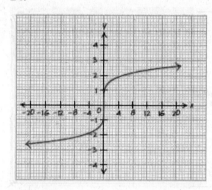

25. $\sqrt{121} = 11$

26. $\sqrt{49} = 7$

27.
$$\sqrt{24x} = \sqrt{4 \cdot 6x}$$
$$= \sqrt{2^2}\sqrt{6x}$$
$$= 2\sqrt{6x}$$

28.
$$\sqrt{72m} = \sqrt{36 \cdot 2m}$$
$$= \sqrt{6^2}\sqrt{2m}$$
$$= 6\sqrt{2m}$$

29.
$$\sqrt{144x^2y^4} = \sqrt{12^2}\sqrt{x^2}\sqrt{y^4}$$
$$= 12xy^2$$

30.
$$\sqrt{100a^3b} = \sqrt{10^2 a^2}\sqrt{ab}$$
$$= 10a\sqrt{ab}$$

31.
$$5mn^3\sqrt{180m^4n^5} = 5mn^3\sqrt{36\cdot 5\cdot m^4n^4n}$$
$$= 5mn^3\sqrt{6^2m^4n^4}\sqrt{5n}$$
$$= 5mn^3\cdot 6m^2n^2\sqrt{5n}$$
$$= 30m^3n^5\sqrt{5n}$$

32.
$$2c^2d\sqrt{220c^7d^6} = 2c^2d\sqrt{4\cdot 55c^6cd^6}$$
$$= 2c^2d\sqrt{2^2c^6d^6}\sqrt{55c}$$
$$= 2c^2d\cdot 2c^3d^3\sqrt{55c}$$
$$= 4c^5d^4\sqrt{55c}$$

33.
$$\sqrt[3]{-8x^3y^7} = \sqrt[3]{(-2)^3x^3y^6}\sqrt[3]{y}$$
$$= -2xy^2\sqrt[3]{y}$$

34.
$$\sqrt[3]{64x^3y^5} = \sqrt[3]{4^3x^3y^3}\sqrt[3]{y^2}$$
$$= 4xy\sqrt[3]{y^2}$$

35.
$$\sqrt[4]{56a^2b^{10}} = \sqrt[4]{56a^2b^8b^2}$$
$$= \sqrt[4]{b^8}\sqrt[4]{56a^2b^2}$$
$$= b^2\sqrt[4]{56a^2b^2}$$

36.
$$\sqrt[4]{25m^3n^{11}} = \sqrt[4]{25m^3n^8n^3}$$
$$= \sqrt[4]{n^8}\sqrt[4]{25m^3n^3}$$
$$= n^2\sqrt[4]{25m^3n^3}$$

37.
$$\sqrt[5]{32a^5b^{10}c^{30}} = \sqrt[5]{2^5a^5b^{10}c^{30}}$$
$$= 2ab^2c^6$$

38.
$$\sqrt[5]{-40a^7b^3c^5} = \sqrt[5]{-40a^5a^2b^3c^5}$$
$$= \sqrt[5]{a^5c^5}\sqrt[5]{-40a^2b^3}$$
$$= ac\sqrt[5]{-40a^2b^3}$$

39. $5\sqrt{3x} + 2\sqrt{3x} = 7\sqrt{3x}$

40.
$$3x\sqrt{5y} - 6\sqrt{5x^2y} = 3x\sqrt{5y} - 6x\sqrt{5y}$$
$$= -3x\sqrt{5y}$$

41.
$$5x\sqrt{3xy^3} + 7\sqrt{27x^3y^3} + 4\sqrt{3y}$$
$$= 5x\sqrt{3xy^2y} + 7\sqrt{9\cdot 3x^2xy^2y} + 4\sqrt{3y}$$
$$= 5xy\sqrt{3xy} + 21xy\sqrt{3xy} + 4\sqrt{3y}$$
$$= 26xy\sqrt{3xy} + 4\sqrt{3y}$$

42.
$$4\sqrt{3a} + 7\sqrt[3]{3a} + 8\sqrt{3a} = 12\sqrt{3a} + 7\sqrt[3]{3a}$$

43.
$$\sqrt[3]{27a^3b^2} - 5a\sqrt[3]{b^2} = 3a\sqrt[3]{b^2} - 5a\sqrt[3]{b^2}$$
$$= -2a\sqrt[3]{b^2}$$

44.
$$\sqrt[5]{-64a^7b^3c^{10}} + 4ac\sqrt[5]{2a^2b^3c^5} + 19ac^2\sqrt[5]{2a^2b^3}$$
$$= \sqrt[5]{(-2)^5\cdot 2a^5a^2b^3c^{10}} + 4ac\sqrt[5]{2a^2b^3c^5} + 19ac^2\sqrt[5]{2a^2b^3}$$
$$= -2ac^2\sqrt[5]{2a^2b^3} + 4ac^2\sqrt[5]{2a^2b^3} + 19ac^2\sqrt[5]{2a^2b^3}$$
$$= 21ac^2\sqrt[5]{2a^2b^3}$$

45.
$$\sqrt{15}\cdot\sqrt{20} = \sqrt{300}$$
$$= \sqrt{100\cdot 3}$$
$$= 10\sqrt{3}$$

46.
$$\sqrt{3xy}\cdot\sqrt{6x} = \sqrt{18x^2y}$$
$$= \sqrt{9\cdot 2x^2y}$$
$$= 3x\sqrt{2y}$$

47.
$$4\sqrt{5a^3b}\cdot 3\sqrt{2ab} = 12\sqrt{10a^4b^2}$$
$$= 12a^2b\sqrt{10}$$

48.
$$\sqrt[3]{3a^3b^2}\cdot\sqrt[3]{18b^2} = \sqrt[3]{54a^3b^4}$$
$$= \sqrt[3]{3^3\cdot 2a^3b^3b}$$
$$= 3ab\sqrt[3]{2b}$$

49.
$$3x^2y\sqrt[3]{15xy^2z}\cdot 2xy^3\sqrt[3]{18xyz^5} = 6x^3y^4\sqrt[3]{270x^2y^3z^6}$$
$$= 6x^3y^4\sqrt[3]{3^3y^3z^6}\sqrt[3]{10x^2}$$
$$= 18x^3y^5z^2\sqrt[3]{10x^2}$$

50.
$$(9+2\sqrt{5})^2 = 81+36\sqrt{5}+4(5)$$
$$= 81+36\sqrt{5}+20$$
$$= 101+36\sqrt{5}$$

51.
$$(3+5\sqrt{7})(2-3\sqrt{7}) = 6 - 9\sqrt{7} + 10\sqrt{7} - 15(7)$$
$$= 6 + \sqrt{7} - 105$$
$$= -99 + \sqrt{7}$$

52.
$$(3+5\sqrt{6})(3-5\sqrt{6}) = 9 - 15\sqrt{6} + 15\sqrt{6} - 25(6)$$
$$= 9 - 150$$
$$= -141$$

53.
$$\sqrt{\frac{36}{2}} = \sqrt{18}$$
$$= \sqrt{9}\sqrt{2}$$
$$= 3\sqrt{2}$$

54.
$$\frac{\sqrt{24}}{\sqrt{3}} = \sqrt{8}$$
$$= \sqrt{4}\sqrt{2}$$
$$= 2\sqrt{2}$$

55.
$$\frac{\sqrt{5x}}{\sqrt{3x}} = \frac{\sqrt{5}}{\sqrt{3}}$$
$$= \frac{\sqrt{5}}{\sqrt{3}} \cdot \frac{\sqrt{3}}{\sqrt{3}}$$
$$= \frac{\sqrt{15}}{3}$$

56.
$$\sqrt{\frac{7ab}{3ac}} = \frac{\sqrt{7b}}{\sqrt{3c}}$$
$$= \frac{\sqrt{7b} \cdot \sqrt{3c}}{\sqrt{3c} \cdot \sqrt{3c}}$$
$$= \frac{\sqrt{21bc}}{3c}$$

57.
$$\frac{5x}{\sqrt[3]{2x^2 y}} = \frac{5x \cdot \sqrt[3]{2^2 xy^2}}{\sqrt[3]{2x^2 y} \cdot \sqrt[3]{2^2 xy^2}}$$
$$= \frac{5x\sqrt[3]{4xy^2}}{\sqrt[3]{2^3 x^3 y^3}}$$
$$= \frac{5x\sqrt[3]{4xy^2}}{2xy}$$
$$= \frac{5\sqrt[3]{4xy^2}}{2y}$$

58.
$$\frac{5}{2+\sqrt{5}} = \frac{5(2-\sqrt{5})}{(2+\sqrt{5})(2-\sqrt{5})}$$
$$= \frac{10 - 5\sqrt{5}}{4 - 2\sqrt{5} + 2\sqrt{5} - \sqrt{25}}$$
$$= \frac{10 - 5\sqrt{5}}{4 - 5}$$
$$= \frac{10 - 5\sqrt{5}}{-1}$$
$$= -10 + 5\sqrt{5}$$

59.
$$\frac{5+2\sqrt{x}}{3-4\sqrt{x}} = \frac{(5+2\sqrt{x})(3+4\sqrt{x})}{(3-4\sqrt{x})(3+4\sqrt{x})}$$
$$= \frac{15 + 20\sqrt{x} + 6\sqrt{x} + 8\sqrt{x^2}}{9 + 12\sqrt{x} - 12\sqrt{x} - 16\sqrt{x^2}}$$
$$= \frac{15 + 26\sqrt{x} + 8x}{9 - 16x}$$

60.
$$\frac{4-\sqrt{7}}{2+\sqrt{6}} = \frac{(4-\sqrt{7})(2-\sqrt{6})}{(2+\sqrt{6})(2-\sqrt{6})}$$
$$= \frac{8 - 4\sqrt{6} - 2\sqrt{7} + \sqrt{42}}{4 - 2\sqrt{6} + 2\sqrt{6} - \sqrt{36}}$$
$$= \frac{8 - 4\sqrt{6} - 2\sqrt{7} + \sqrt{42}}{4 - 6}$$
$$= \frac{8 - 4\sqrt{6} - 2\sqrt{7} + \sqrt{42}}{-2}$$

61.
$$T(2) = 2\pi\sqrt{\frac{2}{32}}$$
$$T(2) \approx 1.57$$

A 2ft pendulum will have a period of about 1.57 seconds.

Chapter 8 Radical Functions

62.

$$2.5 = 2\pi\sqrt{\frac{L}{32}}$$

$$\frac{2.5}{2\pi} = \frac{2\pi}{2\pi}\sqrt{\frac{L}{32}}$$

$$\frac{2.5}{2\pi} = \sqrt{\frac{L}{32}}$$

$$\left(\frac{2.5}{2\pi}\right)^2 = \left(\sqrt{\frac{L}{32}}\right)^2$$

$$\frac{6.25}{4\pi^2} = \frac{L}{32}$$

$$32\left(\frac{6.25}{4\pi^2}\right) = 32\left(\frac{L}{32}\right)$$

$$L \approx 5.07$$

For a pendulum to have a period of 2.5 seconds, it would have to be about 5ft long.

63.

$$17 = 10.3 + 2\sqrt{m}$$

$$17 - 10.3 = 10.3 - 10.3 + 2\sqrt{m}$$

$$6.7 = 2\sqrt{m}$$

$$\frac{6.7}{2} = \frac{2\sqrt{m}}{2}$$

$$\left(\frac{6.7}{2}\right)^2 = \left(\sqrt{m}\right)^2$$

$$m = 11.2225$$

Big Jim's Mart will have a profit of about $17,000 in the 11th month of the year.

64.

$$60 = 22.37\sqrt[5]{a^3}$$

$$\frac{60}{22.37} = \frac{22.37\sqrt[5]{a^3}}{22.37}$$

$$\frac{60}{22.37} = \sqrt[5]{a^3} = a^{3/5}$$

$$\left(\frac{60}{22.37}\right)^{5/3} = \left(a^{3/5}\right)^{5/3}$$

$$a \approx 5.18$$

A 5 year old alpaca will weigh about 60 kilograms.

65.

$$\sqrt{5+x} = 4$$

$$\left(\sqrt{5+x}\right)^2 = 4^2$$

$$5 + x = 16$$

$$x = 11$$

66.

$$\sqrt{3x-7} = 2$$

$$\left(\sqrt{3x-7}\right)^2 = 2^2$$

$$3x - 7 = 4$$

$$3x = 11$$

$$x = \frac{11}{3}$$

67.

$$-3\sqrt{2x-7} + 14 = -1$$

$$-3\sqrt{2x-7} = -15$$

$$\sqrt{2x-7} = 5$$

$$\left(\sqrt{2x-7}\right)^2 = 5^2$$

$$2x - 7 = 25$$

$$2x = 32$$

$$x = 16$$

68.

$$\sqrt{x-4} = \sqrt{2x+8}$$

$$\left(\sqrt{x-4}\right)^2 = \left(\sqrt{2x+8}\right)^2$$

$$x - 4 = 2x + 8$$

$$-x - 4 = 8$$

$$-x = 12$$

$$x = -12$$

Since the domain of the original equation is $x \geq 4$. The equation has no solution.

69.

$$\sqrt{x+2} = \sqrt{3x-7}$$

$$\left(\sqrt{x+2}\right)^2 = \left(\sqrt{3x-7}\right)^2$$

$$x + 2 = 3x - 7$$

$$-2x + 2 = -7$$

$$-2x = -9$$

$$x = 4.5$$

70.
$$\sqrt{x-5} - \sqrt{x} = -1$$
$$\sqrt{x-5} = \sqrt{x} - 1$$
$$\left(\sqrt{x-5}\right)^2 = \left(\sqrt{x}-1\right)^2$$
$$x-5 = \left(\sqrt{x}-1\right)\left(\sqrt{x}-1\right)$$
$$x-5 = \sqrt{x^2} - \sqrt{x} - \sqrt{x} + 1$$
$$x-5 = x - 2\sqrt{x} + 1$$
$$-6 = -2\sqrt{x}$$
$$\sqrt{x} = 3$$
$$\left(\sqrt{x}\right)^2 = 3^2$$
$$x = 9$$

71.
$$\sqrt{x-4} = 5 + \sqrt{3x}$$
$$\left(\sqrt{x-4}\right)^2 = \left(5+\sqrt{3x}\right)^2$$
$$x-4 = \left(5+\sqrt{3x}\right)\left(5+\sqrt{3x}\right)$$
$$x-4 = 25 + 5\sqrt{3x} + 5\sqrt{3x} + \sqrt{9x^2}$$
$$x-4 = 25 + 10\sqrt{3x} + 3x$$
$$-2x - 29 = 10\sqrt{3x}$$
$$(-2x-29)^2 = \left(10\sqrt{3x}\right)^2$$
$$(-2x-29)(-2x-29) = 100 \cdot 3x$$
$$4x^2 + 116x + 841 = 300x$$
$$4x^2 - 184x + 841 = 0$$
$$x = \frac{184 \pm \sqrt{(-184)^2 - 4(4)(841)}}{2(4)}$$
$$x = \frac{184 \pm \sqrt{20400}}{8}$$
$$x = \frac{184 + \sqrt{20400}}{8}, x = \frac{184 - \sqrt{20400}}{8}$$
$$x \approx 40.85, x \approx 5.15$$

The two answers do not check out, therefore there is no solution.

72.
$$\sqrt{x+5} + \sqrt{3x+4} = 13$$
$$\sqrt{3x+4} = 13 - \sqrt{x+5}$$
$$\left(\sqrt{3x+4}\right)^2 = \left(13-\sqrt{x+5}\right)^2$$
$$3x+4 = \left(13-\sqrt{x+5}\right)\left(13-\sqrt{x+5}\right)$$
$$3x+4 = 169 - 13\sqrt{x+5} - 13\sqrt{x+5} + x + 5$$
$$3x+4 = 174 - 26\sqrt{x+5} + x$$
$$2x - 170 = -26\sqrt{x+5}$$
$$(2x-170)^2 = \left(-26\sqrt{x+5}\right)^2$$
$$(2x-170)(2x-170) = 676(x+5)$$
$$4x^2 - 680x + 28900 = 676x + 3380$$
$$4x^2 - 1356x + 25520 = 0$$
$$x = \frac{1356 \pm \sqrt{(-1356)^2 - 4(4)(25520)}}{2(4)}$$
$$x = \frac{1356 \pm \sqrt{1430416}}{8}$$
$$x = \frac{1356 \pm 1196}{8}$$
$$x = 319, x = 20$$

The answer $x = 319$ does not check out, therefore the solution is $x = 20$.

73.
$$\sqrt[3]{2x} = 3$$
$$\left(\sqrt[3]{2x}\right)^3 = 3^3$$
$$2x = 27$$
$$x = 13.5$$

74.
$$\sqrt[4]{5x+2} = 2$$
$$\left(\sqrt[4]{5x+2}\right)^4 = 2^4$$
$$5x + 2 = 16$$
$$5x = 14$$
$$x = 2.8$$

75.
$$\sqrt{-25} = 5i$$

76.
$$\sqrt{-32} = 4i\sqrt{2}$$

77.
$$\sqrt{-4} + \sqrt{-25} = 2i + 5i$$
$$= 7i$$

78.
$$(2+\sqrt{-5})(3-\sqrt{-10}) = (2+i\sqrt{5})(3-i\sqrt{10})$$
$$= 6 - 2i\sqrt{10} + 3i\sqrt{5} - i^2\sqrt{50}$$
$$= 6 - 2i\sqrt{10} + 3i\sqrt{5} - (-1)\cdot 5\sqrt{2}$$
$$= 6 - 2i\sqrt{10} + 3i\sqrt{5} + 5\sqrt{2}$$
$$= (6+5\sqrt{2}) + (-2\sqrt{10}+3\sqrt{5})i$$

79.
$$(2+3i)+(5+7i) = 7+10i$$

80.
$$(3.5+1.2i)+(2.4-3.6i) = 5.9-2.4i$$

81.
$$(4+15i)-(3+11i) = 4-3+15i-11i$$
$$= 1+4i$$

82.
$$(4.5+2.9i)-(1.6-4.2i) = 4.5-1.6+2.9i+4.2i$$
$$= 2.9+7.1i$$

83.
$$(3+4i)(2-7i) = 6-21i+8i-28i^2$$
$$= 6-13i-28i^2$$
$$= 6-13i-28\cdot(-1)$$
$$= 6-13i+28$$
$$= 34-13i$$

84.
$$(6+5i)(2+7i) = 12+42i+10i+35i^2$$
$$= 12+52i+35i^2$$
$$= 12+52i+35\cdot(-1)$$
$$= 12+52i-35$$
$$= -23+52i$$

85.
$$(2-3i)(2+3i) = 4+6i-6i-9i^2$$
$$= 4-9i^2$$
$$= 4-9\cdot(-1)$$
$$= 4+9$$
$$= 13$$

86.
$$(2.3+4.1i)(3.7-9.2i) = 8.51-21.16i+15.17i-37.72i^2$$
$$= 8.51-5.99i-37.72i^2$$
$$= 8.51-5.99i-(37.72)(-1)$$
$$= 8.51-5.99i+37.72$$
$$= 46.23-5.99i$$

87.
$$\frac{12+7i}{3} = \frac{12}{3}+\frac{7i}{3}$$
$$= 4+\frac{7}{3}i$$

88.
$$\frac{12+9i}{3} = \frac{12}{3}+\frac{9i}{3}$$
$$= 4+3i$$

89.
$$\frac{4+7i}{5i} = \frac{(4+7i)i}{(5i)i}$$
$$= \frac{4i+7i^2}{5i^2}$$
$$= \frac{4i+7(-1)}{5(-1)}$$
$$= \frac{4i-7}{-5}$$
$$= \frac{-7}{-5}+\frac{4i}{-5}$$
$$= 1.4-0.8i$$

90.
$$\frac{7}{2+3i} = \frac{7(2-3i)}{(2+3i)(2-3i)}$$
$$= \frac{14-21i}{4-6i+6i-9i^2}$$
$$= \frac{14-21i}{4-9i^2}$$
$$= \frac{14-21i}{4-9(-1)}$$
$$= \frac{14-21i}{4+9}$$
$$= \frac{14-21i}{13}$$
$$= \frac{14}{13}-\frac{21}{13}i$$

91.

$$\frac{5+7i}{3-4i} = \frac{(5+7i)(3+4i)}{(3-4i)(3+4i)}$$

$$= \frac{15+20i+21i+28i^2}{9+12i-12i-16i^2}$$

$$= \frac{15+41i+28i^2}{9-16i^2}$$

$$= \frac{15+41i+28(-1)}{9-16(-1)}$$

$$= \frac{15+41i-28}{9+16}$$

$$= \frac{-13+41i}{25}$$

$$= -\frac{13}{25} + \frac{41}{25}i$$

92.

$$\frac{2-9i}{4+3i} = \frac{(2-9i)(4-3i)}{(4+3i)(4-3i)}$$

$$= \frac{8-6i-36i+27i^2}{16-12i+12i-9i^2}$$

$$= \frac{8-42i+27i^2}{16-9i^2}$$

$$= \frac{8-42i+27(-1)}{16-9(-1)}$$

$$= \frac{8-42i-27}{16+9}$$

$$= \frac{-19-42i}{25}$$

$$= -\frac{19}{25} - \frac{42}{25}i$$

93.

$$\frac{2.5+6.4i}{3.3+8.2i} = \frac{(2.5+6.4i)(3.3-8.2i)}{(3.3+8.2i)(3.3-8.2i)}$$

$$= \frac{8.25-20.5i+21.12i-52.48i^2}{10.89-27.06i+27.06i-67.24i^2}$$

$$= \frac{8.25+0.62i-52.48i^2}{10.89-67.24i^2}$$

$$= \frac{8.25+0.62i-52.48(-1)}{10.89-67.24(-1)}$$

$$= \frac{8.25+0.62i+52.48}{10.89+67.24}$$

$$= \frac{60.73+0.62i}{78.13}$$

$$= \frac{60.73}{78.13} + \frac{0.62}{78.13}i$$

$$\approx 0.777 + 0.008i$$

94.

$$\frac{1.5+7.25i}{3.25-4.5i} = \frac{(1.5+7.25i)(3.25+4.5i)}{(3.25-4.5i)(3.25+4.5i)}$$

$$= \frac{4.875+6.75i+23.5625i+32.625i^2}{10.5625+14.625i-14.625i-20.25i^2}$$

$$= \frac{4.875+30.3125i+32.625i^2}{10.5625-20.25i^2}$$

$$= \frac{4.875+30.3125i+32.625(-1)}{10.5625-20.25(-1)}$$

$$= \frac{4.875+30.3125i-32.625}{10.5625+20.25}$$

$$= \frac{-27.75+30.3125i}{30.8125}$$

$$= \frac{-27.75}{30.8125} + \frac{30.3125}{30.8125}i$$

$$\approx -0.901 + 0.984i$$

95.

$$x^2 + 25 = 0$$

$$x^2 = -25$$

$$x = \pm\sqrt{-25}$$

$$x = \pm 5i$$

96.

$$m^2 - 4 = -20$$

$$m^2 = -16$$

$$m = \pm\sqrt{-16}$$

$$m = \pm 4i$$

97.

$$2(a-7)^2 + 11 = 5$$

$$2(a-7)^2 = -6$$

$$(a-7)^2 = -3$$

$$\sqrt{(a-7)^2} = \pm\sqrt{-3}$$

$$a - 7 = \pm i\sqrt{3}$$

$$a = 7 \pm i\sqrt{3}$$

98.

$$5(g-9)^2 + 25 = 5$$

$$5(g-9)^2 = -20$$

$$(g-9)^2 = -4$$

$$\sqrt{(g-9)^2} = \pm\sqrt{-4}$$

$$g - 9 = \pm 2i$$

$$g = 9 \pm 2i$$

Chapter 8 Radical Functions

99.

$3x^2 + 12x + 15 = 2$

$3x^2 + 12x + 13 = 0$

$x = \dfrac{-12 \pm \sqrt{12^2 - 4(3)(13)}}{2(3)}$

$x = \dfrac{-12 \pm \sqrt{-12}}{6}$

$x = \dfrac{-12 \pm 2i\sqrt{3}}{6}$

$x = -2 \pm \dfrac{\sqrt{3}}{3}i$

100.

$-b^2 - 4b + 6 = 15$

$0 = b^2 + 4b + 9$

$b = \dfrac{-4 \pm \sqrt{4^2 - 4(1)(9)}}{2(1)}$

$b = \dfrac{-4 \pm \sqrt{-20}}{2}$

$b = \dfrac{-4 \pm 2i\sqrt{5}}{2}$

$b = -2 \pm i\sqrt{5}$

101.

$t^3 - 6t^2 = -13t$

$t^3 - 6t^2 + 13t = 0$

$t(t^2 - 6t + 13) = 0$

$t = 0, t^2 - 6t + 13 = 0$

$t = 0, t = \dfrac{6 \pm \sqrt{(-6)^2 - 4(1)(13)}}{2(1)}$

$t = 0, t = \dfrac{6 \pm \sqrt{-16}}{2}$

$t = 0, t = \dfrac{6 \pm 4i}{2}$

$t = 0, t = 3 \pm 2i$

102.

$h^3 + 8h^2 + 65h = 0$

$h(h^2 + 8h + 65) = 0$

$h = 0, h = \dfrac{-8 \pm \sqrt{8^2 - 4(1)(65)}}{2(1)}$

$h = 0, h = \dfrac{-8 \pm \sqrt{-196}}{2}$

$h = 0, h = \dfrac{-8 \pm 14i}{2}$

$h = 0, h = -4 \pm 7i$

Chapter 8 Test

1.

$v(20) = 20.1\sqrt{273+20} \approx 344.1$

The speed of sound in air is about 344.1 meters per second when the temperature is 20°C.

2.

$$90 = 0.243\sqrt{h}$$
$$\frac{90}{0.243} = \frac{0.243\sqrt{h}}{0.243}$$
$$\left(\frac{90}{0.243}\right)^2 = \left(\sqrt{h}\right)^2$$
$$h \approx 137{,}174.2$$

$T(h) = 90$ when $h \approx 137174.2$, add 1500 to allow time to pull the cord. The plane would have to be at an altitude of about 138,674 feet. This is not reasonable so it is not possible for the daredevil to free-fall for 1.5 minutes.

3.

Since it is a even root there is a restricted domain and range:

$x + 7 \geq 0$
$x + 7 - 7 \geq 0 - 7$
$x \geq -7$

Domain: $x \geq -7$
Range: $(-\infty, 0]$

4.

Since it is an odd root:
Domain: All real numbers
Range: All real numbers

5.

Domain: $x \leq 3$
Range: $[0, \infty)$

6. $f(x) = -\sqrt{10-x}$

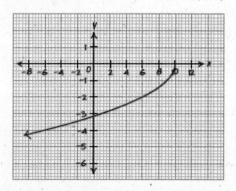

7. $g(x) = \sqrt[3]{x-7}$

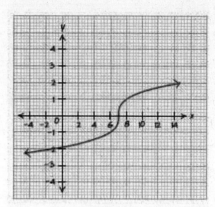

8.

$$\frac{\sqrt{6}}{\sqrt{5x}} = \frac{\sqrt{6}}{\sqrt{5x}} \cdot \frac{\sqrt{5x}}{\sqrt{5x}}$$
$$= \frac{\sqrt{30x}}{5x}$$

9.

$$\sqrt{\frac{8b}{10a}} = \sqrt{\frac{4b}{5a}}$$
$$= \frac{\sqrt{4b}}{\sqrt{5a}}$$
$$= \frac{2\sqrt{b}}{\sqrt{5a}}$$
$$= \frac{2\sqrt{b}}{\sqrt{5a}} \cdot \frac{\sqrt{5a}}{\sqrt{5a}}$$
$$= \frac{2\sqrt{5ab}}{5a}$$

10.

$$\frac{5m}{\sqrt[3]{3mn^2}} = \frac{5m}{\sqrt[3]{3mn^2}} \cdot \frac{\sqrt[3]{3^2 m^2 n}}{\sqrt[3]{3^2 m^2 n}}$$
$$= \frac{5m\sqrt[3]{3^2 m^2 n}}{\sqrt[3]{3^3 m^3 n^3}}$$
$$= \frac{5m\sqrt[3]{9m^2 n}}{3mn}$$
$$= \frac{5\sqrt[3]{9m^2 n}}{3n}$$

Chapter 8 Radical Functions

11.

$$\frac{3-\sqrt{2}}{2+\sqrt{5}} = \frac{(3-\sqrt{2})(2-\sqrt{5})}{(2+\sqrt{5})(2-\sqrt{5})}$$

$$= \frac{6-3\sqrt{5}-2\sqrt{2}+\sqrt{10}}{4-2\sqrt{5}+2\sqrt{5}-5}$$

$$= \frac{6-3\sqrt{5}-2\sqrt{2}+\sqrt{10}}{-1}$$

$$= -(6-3\sqrt{5}-2\sqrt{2}+\sqrt{10})$$

$$= -6+3\sqrt{5}+2\sqrt{2}-\sqrt{10}$$

12.

$$\sqrt{36xy^2} = \sqrt{6^2 y^2} \cdot \sqrt{x}$$

$$= 6y\sqrt{x}$$

13.

$$7\sqrt{120a^2 b^3} = 7\sqrt{2^3 \cdot 3 \cdot 5 \cdot a^2 \cdot b^3}$$

$$= 7 \cdot \sqrt{2^2 a^2 b^2} \cdot \sqrt{2 \cdot 3 \cdot 5 \cdot b}$$

$$= 7 \cdot 2ab \cdot \sqrt{2 \cdot 3 \cdot 5 \cdot b}$$

$$= 14ab\sqrt{30b}$$

14.

$$\sqrt[5]{-32m^3 n^7 p^{10}} = \sqrt[5]{(-2)^5 m^3 n^7 p^{10}}$$

$$= \sqrt[5]{(-2)^5 n^5 p^{10}} \cdot \sqrt[5]{m^3 n^2}$$

$$= -2np^2 \sqrt[5]{m^3 n^2}$$

15.

$$-4(x+5)^2 + 7 = 9$$

$$-4(x+5)^2 = 2$$

$$\frac{-4(x+5)^2}{-4} = \frac{2}{-4}$$

$$(x+5)^2 = -0.5$$

$$x+5 = \pm\sqrt{-0.5}$$

$$x = -5 \pm i\sqrt{0.5}$$

$$x \approx -5 \pm 0.707i$$

16.

$$2.3x^2 + 4.6x + 9 = 5$$

$$2.3x^2 + 4.6x + 4 = 0$$

$$x = \frac{-4.6 \pm \sqrt{4.6^2 - 4(2.3)(4)}}{2(2.3)}$$

$$x = \frac{-4.6 \pm \sqrt{-15.64}}{4.6}$$

$$x \approx -1 \pm 0.86i$$

17.

$$20 = \sqrt{1.5h}$$

$$20^2 = (\sqrt{1.5h})^2$$

$$400 = 1.5h$$

$$\frac{400}{1.5} = \frac{1.5h}{1.5}$$

$$h \approx 266.67$$

To see 20 miles to the horizon a person's eye would have to be about 266.67 ft off the ground.

18.

$$5n\sqrt{6m} - 2\sqrt{24mn^2} = 5n\sqrt{6m} - 2\sqrt{2^3 \cdot 3 \cdot m \cdot n^2}$$

$$= 5n\sqrt{6m} - 2\sqrt{2^2 \cdot n^2} \cdot \sqrt{2 \cdot 3 \cdot m}$$

$$= 5n\sqrt{6m} - 2 \cdot 2n \cdot \sqrt{2 \cdot 3 \cdot m}$$

$$= 5n\sqrt{6m} - 4n\sqrt{6m}$$

$$= n\sqrt{6m}$$

19.

$$2a\sqrt{7ab^5} + 7b\sqrt{28a^3 b^3} + 4b\sqrt{7a}$$

$$= 2a\sqrt{7ab^5} + 7b\sqrt{4 \cdot 7 \cdot a^3 b^3} + 4b\sqrt{7a}$$

$$= 2a\sqrt{b^4} \cdot \sqrt{7ab} + 7b\sqrt{4a^2 b^2} \cdot \sqrt{7ab} + 4b\sqrt{7a}$$

$$= 2a \cdot b^2 \cdot \sqrt{7ab} + 7b \cdot 2ab \cdot \sqrt{7ab} + 4b\sqrt{7a}$$

$$= 2ab^2 \sqrt{7ab} + 14ab^2 \sqrt{7ab} + 4b\sqrt{7a}$$

$$= 16ab^2 \sqrt{7ab} + 4b\sqrt{7a}$$

20.

$$\sqrt{3ab} \cdot 2\sqrt{15ac} = 2\sqrt{3ab} \cdot \sqrt{3 \cdot 5 \cdot ac}$$

$$= 2\sqrt{3^2 \cdot 5 \cdot a^2 bc}$$

$$= 2\sqrt{3^2 a^2} \cdot \sqrt{5bc}$$

$$= 2 \cdot 3a \cdot \sqrt{5bc}$$

$$= 6a\sqrt{5bc}$$

21.

$$\sqrt[3]{4a^4 b} \cdot \sqrt[3]{18b^2} = \sqrt[3]{2^2 \cdot a^4 \cdot b \cdot 2 \cdot 3^2 \cdot b^2}$$

$$= \sqrt[3]{2^3 \cdot 3^2 \cdot a^4 \cdot b^3}$$

$$= \sqrt[3]{2^3 a^3 b^3} \cdot \sqrt[3]{3^2 a}$$

$$= 2ab\sqrt[3]{9a}$$

22.

$2xy\sqrt[3]{18xz^2} \cdot 5xz^2\sqrt[3]{6x^7yz^4}$
$= 2xy \cdot 5xz^2 \cdot \sqrt[3]{2 \cdot 3^2 \cdot xz^2} \cdot \sqrt[3]{2 \cdot 3 \cdot x^7yz^4}$
$= 10x^2yz^2\sqrt[3]{2^2 \cdot 3^3 \cdot x^8 \cdot y \cdot z^6}$
$= 10x^2yz^2 \cdot \sqrt[3]{2^2} \cdot \sqrt[3]{3^3} \cdot \sqrt[3]{x^6} \cdot \sqrt[3]{x^2} \cdot \sqrt[3]{y} \cdot \sqrt[3]{z^6}$
$= 10x^2yz^2 \cdot \sqrt[3]{2^2} \cdot 3 \cdot x^2 \cdot \sqrt[3]{x^2} \cdot \sqrt[3]{y} \cdot z^2$
$= 10x^2yz^2 \cdot 3x^2z^2 \cdot \sqrt[3]{2^2x^2y}$
$= 30x^4yz^4\sqrt[3]{4x^2y}$

23.

$(5+2\sqrt{3})(2-4\sqrt{3}) = 5\cdot 2 - 5\cdot 4\sqrt{3} + 2\cdot 2\sqrt{3} - 2\sqrt{3}\cdot 4\sqrt{3}$
$= 10 - 20\sqrt{3} + 4\sqrt{3} - 24$
$= (10-24) + (-20\sqrt{3} + 4\sqrt{3})$
$= -14 - 16\sqrt{3}$

24.

$2\sqrt{5x+4} - 11 = -3$
$2\sqrt{5x+4} = 8$
$\sqrt{5x+4} = 4$
$\left(\sqrt{5x+4}\right)^2 = 4^2$
$5x+4 = 16$
$5x = 12$
$x = \dfrac{12}{5}$

25.

$\sqrt{x-13} - \sqrt{x} = -1$
$\sqrt{x-13} = \sqrt{x} - 1$
$\left(\sqrt{x-13}\right)^2 = \left(\sqrt{x}-1\right)^2$
$x - 13 = x - 2\sqrt{x} + 1$
$-14 = -2\sqrt{x}$
$\dfrac{-14}{-2} = \dfrac{-2\sqrt{x}}{-2}$
$7 = \sqrt{x}$
$7^2 = \left(\sqrt{x}\right)^2$
$x = 49$

26.

$\sqrt{x-7} + \sqrt{4x-11} = 13$
$\sqrt{4x-11} = 13 - \sqrt{x-7}$
$\left(\sqrt{4x-11}\right)^2 = \left(13-\sqrt{x-7}\right)^2$
$4x-11 = 169 - 26\sqrt{x-7} + x - 7$
$4x-11 = 162 - 26\sqrt{x-7} + x$
$3x - 173 = -26\sqrt{x-7}$
$(3x-173)^2 = \left(-26\sqrt{x-7}\right)^2$
$9x^2 - 1038x + 29{,}929 = 676(x-7)$
$9x^2 - 1038x + 29{,}929 = 676x - 4732$
$9x^2 - 1714x + 34{,}661 = 0$

$x = \dfrac{1714 \pm \sqrt{(-1714)^2 - 4(9)(34{,}661)}}{2(9)}$

$x = \dfrac{1714 + \sqrt{1{,}690{,}000}}{18}, x = \dfrac{1714 - \sqrt{1{,}690{,}000}}{18}$

$x = \dfrac{1507}{9}, x = 23$

Since $x = \dfrac{1507}{9}$ does not check, $x = 23$ is the only solution.

27.

$\sqrt[3]{x+5} = 4$
$\left(\sqrt[3]{x+5}\right)^3 = 4^3$
$x + 5 = 64$
$x = 59$

28.

$(2.7 + 3.4i) + (1.4 - 4.8i) = 4.1 - 1.4i$

29.

$(5+11i) - (4-7i) = 5 + 11i - 4 + 7i$
$= 1 + 18i$

30.

$(7+2i)(3-5i) = 21 - 35i + 6i - 10i^2$
$= 21 - 29i + 10$
$= 31 - 29i$

31.

$(1.5 - 4.5i)(2.25 - 6.5i) = 3.375 - 9.75i - 10.125i + 29.25i^2$
$= 3.375 - 19.875i - 29.25$
$= -25.875 - 19.875i$

Chapter 8 Radical Functions

32.

$$\left(\frac{8}{4+5i}\right)\left(\frac{4-5i}{4-5i}\right) = \frac{32-40i}{16-20i+20i-25i^2}$$

$$= \frac{32-40i}{16+25}$$

$$= \frac{32}{41} - \frac{40}{41}i$$

33.

$$\left(\frac{3+2i}{6-7i}\right)\left(\frac{6+7i}{6+7i}\right) = \frac{18+21i+12i+14i^2}{36+42i-42i-49i^2}$$

$$= \frac{18+33i-14}{36+49}$$

$$= \frac{4+33i}{85}$$

$$= \frac{4}{85} + \frac{33}{85}i$$

34.

a. $v(20) = 58.8\sqrt{273+20} \approx 1006.5$ The speed of sound in helium at 20°C is about 1006.5 meters per second.

b. According to exercise 1, the speed of sound in air (344.1 meters per second at 20°C) is much slower than the speed of sound in helium at 20°C.

Chapter 8 Cumulative Review

1.
$$5(2)^x - 4 = 76$$
$$5(2)^x = 80$$
$$\frac{5(2)^x}{5} = \frac{80}{5}$$
$$(2)^x = 16$$
$$x = 4$$

2.
$$5h + 20 = 3h - 8$$
$$2h + 20 = -8$$
$$2h = -28$$
$$h = -14$$

3.
$$n^2 - 56 = 8$$
$$n^2 = 64$$
$$n = \pm\sqrt{64}$$
$$n = \pm 8$$

4.
$$\frac{4}{a-6} = \frac{7}{a+5}$$
$$\frac{4}{a-6}(a-6)(a+5) = \frac{7}{a+5}(a-6)(a+5)$$
$$4(a+5) = 7(a-6)$$
$$4a + 20 = 7a - 42$$
$$-3a + 20 = -42$$
$$-3a = -66$$
$$a = 22$$

5.
$$3t^2 - 17t = 56$$
$$3t^2 - 17t - 56 = 0$$
$$(3t+7)(t-8) = 0$$
$$t = -\frac{7}{3}, t = 8$$

6.
$$\ln(4x - 9) = 3$$
$$4x - 9 = e^3$$
$$4x = e^3 + 9$$
$$x = \frac{e^3 + 9}{4}$$

7.
$$2\sqrt{x+5} - 12 = 4$$
$$2\sqrt{x+5} = 16$$
$$\frac{2\sqrt{x+5}}{2} = \frac{16}{2}$$
$$\sqrt{x+5} = 8$$
$$\left(\sqrt{x+5}\right)^2 = (8)^2$$
$$x + 5 = 64$$
$$x = 59$$

8.
$$\frac{6}{x+2} + 4 = \frac{4x}{x-3}$$
$$\frac{6}{x+2}(x+2)(x-3) + 4(x+2)(x-3) = \frac{4x}{x-3}(x+2)(x-3)$$
$$6(x-3) + 4(x+2)(x-3) = 4x(x+2)$$
$$6x - 18 + 4(x^2 - x - 6) = 4x^2 + 8x$$
$$6x - 18 + 4x^2 - 4x - 24 = 4x^2 + 8x$$
$$6x - 18 - 4x - 24 = 8x$$
$$2x - 42 = 8x$$
$$-6x - 42 = 0$$
$$-6x = 42$$
$$x = -7$$

9.
$$7c^3 - 120 = -1632$$
$$7c^3 = -1512$$
$$c^3 = -216$$
$$(c^3)^{\frac{1}{3}} = (-216)^{\frac{1}{3}}$$
$$c = -6$$

10.
$$\log_5(x+8) + \log_5(x+12) = 1$$
$$\log_5\left[(x+8)(x+12)\right] = 1$$
$$5^1 = (x+8)(x+12)$$
$$5 = x^2 + 20x + 96$$
$$0 = x^2 + 20x + 91$$
$$0 = (x+13)(x+7)$$
$$x = -13, x = -7$$

The answer $x = -13$ is not part of the domain.
The final answer is:
$$x = -7$$

11.

$\sqrt{x+5} + \sqrt{2x-3} = 3$

$\sqrt{2x-3} = 3 - \sqrt{x+5}$

$(\sqrt{2x-3})^2 = (3 - \sqrt{x+5})^2$

$2x - 3 = (3 - \sqrt{x+5})(3 - \sqrt{x+5})$

$2x - 3 = 9 - 6\sqrt{x+5} + x + 5$

$2x - 3 = -6\sqrt{x+5} + x + 14$

$x - 17 = -6\sqrt{x+5}$

$(x-17)^2 = (-6\sqrt{x+5})^2$

$(x-17)(x-17) = 36(x+5)$

$x^2 - 34x + 289 = 36x + 180$

$x^2 - 70x + 109 = 0$

$x = \dfrac{70 \pm \sqrt{(-70)^2 - 4(1)(109)}}{2(1)}$

$x = \dfrac{70 \pm \sqrt{4464}}{2}$

$x = 35 \pm \sqrt{1116}$

$x = 35 \pm 6\sqrt{31}$

The answer $35 + 6\sqrt{31}$ does not check out.

The final answer is:

$35 - 6\sqrt{31}$

12.

$|2x+7| = 31$

$2x + 7 = 31, \; 2x + 7 = -31$

$2x = 24, \; 2x = -38$

$x = 12, \; x = -19$

13.

$3.4g^2 - 1.6g - 8.4 = 147$

$3.4g^2 - 1.6g - 155.4 = 0$

$g = \dfrac{1.6 \pm \sqrt{(-1.6)^2 - 4(3.4)(-155.4)}}{2(3.4)}$

$g = \dfrac{1.6 \pm \sqrt{2116}}{6.8}$

$g = \dfrac{1.6 \pm 46}{6.8}$

$g = \dfrac{1.6 + 46}{6.8}, \; g = \dfrac{1.6 - 46}{6.8}$

$g = 7, \; g = -\dfrac{111}{17}$

14.

$\dfrac{4}{w^2 + 2w - 24} = \dfrac{-0.5w}{w^2 - 13w + 36}$

$\dfrac{4}{(w+6)(w-4)} = \dfrac{-0.5w}{(w-9)(w-4)}$

$\dfrac{4}{(w+6)(w-4)}(w+6)(w-4)(w-9)$
$= \dfrac{-0.5w}{(w-9)(w-4)}(w+6)(w-4)(w-9)$

$4(w-9) = -0.5w(w+6)$

$4w - 36 = -0.5w^2 - 3w$

$0.5w^2 + 7w - 36 = 0$

$(0.5w - 2)(w + 18) = 0$

$0.5w - 2 = 0, \; w + 18 = 0$

$w = 4, \; w = -18$

The solution $w = 4$ makes the original denomintor zero.

The final answer is:

$w = -18$

15.

$\dfrac{2}{3}(b+5) = \dfrac{5}{6}b - \dfrac{2}{9}$

$\dfrac{2}{3}b + \dfrac{10}{3} = \dfrac{5}{6}b - \dfrac{2}{9}$

$18\left(\dfrac{2}{3}b + \dfrac{10}{3}\right) = 18\left(\dfrac{5}{6}b - \dfrac{2}{9}\right)$

$12b + 60 = 15b - 4$

$-3b = -64$

$b = \dfrac{64}{3}$

16.

$\sqrt[5]{4x-9} = 4$

$\left(\sqrt[5]{4x-9}\right)^5 = (4)^5$

$4x - 9 = 1024$

$4x = 1033$

$x = 258.25$

17.

$4r^2 - 12r + 36 = 0$

$r = \dfrac{12 \pm \sqrt{(-12)^2 - 4(4)(36)}}{2(4)}$

$r = \dfrac{12 \pm \sqrt{-432}}{8}$

$r = \dfrac{12 \pm 12i\sqrt{3}}{8}$

$r = \dfrac{3}{2} \pm \dfrac{3\sqrt{3}}{2}i$

18.

$-3|2t - 5| = -24$

$|2t - 5| = 8$

$2t - 5 = 8, 2t - 5 = -8$

$2t = 13, 2t = -3$

$t = 6.5, t = -1.5$

19.

$1.5^{x^2 - 8} = 2.25$

$\ln\left(1.5^{x^2 - 8}\right) = \ln(2.25)$

$(x^2 - 8)\ln(1.5) = \ln(2.25)$

$x^2 - 8 = \dfrac{\ln(2.25)}{\ln(1.5)}$

$x^2 - 8 = 2$

$x^2 = 10$

$x = \pm\sqrt{10}$

20.

$2d^3 - 4d^2 - 63d + 124 = 63d + 124$

$2d^3 - 4d^2 - 126d = 0$

$2d(d^2 - 2d - 63) = 0$

$2d(d - 9)(d + 7) = 0$

$2d = 0, d - 9 = 0, d + 7 = 0$

$d = 0, d = 9, d = -7$

21.

$\begin{cases} 3x + 6y = 20 \\ 4y = -2x + 15 \end{cases}$

Solve for y in the second equation:

$\dfrac{4y}{4} = \dfrac{-2x}{4} + \dfrac{15}{4}$

$y = -0.5x + 3.75$

Plug this into the first equation:

$3x + 6(-0.5x + 3.75) = 20$

$3x - 3x + 22.5 = 20$

$22.5 = 20$

No Solution.

22. The solution to the system is $(2, 2)$.

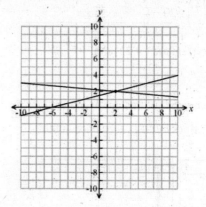

23.

$\begin{cases} y = 3x^2 + 5x - 10 \\ y = \dfrac{1}{4}x - \dfrac{9}{4} \end{cases}$

$3x^2 + 5x - 10 = \dfrac{1}{4}x - \dfrac{9}{4}$

$4(3x^2 + 5x - 10) = 4\left(\dfrac{1}{4}x - \dfrac{9}{4}\right)$

$12x^2 + 20x - 40 = x - 9$

$12x^2 + 19x - 31 = 0$

$(x - 1)(12x + 31) = 0$

$x - 1 = 0, 12x + 31 = 0$

$x = 1, x = -\dfrac{31}{12}$

Substituting x back into either equation solve for y:

$y = \dfrac{1}{4}(1) - \dfrac{9}{4} = -2$

$y = \dfrac{1}{4}\left(-\dfrac{31}{12}\right) - \dfrac{9}{4} = -\dfrac{139}{48}$

The solutions to the system are $(1, -2)$ and $\left(-\dfrac{31}{12}, -\dfrac{139}{48}\right)$.

Chapter 8 Radical Functions

24.

$\begin{cases} y = 2x^2 + 5x - 20 \\ y = -x^2 + 3x + 36 \end{cases}$

$2x^2 + 5x - 20 = -x^2 + 3x + 36$

$3x^2 + 2x - 56 = 0$

$(x-4)(3x+14) = 0$

$x - 4 = 0, 3x + 14 = 0$

$x = 4, x = -\dfrac{14}{3}$

Substituting x back into either equation solve for y:

$y = 2(4)^2 + 5(4) - 20 = 32$

$y = 2\left(-\dfrac{14}{3}\right)^2 + 5\left(-\dfrac{14}{3}\right) - 20 = \dfrac{2}{9}$

The solutions to the system are $(4, 32)$ and $\left(-\dfrac{14}{3}, \dfrac{2}{9}\right)$.

25. Let $M(t)$ be the number of trademark applications for items with stars-and-stripes motif t years since 2000.

a.

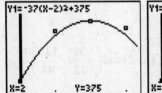

Choose $(2, 375)$ as the vertex and $(0, 227)$ as the additional point used to solve for "a". Then:

$M(t) = a(t-2)^2 + 375$

$227 = a(0-2)^2 + 375$

$227 = a(-2)^2 + 375$

$227 = 4a + 375$

$-148 = 4a$

$\dfrac{-148}{4} = \dfrac{4a}{4}$

$a = -37$

$M(t) = -37(t-2)^2 + 375$

Note: The vertex needed adjustment to better fit the data.

$M(t) = -37(t-2)^2 + 380$

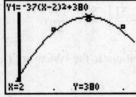

Note: Your answer may vary and yet still represent the trend of the data.

b.

Domain: $[-1, 5]$

$M(-1) = -37(-1-2)^2 + 380 = 47$

$M(2) = -37(2-2)^2 + 380 = 380$

Range: $[47, 380]$

c.

$M(5) = -37(5-2)^2 + 380$

$M(5) = -37(3)^2 + 380$

$M(5) = 47$

In 2005 there were approximately 47 trademark applications for items with a stars-and-stripes motif.

d. The vertex is $(2, 380)$. This means that in the year 2002 there were 380 trademark applications for items with a stars-and-stripes motif.

26. Let $N(t)$ be the net sales (in millions of dollars) for Odyssey Health Care, Inc. t years since 2000.

a.

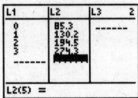

$N(t) = ab^t$

From $(0, 85.3)$ we get $a = 85.3$.

Giving us: $N(t) = 85.3b^t$

To find b use a second point: $(3, 274.3)$

$274.3 = 85.3b^3$

$\dfrac{274.3}{85.3} = \dfrac{85.3b^3}{85.3}$

$\left(\dfrac{274.3}{85.3}\right) = b^3$

$\left(\dfrac{274.3}{85.3}\right)^{1/3} = \left(b^3\right)^{1/3}$

$1.476 \approx b$

Then $N(t) = 85.3(1.476)^t$

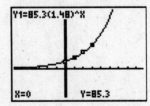

b.
Domain: $[-1,5]$
$N(-1) = 85.3(1.476)^{-1} \approx 57.8$
$N(5) = 85.3(1.476)^5 \approx 597.6$
Range: $[57.8, 597.6]$

c.
$N(-2) = 85.3(1.476)^{-2}$
$N(-2) \approx 39.2$

In 1998, the net sales for Odyssey Health Care were approximately $39.2 million.

d.
$$600 = 85.3(1.476)^t$$
$$\frac{600}{85.3} = \frac{85.3(1.476)^t}{85.3}$$
$$\frac{600}{85.3} = 1.476^t$$
$$\log\left(\frac{600}{85.3}\right) = \log(1.476)^t$$
$$\log\left(\frac{600}{85.3}\right) = t\log(1.476)$$
$$\frac{\log\left(\frac{600}{85.3}\right)}{\log(1.476)} = \frac{t\log(1.476)}{\log(1.476)}$$
$$t \approx 5.0$$

According to the model, the net sales for Odyssey Health Care will reach $600 million in the year 2005.

27. Let $H(t)$ be the average number of hours per year Americans spent listening to recorded music t years since 2000.

a.
Using two points: $(1, 238)$ and $(4, 211)$
Find the slope: $m = \frac{211-238}{4-1} = \frac{-27}{3} = -9$
Vertical Intercept $b = 211 - (-9)4 = 247$
$H(t) = -9t + 247$

b.
$H(-5) = -9(-5) + 247$
$H(-5) = 292$

In 1995 Americans spent an average of 292 hours per year listening to recorded music.

c.
$$150 = -9t + 247$$
$$150 - 247 = -9t + 247 - 247$$
$$-97 = -9t$$
$$\frac{-97}{-9} = \frac{-9t}{-9}$$
$$10.8 \approx t$$

In 2010 Americans will spend an average of 150 hours per year listening to recorded music.

d. The slope is -9. This means that the number of hours per year that Americans spend listening to recorded music is decreasing by 9 hours per year.

e.
Domain: $[-5, 10]$
$H(-5) = -9(-5) + 247 = 292$
$H(10) = -9(10) + 247 = 157$
Range: $[157, 292]$

28.
$$f(x) = \frac{x+2}{x^2 + 6x + 8}$$
$$f(x) = \frac{x+2}{(x+4)(x+2)}$$
Domain: $x \geq -6$ is all real numbers except $x \neq -2$ and $x \neq -4$.

29.
$$\begin{cases} 4x - 6y > 3 \\ y > -2x + 5 \end{cases}$$

Solve for y in the first equation to graph:
$4x - 6y > 3$
$-6y > -4x + 3$
$\frac{-6y}{-6} < \frac{-4x}{-6} + \frac{3}{-6}$
$y < \frac{2}{3}x - \frac{1}{2}$

$$\begin{cases} y < \frac{2}{3}x - \frac{1}{2} \\ y > -2x + 5 \end{cases}$$

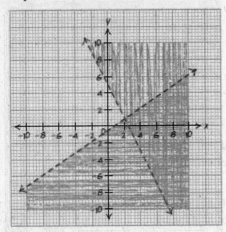

30.
$4x + 20 > 10x - 10$
$-6x + 20 > -10$
$-6x > -30$
$x < 5$

31.
$\frac{1}{4}n + 3 \leq \frac{2}{5}n + \frac{3}{10}$
$20\left(\frac{1}{4}n + 3\right) \leq 20\left(\frac{2}{5}n + \frac{3}{10}\right)$
$5n + 60 \leq 8n + 6$
$-3n + 60 \leq 6$
$-3n \leq -54$
$n \geq 18$

32.
$|2w - 13| \geq 3$
$2w - 13 \geq 3$ or $2w - 13 \leq -3$
$2w \geq 16$ or $2w \leq 10$
$w \geq 8$ or $w \leq 5$

33.
$-2|d + 5| < -10$
$|d + 5| > 5$
$d + 5 > 5$ or $d + 5 < -5$
$d > 0$ or $d < -10$

34.
$|2x + 4| \leq -8$
The absolute value cannot be less than or equal to a negative number.
Therefore, there is no solution

35. $V(t) = 0.275t^2 - 2.77t + 85.73$

a.
$V(15) = 0.275(15)^2 - 2.77(15) + 85.73$
$V(15) \approx 106.06$

In 2005 U.S. amusement parks had approximately 106.06 million visitors.

b.
Vertex: (h, k)

$h = \frac{2.77}{(2)(0.275)} = \frac{2.77}{0.55} \approx 5.04$

$k = V(5.04) = 0.275(5.04)^2 - 2.77(5.04) + 85.73 \approx 78.75$

$(5.04, 78.75)$

In about 1995, U.S amusement parks had their lowest number of visitors with about 78.75 million visitors.

c.
$90 = 0.275t^2 - 2.77t + 85.73$
$0 = 0.275t^2 - 2.77t - 4.27$

$t = \frac{2.77 \pm \sqrt{(-2.77)^2 - 4(0.275)(-4.27)}}{2(0.275)}$

$t \approx 11.4, \; t \approx -1.4$

U.S. amusement parks had about 90 million visitors in 1989 and again in 2001.

36. Let $B(m)$ be the number of salmonella bacteria on the counter m minutes after 2pm. Let n be the number of 10 minute time intervals.

a.

m	$B(m)$
0	4000
10	4000(2)
2(10)	$4000(2)^2$
3(10)	$4000(2)^3$
$n(10)$	$4000(2)^n$

$n(10) = m$
$\frac{n(10)}{10} = \frac{m}{10}$
$n = \frac{m}{10}$
$B(m) = 4000(2)^{m/10}$

b.

$$1{,}000{,}000 = 4000(2)^{m/10}$$

$$\frac{1{,}000{,}000}{4000} = \frac{4000(2)^{m/10}}{4000}$$

$$250 = (2)^{m/10}$$

$$\log_2 250 = \frac{m}{10}$$

$$10\log_2 250 = m$$

$$10\left(\frac{\log 250}{\log 2}\right) = m$$

$$79.7 \approx m$$

After about 80 minutes there will be about 1 million salmonella bacteria on the counter.

c.

6pm → 4 hours = 240 minutes

$$B(240) = 4000(2)^{240/10}$$

$$B(240) \approx 6.71 \times 10^{10}$$

$$B(240) \approx 67{,}100{,}000{,}000$$

When dinner is prepared at 6pm, there will be approximately 67.1 billion salmonella bacteria on the counter.

37.

$$f(x) = \frac{1}{2}x - 5$$

$$y = \frac{1}{2}x - 5$$

$$y + 5 = \frac{1}{2}x$$

$$(2)(y+5) = (2)\left(\frac{1}{2}x\right)$$

$$2y + 10 = x$$

$$2x + 10 = y$$

$$f^{-1}(x) = 2x + 10$$

38.

$$g(x) = -3(5)^x$$

$$y = -3(5)^x$$

$$\frac{y}{-3} = \frac{-3(5)^x}{-3}$$

$$\frac{y}{-3} = (5)^x$$

$$\log_5\left(\frac{y}{-3}\right) = x$$

$$\frac{\log\left(\frac{y}{-3}\right)}{\log 5} = x$$

$$\frac{\log\left(\frac{x}{-3}\right)}{\log 5} = y$$

$$g^{-1}(x) = \frac{\log\left(\frac{x}{-3}\right)}{\log 5}$$

39. $f(x) = -7$

a. $f(9) = -7$

b. Domain: All real numbers   Range: $y = -7$

40. Given $f(x) = 5x + 9$ and $g(x) = -1.4x + 3.2$

a.

$$f(x) + g(x) = (5x+9) + (-1.4x+3.2)$$
$$= 3.6x + 12.2$$

b.

$$f(g(x)) = 5(-1.4x+3.2) + 9$$
$$= -7x + 16 + 9$$
$$= -7x + 25$$

c.

$$f(x)g(x) = (5x+9)(-1.4x+3.2)$$
$$= -7x^2 + 16 - 12.6x + 28.8$$
$$= -7x^2 + 3.4x + 28.8$$

41. Given $f(x) = \frac{1}{2}x - 1$ and $g(x) = -8x + 5$

a.

$$f(6) - g(6) = \left(\frac{1}{2}(6) - 1\right) - (-8(6) + 5)$$
$$= (3 - 1) - (-48 + 5)$$
$$= 2 + 43$$
$$= 45$$

b.
$$g(7) = -8(7) + 5$$
$$= -51$$
$$f(g(7)) = \frac{1}{2}(-51) - 1$$
$$= -26.5$$

c.
$$f(3)g(3) = \left(\frac{1}{2}(3) - 1\right)(-8(3) + 5)$$
$$= (0.5)(-19)$$
$$= -9.5$$

42. Given $f(x) = 2x^2 + 7x - 8$ and $g(x) = 2x - 5$

a.
$$f(x) + g(x) = (2x^2 + 7x - 8) + (2x - 5)$$
$$= 2x^2 + 9x - 13$$

b.
$$f(g(x)) = 2(2x - 5)^2 + 7(2x - 5) - 8$$
$$= 2(4x^2 - 20x + 25) + 14x - 35 - 8$$
$$= 8x^2 - 40x + 50 + 14x - 43$$
$$= 8x^2 - 26x + 7$$

c.
$$f(x)g(x) = (2x^2 + 7x - 8)(2x - 5)$$
$$= 4x^3 - 10x^2 + 14x^2 - 35x - 16x + 40$$
$$= 4x^3 + 4x^2 - 51x + 40$$

43.
$$7.8 = \log\left(\frac{I}{10^{-4}}\right)$$
$$\frac{I}{10^{-4}} = 10^{7.8}$$
$$(10^{-4})\frac{I}{10^{-4}} = (10^{7.8})(10^{-4})$$
$$I = 10^{7.8 - 4} = 10^{3.8}$$
$$I \approx 6309.6$$

An earthquake with a magnitude of 7.8 has an intensity of about 6309.6cm.

44.
Find the slope m:
$$2x + 5y = 11$$
$$5y = -2x + 11$$
$$\frac{5y}{5} = \frac{-2x + 11}{5}$$
$$y = -\frac{2}{5}x + \frac{11}{5}$$
$$m = -\frac{2}{5}$$

This means the perpendicular slope $m_\perp = \frac{5}{2}$

Use $m_\perp = \frac{5}{2}$ and the given point $(-8, 7)$ to find the equation of the line:
$$y - 7 = \frac{5}{2}(x + 8)$$
$$y - 7 = \frac{5}{2}x + 20$$
$$y = \frac{5}{2}x + 27$$

45.
Use two points $(0, -3)$ and $(6, -9)$
$$m = \frac{-9 - (-3)}{6 - 0} = \frac{-6}{6} = -1$$
$$y = -x - 3$$
Dashed line, shaded above uses " > "
$$y > -x - 3$$

46. $3x - 7y = 28$

y-intercept:
$$3(0) - 7y = 28$$
$$0 - 7y = 28$$
$$-7y = 28$$
$$y = -4 \rightarrow (0, -4)$$

x-intercept:
$$3x - 7(0) = 28$$
$$3x + 0 = 28$$
$$3x = 28$$
$$x = \frac{28}{3} \rightarrow (9.33, 0)$$

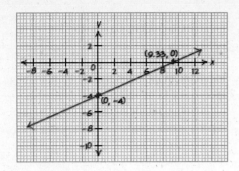

47. $y = \sqrt{x+1}$

y-intercept:
$y = \sqrt{0+1}$
$y = \sqrt{1}$
$y = 1 \to (0,1)$

x-intercept:
$0 = \sqrt{x+1}$
$(0)^2 = \left(\sqrt{x+1}\right)^2$
$0 = x+1$
$x = -1 \to (-1,0)$

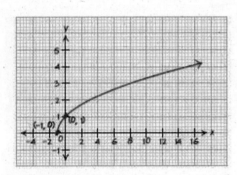

48. $y = 2x^2 - 4x - 18$

vertex: (h,k)
$h = \dfrac{4}{2(2)} = \dfrac{4}{4} = 1$
$k = 2(1)^2 - 4(1) - 18 = -20$
$(1,-20)$

y-intercept:
$y = 2(0)^2 - 4(0) - 18$
$y = 0 - 0 - 18$
$y = -18 \to (0,-18)$

x-intercepts:

$x = \dfrac{4 + \sqrt{(-4)^2 - 4(2)(-18)}}{2(2)} \approx 4.16 \to (4.16, 0)$

$x = \dfrac{4 - \sqrt{(-4)^2 - 4(2)(-18)}}{2(2)} \approx -2.16 \to (-2.16, 0)$

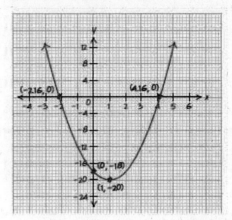

49. $y = 45(0.7)^x$

y-intercept:
$y = 45(0.7)^0$
$y = 45(1)$
$y = 45 \to (0, 45)$

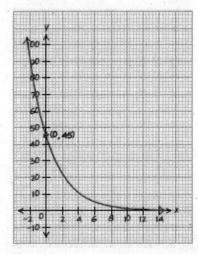

50. $y = -0.3(x-7)^2 + 30$

vertex: (h,k)
$(7, 30)$

y-intercept:
$y = -0.3(0-7)^2 + 30$
$y = 15.3 \to (0, 15.3)$

Chapter 8 Radical Functions

x-intercepts:
$$-0.3(x-7)^2 + 30 = 0$$
$$-0.3(x-7)^2 = -30$$
$$\frac{-0.3(x-7)^2}{-0.3} = \frac{-30}{-0.3}$$
$$(x-7)^2 = 100$$
$$x-7 = \pm\sqrt{100}$$
$$x-7 = \pm 10$$
$$x = 7 \pm 10$$
$$x = 17,\ x = -3$$
$$(17, 0) \text{ and } (-3, 0)$$

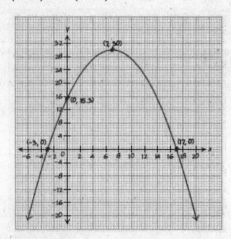

51. $y = \sqrt[3]{5-x}$

y-intercept:
$$y = \sqrt[3]{5-0}$$
$$y = \sqrt[3]{5}$$
$$y \approx 1.7 \to (0, 1.7)$$

x-intercept:
$$0 = \sqrt[3]{5-x}$$
$$(0)^3 = \left(\sqrt[3]{5-x}\right)^3$$
$$0 = 5 - x$$
$$x = 5 \to (5, 0)$$

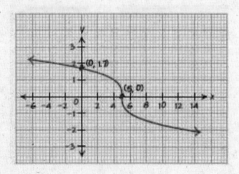

52. $y = 4x - 22$

y-intercept:
$$y = 4(0) - 22$$
$$y = 0 - 22$$
$$y = -22 \to (0, -22)$$

x-intercept:
$$0 = 4x - 22$$
$$22 = 4x$$
$$x = \frac{22}{4}$$
$$x = 5.5 \to (5.5, 0)$$

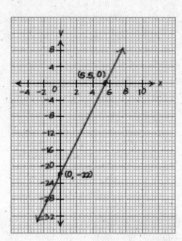

53. $y = 3(4)^x$

y-intercept:
$$y = 3(4)^0$$
$$y = 3(1)$$
$$y = 3 \to (0, 3)$$

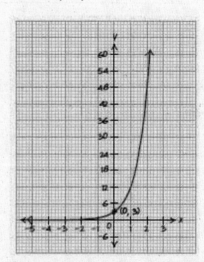

54. $y = \log_7 x$

x-intercept:

$\log_7 x = 0$

$7^0 = x$

$x = 1 \to (1, 0)$

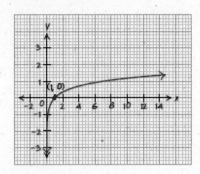

55. $y = -2.5(2)^x - 30$

y-intercept:

$y = -2.5(2)^0 - 30$

$y = -2.5 - 30$

$y = 32.5 \to (0, 32.5)$

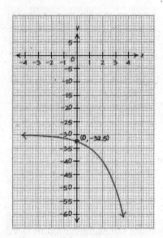

56. $y = -\sqrt{x - 15}$

x-intercept:

$0 = -\sqrt{x - 15}$

$(0)^2 = \left(-\sqrt{x - 15}\right)^2$

$0 = x - 15$

$x = 15 \to (15, 0)$

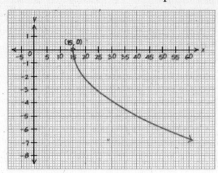

57. $y < x^2 - 3x - 10$

vertex: (h, k)

$h = \dfrac{3}{2} = 1.5$

$k = (1.5)^2 - 3(1.5) - 10 = -12.25$

$(1.5, -12.25)$

y-intercept:

$y = (0)^2 - 3(0) - 10$

$y = -10 \to (0, -10)$

x-intercepts:

$0 = x^2 - 3x - 10$

$x = \dfrac{3 \pm \sqrt{(-3)^2 - 4(1)(-10)}}{2(1)}$

$x = \dfrac{3 \pm \sqrt{49}}{2}$

$x = \dfrac{3 \pm 7}{2}$

$x = \dfrac{3 + 7}{2}, x = \dfrac{3 - 7}{2}$

$x = 5, x = -2$

$(5, 0)$ and $(-2, 0)$

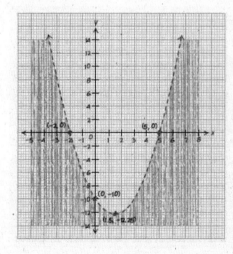

58. $y \geq -0.5x + 6$

y-intercept:
$y = -0.5(0) + 6$
$y = 0 + 6$
$y = 6 \to (0, 6)$

x-intercept:
$0 = -0.5x + 6$
$0.5x = 6$
$x = \dfrac{6}{0.5}$
$x = 12 \to (12, 0)$

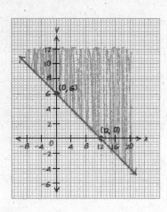

59. $4x - 3y < 12$

y-intercept:
$4(0) - 3y = 12$
$0 - 3y = 12$
$-3y = 12$
$y = -4 \to (0, -4)$

x-intercept:
$4x - 3(0) = 12$
$4x - 0 = 12$
$4x = 12$
$x = 3 \to (3, 0)$

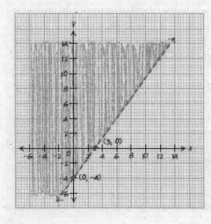

60. $y \geq -\dfrac{1}{2}(x-8)^2 + 4$

vertex: (h, k)
$(8, 4)$

y-intercept:
$y = -\dfrac{1}{2}(0-8)^2 + 4$
$y = -\dfrac{1}{2}(64) + 4$
$y = -28 \to (0, -28)$

x-intercepts:
$-\dfrac{1}{2}(x-8)^2 + 4 = 0$
$-\dfrac{1}{2}(x-8)^2 = -4$
$(x-8)^2 = 8$
$x - 8 = \pm\sqrt{8}$
$x = 8 \pm \sqrt{8}$
$x \approx 10.8,\ x \approx 5.2$
$(10.8, 0)$ and $(5.2, 0)$

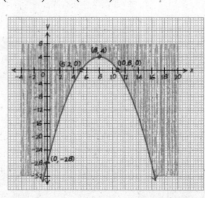

61.

a. Let $P(v)$ be the pressure in pounds per square inch in the syringe when the volume of the syringe is v cubic inches.

$P(v) = \dfrac{k}{v}$ with $P(2) = 4$

$\dfrac{k}{2} = 4$

$k = 8$

$P(v) = \dfrac{8}{v}$

b. $P(1) = \dfrac{8}{1} = 8$

If the volume in the syringe is 1 cubic inch the pressure is 8 pounds per square inch.

62.
$$5\sqrt{x}+7\sqrt{4x}-10 = 5\sqrt{x}+7\cdot 2\sqrt{x}-10$$
$$= 5\sqrt{x}+14\sqrt{x}-10$$
$$= 19\sqrt{x}-10$$

63.
$$(6x^2+8x-14)-(8x^2-3x+5)$$
$$= 6x^2+8x-14-8x^2+3x-5$$
$$= -2x^2+11x-19$$

64.
$$\frac{x+2}{x-8}\cdot\frac{x+5}{x+3} = \frac{(x+2)(x+5)}{(x-8)(x+3)}$$
$$= \frac{x^2+7x+10}{x^2-5x-24}$$

65.
$$(3+\sqrt{2c})(4+\sqrt{18c}) = (3+\sqrt{2c})(4+\sqrt{9\cdot 2c})$$
$$= (3+\sqrt{2c})(4+3\sqrt{2c})$$
$$= 12+9\sqrt{2c}+4\sqrt{2c}+3(2c)$$
$$= 6c+13\sqrt{2c}+12$$

66.

$$\begin{array}{r}8b^2+14b-30\\2b-5\overline{)16b^3-12b^2-130b+150}\\\underline{-(16b^3-40b^2)}\\28b^2-130b\\\underline{-(28b^2-70b)}\\-60b+150\\\underline{-(-60b+150)}\\0\end{array}$$

$$(16b^3-12b^2-130b+150)\div(2b-5)$$
$$= 8b^2+14b-30$$

67.
$$(3x-8)^2 = (3x-8)(3x-8)$$
$$= 9x^2-48x+64$$

68.
$$\frac{4h+7}{h^2+10h+21}\div\frac{h-6}{h^2+3h-28}$$
$$= \frac{4h+7}{h^2+10h+21}\cdot\frac{h^2+3h-28}{h-6}$$
$$= \frac{4h+7}{(h+7)(h+3)}\cdot\frac{(h+7)(h-4)}{h-6}$$
$$= \frac{(4h+7)(h-4)}{(h+3)(h-6)}$$

69.
$$\frac{2x}{x+4}+\frac{3}{x-6} = \left(\frac{x-6}{x-6}\right)\left(\frac{2x}{x+4}\right)+\left(\frac{3}{x-6}\right)\left(\frac{x+4}{x+4}\right)$$
$$= \frac{2x^2-12x}{(x+4)(x-6)}+\frac{3x+12}{(x+4)(x-6)}$$
$$= \frac{2x^2-9x+12}{(x+4)(x-6)}$$

70.

a. $G(10) = 210.9(10)-226.3 = 1882.7$

The National Endowment for the Arts gave about 1883 grants in 2000.

b. $F(10) = 3060(10)^2-55,290(10)+330,910$
$F(10) = 84,010$

The National Endowment for the Arts gave about $84,010 thousand in 2000.

c. $\dfrac{F(10)}{G(10)} = \dfrac{84,010}{1882.7} \approx 44.62$

The average grant given by the National Endowment for the Arts in 2000 was about $44.62 thousand.

d. $\dfrac{F(t)}{G(t)} = \dfrac{3060t^2-55,290t+330,910}{210.9t-226.3}$

e.
$$\frac{3060t^2-55,290t+330,910}{210.9t-226.3} = 50$$
$$3060t^2-55,290t+330,910 = 50(210.9t-226.3)$$
$$3060t^2-55,290t+330,910 = 10,545t-11,315$$
$$\underline{\quad -10,545t+11,315 \quad -10,545t+11,315}$$
$$3060t^2-65,835t+342,225 = 0$$

$$t = \frac{65,835+\sqrt{(-65,835)^2-4(3060)(342,225)}}{2(3060)} \approx 12.7$$

$$t = \frac{65,835-\sqrt{(-65,835)^2-4(3060)(342,225)}}{2(3060)} \approx 8.8$$

The National Endowment for the Arts grants, in 1998 and 2002, averaged $50,000 each.

71.

$$8r^2-128=8(r^2-16)$$
$$=8(r+4)(r-4)$$

72.

$$c^2+10c+25=(c+5)(c+5)$$
$$=(c+5)^2$$

73.

$$8t^2+14t-30=2(4t^2+7t-15)$$

$ac=4\times-15=-60$

Find the factors of -60 that sum to 7:

$-5\times12=-60$

$-5+12=7$

We get:

$$2(4t^2+7t-15)=2(\underline{4t^2+12t}-\underline{5t-15})$$
$$=2(4t(t+3)-5(t+3))$$
$$=2(t+3)(4t-5)$$

74. $t^3-64=(t-4)(t^2+4t+16)$

75.

$$21x^3-68x^2+32x=x(21x^2-68x+32)$$

$ac=21\times32=672$

Find the factors of 672 that sum to -68:

$-56\times(-12)=672$

$-56+(-12)=-68$

We get:

$$x(21x^2-68x+32)=x(\underline{21x^2-12x}-\underline{56x+32})$$
$$=x(3x(7x-4)-8(7x-4))$$
$$=x(7x-4)(3x-8)$$

76. Degree = 2

77.

$f(x)=-2x^2+20x-44$

$f(x)=-2(x^2+10x)-44$

$f(x)=-2(x^2+10x+25-25)-44$

$f(x)=-2(x^2+10x+25)+50-44$

$f(x)=-2(x-5)^2+6$

78.

a. $W(10)=9.3\sqrt[3]{10}\approx20.0$ The average weight of girls who are 10 months old is about 20 pounds.

b.

$9.3\sqrt[3]{m}=25$

$\sqrt[3]{m}=\dfrac{25}{9.3}$

$\left(\sqrt[3]{m}\right)^3=\left(\dfrac{25}{9.3}\right)^3$

$m\approx19.4$

The average weight of girls who are about 19 months old is 25 pounds.

79. $x=1$

80. Domain: $x\le10$ Range: $y\ge0$

81. a is negative since the vertical intercept is negative.

82. b is less than 1 since the graph indicates exponential decay.

83. Domain: All real numbers Range: $(-\infty,0)$

84. a is negative because the parabola is facing downward.

85. $h=-1.5, k=18$ because the vertex pf the parabola is at the point $(-1.5,18)$

86. $f(-3)=10$

87. $x=1, x=-4$

88. Domain: All real numbers Range: $(-\infty,18]$

89. m is negative because the line is decreasing.

90. b is positive because the line intercepts the y-axis above the origin. $b=4$

91. $m=-\dfrac{2}{5}$

92. $x=7.5$

93. Domain: All real numbers Range: All real numbers

94. $\sqrt{100x^4y^8z^2}=10x^2y^4z$

95.

$$\sqrt[3]{32a^4b^8c^6}=\sqrt[3]{2^5a^4b^8c^6}$$
$$=\sqrt[3]{2^3a^3b^6c^6}\cdot\sqrt[3]{2^2a^1b^2}$$
$$=2ab^2c^2\sqrt[3]{4ab^2}$$

96.

$$\frac{\sqrt{2x}}{\sqrt{5xy}}=\frac{\sqrt{2}}{\sqrt{5y}}$$
$$=\frac{\sqrt{2}}{\sqrt{5y}}\cdot\frac{\sqrt{5y}}{\sqrt{5y}}$$
$$=\frac{\sqrt{10y}}{5y}$$

97.

$$\frac{5+\sqrt{3}}{4-\sqrt{7}} = \frac{(5+\sqrt{3})(4+\sqrt{7})}{(4-\sqrt{7})(4+\sqrt{7})}$$

$$= \frac{20+5\sqrt{7}+4\sqrt{3}+\sqrt{21}}{16+4\sqrt{7}-4\sqrt{7}-7}$$

$$= \frac{20+5\sqrt{7}+4\sqrt{3}+\sqrt{21}}{9}$$

98. $(3+2i)+(8-7i) = 11-5i$

99. $(4+3i)-(7-5i) = -3+8i$

100.

$(2+6i)(2-6i) = 4-12i+12i-36i^2$
$\qquad = 4-36i^2$
$\qquad = 4-36(-1)$
$\qquad = 4+36$
$\qquad = 40$

101.

$(2-6i)(4+7i) = 8+14i-24i-42i^2$
$\qquad = 8-10i-42i^2$
$\qquad = 8-10i-42(-1)$
$\qquad = 8-10i+42$
$\qquad = 50-10i$

102.

$$\frac{2+3i}{7i} = \frac{2+3i}{7i} \cdot \frac{i}{i}$$

$$= \frac{2i+3i^2}{7i^2}$$

$$= \frac{2i+3i^2}{7i^2}$$

$$= \frac{2i+3(-1)}{7(-1)}$$

$$= \frac{2i-3}{-7}$$

$$= \frac{3}{7} - \frac{2}{7}i$$

103.

$$\frac{2-5i}{3-8i} = \frac{(2-5i)(3+8i)}{(3-8i)(3+8i)}$$

$$= \frac{6+16i-15i-40i^2}{9+24i-24i-64i^2}$$

$$= \frac{6+i-40i^2}{9-64i^2}$$

$$= \frac{6+i-40(-1)}{9-64(-1)}$$

$$= \frac{6+i+40}{9+64}$$

$$= \frac{46+i}{73}$$

$$= \frac{46}{73} + \frac{1}{73}i$$

104.

$$\frac{5}{4-9i} = \frac{5(4+9i)}{(4-9i)(4+9i)}$$

$$= \frac{20+45i}{16+36i-36i-81i^2}$$

$$= \frac{20+45i}{16-81i^2}$$

$$= \frac{20+45i}{16-81(-1)}$$

$$= \frac{20+45i}{16+81}$$

$$= \frac{20+45i}{97}$$

$$= \frac{20}{97} + \frac{45}{97}i$$

105.

$-x^2 + 4x = 20$
$0 = x^2 - 4x + 20$

$$x = \frac{4 \pm \sqrt{(-4)^2 - 4(1)(20)}}{2(1)}$$

$$x = \frac{4 \pm \sqrt{-64}}{2}$$

$$x = \frac{4 \pm 8i}{2}$$

$x = 2 \pm 4i$

106.

$2x^3 - 20x + 68x = 0$

$2x(x^2 - 10x + 34) = 0$

$2x = 0, x^2 - 10x + 34 = 0$

$x = 0, x = \dfrac{10 \pm \sqrt{(-10)^2 - 4(1)(34)}}{2(1)}$

$x = 0, x = \dfrac{10 \pm \sqrt{-36}}{2}$

$x = 0, x = \dfrac{10 \pm 6i}{2}$

$x = 0, x = 5 \pm 3i$

107.

$x^2 + 9 = 0$

$\quad x^2 = -9$

$\quad x = \pm\sqrt{-9}$

$\quad x = \pm 3i$

108.

$x^2 - 6x + 30 = -4$

$x^2 - 6x + 34 = 0$

$x = \dfrac{6 \pm \sqrt{(-6)^2 - 4(1)(34)}}{2(1)}$

$x = \dfrac{6 \pm \sqrt{-100}}{2}$

$x = \dfrac{6 \pm 10i}{2}$

$x = 3 \pm 5i$

Section 9.1

1.
$y = 3x^2 + 24x - 9$
$a = 3 > 0$ opens up
$h = \dfrac{-24}{2(3)} = -4$
$k = 3(-4)^2 + 24(-4) - 9$ axis of symmetry: $x = -4$
$k = 48 - 96 - 9 = -57$
vertex: $(-4, -57)$

3.
$x = 2y^2 + 12y - 7$
$a = 3 > 0$ opens right
$k = \dfrac{-12}{2(3)} = -2$
$h = 3(-2)^2 + 12(-2) - 7$
$h = 12 - 24 - 7 = -19$
vertex: $(-19, -2)$
axis of symmetry: $y = -2$

5.
$x = -\dfrac{1}{2}y^2 + 4$
$a = -\dfrac{1}{2} < 0$ opens left
vertex: $(4, 0)$
axis of symmetry: $y = 0$

7.
$y = -0.2x^2 + 5x - 3$
$a = -0.2 < 0$ opens down
$h = \dfrac{-5}{2(-0.2)} = 12.5$
$k = -0.2(12.5)^2 + 5(12.5) - 3$
$k = -31.25 + 62.5 - 3 = 28.25$
vertex: $(12.5, 28.25)$
axis of symmetry: $x = 12.5$

9.
$x = \dfrac{y^2}{32}$
$a = \dfrac{1}{32} > 0$ opens right
vertex: $(0, 0)$
axis of symmetry: $y = 0$

11.
$x = y^2 + 10y + 21$
$a = 1 > 0$ opens right
$k = \dfrac{-10}{2(1)} = -5$
$h = (-5)^2 + 10(-5) + 21 = -4$
vertex: $(-4, -5)$
$0 = y^2 + 10y + 21$
$y = \dfrac{-10 \pm \sqrt{10^2 - 4(1)(21)}}{2(1)}$
$y = \dfrac{-10 - \sqrt{16}}{2}$ and $y = \dfrac{-10 + \sqrt{16}}{2}$
$y = -7$ and $y = -3$
y-intercepts: $(0, -7)$ and $(0, -3)$
$x = (0)^2 + 10(0) + 21 = 21$
x-intercept: $(21, 0)$

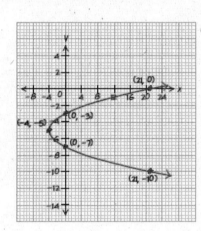

13.
$x = -2y^2 + 15y + 8$
$a = -2 < 0$ opens left
$k = \dfrac{-15}{2(-2)} = 3.75$
$h = -2(3.75)^2 + 15(3.75) + 8 = 36.125$
vertex: $(36.125, 3.75)$

Chapter 9 Conics, Sequences, and Series

$0 = -2y^2 + 15y + 8$

$y = \dfrac{-15 \pm \sqrt{15^2 - 4(-2)(8)}}{2(-2)}$

$y = \dfrac{-15 - \sqrt{289}}{-4}, y = \dfrac{-15 + \sqrt{289}}{-4}$

$y = 8$ and $y = -0.5$

y-intercepts: $(0, -0.5)$ and $(0, 8)$

$x = -2(0)^2 + 15(0) + 8 = 8$

x-intercept: $(8, 0)$

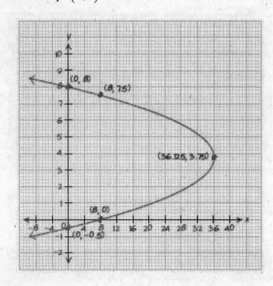

15.

$x = \dfrac{1}{2}(y-4)^2 + 5$

$a = \dfrac{1}{2} > 0$ opens right

vertex: $(5, 4)$

no y-intercepts

$x = \dfrac{1}{2}(0-4)^2 + 5 = 13$

x-intercept: $(13, 0)$

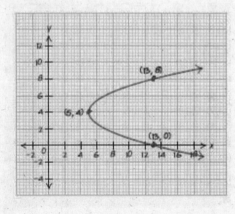

17.

$x = (y+2)^2 - 6$

$a = 1 > 0$ opens right

vertex: $(-6, -2)$

$0 = (y+2)^2 - 6$

$6 = (y+2)^2$

$y + 2 = \pm\sqrt{6}$

$y = -2 - \sqrt{6}, y = -2 + \sqrt{6}$

$y \approx -4.45$ and $y \approx 0.45$

y-intercepts: $(0, -4.45)$ and $(0, 0.45)$

$x = (0+2)^2 - 6 = -2$

x-intercept: $(-2, 0)$

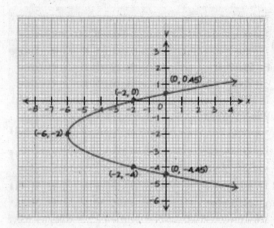

19.

$x = -2(y+5)^2 + 4$

$a = -2 < 0$ opens left

vertex: $(4, -5)$

$0 = -2(y+5)^2 + 4$

$2(y+5)^2 = 4$

$\dfrac{2(y+5)^2}{2} = \dfrac{4}{2}$

$(y+5)^2 = 2$

$y + 5 = \pm\sqrt{2}$

$y = -5 - \sqrt{2}, y = -5 + \sqrt{2}$

$y \approx -6.4$ and $y \approx -3.6$

y-intercepts: $(0, -6.4)$ and $(0, -3.6)$

$x = -2(0+5)^2 + 4 = -46$

x-intercept: $(-46, 0)$

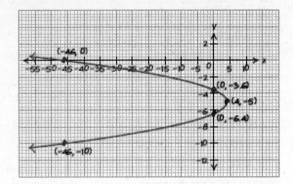

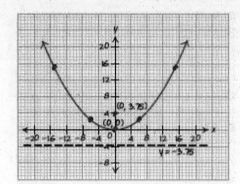

21.

$x = \dfrac{y^2}{36} = \dfrac{y^2}{4p}$

$4p = 36$

$p = 9 > 0$

focus: $(9, 0)$

directrix: $x = -9$

vertex: $(0,0)$	
opens right	
points:	symmetric pairs:
$(1, 6)$	$(1, -6)$
$(4, 12)$	$(4, -12)$

25.

$28x = y^2$

$x = \dfrac{y^2}{28} = \dfrac{y^2}{4p}$

$4p = 28$

$p = 7 > 0$

focus: $(7, 0)$

directrix: $x = -7$

vertex: $(0,0)$	
opens right	
points:	symmetric pairs:
$(2.9, 9)$	$(2.9, -9)$
$(14.3, 20)$	$(14.3, -20)$

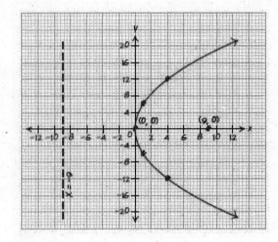

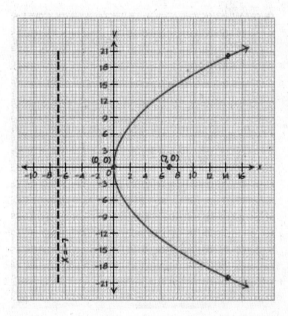

23.

$y = \dfrac{x^2}{15} = \dfrac{x^2}{4p}$

$4p = 15$

$p = \dfrac{15}{4} = 3.75 > 0$

focus: $(0, 3.75)$

directrix: $y = -3.75$

vertex: $(0,0)$	
opens up	
points:	symmetric pairs:
$(6, 2.4)$	$(-6, 2.4)$
$(15, 15)$	$(-15, 15)$

27.

opens up with focus $(0, 1)$ so $p = 1$

$y = \dfrac{x^2}{4(1)} \;\rightarrow\; y = \dfrac{x^2}{4}$

29.

opens right with focus $(6, 0)$ so $p = 6$

$x = \dfrac{y^2}{4(6)} \;\rightarrow\; x = \dfrac{y^2}{24}$

31.

opens left with focus $(-10,0)$ so $p = -10$

$$x = \frac{y^2}{4(-10)} \rightarrow x = -\frac{y^2}{40}$$

33.

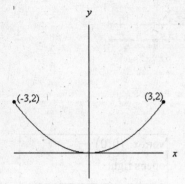

$$x^2 = 4py$$
$$3^2 = 4p(2)$$
$$9 = 8p$$
$$\frac{9}{8} = p = 1\tfrac{1}{8}$$

The receiver should be $1\tfrac{1}{8}$ ft above the center of the satellite along the axis of symmetry.

35.

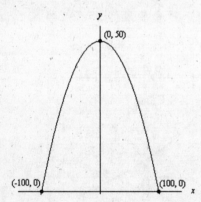

vertex: $(0, 50)$

point on graph: $(100, 0)$

Using: $y = a(x-h)^2 + k$

Solve for a:
$$0 = a(100-0)^2 + 50$$
$$0 = 10000a + 50$$
$$-50 = 10000a$$
$$a = -0.005$$

The equation for the graph is:
$$y = -0.005x^2 + 50$$

Plug in $x = 40$
$$y = -0.005(40)^2 + 50 = 42$$

The arch, 40 feet from the center of the bridge, is 42 feet high.

37.

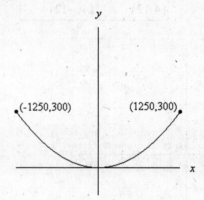

$$x^2 = 4py$$
$$1250^2 = 4p(300)$$
$$1250^2 = 1200p$$
$$\frac{1250^2}{1200} = p$$

so $x^2 = 4 \cdot \dfrac{1250^2}{1200} y$

$$x^2 = \frac{1250^2}{300} y$$

$$x^2 = \frac{15{,}625}{3} y \text{ or } y = \frac{3x^2}{15{,}625}$$

$$y = \frac{3(1000)^2}{15{,}625} = 192$$

The cables, 1000 feet from the center of the bridge, are 192 feet high.

39.

Center $=(0,0)$ and Radius $=6$

$h=0, k=0, r=6$

$(x-0)^2+(y-0)^2=(6)^2$

$x^2+y^2=36$

41.

Center $=(4,9)$ and Radius $=4$

$h=4, k=9, r=4$

$(x-4)^2+(y-9)^2=(4)^2$

$(x-4)^2+(y-9)^2=16$

43.

Center $=(-2,3)$ and Radius $=12$

$h=-2, k=3, r=12$

$(x-(-2))^2+(y-3)^2=(12)^2$

$(x+2)^2+(y-3)^2=144$

45.

Center $=(-1,-4)$ and Radius $=0.5$

$h=-1, k=-4, r=0.5$

$(x-(-1))^2+(y-(-4))^2=(0.5)^2$

$(x+1)^2+(y+4)^2=0.25$

47.

$x^2+y^2=9$

$(x-0)^2+(y-0)^2=(3)^2$

$h=0, k=0, r=3$

Center $=(0,0)$ and Radius $=3$

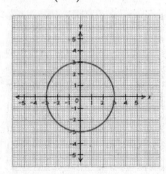

49.

$(x-5)^2+(y-4)^2=25$

$(x-5)^2+(y-4)^2=(5)^2$

$h=5, k=4, r=5$

Center $=(5,4)$ and Radius $=5$

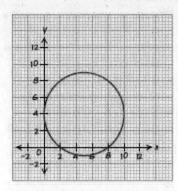

51.

$(x+4)^2+(y+1)^2=1$

$(x-(-4))^2+(y-(-1))^2=(1)^2$

$h=-4, k=-1, r=1$

Center $=(-4,-1)$ and Radius $=1$

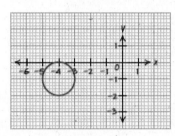

53.

$(x-8)^2+(y+2)^2=36$

$(x-8)^2+(y-(-2))^2=(6)^2$

$h=8, k=-2, r=6$

Center $=(8,-2)$ and Radius $=6$

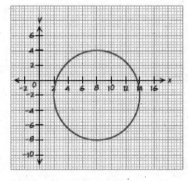

55.
$x^2 + y^2 + 2x - 6y + 9 = 0$
$x^2 + 2x + y^2 - 6y = -9$
$(x^2 + 2x + \_\_) + (y^2 - 6y + \_\_) = -9$
Completing the square: $\left(\frac{2}{2}\right)^2 = 1$ and $\left(\frac{-6}{2}\right)^2 = 9$
$(x^2 + 2x + 1) + (y^2 - 6y + 9) = -9 + 1 + 9$
$(x+1)^2 + (y-3)^2 = 1$
Center $= (-1, 3)$ and Radius $= 1$

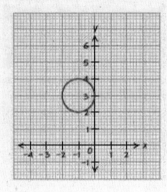

57.
$3x^2 + 3y^2 + 18x - 36y + 27 = 0$
$\dfrac{3x^2 + 3y^2 + 18x - 36y + 27}{3} = \dfrac{0}{3}$
$x^2 + y^2 + 6x - 12y + 9 = 0$
$x^2 + 6x + y^2 - 12y = -9$
$(x^2 + 6x + \_\_) + (y^2 - 12y + \_\_) = -9$
Completing the square: $\left(\frac{6}{2}\right)^2 = 9$ and $\left(\frac{-12}{2}\right)^2 = 36$
$(x^2 + 6x + 9) + (y^2 - 12y + 36) = -9 + 9 + 36$
$(x+3)^2 + (y-6)^2 = 36$
Center $= (-3, 6)$ and Radius $= 6$

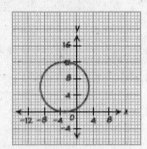

59.
$4x^2 + 4y^2 + 4x - 4y = 34$
$\dfrac{4x^2 + 4y^2 + 4x - 4y}{4} = \dfrac{34}{4}$
$x^2 + y^2 + x - y = \dfrac{34}{4}$
$x^2 + x + y^2 - y = \dfrac{34}{4}$
$\left(x^2 + x + \_\_\right) + \left(y^2 - y + \_\_\right) = \dfrac{34}{4}$
Completing the square: $\left(\frac{1}{2}\right)^2 = \frac{1}{4}$ and $\left(\frac{-1}{2}\right)^2 = \frac{1}{4}$
$\left(x^2 + x + \frac{1}{4}\right) + \left(y^2 - y + \frac{1}{4}\right) = \dfrac{34}{4} + \dfrac{1}{4} + \dfrac{1}{4}$
$\left(x + \frac{1}{2}\right)^2 + \left(y - \frac{1}{2}\right)^2 = 9$
Center $= \left(-\dfrac{1}{2}, \dfrac{1}{2}\right)$ and Radius $= 3$

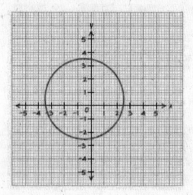

61.
Center $= (0, 2)$ and Radius $= 2$
$h = 0, k = 2, r = 2$
$(x - 0)^2 + (y - 2)^2 = (2)^2$
$x^2 + (y - 2)^2 = 4$

63.
Center $= (-4, 2)$ and Radius $= 3$
$h = -4, k = 2, r = 3$
$(x - (-4))^2 + (y - 2)^2 = (3)^2$
$(x + 4)^2 + (y - 2)^2 = 9$

65.
Center $= (10, 10)$ and Radius $= 5$
$h = 10, k = 10, r = 5$
$(x - 10)^2 + (y - 10)^2 = (5)^2$
$(x - 10)^2 + (y - 10)^2 = 25$

67.

$2x^2 + 2y^2 + 4 = 22$

$2x^2 + 2y^2 = 18$

$\dfrac{2x^2 + 2y^2}{2} = \dfrac{18}{2}$

$x^2 + y^2 = 9$

This is a circle with center $=(0,0)$ and radius $=3$.

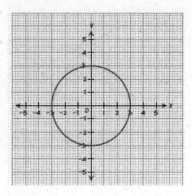

69.

$x^2 - 4x - y - 12 = 0$

$y = x^2 - 4x - 12$

For the vertex:

$x = \dfrac{-(-4)}{2(1)} = 2$

Substitute the x-value into the equation to find y.

$y = (2)^2 - 4(2) - 12$

$y = -16$

This is a parabola with vertex $=(2,-16)$.

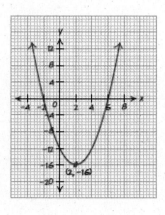

71.

$x - y^2 - 8y = 15$

$x = y^2 + 8y + 15$

For the vertex:

$y = \dfrac{-(8)}{2(1)} = -4$

Substitute the y-value into the equation to find x.

$x = (-4)^2 + 8(-4) + 15$

$x = -1$

This is a parabola with vertex $=(-1,-4)$.

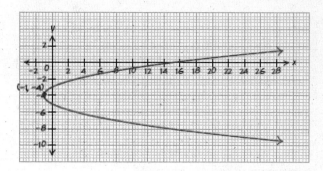

73.

$x^2 = -(y-3)^2 + 49$

$x^2 + (y-3)^2 = 49$

This is a circle with center $=(0,3)$ and radius $=7$.

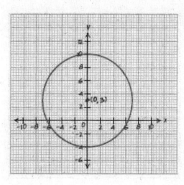

9.1 Exercises

75.

$$x = \frac{y^2}{32}$$

$$x = \frac{1}{32}y^2$$

For the vertex:

$$y = \frac{-(0)}{2\left(\frac{1}{32}\right)} = 0$$

Substitute the y-value into the equation to find x.

$$x = \frac{1}{32}(0)^2$$

$$x = 0$$

This is a parabola with vertex $= (0, 0)$.

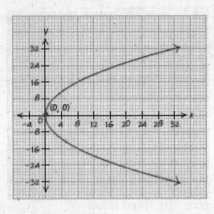

Section 9.2

1.

$\dfrac{x^2}{4}+\dfrac{y^2}{16}=1$

$a^2=4 \to a=2$

$b^2=16 \to b=4$

major axis is vertical

$c^2=16-4$

$c^2=12$

$c=\pm\sqrt{12}\approx \pm 3.46$

foci: $(0,-3.46)$ and $(0,3.46)$

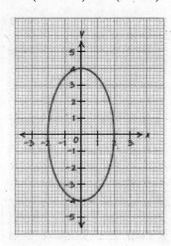

3.

$\dfrac{x^2}{25}+\dfrac{y^2}{9}=1$

$a^2=25 \to a=5$

$b^2=9 \to b=3$

major axis is horizontal

$c^2=25-9$

$c^2=16$

$c=\pm\sqrt{16}=\pm 4$

foci: $(-4,0)$ and $(4,0)$

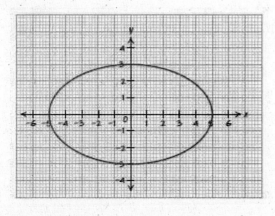

5.

$x^2+\dfrac{y^2}{9}=1$

$a^2=1 \to a=1$

$b^2=9 \to b=3$

major axis is vertical

$c^2=9-1$

$c^2=8$

$c=\pm\sqrt{8}\approx \pm 2.83$

foci: $(0,-2.83)$ and $(0,2.83)$

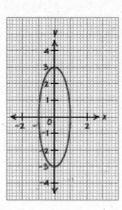

7.

$\dfrac{x^2}{10}+\dfrac{y^2}{49}=1$

$a^2=10 \to a\approx 3.16$

$b^2=49 \to b=7$

major axis is vertical

$c^2=49-10$

$c^2=39$

$c=\pm\sqrt{39}\approx \pm 6.24$

foci: $(0,-6.24)$ and $(0,6.24)$

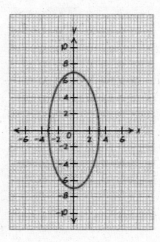

9.

$4x^2 + 9y^2 = 36$

$\dfrac{4x^2}{36} + \dfrac{9y^2}{36} = \dfrac{36}{36}$

$\dfrac{x^2}{9} + \dfrac{y^2}{4} = 1$

$a^2 = 9 \to a = 3$

$b^2 = 4 \to b = 2$

major axis is horizontal

$c^2 = 9 - 4$

$c^2 = 5$

$c = \pm\sqrt{5} \approx \pm 2.24$

foci: $(-2.24, 0)$ and $(2.24, 0)$

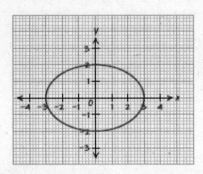

11.

$x^2 + 36y^2 = 36$

$\dfrac{x^2}{36} + \dfrac{36y^2}{36} = \dfrac{36}{36}$

$\dfrac{x^2}{36} + y^2 = 1$

$a^2 = 36 \to a = 6$

$b^2 = 1 \to b = 1$

major axis is horizontal

$c^2 = 36 - 1$

$c^2 = 35$

$c = \pm\sqrt{35} \approx \pm 5.92$

foci: $(-5.92, 0)$ and $(5.92, 0)$

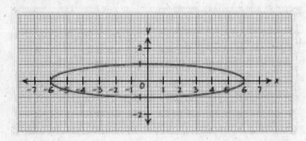

13.

$20x^2 + 9y^2 = 180$

$\dfrac{20x^2}{180} + \dfrac{9y^2}{180} = \dfrac{180}{180}$

$\dfrac{x^2}{9} + \dfrac{y^2}{20} = 1$

$a^2 = 9 \to a = 3$

$b^2 = 20 \to b \approx 4.47$

major axis is vertical

$c^2 = 20 - 9$

$c^2 = 11$

$c = \pm\sqrt{11} \approx \pm 3.32$

foci: $(0, -3.32)$ and $(0, 3.32)$

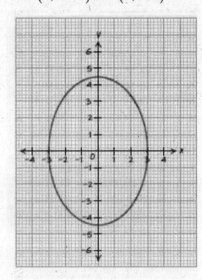

15.

major axis is horizontal:

vertices $(-8, 0)$ and $(8, 0) \to a = 8$

minor axis end points:

$(0, -4)$ and $(0, 4) \to b = 4$

equation: $\dfrac{x^2}{64} + \dfrac{y^2}{16} = 1$

$c^2 = 64 - 16$

$c^2 = 48$

$c = \pm\sqrt{48} \approx \pm 6.93$

foci: $(-6.93, 0)$ and $(6.93, 0)$

17.
major axis is vertical:
vertices $(0,-3)$ and $(0,3) \to b = 3$
minor axis end points:
$(-2,0)$ and $(2,0) \to a = 2$
equation: $\dfrac{x^2}{4} + \dfrac{y^2}{9} = 1$
$c^2 = 9 - 4$
$c^2 = 5$
$c = \pm\sqrt{5} \approx \pm 2.24$
foci: $(0,-2.24)$ and $(0, 2.24)$

19.
major axis is horizontal:
vertices $(-15,0)$ and $(15,0) \to a = 15$
minor axis end points:
$(0,-10)$ and $(0,10) \to b = 10$
equation: $\dfrac{x^2}{225} + \dfrac{y^2}{100} = 1$
$c^2 = 225 - 100$
$c^2 = 125$
$c = \pm\sqrt{125} \approx \pm 11.18$
foci: $(-11.18, 0)$ and $(11.18, 0)$

21.
$a = $ mean distance $= 3671.5$
$c = $ mean distance $-$ perihelion
$c = 3671.5 - 2760 = 911.5$
$c^2 = a^2 - b^2$ so $b^2 = a^2 - c^2$
$b^2 = 3671.5^2 - 911.5^2$
$b^2 = 12{,}649{,}080$
$a^2 = 13{,}479{,}912.25$
$\dfrac{x^2}{13{,}479{,}912.25} + \dfrac{y^2}{12{,}649{,}080} = 1$

23.
aphelion $= 2(\text{mean distance}) - $ perihelion
$= 2(886) - 840$
$= 932$

$a = $ mean distance $= 886$
$c = $ mean distance $-$ perihelion
$c = 886 - 840 = 46$
$c^2 = a^2 - b^2$ so $b^2 = a^2 - c^2$
$b^2 = 886^2 - 46^2$
$b^2 = 782{,}880$
$a^2 = 784{,}996$
$\dfrac{x^2}{784{,}996} + \dfrac{y^2}{782{,}880} = 1$

25.
$\dfrac{w^2}{4556.25} + \dfrac{d^2}{945.5} = 1$
$a^2 = 4556.25 \to a = 67.5$
$b^2 = 945.5 \to b \approx 30.75$
width $= 2(67.5) = 135$ cm
depth $= 30.75$ cm

27. Let w be half the width of the arch and h be the height of the arch.
$a = \dfrac{120}{2} = 60 \to a^2 = 3600$
$b = 35 \to b^2 = 1225$
$\dfrac{w^2}{3600} + \dfrac{h^2}{1225} = 1$

29.
$\dfrac{x^2}{4} - \dfrac{y^2}{9} = 1$
$\dfrac{x^2}{2^2} - \dfrac{y^2}{3^2} = 1$
$a = 2, b = 3$
$c^2 = 2^2 + 3^2$
$c = \pm\sqrt{13}$
Vertices $= (2,0), (-2,0)$
Foci $= (\sqrt{13}, 0), (-\sqrt{13}, 0) \approx (3.6, 0), (-3.6, 0)$

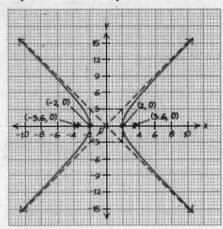

33.
$x^2 - 25y^2 = 25$
$\dfrac{x^2 - 25y^2}{25} = \dfrac{25}{25}$
$\dfrac{x^2}{25} - \dfrac{y^2}{1} = 1$
$\dfrac{x^2}{5^2} - \dfrac{y^2}{1^2} = 1$
$a = 5, b = 1$
$c^2 = 5^2 + 1^2$
$c = \pm\sqrt{26}$
Vertices $= (5,0), (-5,0)$
Foci $= \left(\sqrt{26}, 0\right), \left(-\sqrt{26}, 0\right) \approx (5.1, 0), (-5.1, 0)$

31.
$\dfrac{y^2}{36} - \dfrac{x^2}{4} = 1$
$\dfrac{y^2}{6^2} - \dfrac{x^2}{2^2} = 1$
$a = 2, b = 6$
$c^2 = 2^2 + 6^2$
$c = \pm\sqrt{40}$
Vertices $= (0, 6), (0, -6)$
Foci $= \left(0, \sqrt{40}\right), \left(0, -\sqrt{40}\right) \approx (0, 6.3), (0, -6.3)$

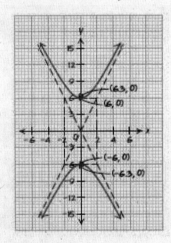

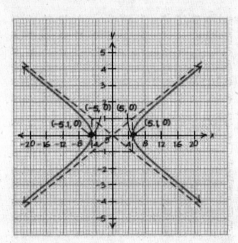

35.
$9y^2 - 81x^2 = 81$
$\dfrac{9y^2 - 81x^2}{81} = \dfrac{81}{81}$
$\dfrac{y^2}{9} - \dfrac{x^2}{1} = 1$
$\dfrac{y^2}{3^2} - \dfrac{x^2}{1^2} = 1$
$a = 1, b = 3$
$c^2 = 1^2 + 3^2$
$c = \pm\sqrt{10}$
Vertices $= (0, 3), (0, -3)$
Foci $= \left(0, \sqrt{10}\right), \left(0, -\sqrt{10}\right) \approx (0, 3.2), (0, -3.2)$

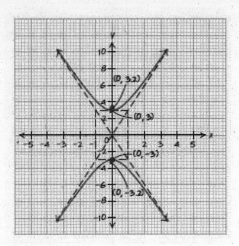

37.
Since hyperbola opens left/right
and $a = 4$ and $b = 3$ then
$c^2 = 4^2 + 3^2$
$c = \pm 5$
$\dfrac{x^2}{4^2} - \dfrac{y^2}{3^2} = 1$
$\dfrac{x^2}{16} - \dfrac{y^2}{9} = 1$
Foci $= (5,0), (-5,0)$

39.
Since hyperbola opens up/down
and $a = 1$ and $b = 2$ then
$c^2 = 1^2 + 2^2$
$c = \pm\sqrt{5}$
$\dfrac{y^2}{2^2} - \dfrac{x^2}{1^2} = 1$
$\dfrac{y^2}{4} - \dfrac{x^2}{1} = 1$
Foci $= (0, \sqrt{5}), (0, -\sqrt{5})$

41.
$3x^2 + 3y^2 = 27 \quad A = 3 \text{ and } C = 3$
Since $A = C$ then this is a circle.

43.
$4x^2 - 9y^2 = 36 \quad A = 4 \text{ and } C = -9$
Since $A \neq C$, $A < 0$ or $C < 0$ then this is a hyperbola.

45.
$6y^2 + 2y + x - 9 = 0 \quad A = 0 \text{ and } C = 6$
Since $A = 0$ or $C = 0$ then this is a parabola.

47.
$64x^2 + 36y^2 = 576 \quad A = 64 \text{ and } C = 36$
Since $A \neq C$, $A > 0$ and $C > 0$ then this is an ellipse.

49. Ellipse

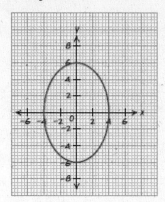

51. Hyperbola

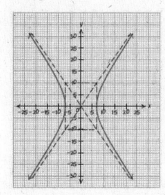

53. Ellipse

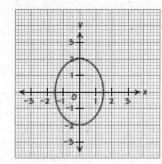

55. Circle

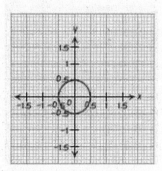

Chapter 9 Conics, Sequences, and Series
57. Circle

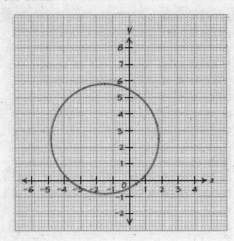

Section 9.3

1.

n	$a_n = 4+(n-1)6$	a_n
1	$a_1 = 4+(1-1)6$	$a_1 = 4$
2	$a_2 = 4+(2-1)6$	$a_2 = 10$
3	$a_3 = 4+(3-1)6$	$a_3 = 16$
4	$a_4 = 4+(4-1)6$	$a_4 = 22$

3.

n	$f(n) = 2n^2$	$f(n)$
1	$f(1) = 2(1)^2$	$f(1) = 2$
2	$f(2) = 2(2)^2$	$f(2) = 8$
3	$f(3) = 2(3)^2$	$f(3) = 18$
4	$f(4) = 2(4)^2$	$f(4) = 32$

5.

n	$a_n = 4(-1)^n$	a_n
1	$a_1 = 4(-1)^1$	$a_1 = -4$
2	$a_2 = 4(-1)^2$	$a_2 = 4$
3	$a_3 = 4(-1)^3$	$a_3 = -4$
4	$a_4 = 4(-1)^4$	$a_4 = 4$
5	$a_5 = 4(-1)^5$	$a_4 = -4$

7.

n	$f(n) = 3^n - 20$
1	$f(1) = 3^1 - 20 = -17$
2	$f(2) = 3^2 - 20 = -11$
3	$f(3) = 3^3 - 20 = 7$
4	$f(4) = 3^4 - 20 = 61$
5	$f(5) = 3^5 - 20 = 223$

The first 5 terms of the sequence are -17, -11, 7, 61, 223.

9.

n	$a_n = \dfrac{4n}{n+7}$
1	$a_1 = \dfrac{4(1)}{(1)+7} = \dfrac{1}{2}$
2	$a_2 = \dfrac{4(2)}{(2)+7} = \dfrac{8}{9}$
3	$a_3 = \dfrac{4(3)}{(3)+7} = \dfrac{6}{5}$
4	$a_4 = \dfrac{4(4)}{(4)+7} = \dfrac{16}{11}$
5	$a_5 = \dfrac{4(5)}{(5)+7} = \dfrac{5}{3}$

The first 5 terms of the sequence are $\dfrac{1}{2}, \dfrac{8}{9}, \dfrac{6}{5}, \dfrac{16}{11}, \dfrac{5}{3}$.

11.

n	$a_n = \dfrac{1}{4^n}$
1	$a_1 = \dfrac{1}{4^1} = \dfrac{1}{4}$
2	$a_2 = \dfrac{1}{4^2} = \dfrac{1}{16}$
3	$a_3 = \dfrac{1}{4^3} = \dfrac{1}{64}$
4	$a_4 = \dfrac{1}{4^4} = \dfrac{1}{256}$
5	$a_5 = \dfrac{1}{4^5} = \dfrac{1}{1024}$

The first 5 terms of the sequence are $\dfrac{1}{4}, \dfrac{1}{16}, \dfrac{1}{64}, \dfrac{1}{256}, \dfrac{1}{1024}$.

13.

n	$a_n = \dfrac{(-1)^n}{n^2}$
1	$a_1 = \dfrac{(-1)^1}{(1)^2} = -1$
2	$a_2 = \dfrac{(-1)^2}{(2)^2} = \dfrac{1}{4}$
3	$a_3 = \dfrac{(-1)^3}{(3)^2} = -\dfrac{1}{9}$
4	$a_4 = \dfrac{(-1)^4}{(4)^2} = \dfrac{1}{16}$
5	$a_5 = \dfrac{(-1)^5}{(5)^2} = -\dfrac{1}{25}$

Chapter 9 Conics, Sequences, and Series

The first 5 terms of the sequence are $-1, \dfrac{1}{4}, -\dfrac{1}{9}, \dfrac{1}{16}, -\dfrac{1}{25}$.

15.
$a_n = 4 + (n-1)6$ where $a_n = 118$
$118 = 4 + (n-1)6$
$118 = 4 + 6n - 6$
$118 = 6n - 2$
$120 = 6n$
$n = 20$

17.
$a_n = 2n^2 + 5$ where $a_n = 397$
$397 = 2n^2 + 5$
$392 = 2n^2$
$196 = n^2$
$n = \pm\sqrt{196}$
$n = \pm 14$
Since n must be a natural number
$n = 14$

19.
$a_n = 2(3)^n - 24$ where $a_n = 354270$
$354270 = 2(3)^n - 24$
$354294 = 2(3)^n$
$177147 = (3)^n$
$\ln(177147) = \ln(3)^n$
$\ln(177147) = n\ln(3)$
$n = \dfrac{\ln(177147)}{\ln(3)}$
$n = 11$

21.
$a_n = \dfrac{2n}{n+8}$ where $a_n = 1.2$
$1.2 = \dfrac{2n}{n+8}$
$1.2(n+8) = \dfrac{2n}{n+8}(n+8)$
$1.2(n+8) = 2n$
$1.2n + 9.6 = 2n$
$9.6 = 0.8n$
$n = 12$

23.
$a_n = 50 + \sqrt{2n}$ where $a_n = 68$
$68 = 50 + \sqrt{2n}$
$18 = \sqrt{2n}$
$324 = 2n$
$n = 162$

25.

n	$a_n = 4n - 5$
1	$a_1 = 4(1) - 5 = -1$
2	$a_2 = 4(2) - 5 = 3$
3	$a_3 = 4(3) - 5 = 7$
4	$a_4 = 4(4) - 5 = 11$
5	$a_5 = 4(5) - 5 = 15$
6	$a_6 = 4(6) - 5 = 19$

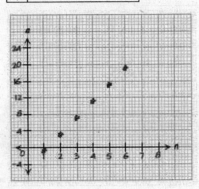

27.

n	$a_n = 4 + (n-1)6$
1	$a_1 = 4 + (1-1)6 = 4$
2	$a_2 = 4 + (2-1)6 = 10$
3	$a_3 = 4 + (3-1)6 = 16$
4	$a_4 = 4 + (4-1)6 = 22$
5	$a_5 = 4 + (5-1)6 = 28$
6	$a_6 = 4 + (6-1)6 = 34$

9.3 Exercises

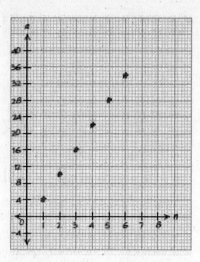

29.

n	$f(n) = 3^n - 20$
1	$f(1) = 3^1 - 20 = -17$
2	$f(2) = 3^2 - 20 = -11$
3	$f(3) = 3^3 - 20 = 7$
4	$f(4) = 3^4 - 20 = 61$
5	$f(5) = 3^5 - 20 = 223$
6	$f(6) = 3^6 - 20 = 709$

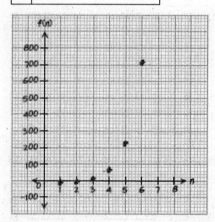

31.

n	$a_n = 4n(-1)^n$
1	$a_1 = 4(1)(-1)^1 = -4$
2	$a_2 = 4(2)(-1)^2 = 8$
3	$a_3 = 4(3)(-1)^3 = -12$
4	$a_4 = 4(4)(-1)^4 = 16$
5	$a_5 = 4(5)(-1)^5 = -20$
6	$a_6 = 4(6)(-1)^6 = 24$

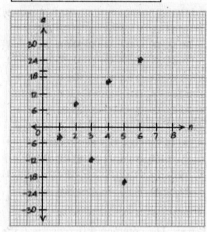

33.

n	$a_n = \dfrac{4n}{n+7}$
1	$a_1 = \dfrac{4(1)}{(1)+7} = \dfrac{1}{2}$
2	$a_2 = \dfrac{4(2)}{(2)+7} = \dfrac{8}{9}$
3	$a_3 = \dfrac{4(3)}{(3)+7} = \dfrac{6}{5}$
4	$a_4 = \dfrac{4(4)}{(4)+7} = \dfrac{16}{11}$
5	$a_5 = \dfrac{4(5)}{(5)+7} = \dfrac{5}{3}$
6	$a_5 = \dfrac{4(6)}{(6)+7} = \dfrac{24}{13}$

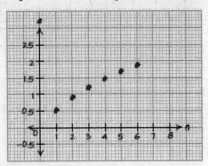

35.
$a_2 - a_1 = 16 - 12 = 4$
$a_3 - a_2 = 20 - 16 = 4$
$a_4 - a_3 = 24 - 20 = 4$
$a_5 - a_4 = 28 - 24 = 4$
$a_6 - a_5 = 32 - 28 = 4$
The sequence is arithmetic and the common difference is 4.

37.
$a_2 - a_1 = -4.3 - (-6.65) = -4.3 + 6.65 = 2.35$
$a_3 - a_2 = -1.95 - (-4.3) = -1.95 + 4.3 = 2.35$
$a_4 - a_3 = 0.4 - (-1.95) = 0.4 + 1.95 = 2.35$
$a_5 - a_4 = 2.75 - 0.4 = 2.35$
$a_6 - a_5 = 5.1 - 2.75 = 2.35$
The sequence is arithmetic. The common difference is 2.35.

39.
$a_2 - a_1 = 80 - 100 = -20$
$a_3 - a_2 = 64 - 80 = -16$
$-20 \ne -16$
The differences between consecutive terms are not the same, so this is not an arithmetic sequence.

41.
$a_2 - a_1 = 10.55 - 14 = -3.45$
$a_3 - a_2 = 7.1 - 10.55 = -3.45$
$a_4 - a_3 = 3.65 - 7.1 = -3.45$
$a_5 - a_4 = 0.2 - 3.65 = -3.45$
$a_6 - a_5 = -3.25 - 0.2 = -3.45$
The sequence is arithmetic. The common difference is -3.45.

43.
$a_2 - a_1 = \dfrac{7}{2} - 4 = -\dfrac{1}{2}$
$a_3 - a_2 = 3 - \dfrac{7}{2} = -\dfrac{1}{2}$
$a_4 - a_3 = \dfrac{5}{2} - 3 = -\dfrac{1}{2}$
$a_5 - a_4 = 2 - \dfrac{5}{2} = -\dfrac{1}{2}$
$a_6 - a_5 = \dfrac{3}{2} - 2 = -\dfrac{1}{2}$
$a_7 - a_6 = 1 - \dfrac{3}{2} = -\dfrac{1}{2}$
The sequence is arithmetic. The common difference is $-\dfrac{1}{2}$.

45.
$a_2 - a_1 = 10 - 4 = 6$
$a_3 - a_2 = 16 - 10 = 6$
$a_4 - a_3 = 22 - 16 = 6$
$a_5 - a_4 = 28 - 22 = 6$
$a_6 - a_5 = 34 - 28 = 6$
The sequence is arithmetic. The common difference is 6.
$a_n = 4 + (n-1)6$
$a_n = 4 + 6n - 6$
$a_n = -2 + 6n$

47.
$a_2 - a_1 = 5.25 - 4 = 1.25$
$a_3 - a_2 = 6.5 - 5.25 = 1.25$
$a_4 - a_3 = 7.75 - 6.5 = 1.25$
$a_5 - a_4 = 9 - 7.75 = 1.25$
$a_6 - a_5 = 10.25 - 9 = 1.25$
The sequence is arithmetic. The common difference is 1.25.
$a_n = 4 + (n-1)1.25$
$a_n = 4 + 1.25n - 1.25$
$a_n = 2.25 + 1.25n$

49.
$a_2 - a_1 = 9 - 3 = 6$
$a_3 - a_2 = -1 - 9 = -10$
$6 \ne -10$
The differences between consecutive terms are not the same, so this is not an arithmetic sequence.

51.

$a_2 - a_1 = \dfrac{10}{3} - 4 = -\dfrac{2}{3}$

$a_3 - a_2 = \dfrac{8}{3} - \dfrac{10}{3} = -\dfrac{2}{3}$

$a_4 - a_3 = 2 - \dfrac{8}{3} = -\dfrac{2}{3}$

$a_5 - a_4 = \dfrac{4}{3} - 2 = -\dfrac{2}{3}$

$a_6 - a_5 = \dfrac{2}{3} - \dfrac{4}{3} = -\dfrac{2}{3}$

The sequence is arithmetic. The common difference is $-\dfrac{2}{3}$.

$a_n = 4 + (n-1)\left(-\dfrac{2}{3}\right)$

$a_n = 4 - \dfrac{2}{3}n + \dfrac{2}{3}$

$a_n = \dfrac{14}{3} - \dfrac{2}{3}n$

53.

a.

$a_1 = 36,000$ and $d = 900$

$a_n = 36,000 + (n-1)900$

$a_n = 36,000 + 900n - 900$

$a_n = 35,100 + 900n$

b.

$a_{12} = 35,100 + 900(12)$

$a_{12} = 45,900$

Your salary will be $45,900 in the 12th year of employment.

c.

$60,300 = 35,100 + 900n$

$\underline{-35,100 \quad -35,100}$

$25,200 = 900n$

$\dfrac{25,200}{900} = \dfrac{900n}{900}$

$28 = n$

In your 28th year of employment, your salary will be $60,300.

55.

a.

$a_1 = 48,000$ and $d = -4500$

$a_n = 48,000 + (n-1)(-4500)$

$a_n = 48,000 - 4500n + 4500$

$a_n = 52,500 - 4500n$

b.

$a_{11} = 52,500 - 4500(11)$

$a_{11} = 3000$

The car will be worth $3000 in the 11th year.

c.

$21,000 = 52,500 - 4500n$

$\underline{-52,000 \quad -52,000}$

$-31,500 = -4500n$

$\dfrac{-31,500}{-4500} = \dfrac{-4500n}{-4500}$

$7 = n$

In the 7th year, the car will be worth $21,000.

57.

a.

$a_1 = 10$ and $d = 1$

$a_n = 10 + (n-1)(1)$

$a_n = 10 + n - 1$

$a_n = 9 + n$

b.

$a_{32} = 9 + 32 = 41$

You will save $41 during the 32nd week.

c.

$61 = 9 + n$

$52 = n$

During week 52, you will save $61.

59.

a.

$a_n = 400 + 0.15n$

$a_{100} = 400 + 0.15(100)$

$a_{100} = 415$

It costs $415 to produce 100 candy bars.

b.

$550 = 400 + 0.15n$

$\underline{-400 \quad -400}$

$150 = 0.15n$

$\dfrac{150}{0.15} = \dfrac{0.15n}{0.15}$

$1000 = n$

One thousand candy bars can be produced for a cost of $550.

c. The common difference is $d = 0.15$. The cost of production increases by $0.15 per candy bar produced.

Chapter 9 Conics, Sequences, and Series

61.
a.
$a_1 = 1$ and $d = 1$
$a_n = 1 + (n-1)$
$a_n = n$

b. $a_{10} = 10$ There are 10 cups in the 10th row.

63.
$\left.\begin{array}{l} a_4 = 9 \\ a_{15} = -13 \end{array}\right\}$ $(4, 9)$ and $(15, -13)$

$d = \dfrac{-13 - 9}{15 - 4} = \dfrac{-22}{11} = -2$

$y - 9 = -2(x - 4)$
$y - 9 = -2x + 8$
$\underline{+9 +9}$
$y = -2x + 17$
so
$a_n = -2n + 17$

65.
$\left.\begin{array}{l} a_7 = 28 \\ a_{21} = 77 \end{array}\right\}$ $(7, 28)$ and $(21, 77)$

$d = \dfrac{77 - 28}{21 - 7} = \dfrac{49}{14} = 3.5$

$y - 28 = 3.5(x - 7)$
$y - 28 = 3.5x - 24.5$
$\underline{+28 +28}$
$y = 3.5x + 3.5$
so
$a_n = 3.5n + 3.5$

67.
$\left.\begin{array}{l} a_{10} = 56.75 \\ a_{32} = 161.25 \end{array}\right\}$ $(10, 56.75)$ and $(32, 161.25)$

$d = \dfrac{161.25 - 56.75}{32 - 10} = \dfrac{104.5}{22} = 4.75$

$y - 56.75 = 4.75(x - 10)$
$y - 56.75 = 4.75x - 47.5$
$\underline{+56.75 +56.75}$
$y = 4.75x + 9.25$
so
$a_n = 4.75n + 9.25$

69.
$\left.\begin{array}{l} a_6 = -36 \\ a_{25} = -32.2 \end{array}\right\}$ $(6, -36)$ and $(25, -32.2)$

$d = \dfrac{-32.2 - (-36)}{25 - 6} = \dfrac{3.8}{19} = 0.2$

$y - (-36) = 0.2(x - 6)$
$y + 36 = 0.2x - 1.2$
$\underline{-36 -36}$
$y = 0.2x - 37.2$
so
$a_n = 0.2n - 37.2$

Section 9.4

1. Yes, the sequence is geometric and the common ratio is 1.5.

$\dfrac{a_2}{a_1} = \dfrac{3}{2} = 1.5$

$\dfrac{a_3}{a_2} = \dfrac{4.5}{3} = 1.5$

$\dfrac{a_4}{a_3} = \dfrac{6.75}{4.5} = 1.5$

$\dfrac{a_5}{a_4} = \dfrac{10.125}{6.75} = 1.5$

$\dfrac{a_6}{a_5} = \dfrac{15.1875}{10.125} = 1.5$

3. Yes, the sequence is geometric and the common ratio is $\dfrac{1}{3}$.

$\dfrac{a_2}{a_1} = \dfrac{90}{270} = \dfrac{1}{3}$

$\dfrac{a_3}{a_2} = \dfrac{30}{90} = \dfrac{1}{3}$

$\dfrac{a_4}{a_3} = \dfrac{10}{30} = \dfrac{1}{3}$

$\dfrac{a_5}{a_4} = \dfrac{\frac{10}{3}}{10} = \dfrac{1}{3}$

$\dfrac{a_6}{a_5} = \dfrac{\frac{10}{9}}{\frac{10}{3}} = \dfrac{1}{3}$

5. No, the sequence is not geometric.

$\dfrac{a_2}{a_1} = \dfrac{-1}{-4} = 0.25$

$\dfrac{a_3}{a_2} = \dfrac{4}{-1} = -4$

$\dfrac{a_4}{a_3} = \dfrac{11}{4} = 2.75$

$\dfrac{a_5}{a_4} = \dfrac{20}{11} \approx 1.818$

$\dfrac{a_6}{a_5} = \dfrac{31}{20} = 1.55$

9.4 Exercises

7. Yes, the sequence is geometric and the common ratio is 2.

$\dfrac{a_2}{a_1} = \dfrac{-12}{-6} = 2$

$\dfrac{a_3}{a_2} = \dfrac{-24}{-12} = 2$

$\dfrac{a_4}{a_3} = \dfrac{-48}{-24} = 2$

$\dfrac{a_5}{a_4} = \dfrac{-96}{-48} = 2$

$\dfrac{a_6}{a_5} = \dfrac{-192}{-96} = 2$

9. Yes, the sequence is geometric and the common ratio is -3.

$\dfrac{a_2}{a_1} = \dfrac{-12}{4} = -3$

$\dfrac{a_3}{a_2} = \dfrac{36}{-12} = -3$

$\dfrac{a_4}{a_3} = \dfrac{-108}{36} = -3$

$\dfrac{a_5}{a_4} = \dfrac{324}{-108} = -3$

$\dfrac{a_6}{a_5} = \dfrac{-972}{324} = -3$

11. Yes, the sequence is geometric.

$\dfrac{a_2}{a_1} = \dfrac{6}{3} = 2$

$\dfrac{a_3}{a_2} = \dfrac{12}{6} = 2$

$\dfrac{a_4}{a_3} = \dfrac{24}{12} = 2$

$\dfrac{a_5}{a_4} = \dfrac{48}{24} = 2$

$\dfrac{a_6}{a_5} = \dfrac{96}{48} = 2$

$r = 2, a_1 = 3$

$a_n = 3(2)^{n-1}$

Chapter 9 Conics, Sequences, and Series

13. Yes, the sequence is geometric.

$\dfrac{a_2}{a_1} = \dfrac{-8}{2} = -4$

$\dfrac{a_3}{a_2} = \dfrac{32}{-8} = -4$

$\dfrac{a_4}{a_3} = \dfrac{-128}{32} = -4$

$\dfrac{a_5}{a_4} = \dfrac{512}{-128} = -4$

$\dfrac{a_6}{a_5} = \dfrac{-2048}{512} = -4$

$r = -4, a_1 = 2$

$a_n = 2(-4)^{n-1}$

15. Yes, the sequence is geometric.

$\dfrac{a_2}{a_1} = \dfrac{125}{625} = 0.2$

$\dfrac{a_3}{a_2} = \dfrac{25}{125} = 0.2$

$\dfrac{a_4}{a_3} = \dfrac{5}{25} = 0.2$

$\dfrac{a_5}{a_4} = \dfrac{1}{5} = 0.2$

$\dfrac{a_6}{a_5} = \dfrac{0.2}{1} = 0.2$

$r = 0.2, a_1 = 625$

$a_n = 625(0.2)^{n-1}$

17. No, the sequence is not geometric.

$\dfrac{a_2}{a_1} = \dfrac{0}{-4.5} = 0$

$\dfrac{a_3}{a_2} = \dfrac{4.5}{0} = undefined$

$\dfrac{a_4}{a_3} = \dfrac{9}{4.5} = 2$

$\dfrac{a_5}{a_4} = \dfrac{13.5}{9} = 1.5$

$\dfrac{a_6}{a_5} = \dfrac{18}{13.5} \approx 1.333$

19. No, the sequence is not geometric.

$\dfrac{a_2}{a_1} = \dfrac{-2}{-4} = 0.5$

$\dfrac{a_3}{a_2} = \dfrac{-1}{-2} = 0.5$

$\dfrac{a_4}{a_3} = \dfrac{-\frac{1}{2}}{-1} = 0.5$

$\dfrac{a_5}{a_4} = \dfrac{-\frac{1}{4}}{-\frac{1}{2}} = 0.5$

$\dfrac{a_6}{a_5} = \dfrac{-\frac{1}{16}}{-\frac{1}{4}} = 0.25$

21.

a..

$a_n = a_1 r^{n-1}$

$a_1 = 56,000$

$r = 1 + 0.015 = 1.015$

$a_n = 56,000(1.015)^{n-1}$

b.

$a_7 = 56,000(1.015)^{7-1}$

$a_7 = 61,232.82$

Your salary will be $61,232.82 in the 7th year of employment.

c.

$80,052 = 56,000(1.015)^{n-1}$

$\dfrac{80,052}{56,000} = \dfrac{56,000(1.015)^{n-1}}{56,000}$

$1.4295 = 1.015^{n-1}$

$n - 1 = \log_{1.015}(1.4295)$

$n = \log_{1.015}(1.4295) + 1$

$n = \dfrac{\log 1.4295}{\log 1.015} + 1$

$n \approx 25$

In your 25th year of employment, your salary will be $80,052.

23.
a.
$a_n = a_1 r^{n-1}$
$a_1 = 8$ and $r = \dfrac{1}{2}$
$a_n = 8\left(\dfrac{1}{2}\right)^{n-1}$

b.
$a_3 = 8\left(\dfrac{1}{2}\right)^{3-1}$
$a_3 = 8\left(\dfrac{1}{2}\right)^2 = 8\left(\dfrac{1}{4}\right)$
$a_3 = 2$

The ball bounces to a height of 2 ft on the 3rd bounce.

25.
a.
$a_n = a_1 r^{n-1}$
$a_1 = 5$ and $r = 2$
$a_n = 5(2)^{n-1}$

b.
$a_7 = 5(2)^{7-1}$
$a_7 = 5(2)^6 = 320$

On day 7, there were 320 people with the flu.

27.
a.
$a_n = a_1 r^{n-1}$
$a_1 = 21,000$
$r = 1 - 0.07 = 0.93$
$a_n = 21,000(0.93)^{n-1}$

b.
$a_4 = 21,000(0.93)^{4-1}$
$a_4 = 21,000(0.93)^3$
$a_4 = 16,891.50$

A four year old Honda Civic hybrid is worth $16,891.50.

29.
a.
$\left.\begin{array}{l}a_0 = 10 \\ a_2 = 5\end{array}\right\} a_n = a_1 r^n$
$a_2 = 10r^2$
$5 = 10r^2$
$\dfrac{5}{10} = r^2$
$\left(\dfrac{1}{2}\right)^{1/2} = (r^2)^{1/2}$
$\left(\dfrac{1}{2}\right)^{1/2} = r$
$a_n = 10\left(\dfrac{1}{2}\right)^{\frac{n}{2}}$

b.
$a_6 = 10\left(\dfrac{1}{2}\right)^{\frac{6}{2}}$
$a_6 = 10\left(\dfrac{1}{2}\right)^3 \approx 1.25$

After 6 hours, there is 1.25 mg of morphine left in the patient.

31.
$a_4 = 250$ and $a_9 = 781,250$
$250 = a_1(r)^{4-1}$ and $781250 = a_1(r)^{9-1}$
$250 = a_1(r)^3$ and $781250 = a_1(r)^8$
$\dfrac{781250}{250} = \dfrac{a_1(r)^8}{a_1(r)^3}$
$3125 = r^5$
$(3125)^{\frac{1}{5}} = (r^5)^{\frac{1}{5}}$
$r = 5$
$250 = a_1(5)^{4-1}$
$250 = 125 a_1$
$a_1 = 2$
$a_n = 2(5)^{n-1}$

33.

$a_4 = 256$ and $a_{10} = \dfrac{1}{16}$

$256 = a_1(r)^{4-1}$ and $\dfrac{1}{16} = a_1(r)^{10-1}$

$256 = a_1(r)^3$ and $\dfrac{1}{16} = a_1(r)^9$

$\dfrac{\frac{1}{16}}{256} = \dfrac{a_1(r)^9}{a_1(r)^3}$

$\dfrac{1}{4096} = r^6$

$\left(\dfrac{1}{4096}\right)^{\frac{1}{6}} = (r^6)^{\frac{1}{6}}$

$r = \dfrac{1}{4}$

$256 = a_1\left(\dfrac{1}{4}\right)^{4-1}$

$256 = \dfrac{1}{64}a_1$

$a_1 = 16384$

$a_n = 16384\left(\dfrac{1}{4}\right)^{n-1}$

35.

$a_4 = -24$ and $a_{11} = 3072$

$-24 = a_1(r)^{4-1}$ and $3072 = a_1(r)^{11-1}$

$-24 = a_1(r)^3$ and $3072 = a_1(r)^{10}$

$\dfrac{3072}{-24} = \dfrac{a_1(r)^{10}}{a_1(r)^3}$

$-128 = r^7$

$(-128)^{\frac{1}{7}} = (r^7)^{\frac{1}{7}}$

$r = -2$

$-24 = a_1(-2)^{4-1}$

$-24 = -8a_1$

$a_1 = 3$

$a_n = 3(-2)^{n-1}$

37.

$a_5 = 16$ and $a_{12} = -\dfrac{1}{8}$

$16 = a_1(r)^{5-1}$ and $-\dfrac{1}{8} = a_1(r)^{12-1}$

$16 = a_1(r)^4$ and $-\dfrac{1}{8} = a_1(r)^{11}$

$\dfrac{-\frac{1}{8}}{16} = \dfrac{a_1(r)^{11}}{a_1(r)^4}$

$-\dfrac{1}{128} = r^7$

$\left(-\dfrac{1}{128}\right)^{\frac{1}{7}} = (r^7)^{\frac{1}{7}}$

$r = -\dfrac{1}{2}$

$16 = a_1\left(-\dfrac{1}{2}\right)^{5-1}$

$16 = \dfrac{1}{16}a_1$

$a_1 = 256$

$a_n = 256\left(-\dfrac{1}{2}\right)^{n-1}$

39.

n	a_n	Ratio
1	11	
2	22	$\dfrac{a_2}{a_1} = \dfrac{22}{11} = 2$
3	44	$\dfrac{a_3}{a_2} = \dfrac{44}{22} = 2$
4	88	$\dfrac{a_4}{a_3} = \dfrac{88}{44} = 2$
5	176	$\dfrac{a_5}{a_4} = \dfrac{176}{88} = 2$
6	352	$\dfrac{a_6}{a_5} = \dfrac{352}{176} = 2$

This sequence has a common ratio of $r = 2$, so it is geometric.

$a_n = a_1(r)^{n-1}$

$a_n = 11(2)^{n-1}$

41.

n	a_n	Difference
1	14	
2	20	$a_2 - a_1 = 20 - 14 = 6$
3	26	$a_3 - a_2 = 26 - 20 = 6$
4	32	$a_4 - a_3 = 32 - 26 = 6$
5	38	$a_5 - a_4 = 38 - 32 = 6$
6	44	$a_6 - a_5 = 44 - 38 = 6$

This sequence has a common difference of $d = 6$, so it is arithmetic.

$a_n = a_1 + (n-1)d$
$a_n = 14 + (n-1)6$
$a_n = 14 + 6n - 6$
$a_n = 8 + 6n$

43.

n	a_n	Ratio
1	7776	
2	3888	$\dfrac{a_2}{a_1} = \dfrac{3888}{7776} = \dfrac{1}{2}$
3	1944	$\dfrac{a_3}{a_2} = \dfrac{1944}{3888} = \dfrac{1}{2}$
4	972	$\dfrac{a_4}{a_3} = \dfrac{972}{1944} = \dfrac{1}{2}$
5	486	$\dfrac{a_5}{a_4} = \dfrac{486}{972} = \dfrac{1}{2}$
6	243	$\dfrac{a_6}{a_5} = \dfrac{243}{486} = \dfrac{1}{2}$

This sequence has a common ratio of $r = \dfrac{1}{2}$, so it is geometric.

$a_n = a_1 (r)^{n-1}$
$a_n = 7776 \left(\dfrac{1}{2}\right)^{n-1}$

45.

n	a_n	Difference
1	−38	
2	−34.75	$a_2 - a_1 = -34.75 - (-38) = 3.25$
3	−31.5	$a_3 - a_2 = -31.5 - (-34.75) = 3.25$
4	−28.25	$a_4 - a_3 = -28.25 - (-31.5) = 3.25$
5	−25	$a_5 - a_4 = -25 - (-28.25) = 3.25$
6	−21.75	$a_6 - a_5 = -21.75 - (-25) = 3.25$

This sequence has a common difference of $d = 3.25$, so it is arithmetic.

$a_n = a_1 + (n-1)d$
$a_n = -38 + (n-1)3.25$
$a_n = -38 + 3.25n - 3.25$
$a_n = -41.25 + 3.25n$

47.

a. $180, 360, 540, \ldots$

$a_2 - a_1 = 360 - 180 = 180$
$a_3 - a_2 = 540 - 360 = 180$

This is an arithmetic sequence with $d = 180$.

$a_n = 180 + (n-1)180$
$a_n = 180 + 180n - 180$
$a_n = 180n$

b.

$a_{10} = 180(10) = 1800$ The sum of the interior angles of a 12 sided polygon is $1800°$.

49.

a. $a_n = 200{,}000 - 8000n$

b.

$a_{10} = 200{,}000 - 8000(10)$
$a_{10} = 120{,}000$

When the machine is 10 years old it will be worth $120,000.

Chapter 9 Conics, Sequences, and Series

51.

a.
$a_n = a_1 + (n-1)d$
$a_n = 15 + (n-1)5$
$a_n = 15 - 5n - 5$
$a_n = 10 + 5n$

b.
$a_5 = 10 + 5(5) = 35$

You should be jogging 35 minutes during the 5th week after surgery.

Section 9.5

1.

a.
$20.5+156+198+162+125+248+244$
$+209+165+262+214+60.6+176+172$
$+161+237+83+182+164+203+14.4$

b.
$20.5+156+198+162+125+248+244$
$+209+165+262+214+60.6+176+172$
$+161+237+83+182+164+203+14.4 = 3456.5$

The 2009 Giro d' Italia professional bike race is 3456.5 km in length.
The 2009 Tour de France professional bike race is 3445 km in length.

3.
$$\sum_{n=1}^{5}(2n+7) = (2(1)+7)+(2(2)+7)+(2(3)+7)$$
$$+(2(4)+7)+(2(5)+7) = 65$$

5.
$$\sum_{n=1}^{4}(4^n) = 4+4^2+4^3+4^4 = 340$$

7.
$$\sum_{n=1}^{6}(5n(-1)^n) = (5(-1))+(10(-1)^2)+(15(-1)^3)$$
$$+(20(-1)^4)+(25(-1)^5)+(30(-1)^6) = 15$$

9.
$$\sum_{n=1}^{5}\left(\frac{1}{n}\right) = 1+\frac{1}{2}+\frac{1}{3}+\frac{1}{4}+\frac{1}{5} = \frac{137}{60}$$

11.
$$\sum_{n=1}^{8}(n^2+5n) = (1+5)+(2^2+10)+(3^2+15)$$
$$+(4^2+20)+(5^2+25)+(6^2+30)$$
$$+(7^2+35)+(8^2+40) = 384$$

13.
$$\sum_{n=1}^{8}(3(-2)^n) = (3(-2))+(3(-2)^2)+(3(-2)^3)$$
$$+(3(-2)^4)+(3(-2)^5)+(3(-2)^6)$$
$$+(3(-2)^7)+(3(-2)^8) = 510$$

15.

a.
$a_n = a_1 + (n-1)d$
$a_1 = 42,000$ and $d = 800$
$a_n = 42,000 + (n-1)800$
$a_n = 42,000 + 800n - 800$
$a_n = 41,200 + 800n$

b.
$a_{15} = 41,200 + 800(15)$
$a_{15} = 53,200$

Your salary will be \$53,200 in the 15th year of employment.

c. Use $S_n = \dfrac{n(a_1+a_n)}{2}$

$S_{15} = \dfrac{15(42,000+53,200)}{2}$

$S_{15} = 714,000$

You will have earned a total of \$714,000 over a 15 year period of employment.

17.

a.
$a_n = a_1 + (n-1)d$
$a_1 = 8$ and $d = 2$
$a_n = 8 + (n-1)2$
$a_n = 8 + 2n - 2$
$a_n = 6 + 2n$

b. $a_8 = 6 + 2(8) = 22$ You will save \$22 during the 8th week.

c.

Use $S_n = \dfrac{n(a_1+a_n)}{2}$

$a_{26} = 6 + 2(26) = 58$

$S_{26} = \dfrac{26(8+58)}{2}$

$S_{25} = 858$

You will have saved a total of \$858 during the first 26 weeks.

19.

a.
$a_n = a_1 + (n-1)d$
$a_1 = 15$ and $d = 1$
$a_n = 15 + (n-1)1$
$a_n = 15 + n - 1$
$a_n = 14 + n$

Chapter 9 Conics, Sequences, and Series

b.

Use $S_n = \dfrac{n(a_1 + a_n)}{2}$

$a_{12} = 14 + 12 = 26$

$S_{12} = \dfrac{12(15 + 26)}{2}$

$S_{12} = 246$

A total of 246 blocks are needed to build a wall with 12 rows.

21.

$a_n = 4n + 1$

$a_1 = 4(1) + 1$

$a_1 = 5$

$a_{30} = 4(30) + 1$

$a_{30} = 121$

$S_{30} = \dfrac{30(5 + 121)}{2}$

$S_{30} = 1{,}890$

23.

$a_n = 20 + (n-1)5$

$a_1 = 20 + (1-1)5$

$a_1 = 20$

$a_{15} = 20 + (15-1)5$

$a_{15} = 90$

$S_{15} = \dfrac{15(20 + 90)}{2}$

$S_{15} = 825$

25.

$a_n = -9n - 2$

$a_1 = -9(1) - 2$

$a_1 = -11$

$a_{40} = -9(40) - 2$

$a_{40} = -362$

$S_{40} = \dfrac{40(-11 + (-362))}{2}$

$S_{40} = -7{,}460$

27.

$a_n = 2.4n + 6.2$

$a_1 = 2.4(1) + 6.2$

$a_1 = 8.6$

$a_{52} = 2.4(52) + 6.2$

$a_{52} = 131$

$S_{52} = \dfrac{52(8.6 + 131)}{2}$

$S_{52} = 3{,}629.6$

29.

$a_n = \dfrac{1}{3}n + 4$

$a_1 = \dfrac{1}{3}(1) + 4$

$a_1 = \dfrac{13}{3}$

$a_{27} = \dfrac{1}{3}(27) + 4$

$a_{27} = 13$

$S_{27} = \dfrac{27\left(\dfrac{13}{3} + 13\right)}{2}$

$S_{27} = 234$

31.

a.

$a_n = a_1 r^{n-1}$

$a_1 = 42{,}000$

$r = 1 + 0.02 = 1.02$

$a_n = 42{,}000(1.02)^{n-1}$

b.

$a_{15} = 42{,}000(1.02)^{15-1}$

$a_{15} = 42{,}000(1.02)^{14}$

$a_{15} = 55{,}418.11$

Your salary will be $55,418.11 during the 15$^\text{th}$ year of employment.

c. Use $S_n = \dfrac{a_1(1 - r^n)}{1 - r}$

$S_{15} = \dfrac{42{,}000(1 - 1.02^{15})}{1 - 1.02}$

$S_{15} = 726{,}323.51$

You will have earned a total of $726,323.51 over a 15 year career.

33.
a.
$a_n = a_1 r^{n-1}$
$a_1 = 10$
$r = 1 + 0.10 = 1.10$
$a_n = 10(1.10)^{n-1}$

b.
$a_8 = 10(1.10)^{8-1}$
$a_8 = 10(1.10)^7 \approx 19.49$

You will save $19.49 during the 8th week.

c.
Use $S_n = \dfrac{a_1(1-r^n)}{1-r}$

$S_{26} = \dfrac{10(1-1.10^{26})}{1-1.10}$

$S_{26} \approx 1091.82$

You will have saved a total of $1091.82 during the first 26 weeks

35.
a.
$a_n = a_1 r^{n-1}$
$a_1 = 3$ and $r = 2$
$a_n = 3(2)^{n-1}$

b.
$a_5 = 3(2)^{5-1}$
$a_5 = 3(2)^4 = 48$

There are 48 people diagnosed with the flu on the 5th day.

c.
Use $S_n = \dfrac{a_1(1-r^n)}{1-r}$

$S_{14} = \dfrac{3(1-2^{14})}{1-2}$

$S_{14} = 49,149$

There were 49,149 people diagnosed with the flu during the first two weeks.

37.
$a_n = 3^n$
$a_1 = 3^1 = 3$
$a_2 = 3^2 = 9$
$r = \dfrac{9}{3} = 3$

$S_9 = \dfrac{3(1-3^9)}{1-3}$

$S_9 = 29,523$

39.
$a_n = 2(7)^n$
$a_1 = 2(7)^1 = 14$
$a_2 = 2(7)^2 = 98$
$r = \dfrac{98}{14} = 7$

$S_6 = \dfrac{14(1-7^6)}{1-7}$

$S_6 = 274,512$

41.
$a_n = 2(-6)^n$
$a_1 = 2(-6)^1 = -12$
$a_2 = 2(-6)^2 = 72$
$r = \dfrac{72}{-12} = -6$

$S_{11} = \dfrac{-12(1-(-6)^{11})}{1-(-6)}$

$S_{11} = -621,937,812$

43.
$a_n = 900\left(\dfrac{1}{3}\right)^n$

$a_1 = 900\left(\dfrac{1}{3}\right)^1 = 300$

$a_2 = 900\left(\dfrac{1}{3}\right)^2 = 100$

$r = \dfrac{100}{300} = \dfrac{1}{3}$

$S_{12} = \dfrac{300\left(1-\left(\dfrac{1}{3}\right)^{12}\right)}{1-\left(\dfrac{1}{3}\right)}$

$S_{12} \approx 449.999$

Chapter 9 Conics, Sequences, and Series

45.
$a_n = -1000\left(\dfrac{2}{5}\right)^n$

$a_1 = -1000\left(\dfrac{2}{5}\right)^1 = -400$

$a_2 = -1000\left(\dfrac{2}{5}\right)^2 = -160$

$r = \dfrac{-160}{-400} = \dfrac{2}{5}$

$S_{10} = \dfrac{-400\left(1-\left(\dfrac{2}{5}\right)^{10}\right)}{1-\left(\dfrac{2}{5}\right)}$

$S_{10} \approx -666.597$

47. Geometric

$a_n = 5(2)^n$

$a_1 = 5(2)^1 = 10$

$a_2 = 5(2)^2 = 20$

$r = \dfrac{20}{10} = 2$

$S_{10} = \dfrac{10(1-2^{10})}{1-2}$

$S_{10} = 10,230$

49. Geometric

$a_n = 2(-3)^n$

$a_1 = 2(-3)^1 = -6$

$a_2 = 2(-3)^2 = 18$

$r = \dfrac{18}{-6} = -3$

$S_6 = \dfrac{-6\left(1-(-3)^6\right)}{1-(-3)}$

$S_6 = 1,092$

51. Arithmetic

$a_n = 8n + 25$

$a_1 = 8(1) + 25 = 33$

$a_{23} = 8(23) + 25 = 209$

$S_{23} = \dfrac{23(33+209)}{2}$

$S_{23} = 2,783$

53. Geometric

$a_n = 900(0.94)^n$

$a_1 = 900(0.94)^1 = 846$

$a_2 = 900(0.94)^2 = 795.24$

$r = \dfrac{795.24}{846} = 0.94$

$S_{12} = \dfrac{846\left(1-(0.94)^{12}\right)}{1-(0.94)}$

$S_{12} \approx 7,389.524$

55. Arithmetic

$a_n = -\dfrac{1}{2}n + 30$

$a_1 = -\dfrac{1}{2}(1) + 30 = 29.5$

$a_{10} = -\dfrac{1}{2}(10) + 30 = 25$

$S_{10} = \dfrac{10(29.5+25)}{2}$

$S_{10} = 272.5$

Chapter 9 Review

1.
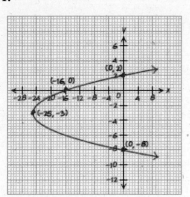

2.
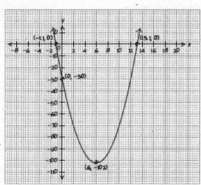

3.
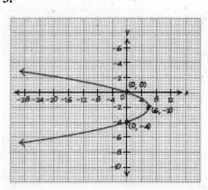

4.
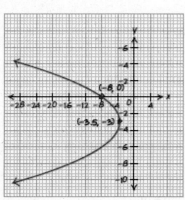

5.
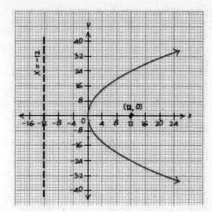

6.
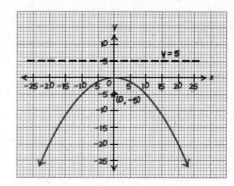

7.
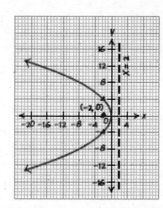

8.
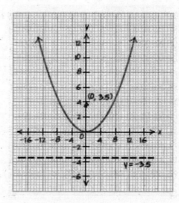

9.
Center $=(2,8)$ and Radius $=3$
$h=2, k=8, r=3$
$(x-2)^2+(y-8)^2=(3)^2$
$(x-2)^2+(y-8)^2=9$

10.
Center $=(5,-2)$ and Radius $=2.5$
$h=5, k=-2, r=2.5$
$(x-5)^2+(y-(-2))^2=(2.5)^2$
$(x-5)^2+(y+2)^2=6.25$

11.
$x^2+y^2=16$
$(x-0)^2+(y-0)^2=(4)^2$
$h=0, k=0, r=4$
Center $=(0,0)$ and Radius $=4$

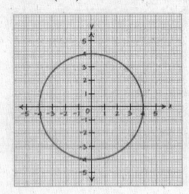

12.
$(x-5)^2+(y-2)^2=4$
$(x-5)^2+(y-2)^2=(2)^2$
$h=5, k=2, r=2$
Center $=(5,2)$ and Radius $=2$

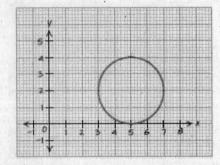

13.
$(x+4)^2+(y-6)^2=1$
$(x-(-4))^2+(y-6)^2=(1)^2$
$h=-4, k=6, r=1$
Center $=(-4,6)$ and Radius $=1$

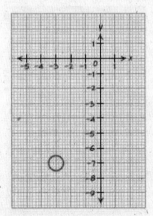

14.
$(x+3)^2+(y+7)^2=0.25$
$(x-(-3))^2+(y-(-7))^2=(0.5)^2$
$h=-3, k=-7, r=0.5$
Center $=(-3,-7)$ and Radius $=0.5$

15.
$x^2+y^2+12x-10y-39=0$
$x^2+12x+y^2-10y=39$
$(x^2+12x+\underline{})+(y^2-10y+\underline{})=39$
Completing the square: $\left(\dfrac{12}{2}\right)^2=36$ and $\left(\dfrac{-10}{2}\right)^2=25$
$(x^2+12x+36)+(y^2-10y+25)=39+36+25$
$(x+6)^2+(y-5)^2=100$
Center $=(-6,5)$ and Radius $=10$

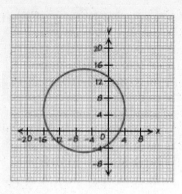

16.
$3x^2 + 3y^2 - 24x + 12y - 15 = 0$
$\dfrac{3x^2 + 3y^2 - 24x + 12y - 15}{3} = \dfrac{0}{3}$
$x^2 + y^2 - 8x + 4y - 5 = 0$
$x^2 - 8x + y^2 + 4y = 5$
$(x^2 - 8x + \underline{}) + (y^2 + 4y + \underline{}) = 5$
Completing the square: $\left(\dfrac{-8}{2}\right)^2 = 16$ and $\left(\dfrac{4}{2}\right)^2 = 4$
$(x^2 - 8x + 16) + (y^2 + 4y + 4) = 5 + 16 + 4$
$(x-4)^2 + (y+2)^2 = 25$
Center $=(4, -2)$ and Radius $= 5$

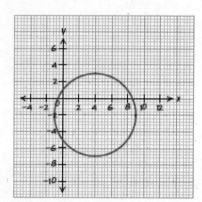

17.
Center $=(2, 6)$ and Radius $= 4$
$h = 2, k = 6, r = 4$
$(x-2)^2 + (y-6)^2 = (4)^2$
$(x-2)^2 + (y-6)^2 = 16$

18.
Center $=(-4, 2)$ and Radius $= 6$
$h = -4, k = 2, r = 6$
$(x-(-4))^2 + (y-2)^2 = (6)^2$
$(x+4)^2 + (y-2)^2 = 36$

19.
Focus $=(5, 0)$
$p = 5$
$x = \dfrac{y^2}{4(5)}$
$x = \dfrac{y^2}{20}$

20.
Focus $=(0, -3)$
$p = -3$
$y = \dfrac{x^2}{4(-3)}$
$y = \dfrac{x^2}{-12}$

21.
$\dfrac{x^2}{16} + \dfrac{y^2}{25} = 1$

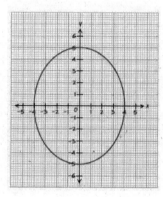

22.
$\dfrac{x^2}{0.25} + y^2 = 1$
$\dfrac{x^2}{0.25} + \dfrac{y^2}{1} = 1$

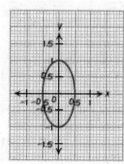

23.

$100x^2 + 4y^2 = 400$

$\dfrac{100x^2}{400} + \dfrac{4y^2}{400} = \dfrac{400}{400}$

$\dfrac{x^2}{4} + \dfrac{y^2}{100} = 1$

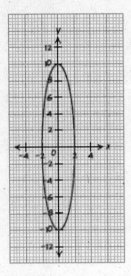

24.

$10x^2 + 2y^2 = 40$

$\dfrac{10x^2}{40} + \dfrac{2y^2}{40} = \dfrac{40}{40}$

$\dfrac{x^2}{4} + \dfrac{y^2}{20} = 1$

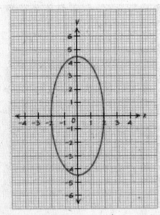

25.

$\dfrac{x^2}{16} - \dfrac{y^2}{25} = 1$

$\dfrac{x^2}{4^2} - \dfrac{y^2}{5^2} = 1$

$a = 4, b = 5$

$c^2 = 4^2 + 5^2$

$c = \pm\sqrt{41}$

Vertices $= (4,0), (-4,0)$

Foci $= \left(\sqrt{41},0\right), \left(-\sqrt{41},0\right) \approx (6.4,0), (-6.4,0)$

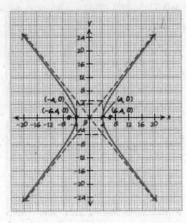

26.

$\dfrac{y^2}{100} - \dfrac{x^2}{4} = 1$

$\dfrac{y^2}{10^2} - \dfrac{x^2}{2^2} = 1$

$a = 2, b = 10$

$c^2 = 2^2 + 10^2$

$c = \pm\sqrt{104}$

Vertices $= (0,10), (0,-10)$

Foci $= \left(0,\sqrt{104}\right), \left(0,-\sqrt{104}\right) \approx (0,10.2), (0,-10.2)$

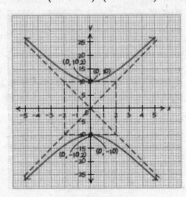

27.

$x^2 - 16y^2 = 16$

$\dfrac{x^2}{16} - \dfrac{16y^2}{16} = \dfrac{16}{16}$

$\dfrac{x^2}{16} - \dfrac{y^2}{1} = 1$

$\dfrac{x^2}{4^2} - \dfrac{y^2}{1^2} = 1$

$a = 4, b = 1$

$c^2 = 4^2 + 1^2$

$c = \pm\sqrt{17}$

Vertices $= (4,0), (-4,0)$

Foci $= \left(\sqrt{17}, 0\right), \left(-\sqrt{17}, 0\right) \approx (4.1, 0), (-4.1, 0)$

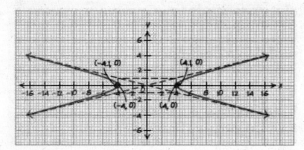

28.

$9y^2 - 25x^2 = 225$

$\dfrac{9y^2}{225} - \dfrac{25x^2}{225} = \dfrac{225}{225}$

$\dfrac{y^2}{25} - \dfrac{x^2}{9} = 1$

$\dfrac{y^2}{5^2} - \dfrac{x^2}{3^2} = 1$

$a = 3, b = 5$

$c^2 = 3^2 + 5^2$

$c = \pm\sqrt{34}$

Vertices $= (0, 5), (0, -5)$

Foci $= \left(0, \sqrt{34}\right), \left(0, -\sqrt{34}\right) \approx (0, 5.8), (0, -5.8)$

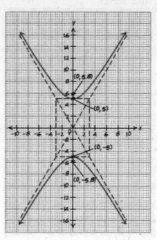

29.

This is an ellipse.

Vertices $= (10, 0), (-10, 0), (0, 15), (0, -15)$

$a = 10, b = 15$

$\dfrac{x^2}{10^2} + \dfrac{y^2}{15^2} = 1$

$\dfrac{x^2}{100} + \dfrac{y^2}{225} = 1$

30.

This is an ellipse.

Vertices $= (20, 0), (-20, 0), (0, 5), (0, -5)$

$a = 20, b = 5$

$\dfrac{x^2}{20^2} + \dfrac{y^2}{5^2} = 1$

$\dfrac{x^2}{400} + \dfrac{y^2}{25} = 1$

31.

This is a hyperbola that opens up and down.

$a = 15, b = 10$

$\dfrac{y^2}{10^2} - \dfrac{x^2}{15^2} = 1$

$\dfrac{y^2}{100} - \dfrac{x^2}{225} = 1$

32.

This is a hyperbola that opens left and right.

$a = 10, b = 8$

$\dfrac{x^2}{10^2} - \dfrac{y^2}{8^2} = 1$

$\dfrac{x^2}{100} - \dfrac{y^2}{64} = 1$

33.

$4x^2 + 36y^2 = 36$ A = 4 and C = 36

Since A ≠ C, A > 0 and C > 0 then this is an ellipse.

$\dfrac{4x^2}{36} + \dfrac{36y^2}{36} = \dfrac{36}{36}$

$\dfrac{x^2}{9} + \dfrac{y^2}{1} = 1$

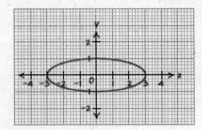

34.

$(x-9)^2 + (y-6)^2 = 9$

This is a circle in standard form.

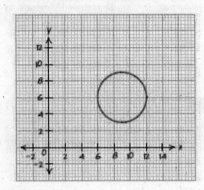

35.

$2y^2 - 50x^2 = 50$ A = −50 and C = 2

Since A ≠ C, A < 0 or C < 0 then this is a hyperbola.

$\dfrac{2y^2}{50} - \dfrac{50x^2}{50} = \dfrac{50}{50}$

$\dfrac{y^2}{25} - \dfrac{x^2}{1} = 1$

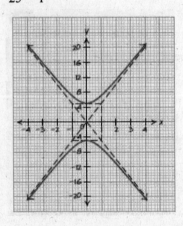

36.

$-2y^2 + 3x + 6y - 21 = 0$ A = 0 and C = −2

Since A = 0 or C = 0 then this is a parabola.

$-2y^2 + 3x + 6y - 21 = 0$

$3x = 2y^2 - 6y + 21$

$\dfrac{3x}{3} = \dfrac{2y^2 - 6y + 21}{3}$

$x = \dfrac{2}{3}y^2 - 2y + 7$

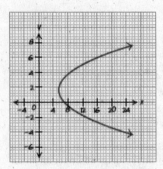

37.

$4x^2 + 4y^2 + 24x = 0$ A = 4 and C = 4

Since A = C then this is a circle.

$\dfrac{4x^2 + 4y^2 + 24x}{4} = \dfrac{0}{4}$

$x^2 + y^2 + 6x = 0$

$(x^2 + 6x + \underline{}) + y^2 = 0$

Completing the square: $\left(\dfrac{6}{2}\right)^2 = 9$

$(x^2 + 6x + 9) + y^2 = 0 + 9$

$(x+3)^2 + (y-0)^2 = 9$

Center = $(-3, 0)$ and Radius = 3

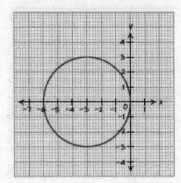

38. Hyperbola

$$9x^2 - 36y^2 = 324$$
$$\frac{9x^2}{324} - \frac{36y^2}{324} = \frac{324}{324}$$
$$\frac{x^2}{36} - \frac{y^2}{9} = 1$$

$a^2 = 36 \rightarrow a = 6$
$b^2 = 9 \rightarrow b = 3$
$c = \pm\sqrt{36+9} = \pm\sqrt{45} \approx \pm 6.7$
Foci: $(-6.7, 0)$ and $(6.7, 0)$
Vertices: $(-6, 0)$ and $(6, 0)$

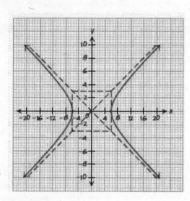

39. Parabola

$$0.5x^2 + 4x - y = 6$$
$$0.5x^2 + 4x - 6 = y$$

$\left. \begin{array}{l} h = \dfrac{-4}{2(0.5)} = -4 \\ k = 0.5(4)^2 + 4(-4) - 6 = -14 \end{array} \right\}$

vertex $(-4, -14)$

$$0.5(x+4)^2 = y + 14$$
$$\frac{0.5(x+4)^2}{0.5} = \frac{y+14}{0.5}$$
$$(x+4)^2 = 2(y+14)$$
$$4p = 2$$
$$p = 0.5$$

Focus: $(-4, -14 + 0.5) \rightarrow (-4, -13.5)$
Directrix: $y = -14 - 0.5 = -14.5$

x-intercepts: $x = \dfrac{-4 \pm \sqrt{(4)^2 - 4(0.5)(-6)}}{2(0.5)}$

$x \approx -9.3$ and $x \approx 1.3$

$(-9.3, 0)$ and $(1.3, 0)$

y-intercept: $(0, -6)$

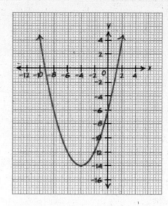

40. Ellipse

$$x^2 + 49y^2 = 49$$
$$\frac{x^2}{49} + \frac{49y^2}{49} = \frac{49}{49}$$
$$\frac{x^2}{49} + y^2 = 1$$

$a^2 = 49 \rightarrow a = 7$
$b^2 = 1 \rightarrow b = 1$
$c = \pm\sqrt{49-1} = \pm\sqrt{48} \approx \pm 6.9$
Foci: $(-6.9, 0)$ and $(6.9, 0)$
Vertices: $(-7, 0)$ and $(7, 0)$

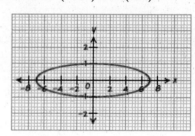

41. $a_n = 7n - 20$

$a_1 = 7(1) - 20 = -13$
$a_2 = 7(2) - 20 = -6$
$a_3 = 7(3) - 20 = 1$
$a_4 = 7(4) - 20 = 8$
$a_5 = 7(5) - 20 = 15$

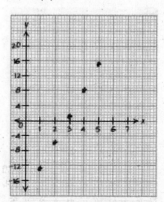

42. $a_n = 4(1.5)^n$

$a_1 = 4(1.5)^1 = 6$

$a_2 = 4(1.5)^2 = 9$

$a_3 = 4(1.5)^3 = 13.5$

$a_4 = 4(1.5)^4 = 20.25$

$a_5 = 4(1.5)^5 = 30.375$

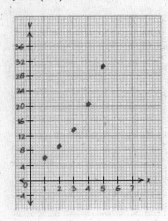

44. $a_n = \dfrac{2n}{n+8}$

$a_1 = \dfrac{2(1)}{1+8} = \dfrac{2}{9} \approx 0.22$

$a_2 = \dfrac{2(2)}{2+8} = \dfrac{4}{10} = 0.4$

$a_3 = \dfrac{2(3)}{3+8} = \dfrac{6}{11} \approx 0.55$

$a_4 = \dfrac{2(4)}{4+8} = \dfrac{8}{12} \approx 0.66$

$a_5 = \dfrac{2(5)}{5+8} = \dfrac{10}{13} \approx 0.77$

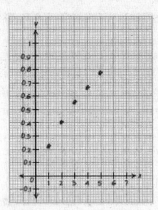

43. $a_n = \dfrac{(-1)^n}{8n}$

$a_1 = \dfrac{(-1)^1}{8(1)} = -\dfrac{1}{8}$

$a_2 = \dfrac{(-1)^2}{8(2)} = \dfrac{1}{16}$

$a_3 = \dfrac{(-1)^3}{8(3)} = -\dfrac{1}{24}$

$a_4 = \dfrac{(-1)^4}{8(4)} = \dfrac{1}{32}$

$a_5 = \dfrac{(-1)^5}{8(5)} = -\dfrac{1}{40}$

45.

$a_n = 50 + 2(n-1)$

$78 = 50 + 2(n-1)$

$28 = 2(n-1)$

$\dfrac{28}{2} = \dfrac{2(n-1)}{2}$

$14 = n-1$

$n = 15$

46.

$a_n = 4n^2 + 3n$

$351 = 4n^2 + 3n$

$0 = 4n^2 + 3n - 351$

$n = \dfrac{-3 \pm \sqrt{3^2 - 4(4)(-351)}}{2(4)}$

$n = 9$ and $n = -9.75$

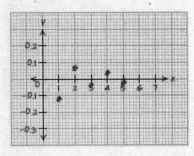

47.

$$a_n = 0.25(2)^n$$
$$1,048,576 = 0.25(2)^n$$
$$\frac{1,048,576}{0.25} = \frac{0.25(2)^n}{0.25}$$
$$4,194,304 = 2^n$$
$$n = \log_2(4,194,304)$$
$$n = \frac{\log(4,194,304)}{\log 2}$$
$$n \approx 22$$

48.

$$a_n = \sqrt{6n+4} + 48$$
$$68 = \sqrt{6n+4} + 48$$
$$20 = \sqrt{6n+4}$$
$$(20)^2 = \left(\sqrt{6n+4}\right)^2$$
$$400 = 6n+4$$
$$396 = 6n$$
$$\frac{396}{6} = \frac{6n}{6}$$
$$n = 66$$

49.

$$a_2 - a_1 = 79 - 82 = -3$$
$$a_3 - a_2 = 76 - 79 = -3$$
$$a_4 - a_3 = 73 - 76 = -3$$
$$a_5 - a_4 = 70 - 73 = -3$$
$$a_6 - a_5 = 67 - 70 = -3$$

The sequence is arithmetic. The common difference is -3.

$$a_n = 82 + (n-1)(-3)$$
$$= 82 - 3n + 3$$
$$a_n = 85 - 3n$$

50.

$$a_2 - a_1 = -1.5 - (-4) = 2.5$$
$$a_3 - a_2 = 1 - (-1.5) = 2.5$$
$$a_4 - a_3 = 3.5 - 1 = 2.5$$
$$a_5 - a_4 = 6 - 3.5 = 2.5$$
$$a_6 - a_5 = 8.5 - 6 = 2.5$$

The sequence is arithmetic. The common difference is 2.5.

$$a_n = -4 + (n-1)2.5$$
$$= -4 + 2.5n - 2.5$$
$$a_n = -6.5 + 2.5n$$

51.

$$a_2 - a_1 = 72 - 48 = 24$$
$$a_3 - a_2 = 108 - 72 = 36$$
$$24 \neq 36$$

The differences between consecutive terms are not the same, so this is not an arithmetic sequence.

52.

$$a_2 - a_1 = \frac{7}{2} - 6 = -\frac{5}{2}$$
$$a_3 - a_2 = 1 - \frac{7}{2} = -\frac{5}{2}$$
$$a_4 - a_3 = -\frac{3}{2} - 1 = -\frac{5}{2}$$
$$a_5 - a_4 = -4 + \frac{3}{2} = -\frac{5}{2}$$
$$a_6 - a_5 = -\frac{13}{2} + 4 = -\frac{5}{2}$$

The sequence is arithmetic. The common difference is $-\frac{5}{2}$.

$$a_n = 6 + (n-1)\left(-\frac{5}{2}\right)$$
$$= 6 - \frac{5}{2}n + \frac{5}{2}$$
$$a_n = \frac{17}{2} - \frac{5}{2}n$$

53.

a.

$$a_1 = 4000, a_2 = 4200, a_3 = 4400, \ldots$$
$$a_n = 4000 + (n-1)200$$
$$a_n = 4000 + 200n - 200$$
$$a_n = 3800 + 200n$$

b.

$$a_{15} = 3800 + 200(15)$$
$$a_{15} = 6800$$

In 2015, the attendance on Easter Sunday at North Coast Church will be 6800.

c.

$$3800 + 200n = 6000$$
$$\underline{-3800 \qquad -3800}$$
$$200n = 2200$$
$$\frac{200n}{200} = \frac{2200}{200}$$
$$n = 11$$

The attendance on Easter Sunday at North Coast Church will be 6000 in the year 2011.

Chapter 9 Conics, Sequences, and Series

54.

a.

$a_1 = 27,000$

$d = -2500$

$a_n = 27,000 + (n-1)(-2500)$

$a_n = 27,000 - 2500n + 2500$

$a_n = 29,500 - 2500n$

b.

$a_6 = 29,500 - 2500(6)$

$a_6 = 14,500$

The car will be worth $14,500 in the sixth year.

c.

$29,500 - 2500n = 2000$

$\underline{-29,500 \qquad -29,500}$

$-2500n = -27,500$

$\dfrac{-2500n}{-2500} = \dfrac{-27,500}{-2500}$

$n = 11$

The car will be 11 years old when it is worth $2000.

55.

$\left.\begin{array}{l} a_5 = 66 \\ a_{17} = 144 \end{array}\right\} d = \dfrac{144 - 66}{17 - 5} = \dfrac{78}{12} = 6.5$

$y - 66 = 6.5(x - 5)$

$y - 66 = 6.5x - 32.5$

$\underline{+66 \qquad +66}$

$y = 6.5x + 33.5$

$a_n = 6.5n + 33.5$

56.

$\left.\begin{array}{l} a_4 = -12 \\ a_{16} = -108 \end{array}\right\} d = \dfrac{-108 + 12}{16 - 4} = \dfrac{-96}{12} = -8$

$y + 12 = -8(x - 4)$

$y + 12 = -8x + 32$

$\underline{-12 \qquad -12}$

$y = -8x + 20$

$a_n = -8n + 20$

57.

$\dfrac{a_2}{a_1} = \dfrac{12}{3} = 4, \dfrac{a_3}{a_2} = \dfrac{48}{12} = 4, \dfrac{a_4}{a_3} = \dfrac{192}{48} = 4,$

$\dfrac{a_5}{a_4} = \dfrac{768}{192} = 4, \dfrac{a_6}{a_5} = \dfrac{3072}{768} = 4$

This sequence is geometric with common ratio 4.

$a_n = 3(4)^{n-1}$

58.

$\dfrac{a_2}{a_1} = \dfrac{12}{36} = \dfrac{1}{3}, \dfrac{a_3}{a_2} = \dfrac{4}{12} = \dfrac{1}{3},$

$\dfrac{a_4}{a_3} = \dfrac{4/3}{4} = \dfrac{1}{3}, \dfrac{a_5}{a_4} = \dfrac{4/9}{4/3} = \dfrac{1}{3},$

$\dfrac{a_6}{a_5} = \dfrac{4/27}{4/9} = \dfrac{1}{3}$

This sequence is geometric with common ratio $\dfrac{1}{3}$.

$a_n = 36\left(\dfrac{1}{3}\right)^{n-1}$

59.

$\dfrac{a_2}{a_1} = \dfrac{-1}{0.5} = -2, \dfrac{a_3}{a_2} = \dfrac{2}{-1} = -2, \dfrac{a_4}{a_3} = \dfrac{-4}{2} = -2,$

$\dfrac{a_5}{a_4} = \dfrac{8}{-4} = -2, \dfrac{a_6}{a_5} = \dfrac{-16}{8} = -2$

This sequence is geometric with common ratio -2.

$a_n = 0.5(-2)^{n-1}$

60.

$\dfrac{a_2}{a_1} = \dfrac{34}{6} = \dfrac{17}{3}, \dfrac{a_3}{a_2} = \dfrac{64}{34} = \dfrac{32}{17},$

This sequence is not geometric.

61.

a.

$a_1 = 4000$

$r = 1 + 0.02 = 1.02$

$a_n = 4000(1.02)^n$

b.

$a_8 = 4000(1.02)^8$

$a_8 = 4686.64$

After eight years, the product will cost $4686.64.

62.

a.

$a_1 = 42,000$

$r = 1 + 0.0125 = 1.0125$

$a_n = 42,000(1.0125)^n$

b.

$a_5 = 42,000(1.0125)^5$

$a_5 = 44,691.45$

After five years of employment, your salary will be $44,691.45.

c.

$50,603 = 42,000(1.0125)^n$

$\dfrac{50,603}{42,000} = \dfrac{42,000(1.0125)^n}{42,000}$

$\dfrac{50,603}{42,000} = (1.0125)^n$

$\log_{1.0125}\left(\dfrac{50,603}{42,000}\right) = n$

$\dfrac{\log\left(\dfrac{50,603}{42,000}\right)}{\log(1.0125)} = n$

$n \approx 15$

After 15 years of employment, your salary will be approximately $50,603.

63.

$\left.\begin{array}{l}a_3 = 72\\a_6 = 243\end{array}\right\} \rightarrow \begin{array}{l}72 = a_1 r^{3-1}\\243 = a_1 r^{6-1}\end{array}$

Solve for r:

$\dfrac{243}{72} = \dfrac{\cancel{a_1}r^5}{\cancel{a_1}r^2} \rightarrow 3.375 = r^3$

$(3.375)^{\frac{1}{3}} = (r^3)^{\frac{1}{3}}$

$1.5 = r$

Solve for a_1:

$72 = a_1(1.5)^2$

$\dfrac{72}{(1.5)^2} = \dfrac{a_1(1.5)^2}{(1.5)^2}$

$a_1 = 32$

$a_n = 32(1.5)^{n-1}$

64.

$\left.\begin{array}{l}a_2 = 27\\a_5 = 1\end{array}\right\} \rightarrow \begin{array}{l}27 = a_1 r^{2-1}\\1 = a_1 r^{5-1}\end{array}$

Solve for r:

$\dfrac{1}{27} = \dfrac{\cancel{a_1}r^4}{\cancel{a_1}r^1} \rightarrow \dfrac{1}{27} = r^3$

$\left(\dfrac{1}{27}\right)^{\frac{1}{3}} = (r^3)^{\frac{1}{3}}$

$\dfrac{1}{3} = r$

Solve for a_1:

$27 = a_1\left(\dfrac{1}{3}\right)$

$(3)(27) = a_1\left(\dfrac{1}{3}\right)(3)$

$a_1 = 81$

$a_n = 81\left(\dfrac{1}{3}\right)^{n-1}$

65.

$\displaystyle\sum_{n=1}^{7}(4n-20) = (4(1)-20)+(4(2)-20)+(4(3)-20)$
$+(4(4)-20)+(4(5)-20)+(4(6)-20)$
$+(4(7)-20) = -28$

66.

$\displaystyle\sum_{n=1}^{6}(5(-2)^n) = (5(-2))+(5(-2)^2)+(5(-2)^3)$
$+(5(-2)^4)+(5(-2)^5)+(5(-2)^6) = 210$

67.

a.

$a_1 = 38,000$, $d = 900$

$a_n = 38,000 + (n-1)900$

$a_n = 38,000 + 900n - 900$

$a_n = 37,100 + 900n$

b.

$a_{15} = 37,100 + 900(15)$

$a_{15} = 50,600$

Your salary will be $50,600 in the 15th year of employment.

c.

$\displaystyle\sum_{n=1}^{15}(37,100 + 900n)$

$S_{15} = \dfrac{15(38,000 + 50,600)}{2}$

$S_{15} = \dfrac{15(88,600)}{2}$

$S_{15} = 664,500$

You will have earned a total of $664,500 over a 15 year period of employment.

68.

a.

$a_1 = 5$, $d = 3$

$a_n = 5 + (n-1)3$

$a_n = 5 + 3n - 3$

$a_n = 2 + 3n$

b. $a_5 = 2 + 3(5) = 17$

On the 5th day you will pick 17 apricots.

c.

$a_{14} = 2 + 3(14) = 44$

$\sum_{n=1}^{14}(2+3n)$

$S_{14} = \dfrac{14(5+44)}{2}$

$S_{14} = 343$

You will pick a total of 343 ripe apricots during the first two weeks.

69.

$S_{15} = \dfrac{42,000(1-1.025^{15})}{1-1.025}$

$S_{15} = 688,226.05$

During a 15 year career, you will have earned a total of $688,226.05.

70.

a.

$a_1 = 10,000$

$r = 1 + 0.2 = 1.2$

$a_n = 10,000(1.2)^{n-1}$

b.

$S_7 = \dfrac{10,000(1-1.2^7)}{1-1.2}$

$S_7 = 129,159$

During the seven day event, 129,159 people attended the car show.

71.

$a_n = 4n + 7$

$a_1 = 11$, $a_{14} = 4(14) + 7 = 63$

$S_{14} = \dfrac{14(11+63)}{2}$

$S_{14} = 518$

72.

$a_n = 300 - 2.5(n-1)$

$a_1 = 300$

$a_{23} = 300 - 2.5(22) = 245$

$S_{23} = \dfrac{23(300+245)}{2}$

$S_{23} = 6267.5$

73.

$a_n = 7(1.8)^n$

$a_1 = 7(1.8)^1 = 12.6$

$S_{11} = \dfrac{12.6(1-1.8^{11})}{1-1.8}$

$S_{11} = 10,106.52459$

74.

$a_n = 48\left(-\dfrac{1}{2}\right)^n$

$a_1 = 48(-0.5)^1 = -24$

$S_{20} = \dfrac{-24(1+0.5^{20})}{1+0.5}$

$S_{20} = -15.99998$

Chapter 9 Test

1.

Center $= (-7, 4)$ and Radius $= 8$

$h = -7, k = 4, r = 8$

$(x-(-7))^2 + (y-4)^2 = (8)^2$

$(x+7)^2 + (y-4)^2 = 64$

2.

$x = \dfrac{y^2}{36}$

$x = \dfrac{y^2}{4(9)}$

$p = 9$

Focus $= (9, 0)$, Directrix: $x = -9$

3.

$\dfrac{x^2}{100} - \dfrac{y^2}{36} = 1$

This is a hyperbola in standard form.

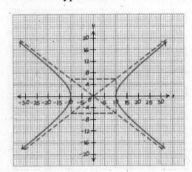

4.

$y^2 + 8 + x - 20 = 0$ $A = 0$ and $C = 1$

Since $A = 0$ or $C = 0$ then this is a parabola.

$y^2 + 8 + x - 20 = 0$

$x = -y^2 + 12$

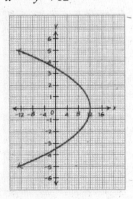

5.

$49x^2 + 4y^2 = 196$ $A = 49$ and $C = 4$

Since $A \neq C$, $A > 0$ and $C > 0$ then this is an ellipse.

$\dfrac{49x^2}{196} + \dfrac{4y^2}{196} = \dfrac{196}{196}$

$\dfrac{x^2}{4} + \dfrac{y^2}{49} = 1$

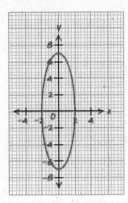

6.

$(x-3)^2 + (y-7)^2 = 81$

This is a circle in standard form.

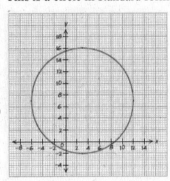

7.

Focus $= (-2, 0)$

$p = -2$

$x = \dfrac{y^2}{4(-2)}$

$x = \dfrac{y^2}{-8}$

8.

Vertices $= (2,0), (-2,0), (0,6), (0,-6)$

$a = 2, b = 6$

$\dfrac{x^2}{2^2} + \dfrac{y^2}{6^2} = 1$

$\dfrac{x^2}{4} + \dfrac{y^2}{36} = 1$

Chapter 9 Conics, Sequences, and Series

9. An arithmetic sequence has a common difference therefore the difference between successive terms is constant. A geometric sequence has a common ratio therefore the ratio between successive terms is constant.

10.

$a_n = 4(-3)^n$

$a_1 = 4(-3)^1 = -12$

$a_2 = 4(-3)^2 = 36$

$a_3 = 4(-3)^3 = -108$

$a_4 = 4(-3)^4 = 324$

$a_5 = 4(-3)^5 = -972$

The first five terms are $-12, 36, -108, 324, -972$.

11.

$a_n = \dfrac{6n}{n+2}$

$a_1 = \dfrac{6(1)}{1+2} = \dfrac{6}{3} = 2$

$a_2 = \dfrac{6(2)}{2+2} = \dfrac{12}{4} = 3$

$a_3 = \dfrac{6(3)}{3+2} = \dfrac{18}{5}$

$a_4 = \dfrac{6(4)}{4+2} = \dfrac{24}{6} = 4$

$a_5 = \dfrac{6(5)}{5+2} = \dfrac{30}{7}$

The first five terms are $2, 3, \dfrac{18}{5}, 4, \dfrac{30}{7}$.

12.

$a_n = 400 - 11(n-1) = 158$

$400 - 11n + 11 = 158$

$411 - 11n = 158$

$\underline{-411 \qquad -411}$

$-11n = -253$

$\dfrac{-11n}{-11} = \dfrac{-253}{-11}$

$n = 23$

13.

$a_n = 20 + \sqrt{n+8} = 38$

$20 + \sqrt{n+8} = 38$

$\underline{-20 \qquad\quad -20}$

$\sqrt{n+8} = 18$

$\left(\sqrt{n+8}\right)^2 = 18^2$

$n + 8 = 324$

$n = 316$

14.

$\dfrac{a_2}{a_1} = \dfrac{50}{250} = \dfrac{1}{5}, \dfrac{a_3}{a_2} = \dfrac{10}{50} = \dfrac{1}{5}, \dfrac{a_4}{a_3} = \dfrac{2}{10} = \dfrac{1}{5},$

$\dfrac{a_5}{a_4} = \dfrac{0.4}{2} = \dfrac{1}{5}, \dfrac{a_6}{a_5} = \dfrac{0.08}{0.4} = \dfrac{1}{5}$

This sequence is geometric with common ratio $\dfrac{1}{5}$.

$a_n = 250\left(\dfrac{1}{5}\right)^{n-1}$

15.

$a_2 - a_1 = 16.5 - 24 = -7.5$

$a_3 - a_2 = 9 - 16.5 = -7.5$

$a_4 - a_3 = 1.5 - 9 = -7.5$

$a_5 - a_4 = -6 - 1.5 = -7.5$

$a_6 - a_5 = -13.5 - (-6) = -7.5$

This sequence is arithmetic with common difference $d = -7.5$.

$a_1 = 24$

$d = -7.5$

$a_n = 24 + (n-1)(-7.5)$

$a_n = 24 - 7.5n + 7.5$

$a_n = 31.5 - 7.5n$

16.

a.

$a_1 = 5$

$d = 3$

$a_n = 5 + (n-1)3$

$a_n = 5 + 3n - 3$

$a_n = 2 + 3n$

b.

$a_3 = 2 + 3(3)$

$a_3 = 11$

By week three, you should be stretching your injured knee for 11 minutes at one time.

c.

$a_n = 2 + 3n$

$2 + 3n = 30$

$\underline{-2 \qquad -2}$

$3n = 28$

$n = \dfrac{28}{3} = 9\dfrac{1}{3}$

In the tenth week, you will be stretching your injured knee for half an hour at one time.

17.

a.

$a_1 = 60,000$

$r = 1 + 0.025 = 1.025$

$a_n = 60,000(1.025)^{n-1}$

b.

$a_{20} = 60,000(1.025)^{20-1} = 95,919.01$

Your salary will be \$95,919.01 in the 20th year.

c.

$S_{20} = \dfrac{60,000\left(1 - 1.025^{20}\right)}{1 - 1.025} = 1,532,679.46$

You will have earned a total of \$1,532,679.46 in salary over a 20 year career.

18.

$\sum_{n=1}^{6}\left(2n^2\right) = 2 + 2(2)^2 + 2(3)^2 + 2(4)^2$

$\qquad\qquad + 2(5)^2 + 2(6)^2 = 182$

19.

$a_n = 600 - 18(n-1)$

$a_1 = 600$ and $d = -18$

$a_{15} = 600 - 18(14) = 348$

$S_{15} = \dfrac{15(600 + 348)}{2}$

$S_{15} = 7110$

20.

$a_n = 20(1.75)^n$

$a_1 = 20(1.75)^1 = 35$

$r = 1.75$

$S_{12} = \dfrac{35\left(1 - 1.75^{12}\right)}{1 - 1.75}$

$S_{12} = 38,453.56699$

Chapter 9 Conics, Sequences, and Series
Chapter 1-9 Cumulative Review

1.
$$7a^3 + 18 = 1530$$
$$7a^3 = 1512$$
$$a^3 = 216$$
$$(a^3)^{\frac{1}{3}} = 216^{\frac{1}{3}}$$
$$a = 6$$

2.
$$-2(t+3)^2 - 8 = -80$$
$$-2(t+3)^2 = -72$$
$$(t+3)^2 = 36$$
$$\sqrt{(t+3)^2} = \pm\sqrt{36}$$
$$t+3 = \pm 6$$
$$t+3 = 6, t+3 = -6$$
$$t = 3, t = -9$$

3.
$$6x + 5 = 2(4x - 9)$$
$$6x + 5 = 8x - 18$$
$$-2x + 5 = -18$$
$$-2x = -23$$
$$x = \frac{23}{2}$$

4.
$$\frac{2}{m-3} = \frac{5}{m+8}$$
$$\frac{2}{m-3}(m-3)(m+8) = \frac{5}{m+8}(m-3)(m+8)$$
$$2(m+8) = 5(m-3)$$
$$2m + 16 = 5m - 15$$
$$-3m + 16 = -15$$
$$-3m = -31$$
$$m = \frac{31}{3}$$

5.
$$-2\sqrt{3h+7} + 10 = -4$$
$$-2\sqrt{3h+7} = -14$$
$$\sqrt{3h+7} = 7$$
$$(\sqrt{3h+7})^2 = 7^2$$
$$3h + 7 = 49$$
$$3h = 42$$
$$h = 14$$

6.
$$40\left(\frac{4}{5}\right)^n + 300 = 320.48$$
$$40\left(\frac{4}{5}\right)^n = 20.48$$
$$\left(\frac{4}{5}\right)^n = 0.512$$
$$\ln\left(\frac{4}{5}\right)^n = \ln(0.512)$$
$$n\ln\left(\frac{4}{5}\right) = \ln(0.512)$$
$$n = \frac{\ln(0.512)}{\ln\left(\frac{4}{5}\right)}$$
$$n = 3$$

7.
$$\ln(2x-5) = 4$$
$$e^{\ln(2x-5)} = e^4$$
$$2x - 5 = e^4$$
$$2x = e^4 + 5$$
$$x = \frac{e^4 + 5}{2}$$
$$x \approx 29.799$$

8.
$$\frac{4}{b+3} + 2 = \frac{-5}{b-7}$$
$$\frac{4}{b+3}(b+3)(b-7) + 2(b+3)(b-7) = \frac{-5}{b-7}(b+3)(b-7)$$
$$4(b-7) + 2(b+3)(b-7) = -5(b+3)$$
$$4b - 28 + 2(b^2 - 4b - 21) = -5b - 15$$
$$4b - 28 + 2b^2 - 8b - 42 = -5b - 15$$
$$2b^2 + b - 55 = 0$$
$$(2b+11)(b-5) = 0$$
$$2b + 11 = 0, b - 5 = 0$$
$$b = -\frac{11}{2}, b = 5$$

9.
$$|5h - 8| = 32$$
$$5h - 8 = 32, 5h - 8 = -32$$
$$5h = 40, 5h = -24$$
$$h = 8, h = -\frac{24}{5}$$

10.
$$\log_2(x+3)+\log_2(x+5)=3$$
$$\log_2(x+3)(x+5)=3$$
$$2^3=(x+3)(x+5)$$
$$8=x^2+8x+15$$
$$0=x^2+8x+7$$
$$0=(x+1)(x+7)$$
$$x+1=0, x+7=0$$
$$x=-1, x=-7$$

The answer $x=-7$ is not part of the domain.
The final answer is:
$$x=-1$$

11.
$$\sqrt{t+12}+\sqrt{2t+1}=7$$
$$\sqrt{2t+1}=7-\sqrt{t+12}$$
$$\left(\sqrt{2t+1}\right)^2=\left(7-\sqrt{t+12}\right)^2$$
$$2t+1=\left(7-\sqrt{t+12}\right)\left(7-\sqrt{t+12}\right)$$
$$2t+1=49-14\sqrt{t+12}+t+12$$
$$t-60=-14\sqrt{t+12}$$
$$(t-60)^2=\left(-14\sqrt{t+12}\right)^2$$
$$(t-60)(t-60)=196(t+12)$$
$$t^2-120t+3600=196t+2352$$
$$t^2-316t+1248=0$$
$$(t-312)(t-4)=0$$
$$t-312=0, t-4=0$$
$$\cancel{t=312}, t=4$$

12.
$$30(2^n)-8400=6960$$
$$30(2^n)=15360$$
$$2^n=512$$
$$2^n=2^9$$
$$n=9$$

13.
$$3.2g^2+2.4g-16.2=59.4$$
$$3.2g^2+2.4g-75.6=0$$
$$g=\frac{-2.4\pm\sqrt{2.4^2-4(3.2)(-75.6)}}{2(3.2)}$$
$$g=\frac{-2.4\pm\sqrt{973.44}}{6.4}$$
$$g=\frac{-2.4\pm 31.2}{6.4}$$
$$g=\frac{-2.4+31.2}{6.4}, g=\frac{-2.4-31.2}{6.4}$$
$$g=4.5, g=-5.25$$

14.
$$\frac{44}{a^2+3a-28}=\frac{7a}{a^2-a-12}$$
$$\frac{44}{(a+7)(a-4)}=\frac{7a}{(a+3)(a-4)}$$
$$\frac{44}{\cancel{(a+7)(a-4)}}(a+7)(a-4)(a+3)=\frac{7a}{\cancel{(a+3)(a-4)}}(a+7)(a-4)(a+3)$$
$$44(a+3)=7a(a+7)$$
$$44a+132=7a^2+49a$$
$$0=7a^2+5a-132$$
$$0=(7a+33)(a-4)$$
$$7a+33=0, a-4=0$$
$$a=-\frac{33}{7}, \cancel{a=4}$$

15.
$$\frac{1}{5}(x+4)=\frac{3}{10}x-\frac{5}{4}$$
$$\frac{1}{5}x+\frac{4}{5}=\frac{3}{10}x-\frac{5}{4}$$
$$20\left(\frac{1}{5}x+\frac{4}{5}\right)=20\left(\frac{3}{10}x-\frac{5}{4}\right)$$
$$4x+16=6x-25$$
$$-2x+16=-25$$
$$-2x=-41$$
$$x=\frac{41}{2}$$

Chapter 9 Conics, Sequences, and Series

16.
$$\sqrt[5]{7-3m} = 3$$
$$\left(\sqrt[5]{7-3m}\right)^5 = 3^5$$
$$7 - 3m = 243$$
$$-3m = 236$$
$$m = -\frac{236}{3}$$

17.
$$3r^3 + 9r^2 - 120r = 0$$
$$3r(r^2 + 3r - 40) = 0$$
$$3r(r+8)(r-5) = 0$$
$$3r = 0, r+8 = 0, r-5 = 0$$
$$r = 0, r = -8, r = 5$$

18.
$$-5|6h+4| = -30$$
$$|6h+4| = 6$$
$$6h+4 = 6, 6h+4 = -6$$
$$6h = 2, 6h = -10$$
$$h = \frac{1}{3}, h = -\frac{5}{3}$$

19.
$$2^{4x+5} = 0.125$$
$$\ln(2^{4x+5}) = \ln(0.125)$$
$$(4x+5)\ln(2) = \ln(0.125)$$
$$4x+5 = \frac{\ln(0.125)}{\ln(2)}$$
$$4x+5 = -3$$
$$4x = -8$$
$$x = -2$$

20.
$$5t^4 + 16 = 1269$$
$$5t^4 = 1253$$
$$t^4 = 250.6$$
$$(t^4)^{\frac{1}{4}} = \pm 250.6^{\frac{1}{4}}$$
$$t = \pm 3.98$$

21.
$$\begin{cases} 6x + 8y = 15 \\ 4y = -3x + 15 \end{cases}$$
Solve for y in the second equation:
$$\frac{4y}{4} = \frac{-3x}{4} + \frac{15}{4}$$
$$y = -0.75x + 3.75$$
Plug this into the first equation:
$$6x + 8(-0.75x + 3.75) = 15$$
$$6x - 6x + 30 = 15$$
$$30 = 15$$
No Solution.

22.
$$\begin{cases} 2x + 7y = 29 \\ 10x - 3y = -7 \end{cases}$$
Solve for x in the first equation:
$$\frac{2x}{2} + \frac{7y}{2} = \frac{29}{2}$$
$$x + 3.5y = 14.5$$
$$x = 14.5 - 3.5y$$
Plug this into the second equation:
$$10(14.5 - 3.5y) - 3y = -7$$
$$145 - 35y - 3y = -7$$
$$145 - 38y = -7$$
$$-38y = -152$$
$$y = 4$$
Substituting y back into either equation solve for x:
$$x = 14.5 - 3.5(4) = 0.5$$
The solution to the system is $(0.5, 4)$.

23.

$\begin{cases} y = 4x^2 - 6x + 5 \\ y = \dfrac{1}{2}x + 8 \end{cases}$

$4x^2 - 6x + 5 = \dfrac{1}{2}x + 8$

$2(4x^2 - 6x + 5) = 2\left(\dfrac{1}{2}x + 8\right)$

$8x^2 - 12x + 10 = x + 16$

$8x^2 - 13x - 6 = 0$

$(8x + 3)(x - 2) = 0$

$8x + 3 = 0, x - 2 = 0$

$x = -\dfrac{3}{8}, x = 2$

Substituting x back into either equation solve for y:

$y = \dfrac{1}{2}\left(-\dfrac{3}{8}\right) + 8 = \dfrac{125}{16}$

$y = \dfrac{1}{2}(2) + 8 = 9$

The solutions to the system are $\left(-\dfrac{3}{8}, \dfrac{125}{16}\right)$ and $(2, 9)$.

24.

$\begin{cases} y = -2x^2 + 3x + 10 \\ y = x^2 - 5x - 25 \end{cases}$

$-2x^2 + 3x + 10 = x^2 - 5x - 25$

$0 = 3x^2 - 8x - 35$

$0 = (3x + 7)(x - 5)$

$3x + 7 = 0, x - 5 = 0$

$x = -\dfrac{7}{3}, x = 5$

Substituting x back into either equation solve for y:

$y = \left(-\dfrac{7}{3}\right)^2 - 5\left(-\dfrac{7}{3}\right) - 25 = -\dfrac{71}{9}$

$y = (5)^2 - 5(5) - 25 = -25$

The solutions to the system are $\left(-\dfrac{7}{3}, -\dfrac{71}{9}\right)$ and $(5, -25)$.

25. Let $P(n)$ be the profit in dollars from manufacturing and selling n customs mp3 players.

a.

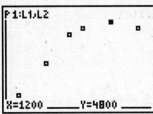

Choose $(1200, 4800)$ as the vertex, and $(200, 800)$ as the additional point used to solve for "a".

$P(n) = a(n - 1200)^2 + 4800$

$800 = a(200 - 1200)^2 + 4800$

$800 = a(1000)^2 + 4800$

$800 = 1{,}000{,}000a + 4800$

$\underline{-4800 \qquad\qquad -4800}$

$-4000 = 1{,}000{,}000a$

$\dfrac{-4000}{1{,}000{,}000} = \dfrac{1{,}000{,}000a}{1{,}000{,}000} \Bigg\} \; a \approx -\dfrac{1}{250}$

$P(n) = -\dfrac{1}{250}(n - 1200)^2 + 4800$

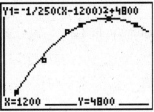

Note: Your answer may vary and yet still represent the trend of the data.

b. Domain: $[150, 2000]$. Within this domain, the maximum value of 4800 comes from the vertex. The minimum value comes from $n = 150$.

$P(150) = -\dfrac{1}{250}(150 - 1200)^2 + 4800$

Range: $[390, 4800]$

c.

$P(1000) = -\dfrac{1}{250}(1000 - 1200)^2 + 4800$

$P(1000) = 4640$

The profit from manufacturing and selling 1000 custom mp3 players is $4640.

d. Vertex: $(1200, 4800)$ The profit from manufacturing and selling 1200 custom mp3 players is $4800.

26. Let $R(t)$ be the primary revenue (in millions of dollars) t years since 2000.

a.

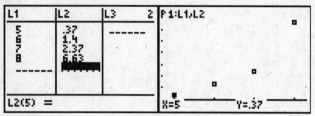

An exponential model is appropriate so choose a point from the flatter end and one from the steeper end.

Note: Your model may differ, but it must follow the trend of the data.

$R(t) = ab^t$

To find a and b choose two points $(5, 0.37)$ and $(8, 6.63)$.

$6.63 = ab^8$ and $0.37 = ab^5$

$$\frac{6.63}{0.37} = \frac{ab^8}{ab^5}$$

$$\frac{6.63}{0.37} = b^3$$

$$\left(\frac{6.63}{0.37}\right)^{\frac{1}{3}} = (b^3)^{\frac{1}{3}}$$

$b \approx 2.62$

Use b and one of the above equations to find a.

$0.37 = a(2.62)^5$

$$\frac{0.37}{(2.62)^5} = \frac{a(2.62)^5}{(2.62)^5}$$

$a \approx 0.003$

Then $R(t) = 0.003(2.62)^t$

b.

Domain: $[2, 10]$

$R(2) = 0.003(2.62)^2 \approx 0.021$

$R(10) = 0.003(2.62)^{10} \approx 45.723$

Range: $[0.021, 45.723]$

c.

$R(10) = 0.003(2.62)^{10} \approx 45.723$

The primary revenue for the Wikimedia Foundation was approximately $45.7 million in 2010.

d.

$50 = 0.003(2.62)^t$

$$\frac{50}{0.003} = \frac{0.003(2.62)^t}{0.003}$$

$$\frac{50}{0.003} = (2.62)^t$$

$$\log_{2.62}\left(\frac{50}{0.003}\right) = t$$

$$t = \frac{\ln\left(\frac{50}{0.003}\right)}{\ln(2.62)} \approx 10.09$$

According to this model, the Wikimedia Foundation will reach $50 million in annual primary revenue in the year 2010.

27. Let $N(t)$ be the total circulation of U.S. Sunday newspapers (in millions) t years since 2000.

a.

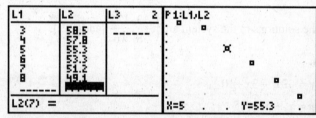

$N(t) = mt + b$

Using two points: $(5, 55.3)$ and $(7, 51.2)$

Find the slope: $m = \dfrac{51.2 - 55.3}{7 - 5} = -2.05$

$N(t) = -2.05t + b$

$b = 55.3 - (-2.05)(5)$

$b = 65.55$

$N(t) = -2.05t + 65.55$

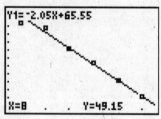

b.

$N(10) = -2.05(10) + 65.55$

$N(10) = 45.05$

In 2010, the total circulation of U.S. Sunday newspapers was 45.05 million.

c.

$25 = -2.05t + 65.55$

$\underline{-65.55 \qquad -65.55}$

$-40.55 = -2.05t$

$\dfrac{-40.55}{-2.05} = \dfrac{-2.05t}{-2.05}$

$19.78 \approx t$

The model predicts that the total circulation of U.S. Sunday newspapers will be only 25 million in the year 2019.

d. Slope $= -2.05$ Each year the total U.S. Sunday newspaper circulation decreases by 2.05 million.

e.

Domain: $[-5, 20]$

$N(-5) = -2.05(-5) + 65.55 = 75.8$

$N(20) = -2.05(20) + 65.55 = 24.55$

Range: $[24.55, 75.8]$

28.

$f(x) = \dfrac{x-2}{x^2 + 7x - 18}$

$f(x) = \dfrac{x-2}{(x+9)(x-2)}$

Domain: All real numbers except for -9 and 2.

29.

$2x - 3y > 6$

$y < -x + 4$

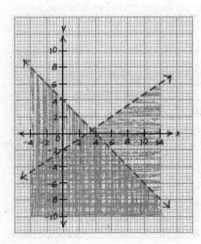

30.

$6a - 15 < 11a + 20$

$\underline{-11a + 15 \quad -11a + 15}$

$-5a < 35$

$\dfrac{-5a}{-5} > \dfrac{35}{-5}$

Remember to reverse the symbol when dividing an inequality by a negative.

$a > -7$

31.

$\dfrac{1}{2}n + 7 \leq \dfrac{3}{4}n + \dfrac{5}{12}$

Multiply both sides by LCD $= 12$.

$(12)\left(\dfrac{1}{2}n + 7\right) \leq (12)\left(\dfrac{3}{4}n + \dfrac{5}{12}\right)$

$6n + 84 \leq 9n + 5$

$\underline{-9n - 84 \quad -9n - 84}$

$-3n \leq -79$

$\dfrac{-3n}{-3} \geq \dfrac{-79}{-3}$

Remember to reverse the symbol when dividing an inequality by a negative.

$n \geq \dfrac{79}{3}$

32.

$|4h - 9| \geq 7$

$4h - 9 \leq -7$ or $4h - 9 \geq 7$

$\underline{+9 \quad +9} \qquad \underline{+9 \quad +9}$

$4h \leq 2$ or $4h \geq 16$

$\dfrac{4h}{4} \leq \dfrac{2}{4}$ or $\dfrac{4h}{4} \geq \dfrac{16}{4}$

$h \leq \dfrac{1}{2}$ or $h \geq 4$

33.

$-6|g - 4| < 30$

$\dfrac{-6|g - 4|}{-6} > \dfrac{30}{-6}$

Remember to reverse the symbol when dividing an inequality by a negative.

$|g - 4| > -5$ which is always true.

Therefore, the solution is all real numbers.

Chapter 9 Conics, Sequences, and Series

34.

$-4|-5x+7| \leq -10$

$\dfrac{-4|-5x+7|}{-4} \geq \dfrac{-10}{-4}$

Remember to reverse the symbol when dividing an inequality by a negative.

$|-5x+7| \geq 2.5$

$-5x+7 \leq -2.5$ or $-5x+7 \geq 2.5$

$ -7 -7$

$-5x \leq -9.5$ or $-5x \geq -4.5$

$\dfrac{-5x}{-5} \geq \dfrac{-9.5}{-5}$ or $\dfrac{-5x}{-5} \leq \dfrac{-4.5}{-5}$

$x \leq 0.9$ or $x \geq 1.9$

35. $A(t) = 1.4t^2 - 14.3t + 202.5$

a.

$A(18) = 1.4(18)^2 - 14.3(18) + 202.5$

$A(18) = 398.7$

In 2008, there were 398.7 thousand apprenticeship training registrations for all major trade groups in Canada.

b.

Vertex: (h, k)

$h = \dfrac{14.3}{2(1.4)} = \dfrac{143}{28} \approx 5.1$

$k = A\left(\dfrac{143}{28}\right) = 1.4\left(\dfrac{143}{28}\right)^2 - 14.3\left(\dfrac{143}{28}\right) + 202.5$

$k \approx 165.98$

Vertex: $(5.1, 165.98)$

In 1995, there were 165.98 thousand apprenticeship training registrations for all major trade groups in Canada.

c.

$300 = 1.4t^2 - 14.3t + 202.5$

$-300 -300$

$0 = 1.4t^2 - 14.3t - 97.5$

$t = \dfrac{14.3 \pm \sqrt{(-14.3)^2 - 4(1.4)(-97.5)}}{2(1.4)}$

$t = \dfrac{14.3 + \sqrt{750.49}}{2.8} \approx 14.89 \to$ In the year 2004.

$t = \dfrac{14.3 - \sqrt{750.49}}{2.8} \approx -4.68 \to$ In the year 1985.

In 1985 and again in 2004, there were 300 thousand apprenticeship training registrations for all major trade groups in Canada.

36. Let $B(h)$ be the number of bacteria present after h hours.

a.

h	$B(h)$
0	2000
1	$2000(2)(2)(2) = 2000(2)^3$
2	$2000(2)^3(2)^3 = 2000(2)^{3 \cdot 2}$
3	$2000(2)^3(2)^3(2)^3 = 2000(2)^{3 \cdot 3}$
h	$2000(2)^{3 \cdot h}$

$B(h) = 2000(2)^{3h}$

b.

$1{,}000{,}000 = 2000(2)^{3h}$

$\dfrac{1{,}000{,}000}{2000} = \dfrac{2000(2)^{3h}}{2000}$

$500 = (2)^{3h} = 8^h$

$\log_8(500) = h$

$h = \dfrac{\ln(500)}{\ln(8)} \approx 2.99$

After approximately 3 hours, there will be 1 million bacteria.

c.

$B(5) = 2000(2)^{3(5)}$

$B(5) = 65{,}536{,}000$

There will 65,536,000 bacteria present after 5 hours.

37.

$f(x) = \dfrac{2}{3}x - 8$ Rewrite without function notation.

$y = \dfrac{2}{3}x - 8$ Switch x and y.

$x = \dfrac{2}{3}y - 8$ Solve for y.

$x + 8 = \dfrac{2}{3}y$

$\left(\dfrac{3}{2}\right)(x + 8) = \left(\dfrac{3}{2}\right)\left(\dfrac{2}{3}y\right)$

$\dfrac{3}{2}x + \dfrac{24}{2} = y$

$f^{-1}(x) = \dfrac{3}{2}x + 12$

38.

$g(x) = 4(3)^x$ Rewrite without function notation.

$y = 4(3)^x$ Switch x and y.

$x = 4(3)^y$ Solve for y.

$\dfrac{x}{4} = \dfrac{4(3)^y}{4}$

$\dfrac{x}{4} = (3)^y$ Write in logaritmic form.

$\log_3\left(\dfrac{x}{4}\right) = y$ Use change of base formula.

$g^{-1}(x) = \dfrac{\log\left(\dfrac{x}{4}\right)}{\log(3)}$

39.

a. $f(3) = 2.5$

b.

Domain: $(-\infty, \infty)$

Range: 2.5

40.

a.

$f(x) + g(x) = (2x - 7) + (-3.4x + 2.7)$
$\qquad = 2x - 7 - 3.4x + 2.7$
$\qquad = -1.4x - 4.3$

b.

$f(g(x)) = 2(-3.4x + 2.7) - 7$
$\qquad = -6.8x + 5.4 - 7$
$\qquad = -6.8x - 1.6$

c.

$f(x)g(x) = (2x - 7)(-3.4x + 2.7)$
$\qquad = -6.8x^2 + 5.4x + 23.8x - 18.9$
$\qquad = -6.8x^2 + 29.2x - 18.9$

41.

a.

$f(2) - g(2) = \left(\dfrac{3}{4}(2) - 10\right) - (-12(2) + 5)$
$\qquad = 1.5 - 10 + 24 - 5$
$\qquad = 10.5$

b.

$(f \circ g)(7) = f(g(7))$

$g(7) = -12(7) + 5 = -79$

$f(g(7)) = f(-79)$

$f(-79) = \dfrac{3}{4}(-79) - 10 = -69.25$

$(f \circ g)(7) = f(g(7)) = f(-79) = -69.25$

c.

$f(2)g(2) = 161.5$

42.

a.

$f(x) + g(x) = (3x^2 + 4x - 6) + (3x - 4)$
$\qquad = 3x^2 + 4x - 6 + 3x - 4$
$\qquad = 3x^2 + 7x - 10$

b.

$f(g(x)) = 3(3x - 4)^2 + 4(3x - 4) - 6$
$\qquad = 3(9x^2 - 24x + 16) + 12x - 16 - 6$
$\qquad = 27x^2 - 72x + 48 + 12x - 22$
$\qquad = 27x^2 - 60x + 26$

c.

$f(x)g(x) = (3x^2 + 4x - 6)(3x - 4)$
$\qquad = 9x^3 - 12x^2 + 12x^2 - 16x - 18x + 24$
$\qquad = 9x^3 - 34x + 24$

43.

$6.5 = \log\left(\dfrac{I}{10^{-4}}\right)$

$10^{6.5} = \dfrac{I}{10^{-4}}$

$(10^{-4})(10^{6.5}) = (10^{-4})\left(\dfrac{I}{10^{-4}}\right)$

$10^{2.5} = I$

$I \approx 316.228$

The intensity of this earthquake is 316.228 cm.

44. Since the lines are perpendicular, $m_1 = -\dfrac{1}{m_2}$.

$-3x + 7y = 4$
$7y = 3x + 4$
$y = \dfrac{3}{7}x + \dfrac{4}{7} \rightarrow \text{slope} = \dfrac{3}{7}$

Using $m = -\dfrac{7}{3}$ and the point $(-5, 2)$

$y - 2 = -\dfrac{7}{3}(x + 5)$

$y = -\dfrac{7}{3}x - \dfrac{35}{3} + 2$

$y = -\dfrac{7}{3}x - \dfrac{29}{3}$

45.
$m = \dfrac{\text{rise}}{\text{run}} = \dfrac{6}{10} = \dfrac{3}{5}$
$y\text{-intercept} = (0, -6)$
Dashed line, shaded above so
$y > \dfrac{3}{5}x - 6$

46. $y = \dfrac{1}{4}x - 6$

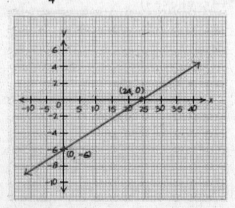

47. $y = -3x^2 - 24x + 2$

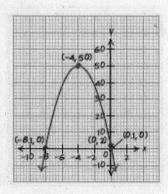

48. $y = -1.5(2.5)^x$

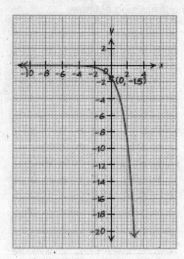

49. $y = \sqrt{x + 17}$

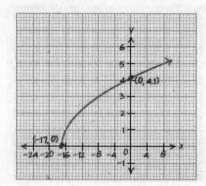

50. $y = \dfrac{1}{3}(x + 4)^2 - 10$

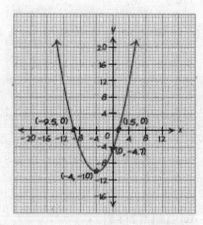

51.
$5x - 30y = 180$
$y = \frac{1}{6}x - 6$

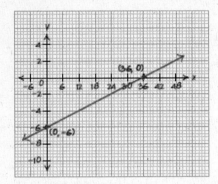

52. $y = 800(0.2)^x$

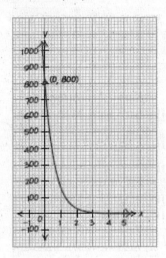

53. $y = \log_9 x$

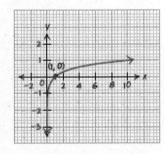

54. $y = \sqrt[3]{x+9}$

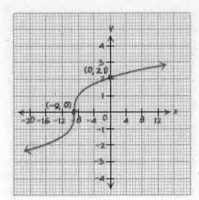

55. $y = 3(1.6)^x + 20$

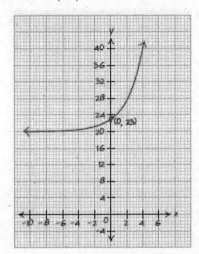

56. $y < x^2 - 5x - 14$

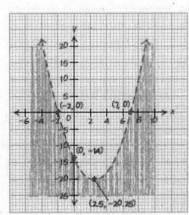

57. $y \geq 2x+3$

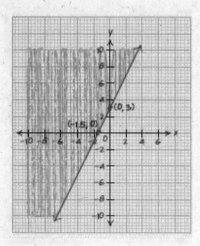

58.
$5x - 6y < 6$
$y > \dfrac{5}{6}x - 1$

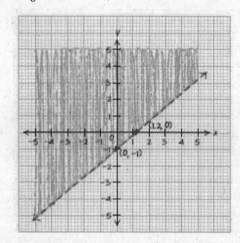

59. $y = -3(x-1)^2 + 8$

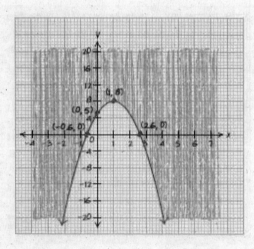

60. Let P be the profit in millions of dollars and t be the time in seconds that it takes to produce a product.

a.

$P = \dfrac{k}{t}$

$1.2 = \dfrac{k}{12} \to k = 12(1.2)$

$\qquad k = 14.4$

$P = \dfrac{14.4}{t}$

b.

$P = \dfrac{14.4}{10} = 1.44$

The profit is $1.44 million if it takes 10 seconds each to produce the product.

61.

$7\sqrt{12a} + 8\sqrt{3a} - 10\sqrt{a}$

$7\sqrt{4 \cdot 3a} + 8\sqrt{3a} - 10\sqrt{a}$

$14\sqrt{3a} + 8\sqrt{3a} - 10\sqrt{a}$

$22\sqrt{3a} - 10\sqrt{a}$

62.

$(14m^3 + 3m^2 - 2m + 15) - (5m^3 - 8m^2 + 6)$

$14m^3 + 3m^2 - 2m + 15 - 5m^3 + 8m^2 - 6$

$9m^3 + 11m^2 - 2m + 9$

63.

$\dfrac{x^2 - 16}{x-3} \cdot \dfrac{x+2}{x+4}$

$\dfrac{\cancel{(x+4)}(x-4)(x+2)}{(x-3)\cancel{(x+4)}}$

$\dfrac{(x-4)(x+2)}{(x-3)}$

64.

$(7 + \sqrt{3w})(-2 + \sqrt{12w})$

$(7 + \sqrt{3w})(-2 + 2\sqrt{3w})$

$-14 + 14\sqrt{3w} - 2\sqrt{3w} + 2(3w)$

$-14 + 12\sqrt{3w} + 6w$

65.

$$\begin{array}{r}4a^2+27a+35\\3a-4{\overline{\smash{\big)}\,12a^3+65a^2-3a-140}}\\\underline{-(12a^3-16a^2)}\\81a^2-3a\\\underline{-(81a^2-108a)}\\105a-140\\\underline{-(105a-140)}\\0\end{array}$$

$(12a^3+65a^2-3a-140)\div(3a-4)=4a^2+27a+35$

$4a^2+27a+35$

66.

$(5x-6)^2$

$(5x-6)(5x-6)$

$25x^2-60x+36$

67.

$\dfrac{3h-4}{h^2-5h+6}\div\dfrac{h+2}{h^2+4h-21}$

$\dfrac{(3h-4)}{(h-3)(h-2)}\cdot\dfrac{(h+7)(h-3)}{h+2}$

$\dfrac{(3h-4)(h+7)}{(h-2)(h+2)}$

68.

$\dfrac{5m}{m-3}+\dfrac{4}{m+7}$

Use LCD $=(m+7)(m-3)$

$\left(\dfrac{5m}{m-3}\right)\left(\dfrac{m+7}{m+7}\right)+\left(\dfrac{4}{m+7}\right)\left(\dfrac{m-3}{m-3}\right)$

$\dfrac{5m^2+35m}{(m+7)(m-3)}+\dfrac{4m-12}{(m+7)(m-3)}$

$\dfrac{5m^2+39m-12}{(m+7)(m-3)}$

69.

a.

$G(12)=3400+155(12)$

$G(12)=5260$

There will be 5260 graduating students at this college in 2012.

b.

$P(12)=0.86-0.03(12)$

$P(12)=0.5$

In 2012, fifty percent of the graduating students at this college will attend graduating ceremonies.

c.

$G(12)P(12)=(5260)(0.5)$

$=2630$

There will be 2630 graduating students attending graduating ceremonies at this college in 2012.

d. Let $A(t)$ be the number of graduating students attending graduating ceremonies at this college t years since 2000.

$A(t)=G(t)P(t)$

$=(3400+155t)(0.86-0.03t)$

$=2924-102t+133.3t-4.65t^2$

$=2924+31.3t-4.65t^2$

e.

$A(15)=2924+31.3(15)-4.65(15)^2$

$A(15)=2347.25$

There will be approximately 2347 graduating students attending graduating ceremonies at this college in 2015.

70.

$2x^2-x-28$

$ac=2\times-28=-56$

Find the factors of -56 that sum to -1:

$-8\times 7=-56$

$-8+7=-1$

We get:

$2x^2-x-28$

$2x^2+7x-8x-28$

$x(2x+7)-4(2x+7)$

$(2x+7)(x-4)$

71.

$20a^2-45b^2$

$5(4a^2-9b^2)$

$5((2a)^2-(3b)^2)$

$5(2a+3b)(2a-3b)$

72.

h^3+125

h^3+5^3

$(h+5)(h^2-5h+25)$

Chapter 9 Conics, Sequences, and Series

73.
$21m^3 - 163m^2 - 48m$
$3m(7m^2 - 54m - 16)$
$ac = 7 \times -16 = -112$
Find the factors of -112 that sum to -54:
$-56 \times 2 = -112$
$-56 + 2 = -54$
We get:
$3m(7m^2 - 54m - 16)$
$3m(7m^2 + 2m - 56m - 16)$
$3m(m(7m + 2) - 8(7m + 2))$
$3m(7m + 2)(m - 8)$

74.
$60x^2 - 265x + 280$
$5(12x^2 - 53x + 56)$
Factors of 12:
$1 \cdot 12$
~~$2 \cdot 6$~~ Would create common factor of 2.
$\boxed{3 \cdot 4} \rightarrow$ Use $3x$ and $4x$
Factors of 56:
$-1 \cdot -56$
~~$-2 \cdot -28$~~ Would create common factor of 2.
~~$-4 \cdot -14$~~ Would create common factor of 2.
$\boxed{-7 \cdot -8} \rightarrow$ Use -7 and -8
We get:
$5(3x - 8)(4x - 7)$

75. $21m^3 - 163m^2 - 48m$ has degree 3.

76.
$f(x) = 4x^2 - 48x + 126$
$f(x) = 4(x^2 - 12x) + 126$
$f(x) = 4(x^2 - 12x + \square - \square) + 126$
Completing the square: $\left(\dfrac{-12}{2}\right)^2 = (-6)^2 = 36$
$f(x) = 4(x^2 - 12x + \boxed{36} - \boxed{36}) + 126$
$f(x) = 4(x^2 - 12x + 36) + 4(-36) + 126$
$f(x) = 4(x - 6)(x - 6) - 144 + 126$
$f(x) = 4(x - 6)^2 - 18$

77. $P(m) = 6.5 + 3\sqrt{m}$

a.
$P(12) = 6.5 + 3\sqrt{12}$
$P(12) \approx 16.89$
The profit for Beach Shack Rentals one year after it opens was $16.89 thousand.

b.
$25 = 6.5 + 3\sqrt{m}$
$25 - 6.5 = 3\sqrt{m}$
$18.5 = 3\sqrt{m}$
$(18.5)^2 = (3\sqrt{m})^2$
$342.25 = 9m$
$\dfrac{342.25}{9} = \dfrac{9m}{9}$
$m \approx 38.028$
After approximately 38 months, the profit for Beach Shack Rentals will be $25,000.

78. $x = -6$

79.
Domain: $(-\infty, 10]$
Range: $[0, \infty)$

80. The value of "a" is negative. The y-intercept is below the x axis.

81. The value of "b" is less than 1. This is an increasing function with $a < 0$.

82.
Domain: $(-\infty, \infty)$
Range: $(-\infty, 0)$

83. The value of a is positive because the parabola opens upward.

84. The vertex of the parabola is $(-2, -16)$ therefore $h = -2$ and $k = -16$.

85. $f(-4) = -12$

86. $x = -6$ and $x = 2$ when $f(x) = 0$.

87.
Domain: $(-\infty, \infty)$
Range: $(-16, \infty)$

88. The value of m is positive because the slope of the line is positive.

89. $b = -3$

90. $m = \dfrac{2}{5}$

91. $x = -5$ when $f(x) = -5$.

92.
Domain: $(-\infty, \infty)$
Range: $(-\infty, \infty)$

93.
$$\sqrt{36a^4b^6c^2} = \sqrt{6^2 a^4 b^6 c^2}$$
$$= 6a^2b^3c$$

94.
$$\sqrt[3]{54x^3y^5} = \sqrt[3]{2 \cdot 3^3 x^3 y^5}$$
$$= \sqrt[3]{3^3 x^3 y^3} \cdot \sqrt[3]{2y^2}$$
$$= 3xy\sqrt[3]{2y^2}$$

95.
$$\frac{\sqrt{7a}}{\sqrt{2b}} = \frac{\sqrt{7a}}{\sqrt{2b}} \cdot \frac{\sqrt{2b}}{\sqrt{2b}}$$
$$= \frac{\sqrt{14ab}}{2b}$$

96.
$$\frac{6+\sqrt{5}}{2-\sqrt{3}} = \frac{(6+\sqrt{5})}{(2-\sqrt{3})} \cdot \frac{(2+\sqrt{3})}{(2+\sqrt{3})}$$
$$= \frac{12 - 6\sqrt{3} + 2\sqrt{5} + \sqrt{5}\cdot\sqrt{3}}{4 + 2\sqrt{3} - 2\sqrt{3} - \sqrt{3}\cdot\sqrt{3}}$$
$$= \frac{12 - 6\sqrt{3} + 2\sqrt{5} + \sqrt{15}}{4-3}$$
$$= \frac{12 - 6\sqrt{3} + 2\sqrt{5} + \sqrt{15}}{1}$$
$$= 12 - 6\sqrt{3} + 2\sqrt{5} + \sqrt{15}$$

97.
$$(7+4i) + (2-7i) = 7 + 2 + 4i - 7i$$
$$= 9 - 3i$$

98.
$$(6+2i) - (10-6i) = 6 + 2i - 10 + 6i$$
$$= 6 - 10 + 2i + 6i$$
$$= -4 + 8i$$

99.
$$(4+9i)(4-9i) = 16 - 36i + 36i - 81i^2$$
$$= 16 - 81(-1)$$
$$= 16 + 81$$
$$= 97$$

100.
$$(11-3i)(8+2i) = 88 + 22i - 24i - 6i^2$$
$$= 88 - 2i - 6(-1)$$
$$= 88 - 2i + 6$$
$$= 94 - 2i$$

101.
$$\frac{6+5i}{4i} = \frac{(6+5i)}{4i} \cdot \frac{i}{i}$$
$$= \frac{6i + 5i^2}{4i^2}$$
$$= \frac{6i + 5(-1)}{4(-1)}$$
$$= \frac{6i - 5}{-4}$$
$$= -\frac{6}{4}i + \frac{5}{4}$$
$$= \frac{5}{4} - \frac{3}{2}i$$

102.
$$\frac{8-6i}{2-5i} = \frac{(8-6i)}{(2-5i)} \cdot \frac{(2+5i)}{(2+5i)}$$
$$= \frac{16 + 40i - 12i - 30i^2}{4 + 10i - 10i - 25i^2}$$
$$= \frac{16 + 40i - 12i - 30(-1)}{4 - 25(-1)}$$
$$= \frac{16 + 28i + 30}{4 + 25}$$
$$= \frac{46 + 28i}{29}$$
$$= \frac{46}{29} + \frac{28}{29}i$$

103.
$$\frac{10}{3+4i} = \frac{10}{(3+4i)} \cdot \frac{(3-4i)}{(3-4i)}$$
$$= \frac{30 - 40i}{9 - 12i + 12i - 16i^2}$$
$$= \frac{30 - 40i}{9 - 16(-1)}$$
$$= \frac{30 - 40i}{9 + 16}$$
$$= \frac{30 - 40i}{25}$$
$$= \frac{30}{25} - \frac{40}{25}i$$
$$= \frac{6}{5} - \frac{8}{5}i$$

Chapter 9 Conics, Sequences, and Series

104.
$x^2 + 4x + 29 = 0$
$x = \dfrac{-4 \pm \sqrt{4^2 - 4(1)(29)}}{2(1)}$
$x = \dfrac{-4 \pm \sqrt{-100}}{2}$
$x = \dfrac{-4 \pm 10i}{2}$
$x = -2 \pm 5i$

105.
$x^3 + 40x = -12x^2$
$x^3 + 12x^2 + 40x = 0$
$x(x^2 + 12x + 40) = 0$
$x = 0, x^2 + 12x + 40 = 0$
$x = \dfrac{-12 \pm \sqrt{12^2 - 4(1)(40)}}{2(1)}$
$x = \dfrac{-12 \pm \sqrt{-16}}{2}$
$x = \dfrac{-12 \pm 4i}{2}$
$x = -6 \pm 2i$

106.
$x^2 + 70 = 0$
$x^2 = -70$
$x = \pm\sqrt{-70}$
$x = \pm i\sqrt{70}$

107.
$-3(x-6)^2 + 20 = 32$
$-3(x-6)^2 = 12$
$(x-6)^2 = -4$
$x - 6 = \pm\sqrt{-4}$
$x - 6 = \pm 2i$
$x = 6 \pm 2i$

108.
Center $= (3, -9)$ and Radius $= 12$
$h = 3, k = -9, r = 12$
$(x-3)^2 + (y-(-9))^2 = (12)^2$
$(x-3)^2 + (y+9)^2 = 144$

109.
$x = \dfrac{y^2}{-44}$
$x = \dfrac{y^2}{4(-11)}$
$p = -11$
Focus $= (-11, 0)$, Directrix: $x = 11$

110. $\dfrac{x^2}{144} - \dfrac{y^2}{64} = 1$ hyperbola

The $\dfrac{x^2}{a^2}$ term is first so opens left and right.
$a^2 = 144 \rightarrow a = 12$
$b^2 = 64 \rightarrow b = 8$
Vertices at $(-12, 0)$ and $(12, 0)$.
The fundamental rectangle will have vertical sides at $x = \pm 12$ and horizontal sides at $y = \pm 8$.
$c^2 = a^2 + b^2$
$c^2 = 144 + 64 = 208$
$c = \pm\sqrt{208} \approx \pm 14.4$
Foci at $(\pm 14.4, 0)$

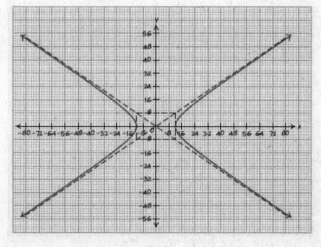

111. $121x^2 + 4y^2 = 484$ ellipse
$\dfrac{121x^2}{484} + \dfrac{4y^2}{484} = \dfrac{484}{484}$
$\dfrac{x^2}{4} + \dfrac{y^2}{121} = 1$
The bigger number is under the y so the major axis is vertical.
$a^2 = 4 \rightarrow a = 2$
$b^2 = 121 \rightarrow b = 11$
Vertices at $(0, -11)$ and $(0, 11)$.

$c^2 = b^2 - a^2$
$c^2 = 121 + 4 = 117$
$c = \pm\sqrt{117} \approx \pm 10.8$
Foci at $(0, \pm 10.8)$

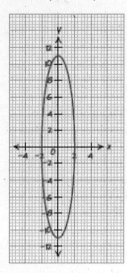

112. $y^2 + 2y + x - 10 = 0$ parabola
$x = -y^2 - 2y + 10$
Opens to the left $a = -1 < 0$.
Vertex (h, k).
$k = \dfrac{2}{2(-1)} = -1$
$h = -(-1)^2 - 2(-1) + 10 = 11$
$(h, k) = (11, -1)$
x-intercept: $(10, 0)$
y-intercept:
$0 = -y^2 - 2y + 10$
$y = \dfrac{2 \pm \sqrt{(-2)^2 - 4(-1)(10)}}{2(-1)}$
$y = \dfrac{2 + \sqrt{44}}{-2} \approx -4.3 \to (0, -4.3)$
$y = \dfrac{2 - \sqrt{44}}{-2} \approx 2.3 \to (0, 2.3)$

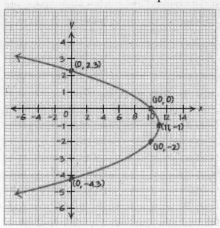

113. $(x+8)^2 + (y-3)^2 = 36$ circle
Center at $(-8, 3)$, radius $r = 6$

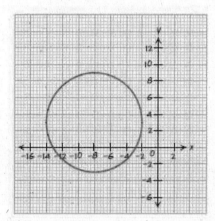

114.
The parabola opens to the left so it is of the form $y^2 = 4px$.
Use point $(-6, -12)$ to determine p.
$(-12)^2 = 4p(-6)$
$144 = -24p$
$\dfrac{144}{-24} = \dfrac{-24p}{-24}$
$-6 = p$

Therefore the equation is $y^2 = -24x$.

115.
The major axis is horizontal and it is centered at the origin so it is of the form $\dfrac{x^2}{a^2} + \dfrac{y^2}{b^2} = 1$.

$a = 12 \to a^2 = 144$
$b = 5 \to b^2 = 25$
$\dfrac{x^2}{144} + \dfrac{y^2}{25} = 1$

116.

$a_n = 34 + (n-1)11$
$a_n = 34 + 11n - 11$
$a_n = 23 + 11n$

117.

$a_1 = 5(-2)^1 = -10$
$a_2 = 5(-2)^2 = 20$
$a_3 = 5(-2)^3 = -40$
$a_4 = 5(-2)^4 = 80$
$a_5 = 5(-2)^5 = -160$
$\{-10, 20, -40, 80, -160, ...\}$

118.

$a_1 = \dfrac{-12(1)}{3(1)+5} = \dfrac{-12}{8} = -\dfrac{3}{2}$

$a_2 = \dfrac{-12(2)}{3(2)+5} = -\dfrac{24}{11}$

$a_3 = \dfrac{-12(3)}{3(3)+5} = \dfrac{-36}{14} = -\dfrac{18}{7}$

$a_4 = \dfrac{-12(4)}{3(4)+5} = -\dfrac{48}{17}$

$a_5 = \dfrac{-12(5)}{3(5)+5} = \dfrac{-60}{20} = -3$

$\left\{-\dfrac{3}{2}, -\dfrac{24}{11}, -\dfrac{18}{7}, -\dfrac{48}{17}, -3, ...\right\}$

119.

$26 + 7(n-1) = 607$
$26 + 7n - 7 = 607$
$7n = 588$
$\dfrac{7n}{7} = \dfrac{588}{7}$
$n = 84$

120.

$-45 + \sqrt{n+20} = -37$
$\sqrt{n+20} = 8$
$\left(\sqrt{n+20}\right)^2 = (8)^2$
$n + 20 = 64$
$n = 44$

121.

n	a_n	Common Difference
1	48	
2	45	$a_2 - a_1 = 45 - 48 = -3$
3	42	$a_3 - a_2 = 42 - 45 = -3$
4	39	$a_4 - a_3 = 39 - 42 = -3$
5	36	$a_5 - a_4 = 36 - 39 = -3$
6	33	$a_6 - a_5 = 33 - 36 = -3$

This is an arithmetic sequence with common difference $d = -3$.

$a_n = 48 + (n-1)(-3)$
$a_n = 48 - 3n + 3$
$a_n = 51 - 3n$

122.

n	a_n	Common Ratio
1	96	
2	384	$\dfrac{a_2}{a_1} = \dfrac{384}{96} = 4$
3	1536	$\dfrac{a_3}{a_2} = \dfrac{1536}{384} = 4$
4	6144	$\dfrac{a_4}{a_3} = \dfrac{6144}{1536} = 4$
5	24,576	$\dfrac{a_5}{a_4} = \dfrac{24,576}{6144} = 4$
6	98,304	$\dfrac{a_6}{a_5} = \dfrac{98,304}{24,576} = 4$

This is a geometric sequence with common ratio $r = 4$.

$a_n = 96(4)^{n-1}$

123. $a_1 = 55,000$ and $d = 825$

a.

$a_n = 55,000 + (n-1)(825)$
$a_n = 55,000 + 825n - 825$
$a_n = 54,175 + 825n$

b.

$a_{10} = 54,175 + 825(10)$
$a_{10} = 62,425$

Your salary will be $62,425 in the 10th year of employment.

c.

$$S_{10} = \frac{10(55,000 + 62,425)}{2}$$

$$S_{10} = 587,125$$

You will have earned a total of $587,125 in the first ten year period of employment.

124. $a_1 = 55,000$ and $r = 1 + 0.015 = 1.015$

a. $a_n = 55,000(1.015)^{n-1}$

b.

$$a_{10} = 55,000(1.015)^{10-1}$$

$$a_{10} = 55,000(1.015)^9$$

$$a_{10} \approx 62,886.45$$

Your salary will be $62,886.45 in the 10th year of employment.

c.

$$S_{10} = \frac{55,000(1-1.015^{10})}{1-1.015}$$

$$S_{10} \approx 588,649.69$$

You will have earned a total of $588,649.69 in the first ten year period of employment.

125.

$$\sum_{n=1}^{6}(2n^3) = 2 + 2(2)^3 + 2(3)^3 + 2(4)^3$$
$$+ 2(5)^3 + 2(6)^3 = 882$$

126.

$$a_1 = 75$$

$$a_{15} = 75 - 6(15-1)$$

$$a_{15} = -9$$

$$S_{15} = \frac{15(75-9)}{2}$$

$$S_{15} = 495$$

127.

$$a_1 = 300(1.3)$$

$$a_1 = 390$$

$$S_{12} = \frac{390(1-1.3^{12})}{1-1.3}$$

$$S_{12} \approx 28,987.511$$

Basic Algebra Review
Appendix A

1. Natural Numbers, Whole Numbers, Integers, Rational Numbers, and Real Numbers.
3. Whole Numbers, Integers, Rational Numbers, and Real Numbers.
5. Irrational Numbers and Real Numbers.
7. Rational Numbers and Real Numbers.
9. Natural Numbers, Whole Numbers, Integers, Rational Numbers, and Real Numbers.
11. – 19.

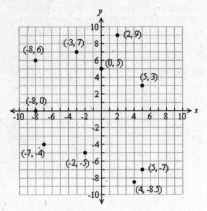

21. $7+(-25)=-18$

When adding two integers with different signs use subtraction. The difference between 25 and 7 is 18. The sign on the answer is negative because the $|-25|>|7|$.

23. $-5+(-9)=-14$

When adding two integers with the same sign use addition. The sum of 5 and 9 is 14. The sign on the answer is negative because when adding two negative numbers the sum is negative.

25. $-14-(-12)=-14+12=-2$

When adding two integers with different signs use subtraction. The difference between 14 and 12 is 2. The sign on the answer is negative because the $|-14|>|12|$.

27. $2(-8)=-16$

When multiplying two integers with different signs the product is negative.

29. $-5(-4)=20$

When multiplying two integers with the same sign the product is positive.

31.
$$\frac{30}{-5}=-6$$
When dividing two integers with different signs the quotient is negative.

33.
$$\frac{-100}{-25}=4$$
When dividing two integers with the same sign the quotient is positive.

35.
$$\frac{144}{6}=24$$
When dividing two integers with the same sign the quotient is positive.

37.
$$\frac{-85}{5}=-17$$
When dividing two integers with different signs the quotient is negative.

39.
$$\frac{2}{3}+\frac{7}{3}=\frac{2+7}{3}$$
$$=\frac{9}{3}$$
$$=3$$

41.
$$\frac{15}{34}-\frac{8}{34}=\frac{15-8}{34}$$
$$=\frac{7}{34}$$

43.
$$\frac{2}{3}+\frac{5}{6}=\frac{4}{6}+\frac{5}{6}$$
$$=\frac{4+5}{6}$$
$$=\frac{9}{6}$$
$$=\frac{3}{2}$$

45.
$$\frac{2}{5}\cdot\frac{3}{7}=\frac{2\cdot3}{5\cdot7}$$
$$=\frac{6}{35}$$

47.
$$\frac{4}{15}+\frac{7}{25}=\frac{20}{75}+\frac{21}{75}$$
$$=\frac{20+21}{75}$$
$$=\frac{41}{75}$$

49.
$$\frac{3}{4}\div\frac{7}{5}=\frac{3}{4}\cdot\frac{5}{7}$$
$$=\frac{3\cdot 5}{4\cdot 7}$$
$$=\frac{15}{28}$$

51.
$$\frac{3}{20}\div\frac{6}{7}=\frac{3}{20}\cdot\frac{7}{6}$$
$$=\frac{3\cdot 7}{20\cdot 6}$$
$$=\frac{21}{120}$$
$$=\frac{7}{40}$$

53.
$$\frac{8}{26}\cdot\frac{3}{15}=\frac{8\cdot 3}{26\cdot 15}$$
$$=\frac{24}{390}$$
$$=\frac{4}{65}$$

55.
$$7-2(4+5)=7-2(9)$$
$$=7-18$$
$$=-11$$

57.
$$2+3^2(5+7)=2+3^2(12)$$
$$=2+9(12)$$
$$=2+108$$
$$=110$$

59.
$$3^2+6(13-8)^2=3^2+6(5)^2$$
$$=9+6\cdot 25$$
$$=9+150$$
$$=159$$

61.
$$\frac{2}{3}\left(\frac{5}{2}+\frac{7}{3}\right)-\left(\frac{2}{3}\right)^2=\frac{2}{3}\left(\frac{29}{6}\right)-\left(\frac{2}{3}\right)^2$$
$$=\frac{2}{3}\left(\frac{29}{6}\right)-\frac{4}{9}$$
$$=\frac{29}{9}-\frac{4}{9}$$
$$=\frac{25}{9}$$

63.
$$5+7\left[3-5(6+2)^2\right]=5+7\left[3-5(8)^2\right]$$
$$=5+7[3-5\cdot 64]$$
$$=5+7[3-320]$$
$$=5+7[-317]$$
$$=5+(-2219)$$
$$=-2214$$

65.
$$2^3+5\left[6^2-4(2+7)\right]=2^3+5\left[6^2-4(9)\right]$$
$$=2^3+5[36-4(9)]$$
$$=2^3+5[36-36]$$
$$=2^3+5[0]$$
$$=8+5[0]$$
$$=8+0$$
$$=8$$

67.
$$\frac{14+10}{5}-\frac{6}{3}+4=\frac{24}{5}-\frac{6}{3}+4$$
$$=\frac{72}{15}-\frac{30}{15}+\frac{60}{15}$$
$$=\frac{102}{15}$$
$$=\frac{34}{5}$$

69.
$$5+3^2\left[6-\left(4+\frac{12+3}{6-3}\right)\right]=5+3^2\left[6-\left(4+\frac{15}{3}\right)\right]$$
$$=5+3^2[6-(4+5)]$$
$$=5+3^2[6-(9)]$$
$$=5+3^2[-3]$$
$$=5+9[-3]$$
$$=5+(-27)$$
$$=-22$$

Basic Algebra Review

71.
$$3x + 8 = 20$$
$$3x + 8 - 8 = 20 - 8$$
$$3x = 12$$
$$x = 4$$

73.
$$\frac{2}{3}x - 5 = 4$$
$$\frac{2}{3}x - 5 + 5 = 4 + 5$$
$$\frac{2}{3}x = 9$$
$$\frac{3}{2}\left(\frac{2}{3}x\right) = \frac{3}{2}(9)$$
$$x = \frac{27}{2}$$

75.
$$2(m-6) + 11 = 61$$
$$2m - 12 + 11 = 61$$
$$2m - 1 = 61$$
$$2m - 1 + 1 = 61 + 1$$
$$2m = 62$$
$$m = 31$$

77.
$$\frac{h+6}{5} = 17$$
$$5\left(\frac{h+6}{5}\right) = 5(17)$$
$$h + 6 = 85$$
$$h + 6 - 6 = 85 - 6$$
$$h = 79$$

79.
$$\frac{k-5}{4} + 9 = 21$$
$$\frac{k-5}{4} + 9 - 9 = 21 - 9$$
$$\frac{k-5}{4} = 12$$
$$4\left(\frac{k-5}{4}\right) = 4(12)$$
$$k - 5 = 48$$
$$k - 5 + 5 = 48 + 5$$
$$k = 53$$

81.
$$\frac{4}{5}b - 5 = 6$$
$$\frac{4}{5}b - 5 + 5 = 6 + 5$$
$$\frac{4}{5}b = 11$$
$$\frac{5}{4}\left(\frac{4}{5}b\right) = \frac{5}{4}(11)$$
$$b = \frac{55}{4}$$

83.
$$\frac{2}{9}w - 3 = \frac{4}{7}$$
$$63\left(\frac{2}{9}w - 3\right) = 63\left(\frac{4}{7}\right)$$
$$14w - 189 = 36$$
$$14w - 189 + 189 = 36 + 189$$
$$14w = 225$$
$$w = \frac{225}{14}$$

85.
$$\frac{2}{7}m + 4 = \frac{3}{7}(5) + 7$$
$$\frac{2}{7}m + 4 = \frac{15}{7} + 7$$
$$7\left(\frac{2}{7}m + 4\right) = 7\left(\frac{15}{7} + 7\right)$$
$$2m + 28 = 15 + 49$$
$$2m + 28 = 64$$
$$2m + 28 - 28 = 64 - 28$$
$$2m = 36$$
$$m = 18$$

87.
$$\frac{7}{10}x - \frac{4}{5} = \frac{5}{6}(x+2)$$
$$\frac{7}{10}x - \frac{4}{5} = \frac{5}{6}x + \frac{10}{6}$$
$$30\left(\frac{7}{10}x - \frac{4}{5}\right) = 30\left(\frac{5}{6}x + \frac{10}{6}\right)$$
$$21x - 24 = 25x + 50$$
$$-4x - 24 = 50$$
$$-4x - 24 + 24 = 50 + 24$$
$$-4x = 74$$
$$x = -18.5$$

89. $2.74 \times 10^8 = 274{,}000{,}000$

91. $-4.63 \times 10^{12} = -4{,}630{,}000{,}000{,}000$

93. $1.28 \times 10^{-5} = 0.0000128$

95. $-9.6 \times 10^{-13} = -0.00000000000096$

97. $14,000,000,000 = 1.4 \times 10^{10}$

99. $-547,000,000 = -5.47 \times 10^{8}$

101. $0.000000078 = 7.8 \times 10^{-8}$

103. $-0.00000000000751 = -7.51 \times 10^{-12}$

105.

$[-6, 4]$

Use brackets because the interval includes the endpoints.

107.

$(-5, 10]$

Use a parenthesis on one end because the interval does not include the endpoint and use a bracket on the other end because the interval does include the other endpoint.

109.

$[-8, \infty)$

Use a bracket on one end because the interval does include the endpoint and use a parenthesis on the other end because the interval extends forever.

111.

$(4, \infty)$

Use a parenthesis on one end because the interval does not include the endpoint and use a parenthesis on the other end because the interval extends forever.

Matrices
Appendix B

1.

a.

Substitute in $(x, y, z) = (0, 5, 1)$

$x + y + z = -2$
$2x - y + 4z = -17$
$x + 3y - z = 18$

Simplify and check

$0 + 5 + 1 \neq -2$
$2(0) - 5 + 4(1) \neq -17$
$0 + 3(5) - 1 \neq 18$

Since at least one of the equations is not true, the point $(x, y, z) = (0, 5, 1)$ is not a solution of the system.

b.

Substitute in $(x, y, z) = (0, 5, -3)$

$x + y + z = -2$
$2x - y + 4z = -17$
$x + 3y - z = 18$

Simplify and check

$0 + 5 + (-3) \neq -2$
$2(0) - 5 + 4(-3) = -17$
$0 + 3(5) - (-3) = 18$

Since at least one of the equations is not true, the point $(x, y, z) = (0, 5, -3)$ is not a solution of the system.

3.

a.

Substitute in $(x, y, z) = (-5, 0, 1)$

$x + y + z = 7$
$-2x + 3y - 2z = -14$
$-x - y + 3z = -19$

Simplify and check

$-5 + 0 + 1 \neq 7$
$-2(-5) + 3(0) - 2(1) \neq -14$
$-(5) - 0 + 3(1) \neq -19$

Since at least one of the equations is not true, the point $(x, y, z) = (-5, 0, 1)$ is not a solution of the system.

b.

Substitute in $(x, y, z) = (10, 0, -3)$

$x + y + z = 7$
$-2x + 3y - 2z = -14$
$-x - y + 3z = -19$

Simplify and check

$10 + 0 + (-3) = 7$
$-2(10) + 3(0) - 2(-3) = -14$
$-10 - 0 + 3(-3) = -19$

Since all of the equations are true, the point $(x, y, z) = (10, 0, -3)$ is a solution of the system.

5.

a.

Substitute in $(x, y, z) = (16, 1, 0)$

$x - y + z = -15$
$y + z = -1$
$3x + 2y = 14$

Simplify and check

$16 - 1 + 0 \neq -15$
$1 + 0 \neq -1$
$3(16) + 2(1) \neq 14$

Since at least one of the equations is not true, $x = 16, y = 1, z = 0$ is not a solution of the system.

b.

Substitute in $(x, y, z) = (0, 7, -8)$

$x - y + z = -15$
$y + z = -1$
$3x + 2y = 14$

Simplify and check

$0 - 7 + (-8) = -15$
$7 + (-8) = -1$
$3(0) + 2(7) = 14$

Since all of the equations are true, $x = 0, y = 7, z = -8$ is a solution of the system.

7.

1. $x + y - z = 3$
2. $x - y + 2z = 6$
3. $-x - 2y + 3z = -1$

Plan: Eliminate x

add #1. & #3 label result #4.
1. $x + y - z = 3$
3. $-x - 2y + 3z = -1$
4. $\quad -y + 2z = 2$

add #2. & #3 label result #5.
2. $x - y + 2z = 6$
3. $-x - 2y + 3z = -1$
5. $\quad -3y + 5z = 5$

Plan: Eliminate y

Create $-3(\#4.)$ label this #6.
$(-3)(-y + 2z) = (2)(-3)$
6. $3y - 6z = -6$

add #5. & #6
5. $-3y + 5z = 5$
6. $\quad 3y - 6z = -6$
$\quad\quad -z = -1$
$\quad\quad\; z = 1$

Back substitute to solve for y
4. $-y + 2(1) = 2$
$\quad -y + 2 = 2$
$\quad\quad y = 0$

Back substitute to solve for x
1. $x + 0 - 1 = 3$
$\quad\; x = 4$

The solution to this system is the point $(x, y, z) = (4, 0, 1)$.

Check:
1. $4 + 0 - 1 = 3$
2. $4 - 0 + 2(1) = 6$
3. $-4 - 2(0) + 3(1) = -1$

9.

1. $x + y - 2z = -6$
2. $-x + y + z = 7$
3. $x - 2y + z = -3$

Plan: Eliminate x

add #1. & #2 label result #4.
1. $x + y - 2z = -6$
2. $-x + y + z = 7$
4. $\quad 2y - z = 1$

add #2. & #3 label result #5.
2. $-x + y + z = 7$
3. $x - 2y + z = -3$
5. $\quad -y + 2z = 4$

Plan: Eliminate y

Create $2(\#5.)$ label this #6.
$(2)(-y + 2z) = (4)(2)$
6. $-2y + 4z = 8$

add #4. & #6
4. $2y - z = 1$
6. $-2y + 4z = 8$
$\quad\quad 3z = 9$
$\quad\quad\; z = 3$

Back substitute to solve for y
5. $-y + 2(3) = 4$
$\quad -y + 6 = 4$
$\quad -y = -2$
$\quad\; y = 2$

Back substitute to solve for x
1. $x + 2 - 2(3) = -6$
$\quad x + 2 - 6 = -6$
$\quad x - 4 = -6$
$\quad\; x = -2$

The solution to this system is the point $(x, y, z) = (-2, 2, 3)$.

Matrices
Check:
1. $(-2)+2-2(3)=-6$
2. $-(-2)+2+3=7$
3. $(-2)-2(2)+3=-3$

11.
1. $x-y+2z=6$
2. $2x+y-z=-1$
3. $-x-2y+z=3$

Plan: Eliminate x

add #1. & #3. label result #4.
1. $x-y+2z=6$
3. $-x-2y+z=3$
4. $\quad -3y+3z=9$

Create $2(\#3.)$ label result #5.
$(2)(-x-2y+z)=(3)(2)$
5. $-2x-4y+2z=6$

add #2. & #5. label result #6.
2. $2x+y-z=-1$
5. $-2x-4y+2z=6$
6. $\quad -3y+z=5$

Plan: Eliminate y

Create $-1(\#6.)$ label result #7.
$(-1)(-3y+z)=(5)(-1)$
7. $3y-z=-5$

add #4. & #7.
4. $-3y+3z=9$
7. $3y-z=-5$
$\quad\quad 2z=4$
$\quad\quad z=2$

Back substitute to solve for y
6. $-3y+2=5$
$\quad -3y=3$
$\quad y=-1$

Back substitute to solve for x
1. $x-(-1)+2(2)=6$
$\quad x+5=6$
$\quad x=1$

The solution to this system is the point $(x,y,z)=(1,-1,2)$.

Check:
1. $1-(-1)+2(2)=6$
2. $2(1)+(-1)-2=-1$
3. $-1-2(-1)+2=3$

13.
1. $2x+2y-z=-5$
2. $-x-y+z=4$
3. $3x-y-2z=-1$

Plan: Eliminate z

add #1. & #2. label result #4.
1. $2x+2y-z=-5$
2. $-x-y+z=4$
4. $x+y=-1$

Create $2(\#2.)$ label result #5.
$(2)(-x-y+z)=(4)(2)$
5. $-2x-2y+2z=8$

add #3. & #5. label result #6.
3. $3x-y-2z=-1$
5. $-2x-2y+2z=8$
6. $x-3y=7$

Plan: Eliminate x

Create $-1(\#4.)$ label result #7.
$(-1)(x+y)=(-1)(-1)$
7. $-x-y=1$

add #6. & #7.
6. $x-3y=7$
7. $-x-y=1$
$\quad -4y=8$
$\quad y=-2$

476

Appendix B

Back substitute to solve for x
4. $x + (-2) = -1$
$x = 1$

Back substitute to solve for z
2. $-1 - (-2) + z = 4$
$1 + z = 4$
$z = 3$

The solution to this system is the point
$(x, y, z) = (1, -2, 3)$.

Check:
1. $2(1) + 2(-2) - 3 = -5$
2. $-1 - (-2) + 3 = 4$
3. $3(1) - (-2) - 2(3) = -1$

15. $\begin{bmatrix} 3 & -1 \\ 4 & 7 \end{bmatrix}$

a. This is a 2x2 matrix.
b. -1 is in the (1, 2) position.

17. $\begin{bmatrix} 4 & 0 & -2 \\ 3 & -1 & 5 \end{bmatrix}$

a. This is a 2x3 matrix.
b. -2 is in the (1, 3) position.

19. $\begin{bmatrix} 1 & -2 & 6 \\ 0 & -3 & 2 \\ 0 & 1 & 1 \\ -3 & 2 & 5 \end{bmatrix}$

a. This is a 4x3 matrix.
b. -3 is in the (4, 1) position.

21. $\begin{bmatrix} 2 & -1 & -5 \\ 4 & 6 & -1 \end{bmatrix}$

23. $\begin{bmatrix} 1 & -1 & 0 \\ -3 & 4 & -1 \end{bmatrix}$

25. $\begin{bmatrix} -2 & 3 & -1 & 0 \\ 1 & 1 & -1 & 5 \\ 1 & -1 & 1 & 2 \end{bmatrix}$

27. $\begin{bmatrix} -1 & 1 & 1 & 3 \\ 0 & 3 & -2 & 1 \\ -1 & 0 & 2 & 0 \end{bmatrix}$

29.

$\begin{bmatrix} 3 & 1 & 5 \\ 0 & 2 & -4 \end{bmatrix}$

Rewrite the last row to solve for y:
$0x + 2y = -4$
$2y = -4$
$\dfrac{2y}{2} = \dfrac{-4}{2}$
$y = -2$

Rewrite the first row and substitute
$y = -2$ to solve for x:
$3x + (-2) = 5$
$\quad +2 \quad +2$
$3x = 7$
$\dfrac{3x}{3} = \dfrac{7}{3}$
$x = \dfrac{7}{3}$

The solution to the system is $(x, y) = \left(\dfrac{7}{3}, -2\right)$

31.

$\begin{bmatrix} 4 & -3 & 7 \\ 0 & -1 & 5 \end{bmatrix}$

Rewrite the last row to solve for y:
$0x - y = 5$
$-y = 5$
$y = -5$

Matrices

Rewrite the first row and substitute $y = -5$ to solve for x:

$4x - 3(-5) = 7$
$4x + 15 = 7$
$\underline{-15 \; -15}$
$4x = -8$
$\dfrac{4x}{4} = \dfrac{-8}{4}$
$x = -2$

The solution to the system is $(x, y) = (-2, -5)$.

33.

$\begin{bmatrix} 1 & -1 & 0 & 2 \\ 0 & 2 & 1 & 3 \\ 0 & 0 & 1 & -2 \end{bmatrix}$

Rewrite the third row to solve for z:

$0x + 0y + z = -2$
$z = -2$

Rewrite the second row and substitute $z = -2$ to solve for y:

$0x + 2y + (-2) = 3$
$2y - 2 = 3$
$\underline{+2 \; +2}$
$2y = 5$
$\dfrac{2y}{2} = \dfrac{5}{2}$
$y = \dfrac{5}{2}$

Rewrite the first row and substitute $y = \dfrac{5}{2}$ and $z = -2$ to solve for x:

$x - \dfrac{5}{2} + 0(-2) = 2$
$x - \dfrac{5}{2} + \dfrac{5}{2} = 2 + \dfrac{5}{2}$
$x = \dfrac{4}{2} + \dfrac{5}{2} = \dfrac{9}{2}$

The solution to the system is $(x, y, z) = \left(\dfrac{9}{2}, \dfrac{5}{2}, -2\right)$.

35.

$\begin{bmatrix} 2 & -1 & 1 & 7 \\ 0 & -2 & 1 & -4 \\ 0 & 0 & -3 & 15 \end{bmatrix}$

Rewrite the third row to solve for z:

$0x + 0y - 3z = 15$
$\dfrac{-3z}{-3} = \dfrac{15}{-3}$
$z = -5$

Rewrite the second row and substitute $z = -5$ to solve for y:

$0x - 2y - 5 = -4$
$\underline{+5 \; +5}$
$-2y = 1$
$\dfrac{-2y}{-2} = \dfrac{1}{-2}$
$y = -\dfrac{1}{2}$

Rewrite the first row and substitute $y = -\dfrac{1}{2}$ and $z = -5$ to solve for x:

$2x - \left(-\dfrac{1}{2}\right) + (-5) = 7$
$2x + \dfrac{1}{2} - 5 = 7$
$(2)\left(2x + \dfrac{1}{2} - 5\right) = (7)(2)$
$4x + 1 - 10 = 14$
$4x - 9 = 14$
$\underline{+9 \; +9}$
$4x = 23$
$\dfrac{4x}{4} = \dfrac{23}{4}$
$x = \dfrac{23}{4}$

The solution to the system is $(x, y, z) = \left(\dfrac{23}{4}, -\dfrac{1}{2}, -5\right)$.

37.

$$\begin{bmatrix} 1 & -1 & -5 \\ -2 & 1 & 7 \end{bmatrix}$$

$2 \cdot$ row 1 + row 2. Overwrite row 2.

$$= \begin{bmatrix} 1 & -1 & -5 \\ 2(1)+-2 & 2(-1)+1 & 2(-5)+7 \end{bmatrix}$$

$$= \begin{bmatrix} 1 & -1 & -5 \\ 0 & -1 & -3 \end{bmatrix}$$

39.

$$\begin{bmatrix} 4 & 3 & 7 \\ 4 & 11 & 24 \end{bmatrix}$$

$-1 \cdot$ row 1 + row 2. Overwrite row 2.

$$= \begin{bmatrix} 4 & 3 & 7 \\ -1(4)+4 & -1(3)+11 & -1(7)+24 \end{bmatrix}$$

$$= \begin{bmatrix} 4 & 3 & 7 \\ 0 & 8 & 17 \end{bmatrix}$$

41.

$$\begin{bmatrix} 1 & 1 & 1 & 2 \\ 2 & 3 & 2 & 3 \\ 0 & -2 & 2 & 6 \end{bmatrix}$$

$-2 \cdot$ row 1 + row 2. Overwrite row 2

$$\begin{bmatrix} 1 & 1 & 1 & 2 \\ 0 & 1 & 0 & -1 \\ 0 & -2 & 2 & 6 \end{bmatrix}$$

$2 \cdot$ row 2 + row 3. Overwrite row 3

$$\begin{bmatrix} 1 & 1 & 1 & 2 \\ 0 & 1 & 0 & -1 \\ 0 & 0 & 2 & 4 \end{bmatrix}$$

$3 \cdot$ row 2 + row 3. Overwrite row 3.

$$= \begin{bmatrix} 1 & 1 & 1 & 2 \\ 0 & 1 & 0 & -1 \\ 3(0)+0 & 3(1)-3 & 3(0)+1 & 3(-1)+4 \end{bmatrix}$$

$$= \begin{bmatrix} 1 & 1 & 1 & 2 \\ 0 & 1 & 0 & -1 \\ 0 & 0 & 1 & 1 \end{bmatrix}$$

43.

$$\begin{bmatrix} 3 & 0 & 2 & 9 \\ 4 & 1 & 2 & 8 \\ 3 & 2 & 1 & 2 \end{bmatrix}$$

$-\dfrac{4}{3} \cdot$ row 1 + row 2. Overwrite row 2.

$-1 \cdot$ row 1 + row 3. Overwrite row 3.

$$= \begin{bmatrix} 3 & 0 & 2 & 9 \\ -\dfrac{4}{3}(3)+4 & -\dfrac{4}{3}(0)+1 & -\dfrac{4}{3}(2)+2 & -\dfrac{4}{3}(9)+8 \\ -1(3)+3 & -1(0)+2 & -1(2)+1 & -1(9)+2 \end{bmatrix}$$

$$= \begin{bmatrix} 3 & 0 & 2 & 9 \\ 0 & 1 & -\dfrac{2}{3} & -4 \\ 0 & 2 & -1 & -7 \end{bmatrix}$$

$-2 \cdot$ row 2 + row 3. Overwrite row 3.

$$= \begin{bmatrix} 3 & 0 & 2 & 9 \\ 0 & 1 & -\dfrac{2}{3} & -4 \\ -2(0)+0 & -2(1)+2 & -2\left(-\dfrac{2}{3}\right)-1 & -2(-4)-7 \end{bmatrix}$$

$$= \begin{bmatrix} 3 & 0 & 2 & 9 \\ 0 & 1 & -\dfrac{2}{3} & -4 \\ 0 & 0 & \dfrac{1}{3} & 1 \end{bmatrix}$$

Matrices

45.

$\begin{cases} 2x+y=-1 \\ 4x+y=-7 \end{cases} \rightarrow \begin{bmatrix} 2 & 1 & -1 \\ 4 & 1 & -7 \end{bmatrix}$

$-2 \cdot$ row 1 + row 2. Overwrite row 2.

$= \begin{bmatrix} 2 & 1 & -1 \\ -2(2)+4 & -2(1)+1 & -2(-1)-7 \end{bmatrix}$

$= \begin{bmatrix} 2 & 1 & -1 \\ 0 & -1 & -5 \end{bmatrix}$

Rewrite the last row to solve for y:

$0x - y = -5$
$-y = -5$
$y = 5$

Rewrite the first row and substitute $y = 5$ to solve for x:

$2x + 5 = -1$
$-5-5$
$2x = -6$
$\dfrac{2x}{2} = \dfrac{-6}{2}$
$x = -3$

The solution to the system is $(x, y) = (-3, 5)$

47.

$\begin{cases} -2x+3y=24 \\ 6x+y=-12 \end{cases} \rightarrow \begin{bmatrix} -2 & 3 & 24 \\ 6 & 1 & -12 \end{bmatrix}$

$3 \cdot$ row 1 + row 2. Overwrite row 2.

$= \begin{bmatrix} -2 & 3 & 24 \\ 3(-2)+6 & 3(3)+1 & 3(24)-12 \end{bmatrix}$

$= \begin{bmatrix} -2 & 3 & 24 \\ 0 & 10 & 60 \end{bmatrix}$

Rewrite the last row to solve for y:

$0x + 10y = 60$
$\dfrac{10y}{10} = \dfrac{60}{10}$
$y = 6$

Rewrite the first row and substitute $y = 6$ to solve for x:

$-2x + 3(6) = 24$
$-2x + 18 = 24$
$-18-18$
$-2x = 6$
$\dfrac{-2x}{-2} = \dfrac{6}{-2}$
$x = -3$

The solution to the system is $(x, y) = (-3, 6)$

49.

$\begin{cases} x-4y+6z=-3 \\ -x+5y-2z=-1 \\ 2x+y-z=7 \end{cases} \rightarrow \begin{bmatrix} 1 & -4 & 6 & -3 \\ -1 & 5 & -2 & -1 \\ 2 & 1 & -1 & 7 \end{bmatrix}$

row 1 + row 2. Overwrite row 2.
$-2 \cdot$ row 1 + row 3. Overwrite row 3.

$= \begin{bmatrix} 1 & -4 & 6 & -3 \\ (1)-1 & (-4)+5 & (6)-2 & (-3)-1 \\ -2(1)+2 & -2(-4)+1 & -2(6)-1 & -2(-3)+7 \end{bmatrix}$

$= \begin{bmatrix} 1 & -4 & 6 & -3 \\ 0 & 1 & 4 & -4 \\ 0 & 9 & -13 & 13 \end{bmatrix}$

$-9 \cdot$ row 2 + row 3. Overwrite row 3.

$= \begin{bmatrix} 1 & -4 & 6 & -3 \\ 0 & 1 & 4 & -4 \\ -9(0)+0 & -9(1)+9 & -9(4)-13 & -9(-4)+13 \end{bmatrix}$

$= \begin{bmatrix} 1 & -4 & 6 & -3 \\ 0 & 1 & 4 & -4 \\ 0 & 0 & -49 & 49 \end{bmatrix}$

Rewrite the third row to solve for z:

$0x + 0y - 49z = 49$
$\dfrac{-49z}{-49} = \dfrac{49}{-49}$
$z = -1$

Rewrite the second row and substitute $z = -1$ to solve for y:

$0x + y + 4(-1) = -4$
$y - 4 = -4$
$y = 0$

Rewrite the first row and substitute
$y = 0$ and $z = -1$ to solve for x:
$$x - 4(0) + 6(-1) = -3$$
$$x - 6 = -3$$
$$\underline{+6 \quad +6}$$
$$x = 3$$

The solution to the system is $(x, y, z) = (3, 0, -1)$

51.

$$\begin{cases} 4x + y - z = -1 \\ -2x - y + z = -1 \\ -x + 7y - 2z = 7 \end{cases} \rightarrow \begin{bmatrix} 4 & 1 & -1 & -1 \\ -2 & -1 & 1 & -1 \\ -1 & 7 & -2 & 7 \end{bmatrix}$$

interchange row 1 and row 3

$$= \begin{bmatrix} -1 & 7 & -2 & 7 \\ -2 & -1 & 1 & -1 \\ 4 & 1 & -1 & -1 \end{bmatrix}$$

$2 \cdot$ row 2 + row 3. Overwrite row 3.

$$= \begin{bmatrix} -1 & 7 & -2 & 7 \\ -2 & -1 & 1 & -1 \\ 2(-2)+4 & 2(-1)+1 & 2(1)-1 & 2(-1)-1 \end{bmatrix}$$

$$= \begin{bmatrix} -1 & 7 & -2 & 7 \\ -2 & -1 & 1 & -1 \\ 0 & -1 & 1 & -3 \end{bmatrix}$$

$-2 \cdot$ row 1 + row 2. Overwrite row 2.

$$= \begin{bmatrix} -1 & 7 & -2 & 7 \\ -2(-1)-2 & -2(7)-1 & -2(-2)+1 & -2(7)-1 \\ 0 & -1 & 1 & -3 \end{bmatrix}$$

$$= \begin{bmatrix} -1 & 2 & 1 & -4 \\ 0 & -15 & 5 & -15 \\ 0 & -1 & 1 & -3 \end{bmatrix}$$

$(-\frac{1}{5}) \cdot$ row 2. Overwrite row 2.

$$= \begin{bmatrix} -1 & 7 & -2 & 7 \\ (-\frac{1}{5})0 & (-\frac{1}{5})(-15) & (-\frac{1}{5})(5) & (-\frac{1}{5})(-15) \\ 0 & -1 & 1 & -3 \end{bmatrix}$$

$$= \begin{bmatrix} -1 & 7 & -2 & 7 \\ 0 & 3 & -1 & 3 \\ 0 & -1 & 1 & -3 \end{bmatrix}$$

interchange row 2 and row 3

$$= \begin{bmatrix} -1 & 7 & -2 & 7 \\ 0 & -1 & 1 & -3 \\ 0 & 3 & -1 & 3 \end{bmatrix}$$

$3 \cdot$ row 2 + row 3. Overwrite row 3.

$$= \begin{bmatrix} -1 & 7 & -2 & 7 \\ 0 & -1 & 1 & -3 \\ 3(0)+0 & 3(-1)+3 & 3(1)-1 & 3(-3)+3 \end{bmatrix}$$

$$= \begin{bmatrix} -1 & 7 & -2 & 7 \\ 0 & -1 & 1 & -3 \\ 0 & 0 & 2 & -6 \end{bmatrix}$$

Rewrite the third row to solve for z:
$$0x + 0y + 2z = -6$$
$$2z = -6$$
$$\frac{2z}{2} = \frac{-6}{2}$$
$$z = -3$$

Rewrite the second row and substitute
$z = -3$ to solve for y:
$$0x - y + (-3) = -3$$
$$\underline{+3 \quad +3}$$
$$y = 0$$

Rewrite the first row and substitute
$y = 0$ and $z = -3$ to solve for x:
$$-x + 7(0) - 2(-3) = 7$$
$$-x + 6 = 7$$
$$\underline{-6 \quad -6}$$
$$-x = 1$$
$$x = -1$$

The solution to the system is $(x, y, z) = (-1, 0, -3)$

53.

$$\begin{cases} 2x - y - 2z = 4 \\ -x + 5y = 26 \\ -y + 4z = -10 \end{cases} \rightarrow \begin{bmatrix} 2 & -1 & -2 & 4 \\ -1 & 5 & 0 & 26 \\ 0 & -1 & 4 & -10 \end{bmatrix}$$

Interchange row 1 and row 2.

$$\begin{bmatrix} -1 & 5 & 0 & 26 \\ 2 & -1 & -2 & 4 \\ 0 & -1 & 4 & -10 \end{bmatrix}$$

$2 \cdot$ row 1 + row 2. Overwrite row 2.

$$= \begin{bmatrix} -1 & 5 & 0 & 26 \\ 2(-1)+2 & 2(5)-1 & 2(0)-2 & 2(26)+4 \\ 0 & -1 & 4 & -10 \end{bmatrix}$$

$$= \begin{bmatrix} -1 & 5 & 0 & 26 \\ 0 & 9 & -2 & 56 \\ 0 & -1 & 4 & -10 \end{bmatrix}$$

Matrices

Interchange row 2 and row 3.

$$\begin{bmatrix} -1 & 5 & 0 & 26 \\ 0 & -1 & 4 & -10 \\ 0 & 9 & -2 & 56 \end{bmatrix}$$

9·row 2 + row 3. Overwrite row 3.

$$\begin{bmatrix} -1 & 5 & 0 & 26 \\ 0 & -1 & 4 & -10 \\ 9(0)+0 & 9(-1)+9 & 9(4)-2 & 9(-10)+56 \end{bmatrix}$$

$$\begin{bmatrix} -1 & 5 & 0 & 26 \\ 0 & -1 & 4 & -10 \\ 0 & 0 & 34 & -34 \end{bmatrix}$$

Rewrite the third row to solve for z:

$0x + 0y + 34z = -34$

$\dfrac{34z}{34} = \dfrac{-34}{34}$

$z = -1$

Rewrite the second row and substitute $z = -1$ to solve for y:

$0x - y + 4(-1) = -10$

$-y - 4 = -10$

$+4 +4$

$-y = -6$

$y = 6$

Rewrite the first row and substitute $y = 6$ and $z = -1$ to solve for x:

$-x + 5(6) + 0(-1) = 26$

$-x + 30 = 26$

$-30 -30$

$-x = -4$

$x = 4$

The solution to the system is $(x, y, z) = (4, 6, -1)$.

55. The three points are $(-1, -6)$, $(1, 2)$, and $(2, 3)$.

Substitute in the equation $y = ax^2 + bx + c$ the given x and y coordinates from the points.

Points	Substitute (x, y)
$(-1, -6)$	$a(-1)^2 + b(-1) + c = -6$
$(1, 2)$	$a(1)^2 + b(1) + c = 2$
$(2, 3)$	$a(2)^2 + b(2) + c = 3$

System of equations Augmented matrix

$\begin{cases} a - b + c = -6 \\ a + b + c = 2 \\ 4a + 2b + c = 3 \end{cases}$ $\begin{bmatrix} 1 & -1 & 1 & -6 \\ 1 & 1 & 1 & 2 \\ 4 & 2 & 1 & 3 \end{bmatrix}$

−1·row 1 + row 2. Overwrite row 2.
−4·row 1 + row 3. Overwrite row 3.

$$= \begin{bmatrix} 1 & -1 & 1 & -6 \\ -1(1)+1 & -1(-1)+1 & -1(1)+1 & -1(-6)+2 \\ -4(1)+4 & -4(-1)+2 & -4(1)+1 & -4(-6)+3 \end{bmatrix}$$

$$= \begin{bmatrix} 1 & -1 & 1 & -6 \\ 0 & 2 & 0 & 8 \\ 0 & 6 & -3 & 27 \end{bmatrix}$$

$\dfrac{1}{2}$ row 2. Overwrite row 2.

$$= \begin{bmatrix} 1 & -1 & 1 & -6 \\ \dfrac{1}{2}\cdot 0 & \dfrac{1}{2}\cdot 2 & \dfrac{1}{2}\cdot 0 & \dfrac{1}{2}\cdot 8 \\ 0 & 6 & -3 & 27 \end{bmatrix}$$

$$= \begin{bmatrix} 1 & -1 & 1 & -6 \\ 0 & 1 & 0 & 4 \\ 0 & 6 & -3 & 27 \end{bmatrix}$$

−6·row 2 + row 3. Overwrite row 3.

$$= \begin{bmatrix} 1 & -1 & 1 & -6 \\ 0 & 1 & 0 & 4 \\ -6(0)+0 & -6(1)+6 & -6(0)-3 & -6(4)+27 \end{bmatrix}$$

$$= \begin{bmatrix} 1 & -1 & 1 & -6 \\ 0 & 1 & 0 & 4 \\ 0 & 0 & -3 & 3 \end{bmatrix}$$

Rewrite the third row to solve for c:

$0a + 0b - 3c = 3$

$\dfrac{-3c}{-3} = \dfrac{3}{-3}$

$c = -1$

Rewrite the second row and substitute $c = -1$ to solve for b:

$0a + b + 0(-1) = 4$

$b = 4$

Rewrite the first row and substitute $b = 4$ and $c = -1$ to solve for a:

$a - (4) + (-1) = -6$

$a - 4 - 1 = -6$

$a - 5 = -6$

$a = -1$

The quadratic function that passes through these three points is $f(x) = -x^2 + 4x - 1$.

57. The three points are $(1,10)$, $(2,18)$, and $(3,28)$.

Substitute in the equation $y = ax^2 + bx + c$ the given x and y coordinates from the points.

Points	Substitute (x,y)
$(1,10)$	$a(1)^2 + b(1) + c = 10$
$(2,18)$	$a(2)^2 + b(2) + c = 18$
$(3,28)$	$a(3)^2 + b(3) + c = 28$

System of equations   Augmented matrix

$\begin{cases} a+b+c=10 \\ 4a+2b+c=18 \\ 9a+3b+c=28 \end{cases}$ $\begin{bmatrix} 1 & 1 & 1 & 10 \\ 4 & 2 & 1 & 18 \\ 9 & 3 & 1 & 28 \end{bmatrix}$

$-4 \cdot$ row 1 + row 2. Overwrite row 2.
$-9 \cdot$ row 1 + row 3. Overwrite row 3.

$= \begin{bmatrix} 1 & 1 & 1 & 10 \\ -4(1)+4 & -4(1)+2 & -4(1)+1 & -4(10)+18 \\ -9(1)+9 & -9(1)+3 & -9(1)+1 & -9(10)+28 \end{bmatrix}$

$= \begin{bmatrix} 1 & 1 & 1 & 10 \\ 0 & -2 & -3 & -22 \\ 0 & -6 & -8 & -62 \end{bmatrix}$

$-\dfrac{1}{2}$ row 2. Overwrite row 2.

$= \begin{bmatrix} 1 & 1 & 1 & 10 \\ -\dfrac{1}{2} \cdot 0 & -\dfrac{1}{2} \cdot -2 & -\dfrac{1}{2} \cdot -3 & -\dfrac{1}{2} \cdot -22 \\ 0 & -6 & -8 & -62 \end{bmatrix}$

$= \begin{bmatrix} 1 & 1 & 1 & 10 \\ 0 & 1 & \dfrac{3}{2} & 11 \\ 0 & -6 & -8 & -62 \end{bmatrix}$

$6 \cdot$ row 2 + row 3. Overwrite row 3.

$= \begin{bmatrix} 1 & 1 & 1 & 10 \\ 0 & 1 & \dfrac{3}{2} & 11 \\ 6(0)+0 & 6(1)-6 & 6\left(\dfrac{3}{2}\right)-8 & 6(11)-62 \end{bmatrix}$

$= \begin{bmatrix} 1 & 1 & 1 & 10 \\ 0 & 1 & \dfrac{3}{2} & 11 \\ 0 & 0 & 1 & 4 \end{bmatrix}$

Rewrite the third row to solve for c:
$0a + 0b + c = 4$
$c = 4$

Rewrite the second row and substitute $c = 4$ to solve for b:

$0a + b + \dfrac{3}{2}(4) = 11$

$b + 6 = 11$

$b = 5$

Rewrite the first row and substitute $b = 5$ and $c = 4$ to solve for a:

$a + (5) + (4) = 10$

$a + 9 = 10$

$a = 1$

The quadratic function that passes through these three points is $f(x) = x^2 + 5x + 4$.

59. The three points are $(1,4)$, $(2,15)$, and $(3,30)$.

Substitute in the equation $y = ax^2 + bx + c$ the given x and y coordinates from the points.

Points	Substitute (x,y)
$(1,4)$	$a(1)^2 + b(1) + c = 4$
$(2,15)$	$a(2)^2 + b(2) + c = 15$
$(3,30)$	$a(3)^2 + b(3) + c = 30$

Matrices

System of equations   Augmented matrix

$\begin{cases} a+b+c=4 \\ 4a+2b+c=15 \\ 9a+3b+c=30 \end{cases}$ $\begin{bmatrix} 1 & 1 & 1 & 4 \\ 4 & 2 & 1 & 15 \\ 9 & 3 & 1 & 30 \end{bmatrix}$

$-4 \cdot$ row 1 + row 2. Overwrite row 2.
$-9 \cdot$ row 1 + row 3. Overwrite row 3.

$= \begin{bmatrix} 1 & 1 & 1 & 4 \\ -4(1)+4 & -4(1)+2 & -4(1)+1 & -4(4)+15 \\ -9(1)+9 & -9(1)+3 & -9(1)+1 & -9(4)+30 \end{bmatrix}$

$= \begin{bmatrix} 1 & 1 & 1 & 4 \\ 0 & -2 & -3 & -1 \\ 0 & -6 & -8 & -6 \end{bmatrix}$

$-\dfrac{1}{2}$ row 2. Overwrite row 2.

$= \begin{bmatrix} 1 & 1 & 1 & 4 \\ -\dfrac{1}{2} \cdot 0 & -\dfrac{1}{2} \cdot -2 & -\dfrac{1}{2} \cdot -3 & -\dfrac{1}{2} \cdot -1 \\ 0 & -6 & -8 & -6 \end{bmatrix}$

$= \begin{bmatrix} 1 & 1 & 1 & 4 \\ 0 & 1 & \dfrac{3}{2} & \dfrac{1}{2} \\ 0 & -6 & -8 & -6 \end{bmatrix}$

$6 \cdot$ row 2 + row 3. Overwrite row 3.

$= \begin{bmatrix} 1 & 1 & 1 & 4 \\ 0 & 1 & \dfrac{3}{2} & \dfrac{1}{2} \\ 6(0)+0 & 6(1)-6 & 6\left(\dfrac{3}{2}\right)-8 & 6\left(\dfrac{1}{2}\right)-6 \end{bmatrix}$

$= \begin{bmatrix} 1 & 1 & 1 & 4 \\ 0 & 1 & \dfrac{3}{2} & \dfrac{1}{2} \\ 0 & 0 & 1 & -3 \end{bmatrix}$

Rewrite the third row to solve for c:
$0a+0b+c=-3$
$c=-3$

Rewrite the second row and substitute
$c=-3$ to solve for b:
$0a+b+\dfrac{3}{2}(-3)=\dfrac{1}{2}$
$b-\dfrac{9}{2}=\dfrac{1}{2}$
$b=\dfrac{10}{2}=5$

Rewrite the first row and substitute
$b=5$ and $c=-3$ to solve for a:
$a+(5)+(-3)=4$
$a+2=4$
$a=2$

The quadratic function that passes through these three points is
$f(x)=2x^2+5x-3$.

61. The three points are $(-1,-5)$, $(1,15)$, and $(2,16)$.

Substitute in the equation $y=ax^2+bx+c$ the given x and y coordinates from the points.

Points   Substitute (x,y)
$(-1,-5)$ $a(-1)^2+b(-1)+c=-5$
$(1,15)$ $a(1)^2+b(1)+c=15$
$(2,16)$ $a(2)^2+b(2)+c=16$

System of equations   Augmented matrix

$\begin{cases} a-b+c=-5 \\ a+b+c=15 \\ 4a+2b+c=16 \end{cases}$ $\begin{bmatrix} 1 & -1 & 1 & -5 \\ 1 & 1 & 1 & 15 \\ 4 & 2 & 1 & 16 \end{bmatrix}$

$-1 \cdot$ row 1 + row 2. Overwrite row 2.
$-4 \cdot$ row 1 + row 3. Overwrite row 3.

$= \begin{bmatrix} 1 & -1 & 1 & -5 \\ -1(1)+1 & -1(-1)+1 & -1(1)+1 & -1(-5)+15 \\ -4(1)+4 & -4(-1)+2 & -4(1)+1 & -4(-5)+16 \end{bmatrix}$

$= \begin{bmatrix} 1 & -1 & 1 & -5 \\ 0 & 2 & 0 & 20 \\ 0 & 6 & -3 & 36 \end{bmatrix}$

$\dfrac{1}{2}$ row 2. Overwrite row 2.

$= \begin{bmatrix} 1 & -1 & 1 & -5 \\ \dfrac{1}{2} \cdot 0 & \dfrac{1}{2} \cdot 2 & \dfrac{1}{2} \cdot 0 & \dfrac{1}{2} \cdot 20 \\ 0 & 6 & -3 & 36 \end{bmatrix}$

$= \begin{bmatrix} 1 & -1 & 1 & -5 \\ 0 & 1 & 0 & 10 \\ 0 & 6 & -3 & 36 \end{bmatrix}$

$-6 \cdot$ row 2 + row 3. Overwrite row 3.

$$= \begin{bmatrix} 1 & -1 & 1 & -5 \\ 0 & 1 & 0 & 10 \\ -6(0)+0 & -6(1)+6 & -6(0)-3 & -6(10)+36 \end{bmatrix}$$

$$= \begin{bmatrix} 1 & -1 & 1 & -5 \\ 0 & 1 & 0 & 10 \\ 0 & 0 & -3 & -24 \end{bmatrix}$$

Rewrite the third row to solve for c:
$$0a + 0b - 3c = -24$$
$$c = 8$$

Rewrite the second row and substitute $c = 8$ to solve for b:
$$0a + b + 0(8) = 10$$
$$b = 10$$

Rewrite the first row and substitute $b = 10$ and $c = 8$ to solve for a:
$$a - (10) + (8) = -5$$
$$a - 2 = -5$$
$$a = -3$$

The quadratic function that passes through these three points is $f(x) = -3x^2 + 10x + 8$.